Electric Machines

Electric Machines

Transients, Control Principles, Finite Element Analysis, and Optimal Design with MATLAB®

Second Edition

Ion Boldea
and
Lucian N. Tutelea

CRC Press
Taylor & Francis Group
Boca Raton London New York

CRC Press is an imprint of the
Taylor & Francis Group, an **informa** business

Second edition published 2022
by CRC Press
6000 Broken Sound Parkway NW, Suite 300, Boca Raton, FL 33487-2742

and by CRC Press
2 Park Square, Milton Park, Abingdon, Oxon, OX14 4RN

First edition published by CRC Press 2009

CRC Press is an imprint of Taylor & Francis Group, LLC

Library of Congress Cataloging-in-Publication Data
Names: Boldea, I., author. | Tutelea, Lucian, author.
Title: Electric machines : transients, control principles, finite element
analysis and optimal design with MATLAB / Ion Boldea, Lucian N. Tutelea.
Description: Second edition. | Boca Raton, FL : CRC Press, 2021. | Includes
bibliographical references and index.
Identifiers: LCCN 2021035679 (print) | LCCN 2021035680 (ebook) | ISBN
9780367375652 (hbk) | ISBN 9781032105727 (pbk) | ISBN 9781003216018 (ebk)
Subjects: LCSH: Electric machinery--Design and construction--Data
processing. | MATLAB.
Classification: LCC TK2331 .B584 2021 (print) | LCC TK2331 (ebook) | DDC
621.31/042--dc23
LC record available at https://lccn.loc.gov/2021035679
LC ebook record available at https://lccn.loc.gov/2021035680

ISBN: 978-0-367-37565-2 (hbk)
ISBN: 978-1-032-10572-7 (pbk)
ISBN: 978-1-003-21601-8 (ebk)

DOI: 10.1201/9781003216018

Typeset in Times
by SPi Technologies India Pvt Ltd (Straive)

Contents

Preface

This textbook treats, as a second (and third) semester, advanced issues/subjects in electric machines—rotary and linear—such as:

- Modeling of transients
- Control principles
- Electromagnetic and thermal Finite Element Analysis
- Optimal design (dimensioning),

for line start (constant speed) and variable speed applications with advanced parameters' estimation techniques.

Many numerical examples and case studies are included in all 10 Chapters while 8+3 MATLAB®/SIMULINK® Programs, available online constitute a strong method to deepen the knowledge assimilation by case studies with a strong feeling of magnitudes.

Six of the mentioned Programs illustrate complete FEA and optimal design case studies.

Notable recent knowledge with strong industrialization potential was added to this edition and we mention here:

- Orthogonal models of multiphase a.c. machines (Chapter 4)
- Thermal Finite Element Analysis of (FEA) electric machines (new Chapter 7, with one MATLAB Program online example)
- FEA—based—only optimal design of a PM motor case study (in Chapter 9 with a new (9th) Matlab Computer Program available online)
- Line start synchronizing premium efficiency PM induction machines (in Chapter 6)

The Contents

1: Electric machine circuit models for transients and control
2: Transients and control principles of brush-commutator machines
3: Synchronous machine transients and control principles.
4: Induction machine transients and control principles.
5: Essential of Finite element analysis (FEA) in electromagnetics.
6: FEA of electric machines electromagnetics.
7: Thermal FEA of electric machines.
8: Optimal electromagnetic design of electric machines: basics.
9: Optimal electromagnetic design of PM synchronous machines (PMSM).
10: Optimal electromagnetic design of induction machines.

All chapters follow a rather unique pattern: from principles to applications and case studies with ample graphical (numerical) results.

The detailed contents exhibit the large plethora of issues of interest treated in each Chapter.

How to Use the Textbook

The textbook may be used in the strict order of Chapters with some paragraphs used as individual Projects (homework). In some parts of the world Chapters 1–4 may be treated exhaustively as a Senior B. S. (or required M.S.) course with Chapters 5–10 left for a senior graduate Course as both FEA and Optimal design are crucial engineering tools today and in the future in industry.

The 11 Matlab-Simulink Computer Programs (online) refer to:

- Example 10—No-load transformer grid connection transients
- Example 11—Loaded transformer grid connection
- Example 12—DC-brush PM motor transients
- Example 13—Induction machine transients
- Example 14—Synchronous motor transients
- Example 15—PMSM optimal design by the Hooke–Jeeves method
- Example 16—PMSM optimal design by Genetic Algorithms
- Example 17—IM optimal design by Hooke-Jeeves method
- Example 18—IM optimal design by Genetic Algorithms
- Example 19—BLDC-PM motor FEA-thermal
- Example 20—FEA only optimal design of a one-phase PMSM

Ion Boldea

Lucian N. Tutelea
Timişoara, Romania

Authors

Ion Boldea is a full professor of Electrical Engineering at the University Politechnica of Timişoara, Romania. He has spent approximately 5 years as visiting professor of Electrical Engineering in both Kentucky and Oregon, USA since 1973, when he was a senior Fulbright Scholar for 10 months. He was also a visiting professor in the UK at UMIST and Glasgow University. He is a full member of the Romanian Academy of Technical Sciences, a full member of the European Academy of Sciences and Arts of Salzburg, Austria, and a full member of the Romanian Academy. He has delivered IEEE-IAS Distinguished Lectures since 2008. He has given keynote speeches, tutorial courses, intensive courses, technical consulting in the USA, South America, E.U, South Korea, and China based on his numerous books and IEEE Trans. and Conference papers over the last 45 years in the field of rotating and linear electric machines and drives for renewable energy, vehicular, industrial, and residential applications. Professor Boldea is a life fellow of IEEE. He won the IEEE 2015 Nikola Tesla Award for "contributions to the design and control of rotating and linear electric machines for industry applications."

Lucian N. Tutelea (M'07) received the B.S. and Ph.D. degrees in electrical engineering from the Politehnica University Timişoara, Timişoara, Romania, in 1989 and 1997, respectively. He was a visiting researcher with the Institute of Energy Technology, Aalborg University, Denmark, in 1997, 1999, 2000, and 2006, as well as the Department of Electrical Engineering, Hanyang University, South Korea, in 2004. He is currently a professor with the Department of Electric Engineering, Politehnica University Timisoara. His main research interests include design, modeling, and control of electric machines and drives. Professor Tutelea published more than 80 papers indexed IEEE Xplore or in Web of Science with more than 400 citations.

Authors

1 Electric Machine Circuit Models for Transients and Control

1.1 INTRODUCTION

In the companion book (*Electric Machines: Steady State and Performance*), only electrical machines under steady-state operation were treated. Under steady state, speed, terminal voltages, currents, amplitudes, and their frequency remain constant. In reality, all electric machines undergo speed, voltage, current, amplitude, and frequency variations under transients at (or when connected to) power grid and when associated with PWM static converters for variable speed close-loop control operation both in generator and motor modes.

Transients may also be preferred to estimate electric machine circuit parameters (resistances, inductances, etc.), by the analysis of the corresponding machine response. This chapter introduces a few general electric machine models used to handle machine transients.

There are two main categories of machine models:

- Circuit models
- Field-circuit-coupled models

Another classification of machine models is

- Fundamental frequency models
- Super high-frequency models: when stray capacitors inside the machine are considered—for instance, for switching frequency when a PWM static converter supplies the machine.

The main circuit models are

- Phase-coordinate circuit models
- Orthogonal (dq)—space phasor (complex variable)—models [1–6]

The magnetic field-to-circuit models have gained widespread acceptance:

- Analytical field models: in simplified form, they are used to derive circuit models [7,8] for steady-state operation.

DOI: 10.1201/9781003216018-1

- Finite element models (FEMs) [7,8]: Numerical integral field models, with field-to-circuit coupling to investigate electric machine operation under steady state or transients while considering most secondary effects (such as slot openings, machine load saturation, skin effect). Last part of this book dwells extendedly on FEM as it is paramount in all design refinements of electric machines and drives.
- Multiple magnetic circuit models: The machine zones are divided into region permeances with uniform flux density; they may be 3D and they consider the placement of windings in slots and their connections, slot openings, magnetic saturation, but, in a coarser way, with the advantage of at least 10 times lower computation time than FEM on similar CPUs [9].

Other models, such as spiral vector theory [10] have been proposed but have not gained widespread acceptance, yet.

Of the above models, we will discuss here thoroughly the physical orthogonal model and its spin-off, the space vector (complex variable) model, as they are applied extensively to investigate electric machine transients in power systems and in variable speed power electronics control of electric drives. The dq model may be introduced mathematically, by a transformation of variables, or through a physical model (when the model may be built and tested).

We choose the physical model because it is believed to be more intuitive.

1.2 ORTHOGONAL (dq) PHYSICAL MODEL

The dq physical model of electric machines is shown in Figure 1.1.

We call it a physical model because it can be actually built (in the United Kingdom it was used for teaching until rather recently).

The dq physical model as introduced here is provided with brush–commutator windings on the stator and on the rotor, with brushes along two orthogonal axes d and q. There are two such windings on the stator (d and q) and three such windings (d_r, F, along axis d, and q_r along axis q) on the rotor. The brushes of all windings are aligned to the orthogonal axes d and q. It is known that the armature field axis in brush–commutator machines falls along the brush physical axis for asymmetrical rotor coils.

It follows that the mmfs and the corresponding airgap fields of all windings are mostly along brush axes and are at a standstill with each other during all operation modes.

On the other hand, the coils of stator windings are at a standstill while the rotor coils travel at speed ω_r, where the axes of their fields travel at brush speed ω_b. The brushes are rotated at ω_b by a small fictitious (or real) power drive.

We should mention that in real brush–commutator windings, the airgap field is triangular rather than sinusoidal; only its fundamental is considered here. Let us observe also that for all orthogonal windings, self and mutual inductances are independent of rotor position if and only if the brushes are attached to the machine part (stator and rotor) with magnetic anisotropy (salient poles).

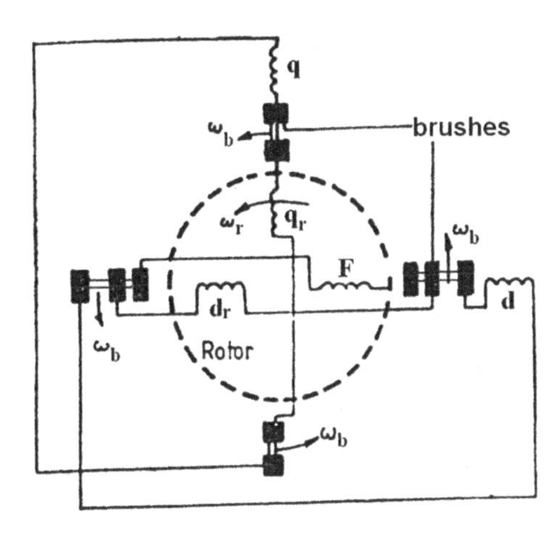

FIGURE 1.1 The dq physical model.

In this way, we obtain a system of differential equations for the dq physical models that have constant coefficients (inductances).

This is an extraordinary simplification in treating transients of electric machines but we have to work out the equivalence between the dq physical model and the actual electric machine.

Let us now pursue the restrictions in terms of brush speed, ω_b, for synchronous, brush–commutator and induction machines.

If the electric machine has a rotor magnetic saliency, such as in synchronous like machines, it is mandatory to have $\omega_b = \omega_r$, in order to obtain rotor position independence of dq model parameters (inductances).

The dq model in Figure 1.1 with $\omega_b = \omega_r$ refers, in fact, directly to the synchronous machine. If the stator shows magnetic saliency—as in brush–commutator machines—$\omega_b = 0$ and all but the d-axis winding in the stator and the q_r winding in the rotor are eliminated.

For machines where the airgap is uniform (induction machines) for any speed of brushes (axes) ω_b, the dq-model inductances will be independent of rotor position. However, $\omega_b = 0$, ω_r, ω_1 (stator frequency) are the main axes speed that are applied extensively.

The brush–commutators in the dq physical model (Figure 1.1) are the physical correspondent of coordinate (or variable) mathematical transformation that leads to same dq model.

When can the dq model properly model actual 3(2)-phase synchronous and induction machines?

As is evident from Figure 1.1, that the d and q windings, on the machine part that is moving in accordance with *dq* axes speed, ω_b, should be symmetric. Also, the magnetic field (not necessarily the mmf) of all windings of the dq model (and of the real machine) should have a sinusoidal spatial distribution along the rotor periphery. Space harmonics of order ν of this field may be treated by adding additional dq windings with their *dq* axes moving at different speeds, $\omega_{b\nu}$.

Can double magnetic saliency machines (switched reluctance machines) be treated via the dq model?

Yes, but only if they have no winding in the rotor and if their stator winding inductances vary sinusoidally with the rotor position (as in synchronous machines, but with almost zero (in three-phase SRM) mutual inductances). In machines with PMs, the PM field should create sinusoidal emfs in the stator windings to make the dq model adequate for the scope.

So the dq model may be applied not only to distributed a.c. winding machines (synchronous and induction) but also to nonoverlapping coil stator winding machines (SRM and PMSM alike), provided there are no rotor windings and the stator winding inductances vary sinusoidally with the rotor position and so do, in time, the PM-produced emfs.

To document the above claims we will investigate the concept of transformer (pulsational) and motion-induced voltages.

1.3 PULSATIONAL AND MOTION-INDUCED VOLTAGES IN dq MODELS

The total induced voltage in coupled electric circuits is generated by two types of voltages: one by pulsational (transformer) action, E_p, and the other by relative motion between conductors and magnetic fields, E_m. Let us consider the flux linkage of a winding $\Psi(\theta,t)$ as dependent on position and time.

The total emf E_t is

$$E_t = -\frac{d_s\psi(\theta,t)}{dt} = E_p + E_m \tag{1.1}$$

$$E_p = -\frac{\partial\psi(\theta,t)}{\partial t}; \quad E_m = -\frac{\partial\psi(\theta,t)}{\partial\theta_r}\cdot\frac{d\theta_r}{dt} \tag{1.2}$$

Note: In linear electric machines, the rotor position angle, θ_r, is replaced by the linear position, x.

Let us now assume that the flux linkage of the windings along the two axes varies sinusoidally with θ_r:

$$\Psi_d(\theta_r) = \Psi_m \sin\theta_\gamma = \frac{\partial\Psi_q}{\partial\theta_\gamma}; \quad \Psi_q(\theta_r) = -\Psi_m \cos\theta_\gamma = -\frac{\partial\Psi_d}{\partial\theta_\gamma} \tag{1.3}$$

Axis q in Figure 1.1 is ahead of axis d along the direction of rotation (of ω_r). Conditions (1.3) lead to the idea that the motion emf in one orthogonal axis is produced by the flux linkage in the other orthogonal axis.

However, both windings have the same maximum flux linkage, so they have to be symmetric as well. This condition is valid only if the *dq* axes are moving in accordance with those of dq windings. It has already been discussed in [5] that only the motion emfs contribute to the torque production.

1.4 dq MODEL OF DC BRUSH PM MOTOR ($\omega_b = 0$)

Let us reconsider here the case of the d.c. brush PM motor (Figure 1.2).

The dq model of this machine is obtained by eliminating all windings in Figure 1.1 but the *q* axis winding in the rotor (the armature winding) and by introducing a PM along axis *d* in the stator, for the magnetic field production.

The magnetic saliency is on the stator (the PM in this case) and thus the brushes are fixed as they are in reality; thus the stator windings do not need to have a brush–commutator. We may now forget about the brushes (Figure 1.2b) by noticing that the axis of the rotor winding field is fixed along axis *q*. The PM induces an emf in the rotor by motion.

We end up with one voltage equation:

$$V_{qr} = R_a I_{qr} + \frac{\partial \Psi_{qr}}{\partial t} + \frac{d\Psi_{qr}}{d\theta_r} \cdot \frac{d\theta_r}{dt} \tag{1.4}$$

But $d\theta_r/dt$ refers to the relative motion between the conductors of the respective windings and the external field:

$$\frac{d\theta_r}{dt} = \omega_r - \omega_b = \omega_r \tag{1.5}$$

Also (1.4)

$$\frac{\partial \Psi_{qr}}{\partial \theta_r} = \Psi_{dr} = \Psi_{PM}; \quad \Psi_{qr} = L_a I_{qr} \tag{1.6}$$

Let us note that Ψ_{dr} corresponds to the flux linkage of a fictitious rotor winding identical to q_r but placed along axis d_r. As this, in fact, does not exist, its flux linkage comes directly from the PM flux produced along axis *d* in the stator.

So Equation 1.4 becomes

$$V_{qr} = R_a I_{qr} + L_a \frac{di_{qr}}{dt} + \Psi_{PM}\omega_r \tag{1.7}$$

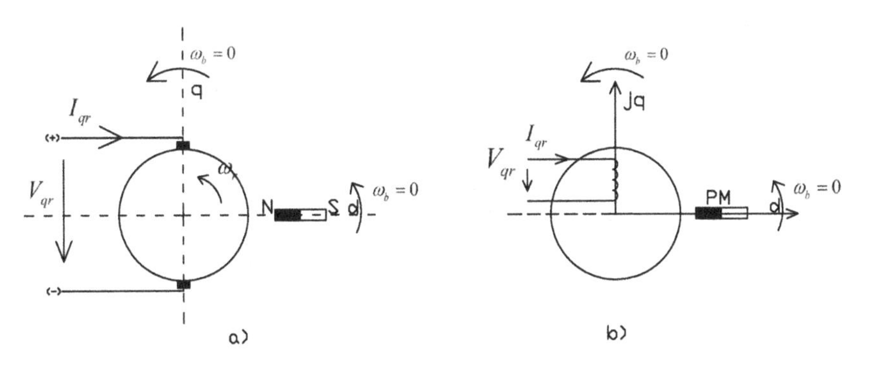

FIGURE 1.2 The d.c. brush PM motor (a) and its dq model (b).

Let us now multiply Equation 1.7 by i_{qr}:

$$V_{qr}I_{qr} = \underset{\substack{\textit{Input}\\ \textit{power}}}{} \quad \underset{\substack{\textit{Copper power}\\ \textit{losses}}}{R_a I_{qr}^2} + \underset{\substack{\textit{Magnetic energy}\\ \textit{variation}}}{L_a I_{qr}\frac{\partial I_{qr}}{\partial t}} + \underset{\substack{\textit{Electromagnetic}\\ \textit{power } P_e}}{\Psi_{PM}\omega_r I_{qr}} \tag{1.8}$$

Note that ω_r is the electrical rotor angular speed $\omega_r = p_1\Omega_1 = p_1 2\pi n$; n is the rotor speed in rps; and p_1 is the pole pairs of the motor.

From the electromagnetic power expression, P_e, the torque, T_e, is straightforward:

$$T_e = \frac{P_e}{\Omega_r} = p_1\Psi_{PM} i_{qr} \tag{1.9}$$

In Equation 4.60 (Vol.1) we had

$$T_e = \frac{p_1 N}{a}\cdot\frac{\Phi_p i_{q1}}{2\pi} = p_1\Psi_{PM} i_{qr} \tag{1.10}$$

N is the total number of conductors in rotor slots, $2a$ is the current path count, and Φ_p is the pole flux.

So

$$\Psi_{PM} = \frac{N}{a}\cdot\frac{\Phi_p}{2\pi} \tag{1.11}$$

The d.c. brush machine is a simplified version of the dq model!

1.5 BASIC dq MODEL OF SYNCHRONOUS MACHINES ($\omega_b = \omega_r$)

For synchronous machines (SMs), in general, the dq axes are fixed to the rotor ($\omega_b = \omega_r$), and thus there will be no motion-induced voltages in the rotor circuits, which are by construction asymmetric (Figure 1.3).

The motion emfs in the stator are proportional to ($-\omega_b$) as this is the speed of the stator coils with respect to the magnetic fields of dq axes.

So the SM equations are straightforward:

$$V_d = R_s I_d + \frac{\partial\Psi_d}{\partial t} - \omega_r\Psi_q; \quad \frac{\partial\Psi_d}{\partial\theta_{er}} = -\Psi_q; \quad \frac{d\theta_{er}}{dt} = \omega_r \tag{1.12}$$

$$V_q = R_s I_q + \frac{\partial\Psi_q}{\partial t} + \omega_r\Psi_d; \quad \frac{\partial\Psi_q}{\partial\theta_{er}} = \Psi_d \tag{1.13}$$

$$V_F = R_F I_F + \frac{\partial\Psi_F}{\partial t} \tag{1.14}$$

$$0 = R_{dr} I_{dr} + \frac{\partial\Psi_{dr}}{\partial t} \tag{1.15}$$

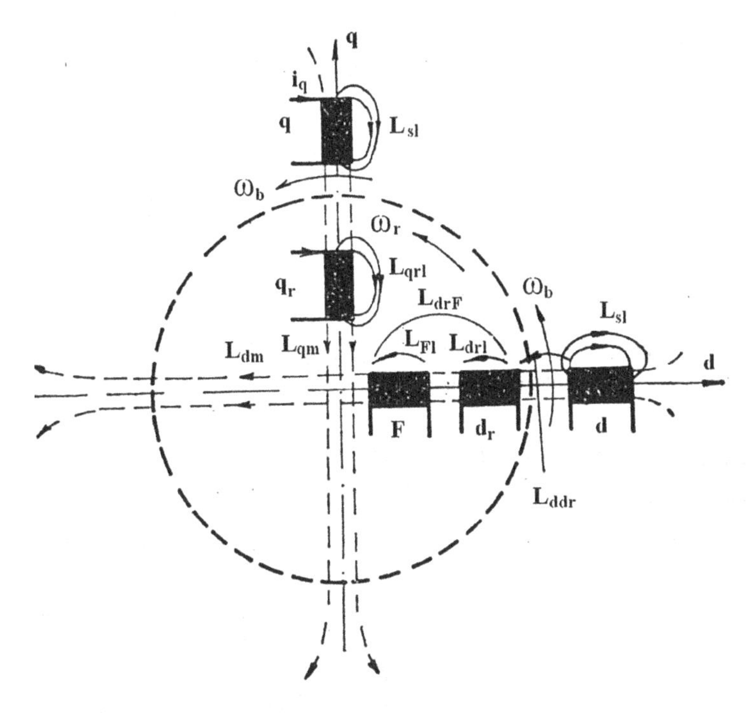

FIGURE 1.3 The dq model of SM.

$$0 = R_{qr} I_{qr} + \frac{\partial \Psi_{qr}}{\partial t} \tag{1.16}$$

After multiplying Equation 1.12 by I_d and Equation 1.13 by I_q, and after adding the two we obtain

$$\underbrace{V_d I_d + V_q I_q}_{\text{Input power}} = \underbrace{R_s\left(I_d^2 + I_q^2\right)}_{\text{Copper losses}} + \underbrace{I_d \frac{\partial \Psi_d}{\partial t} + I_q \frac{\partial \Psi_q}{\partial t}}_{\text{Magnetic energy variation}} + \underbrace{\omega_r\left(\Psi_d I_q - \Psi_q I_d\right)}_{\text{Electromagnetic power}} \tag{1.17}$$

But P_e is

$$P_e = T_e \frac{\omega_1}{p_1}; \quad T_e = p_1\left(\Psi_d I_q - \Psi_q I_d\right) \tag{1.18}$$

We may add the motion equations to complete the set:

$$\frac{J}{p_1} \frac{d\omega_r}{dt} = T_e - T_{load}; \quad \frac{d\theta_{er}}{dt} = \omega_r \tag{1.19}$$

We still have to add the flux linkage–currents relationships. Due to orthogonality of axes, in the absence of magnetic saturation, there is no magnetic coupling between axes d and q and thus

$$\begin{aligned}
&\Psi_d = L_{sl}I_d + \Psi_{dm}; \quad \Psi_{dr} = L_{drl}I_{dr} + \Psi_{dm} + L_{drF}\left(I_{dr} + I_F\right) \\
&\Psi_F = L_{Fl}I_F + \Psi_{dm} + L_{drF}\left(I_{dr} + I_F\right); \quad \Psi_{dm} = L_{dm}\left(I_d + I_{dr} + I_F\right) = L_{dm}I_{dm} \\
&\Psi_q = L_{sl}I_q + \Psi_{qm}; \quad \Psi_{qr} = L_{qrl}I_{qr} + \Psi_{qm}; \quad \Psi_{qm} = L_{qm}\left(I_q + I_{qr}\right) = L_{qm}I_{qm}
\end{aligned} \tag{1.20}$$

From Equations 1.12 through 1.20 we decipher an eighth order system with the variables as six currents, ω_r, and θ_{er} and the inputs as V_d, V_q, V_F, and T_{load}. It is also feasible to use six flux variables with currents as dummy variables, etc.

1.6 BASIC DQ MODEL OF INDUCTION MACHINES ($\omega_b = 0, \omega_r, \omega_1$)

Induction machines with symmetric slots and rotor windings are considered here. They have two stator and two rotor symmetric orthogonal windings (Figure 1.4). By now, we know that two symmetric orthogonal windings are equivalent to three windings at 120°, as in induction machines.

Now the dq axes speed is without constraints, ω_b.

There are motion-induced voltages in the stator (by ω_b) and rotor (by $\omega_b - \omega_r$):

$$\begin{aligned}
V_d &= R_s I_d + \frac{\partial \Psi_d}{\partial t} - \omega_b \Psi_q \\
V_q &= R_s I_q + \frac{\partial \Psi_q}{\partial t} + \omega_b \Psi_d \\
V_{dr} &= R_r I_{dr} + \frac{\partial \Psi_{dr}}{\partial t} - \left(\omega_b - \omega_r\right)\Psi_{qr} \\
V_{qr} &= R_r I_{qr} + \frac{\partial \Psi_{qr}}{\partial t} + \left(\omega_b - \omega_r\right)\Psi_{dr}
\end{aligned} \tag{1.21}$$

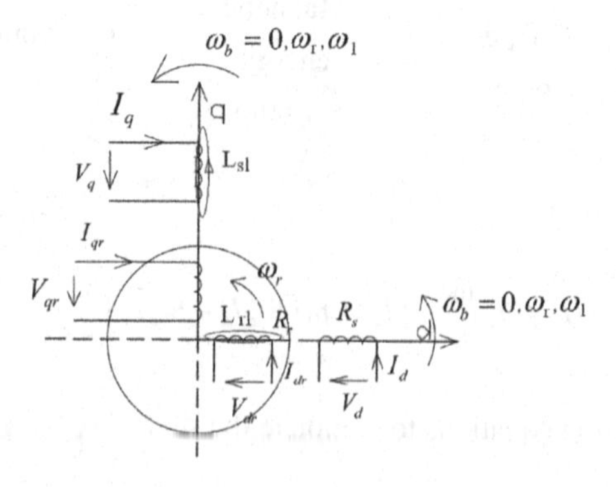

FIGURE 1.4 The basic dq model of IMs.

$$T_e = p_1\left(\Psi_d I_q - \Psi_q I_d\right)$$

$$\frac{J}{p_1}\frac{d\omega_r}{dt} = T_e - T_{load}; \quad \frac{d\theta_{er}}{dt} = \omega_r$$

$$\begin{aligned} \Psi_d &= L_{sl} I_d + \Psi_{dm}; \quad \Psi_{dr} = L_{drl} I_{dr} + \Psi_{dm}; \quad \Psi_{dm} = L_m I_{dm}; \quad I_{dm} = I_d + I_{dr} \\ \Psi_q &= L_{sl} I_q + \Psi_{qm}; \quad \Psi_{qr} = L_{qrl} I_{qr} + \Psi_{qm}; \quad \Psi_{qm} = L_m I_{qm}; \quad I_{qm} = I_q + I_{qr} \end{aligned} \tag{1.22}$$

The torque may also be calculated from rotor equations as

$$T_e = -p_1\left(\Psi_{dr} I_{qr} - \Psi_{qr} I_{dr}\right) \tag{1.23}$$

The sign (−) appears as T_e in Equation 1.23 refers to the torque on the rotor. For the cage rotor, Equation 1.23 is obtained with $V_{dr} = V_{qr} = 0$, while for wound rotor $V_{dr} \neq V_{qr} \neq 0$.

1.7 MAGNETIC SATURATION IN dq MODELS

Flux–current relationships, Equations 1.20 and 1.22, exhibit magnetization inductances L_{dm}, L_{qm}, L_m and leakage inductances L_{sl}, L_{rl}. Magnetic saturation influences both, particularly main inductances, with the exception of the closed-rotor slot configurations or for very large stator (rotor) currents when leakage flux paths saturate as well. Inclusion of magnetic saturation in the dq model, presupposes to notice the type of magnetization in the stator and rotor of main electric machine cores (Table 1.1).

Consequently, magnetic saturation will manifest itself differently in the stator and rotor of various electric machines and so will core losses. However, in the dq model of all machines, as shown in Table 1.1, under steady state, voltages, currents, flux linkages are all d.c. quantities.

This is optimum for a control system design.

Among the many models that include magnetic saturation in the dq model of electric machines, we present here only the model of distinct and unique d, q magnetization curves, $\Psi_{dm}(i_m)$ and $\Psi_{qm}(i_m)$, as shown in Figure 1.5.

In other words, the main (magnetization) inductances, L_{dm}, L_{qm}, in axes d and q depend only on the total magnetization current, I_m:

$$I_m = \sqrt{I_{dm}^2 + I_{qm}^2}; \quad I_{dm} = I_d + I_{dr} + I_F; \quad I_{qm} = I_q + I_{qr} \tag{1.24}$$

TABLE 1.1 Frequences in the Machines

Type of Magnetization	SM	IM	DC Brush Machine
dq model coordinates	$\omega_b = \omega_r$	$\omega_b = \omega_1$	$\omega_b = 0$
Rotor	DC	AC at $\omega_2 = \omega_1 - \omega_r$	AC at $\omega_2 = \omega_r$
Stator	AC at $\omega_1 = \omega_r$	AC at ω_1	DC

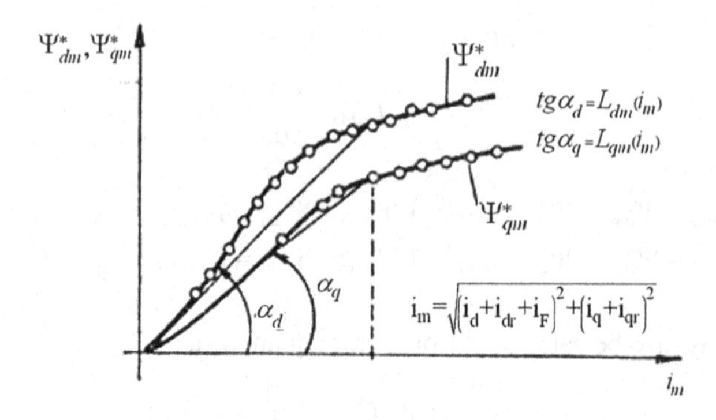

FIGURE 1.5 Distinct unique d,q magnetic curves.

$$\Psi_{dm} = L_{dm}\left(I_m\right)\cdot I_{dm};\quad \Psi_{qm} = L_{qm}\left(I_m\right)\cdot I_{qm} \tag{1.25}$$

with

$$L_{dm} = \frac{\Psi^*_{dm}\left(I_m\right)}{I_m};\quad L_{dm} = \frac{\Psi^*_{qm}\left(I_m\right)}{I_m} \tag{1.26}$$

Ψ^*_{dm}, Ψ^*_{qm} are measured (or calculated) values.

L_{dm}, L_{qm} are related to the normal permeability, μ_n, in the core, as shown in Figure 1.5. They are valid for steady state (exactly so only in the d.c. rotor of SMs).

For transients, the time derivatives of Ψ_{dm} and Ψ_{qm} are required.

From Equations 1.24 through 1.26:

$$\frac{d\Psi_{dm}}{dt} = \frac{d\Psi^*_{dm}}{di_m}\cdot\frac{di_m}{dt}\cdot\frac{i_{dm}}{i_m} + \frac{\Psi^*_{dm}}{i_m^2}\left(i_m\frac{di_m}{dt} - i_{dm}\frac{di_m}{dt}\right) \tag{1.27}$$

$$\frac{d\Psi_{qm}}{dt} = \frac{d\Psi^*_{qm}}{di_m}\cdot\frac{di_m}{dt}\cdot\frac{i_{qm}}{i_m} + \frac{\Psi^*_{qm}}{i_m^2}\left(i_m\frac{di_{qm}}{dt} - i_{qm}\frac{di_m}{dt}\right) \tag{1.28}$$

$$\frac{di_m}{dt} = \frac{i_{qm}}{i_m}\cdot\frac{di_{qm}}{dt} + \frac{i_{dm}}{i_m}\cdot\frac{di_{dm}}{dt} \tag{1.29}$$

So, finally

$$\begin{aligned}\frac{d\Psi_{dm}}{dt} &= L_{ddm}\frac{di_{dm}}{dt} + L_{qdm}\frac{di_{qm}}{dt}\\ \frac{d\Psi_{qm}}{dt} &= L_{qdm}\frac{di_{qm}}{dt} + L_{qqm}\frac{di_{qm}}{dt}\end{aligned} \tag{1.30}$$

with

$$L_{ddm} = L_{dmt}\frac{i_{dm}^2}{i_{m^2}} + L_{dm}\frac{i_{qm}^2}{i_m^2}$$

$$L_{qqm} = L_{qmt}\frac{i_{qm}^2}{i_m^2} + L_{qm}\frac{i_{dm}^2}{i_m^2}$$

$$L_{dqm} = \left(L_{dmt} - L_{dm}\right)\frac{i_{dm}i_{qm}}{i_m^2} = \left(L_{qmt} - L_{qm}\right)\frac{i_{dm}i_{qm}}{i_m^2} = L_{qdm} \tag{1.31}$$

$$L_{dmt} = \frac{d\Psi_{dm}^*}{di_m}; \quad L_{qmt} = \frac{d\Psi_{qm}^*}{di_m}$$

$L_{dqm} = L_{qdm}$; ($L_{dmt} - L_{dm} = L_{qmt} - L_{qm}$) because of reciprocity theorem.
The above equations give rise to the following observations:

- The magnetization flux along each of the d and q axes is influenced by both currents i_d and i_q, both under steady state and transients; but, in this model, unique d and q magnetization curves exist. Only for the underexcited large synchronous machine, the model was found through the FEM [8], less acceptable.
- For transients ($d\Psi_{dm}/dt \neq 0$ and/or $d\Psi_{qm}/dt \neq 0$), there is a kind of magnetic coupling between the two orthogonal axes due to magnetic saturation.
- Zero magnetic coupling inductance ($L_{qdm} = 0$) occurs only in the absence of magnetic saturation (since $L_{dm} = L_{dmt}$) or when either $i_d = 0$ or $i_q = 0$.
- For standstill operations at small a.c. p.u. currents, the transient inductances L_{dmt}, L_{qmt} are replaced by incremental inductances L_{dmi}, L_{qmi}:

$$L_{dmi} = \frac{\Delta\Psi_{dm}^*}{\Delta i}; \quad L_{qmi} = \frac{\Delta\Psi_{qm}^*}{\Delta i} \tag{1.32}$$

  found from the local (small-amplitude) hysteresis cycle incremental permeabilities $\mu_i \approx (120 - 150)\mu_0$.
- Magnetic saturation is worth accounting for in modern, stressed-to-the-limit, electric machines.
- To consider magnetic saturation, we need to apply Equations 1.28 through 1.31 to the dq model of SM or IM and then introduce, perhaps, flux variables, but with Ψ_{dm}, Ψ_{qm} as stator variables instead of Ψ_d, Ψ_q, to handle the solution of the system of equations easily [11,12].

1.8 FREQUENCY (SKIN) EFFECT CONSIDERATION IN dq MODELS

The skin (frequency) effect on rotor cage bars is known. It translates into an increase in rotor resistance R_r by $k_R > 1$ and a decrease in the rotor leakage inductance L_{rl} by $k_X < 1$, as the rotor (slip) frequency, ω_2, increases. Large SGs exhibit strong skin effect in the solid iron rotor core (for nonsalient two-pole configurations) or in the rotor damping cage.

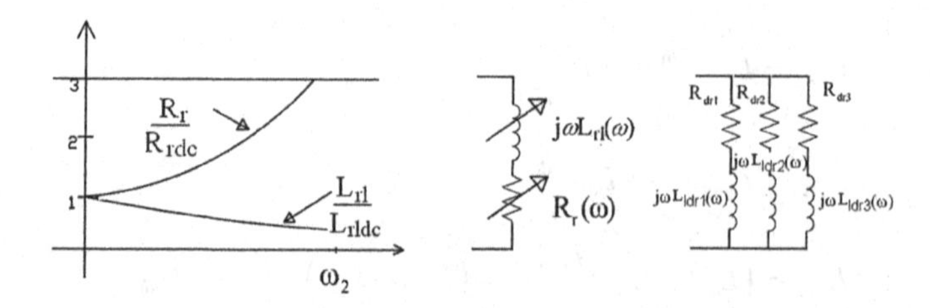

FIGURE 1.6 From single circuits with variable parameters to multiple circuits with constant parameters.

The skin effect in the electric conductors is influenced by magnetic saturation also, but accounting for this influence is not considered here. We thus implicitly separate magnetic saturation from the skin effect.

To include the skin effect in the dq model, we introduce 2,3 fictitious cage constant, R_{dr}, L_{1dr}, parameters circuits in parallel such that their frequency response matches, to an assigned global error, the real machine with variable rotor parameters (Figure 1.6).

The multiple circuit parameters may be calculated by FEM or from the frequency response standstill test by regression methods.

In machines with strong core losses, the latter may be introduced in the dq model by adding two more orthogonal short-circuited stator windings (on the machine side with a.c. magnetization), whose leakage inductance may be neglected and whose resistances are calculated from measured or calculated iron losses in the machine [11].

Note: By including magnetic saturation, frequency effects, and core losses, the dq model has been enhanced considerably.

1.9 EQUIVALENCE BETWEEN DQ MODELS AND AC MACHINES

The dq model is a fair representation of the actual a.c. machine if it behaves like the latter in terms of torque, power, speed, and losses.

Equivalence conditions are thus necessary.

Conservation of mmf fundamentals is an implicitly powerful equivalence condition. Let us project the A, B, C stator phase mmfs along the d and q axes (Figure 1.7):

$$\begin{aligned} F_d &= W_1K_{w1}[i_A\cos(-\theta_{eb}) + i_B\cos(-\theta_{eb}+2\pi/3) + i_C\cos(-\theta_{eb}-2\pi/3) \\ F_q &= W_1K_{w1}[i_A\sin(-\theta_{eb}) + i_B\sin(-\theta_{eb}+2\pi/3) + i_C\sin(-\theta_{eb}-2\pi/3) \end{aligned} \tag{1.33}$$

For

$$\begin{aligned} F_d &= W_dK_{wd}i_d \\ F_q &= W_dK_{wd}iq \end{aligned} \tag{1.34}$$

in practice

$$\frac{W_1K_{w1}}{W_dK_{wd}} = \frac{2}{3} \text{ or } \sqrt{\frac{2}{3}} \tag{1.35}$$

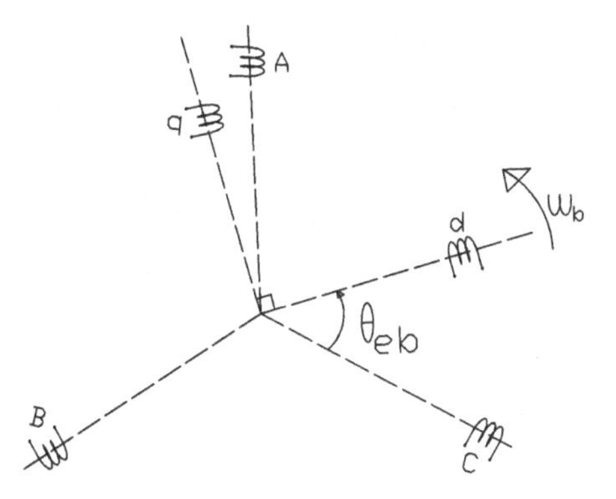

FIGURE 1.7 Three-phase a.c.-dq mmf equivalence.

As can be shown easily, the $\sqrt{\frac{2}{3}}$ ratio conserves the power—from the dq model to the actual machine—while for the ratio $\frac{2}{3}$, the $\frac{3}{2}$ dq model power equals the actual active power of the machine; the same is true for torque equivalence.

Let us consider the $\frac{2}{3}$ ratio, which is extendedly used for electric machine control.

The equivalence between the two-phase dq model and the three-phase a.c. machine requires one more variable: the zero component, or V_0, i_0, Ψ_0:

So, the so-called generalized Park transformation for the three-phase stator, $S(\theta_{eb})$, is the same for currents, voltages, and flux linkages:

$$\begin{vmatrix} i_d \\ i_q \\ i_0 \end{vmatrix} = \left|S_{dq0}\right| \cdot \begin{vmatrix} i_A \\ i_B \\ i_C \end{vmatrix}; \quad \theta_{eb} = \int \omega_b \, dt; \quad \frac{d\theta_{eb}}{dt} = \omega_b \tag{1.36}$$

$$\left|S_{dq0}\right| = \frac{2}{3} \begin{vmatrix} \cos(-\theta_{eb}) & \cos(-\theta_{eb} + 2\pi/3) & \cos(-\theta_{eb} - 2\pi/3) \\ \sin(-\theta_{eb}) & \sin(-\theta_{eb} + 2\pi/3) & \sin(-\theta_{eb} - 2\pi/3) \\ 1/2 & 1/2 & 1/2 \end{vmatrix} \tag{1.37}$$

The zero-sequence component, i_0 in Equation 1.37 is

$$i_0 = \frac{(i_A + i_B + i_C)}{3} \tag{1.38}$$

It does not contribute to the traveling field and thus interacts with the leakage inductance and resistance of stator phases:

$$V_{0s} \approx R_s + L_{sl} \frac{di_{0s}}{dt} \tag{1.39}$$

For symmetric steady state and transients $i_0 = 0$; it is also zero for star phase connection, anyway.

Note: The Park transformation may be extended to m phases ($m_1 = 6, 9, 12 \ldots$) when the angle in Equation 1.37 is $2\pi/m_1$ and the number of columns is m_1.

For every three symmetric phases, a zero sequence component is required for full (mathematical) equivalence, unless a separated star connection is applied for all three symmetric phase modules.

The inverse matrix of $|S_{dq0}|$ is

$$\left[S_{dq0}\right]^{-1} = \frac{3}{2}\left[S_{dq0}\right]^{t} \tag{1.40}$$

Let us calculate the power in the dq0 model:

$$P_{\mathrm{dq0}} = \left[V_{\mathrm{dq0}}\right]^{t}\left[I_{\mathrm{dq0}}\right] = V_d I_d + V_q I_q + V_0 I_0 \tag{1.41}$$

It follows that

$$\begin{aligned}\frac{3}{2}P_{\mathrm{dq0}} &= \frac{3}{2}\left[V_{\mathrm{ABC}}\right]^{\mathrm{T}}\left[S_{\mathrm{dq0}}\right]^{\mathrm{T}}\left[S_{\mathrm{dq0}}\right]\left[I_{\mathrm{ABC}}\right] = \left[V_{\mathrm{ABC}}\right]^{\mathrm{T}}\left[I_{\mathrm{ABC}}\right] = P_{\mathrm{ABC}} \\ &= V_{\mathrm{a}}I_{\mathrm{a}} + V_{\mathrm{b}}I_{\mathrm{b}} + V_{\mathrm{c}}I_{\mathrm{c}}\end{aligned} \tag{1.42}$$

This equivalence refers to the active instantaneous power and thus also to instantaneous torque (see Ref. [5] for a detailed derivation of Equation 1.42):

$$\frac{3}{2}T_{\mathrm{edq0}} = T_{\mathrm{eABC}} = \frac{3}{2}p_1\left(\psi_d i_q - \psi_q i_d\right) \tag{1.43}$$

Note: At steady state in rotor coordinates ($\omega_b = \omega_r$) for SMs and synchronous coordinates ($\omega_b = \omega_1$) for IMs, d, q voltages and currents are d.c. and thus the dq model does not show any reactive power.

However, it may be demonstrated that the reactive power, Q_{ABC}, of the actual a.c. three-phase machine is represented in the dq model by

$$Q_{\mathrm{ABC}} = -\frac{3}{2}\left(V_d I_q - V_q I_d\right) \tag{1.44}$$

Also, stator copper losses are

$$p_{\mathrm{copper}} = 3R_s\left(I_d^2 + I_q^2\right)/2 \tag{1.45}$$

R_s stays the same for the dq0 model as for the three a.c. phases.

The damping cage and the field winding in SMs are similar (orthogonal), both in the actual machine and in the dq0 model.

For the wound rotor induction machine, which has three phases in the rotor, a similar Park transformation as for the stator is applied, but, instead of θ_{es}, the d axis coordinate angle, θ_{ebr}, is

$$\theta_{ebr} = \int(\omega_b - \omega_r)dt; \quad \frac{d\theta_{ebr}}{dt} = (\omega_b - \omega_r) \tag{1.46}$$

because the rotor conductors rotate at $\omega_b - \omega_r$ with respect to the dq axes (coordinates) at a speed ω_b.

So far we have defined the voltage, current, and flux linkage equivalence between three a.c. phase windings, and the dq0 model with the power–torque losses relationships. The only point left out is the relationship between the three-phase a.c. windings and the dq model inductances. This will be done in the following chapters. We will now introduce the space phasor (complex variable) model.

1.10 SPACE PHASOR (COMPLEX VARIABLE) MODEL

The space phasor (or complex variable) model may be derived from the dq0 model because it is based on same assumptions—but it may also be derived directly from the phase-coordinate model [4,6].

Here we make use of the dq0 model because we have already derived it.

For the space phasor model, we introduce the denotations in the stator:

$$\bar{I}_s = I_d + jI_q; \quad \bar{V}_s = V_d + jV_q; \quad \bar{\psi}_s = \psi_d + j\psi_q \tag{1.47}$$

$\bar{I}_s$, $\bar{V}_s$, $\bar{\psi}_s$ are called direct space phasors (vectors) of currents, voltages, and flux linkages.

At least for the flux linkage, $\bar{\psi}_s$refers to a space phasor whose position with respect to phase A in stator axis is that of the traveling magnetic field axis, which rotates at speed ω_b.

Making use of the I_d, I_q expressions in Equation 1.36 we obtain

$$I_d + jI_q = \bar{I}_s = \frac{2}{3}\left[I_A + I_B e^{j\frac{2\pi}{3}} + I_C e^{-j\frac{2\pi}{3}}\right]e^{-j\theta_{eb}}; \quad \frac{d\theta_{eb}}{dt} = \omega_b \tag{1.48}$$

For the rotor, $\theta_{ebr} = \int(\omega_b - \omega_r)dt$.

It is evident that the zero-sequence current is missing in Equation 1.48 and thus has to be added. Equation 1.48 illustrates, in fact, a rotational transformation by angle θ_{eb}:

$$\bar{I}_s = \bar{I}_s^s e^{-j\theta_{eb}} \tag{1.49}$$

where $\bar{I}_s^s s$ is the space phasor in stator coordinates; if we take the transformation backward from Equation 1.49 we find

$$I_A(t) = \mathrm{Re}\left(\overline{I}_s^s\right) + I_0 = \mathrm{Re}\left(\overline{I}_s e^{j\theta_{eb}}\right) + I_0 \tag{1.50}$$

$$I_0 = \frac{1}{3}\left(I_A + I_B + I_C\right) \tag{1.51}$$

The stator voltage dq equations, Equations 1.12 through 1.14, and 1.22, are the same for SMs and IMs, for the *dq* axis rotating at ω_b and can simply be converted into the space phasor (complex variable) formulation:

$$\overline{V}_s = V_d + jV_q = R_s\overline{I}_s + \frac{d\overline{\Psi}_s}{dt} + j\omega_b\overline{\Psi}_s; \quad \overline{\Psi}_s = \Psi_d + j\Psi_q \tag{1.52}$$

For the wound rotor induction machine, the rotor voltage equation (Equation 1.22) is

$$\overline{V}_r = V_{dr} + jV_{qr} = R_r\overline{I}_r + \frac{d\overline{\Psi}_r}{dt} + j\left(\omega_b - \omega_r\right)\overline{\Psi}_r; \quad \overline{\Psi}_r = \Psi_{dr} + j\Psi_{qr} \tag{1.53}$$

The electromagnetic torque (Equation 1.44) is

$$T_e = \frac{3}{2}p_1 \mathrm{Re}\left[j\overline{\Psi}_s\overline{I}_s^*\right] = -\frac{3}{2}p_1 \mathrm{Re}\left[j\overline{\Psi}_r\overline{I}_r^*\right] \tag{1.54}$$

To illustrate phenomenologically the complex variable, let us follow the expression of the symmetric three-phase sinusoidal currents, typical in a.c. machines:

$$I_{ABC} = I\sqrt{2}\cos\left[\omega_1 t - (i-1)\frac{2\pi}{3}\right]; \quad i = 1,2,3 \tag{1.55}$$

For stator coordinates ($\omega_b = 0$, $\theta_{eb} = \theta_{0s} = 0$) from Equations 1.48 and 1.55:

$$\overline{I}_s^s = \frac{3}{2}\left[i_A(t) + i_B(t)e^{j\frac{2\pi}{3}} + i_C(t)e^{-j\frac{2\pi}{3}}\right] = I\sqrt{2}\left[\cos\omega_1 t + j\sin\omega_1 t\right] \tag{1.56}$$

$\theta_{eb} = 0$ means that axis *d* is aligned to stator phase A axis.

For synchronous coordinates ($\omega_b = \omega_1$), with $\theta_{eb} = \omega_b t = \omega_1 t$ from Equation 1.49:

$$\overline{I}_s = \overline{I}_s^s e^{-j\omega_1 t} = I\sqrt{2}\left(\cos\omega_1 t + j\sin\omega_1 t\right)\left(\cos\omega_1 t - j\sin\omega_1 t\right) = I\sqrt{2} = I_d \tag{1.57}$$

So, in stator coordinates ($\omega_b = 0$) $\overline{I}_s^s$ refers to an a.c. vector whose position changes in time at ω_1 speed (Figure 1.8.a). On the other hand, in synchronous coordinates ($\omega_b = \omega_1$), we have a d.c. vector, along axis *d* (in our case, because the initial value of θ_{eb} was chosen as zero), which now rotates together with axis *d* at ω_1 speed (Figure 1.8.b).

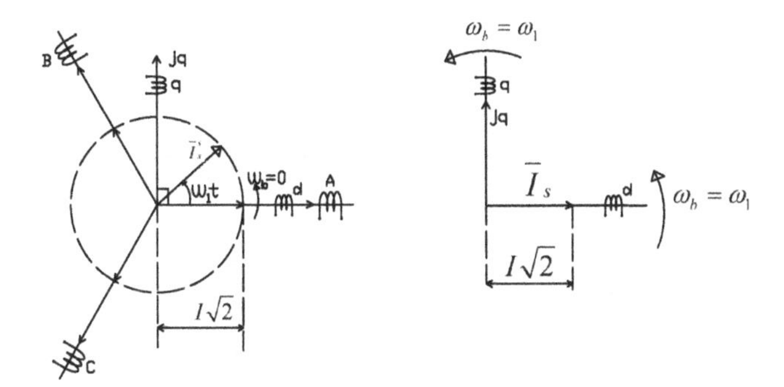

FIGURE 1.8 Steady-state a.c. current space phasor in stator coordinates (a) and synchronous coordinates (b).

Besides the abbreviation in writing the dq0 model equations, the space phasor (complex variable) formalism brings up some more concepts in the flux orientation control of electric machines.

1.11 HIGH-FREQUENCY MODELS FOR ELECTRIC MACHINES

Atmospheric (microsecond front) waves, commutation (tens of microsecond front) voltage pulses, and IGBT (MOSFET)-triggered pulses (0.5–2 μs front waves) are so fast that the machine resistances and inductance influences are small, while the stray capacitors between turns, coils, and from windings to frame, similarly as for transformers (Figure 1.9), take over up to 400 Hz the R,L,E circuit model of electric machines is valid, while above 20 kHz, the high-frequency model is practical (Figure 1.9).

In between, (400 Hz–20 kHz) the electric machines have to be modeled correctly as they interact with the cables that carry the current from the power electronics supply to the motor. There we should use two types of voltage fast pulses to reach the machine terminals: differential mode (DM) and common mode (CM) pulses. The latter refers to the neutral potential pulsation with respect to the ground and the machine response through stray capacitors of windings to frame, and frame (bearing) to ground.

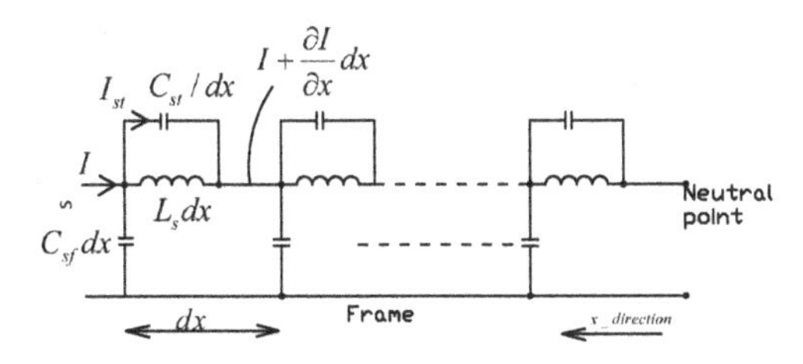

FIGURE 1.9 Distributed high-frequency parameter motor for a.c. stator windings.

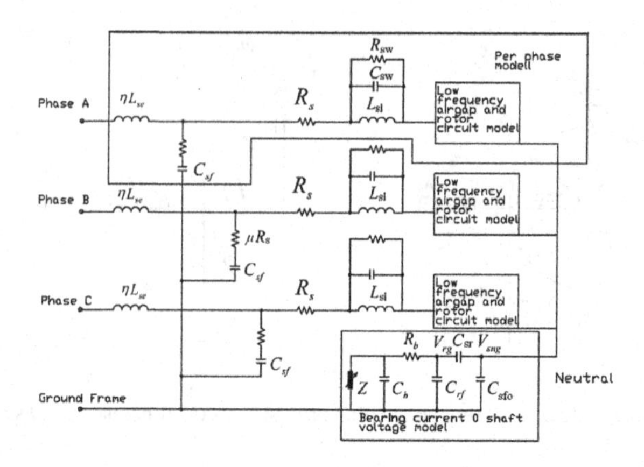

FIGURE 1.10 Universal (low-to-high-frequency) model of a.c. machines.

Recently a universal, low-to-high-frequency model for the IM has been introduced [13,14] (Figure 1.10). It is valid for all a.c. machines if the low-frequency section of the circuit model is the pertinent one.

Such a model is instrumental in explaining voltage tripling, electromagnetic interference, and bearing currents when EMs are associated with PWM converters for energy or motion modern control.

The bearing model in Figure 1.10 consists of common mode (windings to frame) capacitances: stator to frame, C_{sf}, rotor to frame, C_{rf}, stator to rotor, C_{sr}, and bearing balls race resistance, R_b, with a series/parallel combination of capacitance, C_b, and nonlinear impedance, Z, for random charging and discharging of the rotor shaft due to asperity point contacts puncturing the oil film. C_{sf-0} matches the low-frequency response.

The PWM converter intervenes here mainly as a neutral-to-ground zero-sequence voltage source, V_{nsg}. At low frequency, the voltage at machine terminals is distributed uniformly from the terminal point to the neutral point.

In the medium frequency (400–20 kHz) the impact of the airgap—in the rotor part of the low-frequency model—may be neglected. Three capacitors, C_{sf}, C_{sf-0}, C_{sw}, are introduced in Figure 1.10 to match the mid-to-high-frequency response:

- C_{sf} is the stator-to-frame capacitor of the first slot per phase
- C_{sf0} is the capacitance of stator-to-frame in the neutral point
- C_{sw} is the interturn capacitance per phase

The introduction of the universal model here was done for completeness and as a starting point when electromagnetic compatibility (EMC) aspects of PWM converter-controlled generators and motors are of interest.

The recent developments in SiC PWM converters with switching frequencies above 20 Hz make the super high-frequency models of a.c. machine even more important.

1.12 ORTHOGONAL MODELS OF MULTIPHASE A.C. MACHINES

Three-phase a.c. machines—induction and synchronous—are typical. But the advantages in power electronics led to the use of multiphase a.c. machines for variable speed motor/generator drives as the power per phase is reduced, the PWM offers more voltage vectors (lower current THD for given switching frequency) and the system becomes more fault tolerant. The efficiency of the machine is not increased notably but the maximum torque is up (by up to 20%) for given geometry.

Clarke transformation is used to map the phase voltages and currents in a, for example, 6 phase asymmetrical winding (30° electrical degrees phase shift of neighboring phases) into two orthogonal stationary coordinates, α–β and x–y, plus a zero sequence [15] model:

$$\begin{bmatrix} V_\alpha & V_\beta & V_x & V_y & V_{0+} & V_{0-} \end{bmatrix}^T = \begin{bmatrix} T \end{bmatrix} \cdot \begin{bmatrix} V_{a1} & V_{b1} & V_{c1} & V_{a2} & V_{b2} & V_{c2} \end{bmatrix} \quad (1.58)$$

$$\left|T_{6phase}\right| = \frac{1}{\sqrt{3}} \begin{bmatrix} 1 & -\frac{1}{2} & -\frac{1}{2} & \frac{\sqrt{3}}{2} & -\frac{\sqrt{3}}{2} & 0 \\ 0 & \frac{\sqrt{3}}{2} & -\frac{\sqrt{3}}{2} & \frac{1}{2} & \frac{1}{2} & -1 \\ 1 & -\frac{1}{2} & -\frac{1}{2} & -\frac{\sqrt{3}}{2} & \frac{\sqrt{3}}{2} & 0 \\ 0 & -\frac{\sqrt{3}}{2} & \frac{\sqrt{3}}{2} & \frac{1}{2} & \frac{1}{2} & -1 \\ 1 & 1 & 1 & 0 & 0 & 0 \\ 0 & 0 & 0 & 1 & 1 & 1 \end{bmatrix} \quad (1.59)$$

To reach the dq model for the αβ subspace, the Park transform is needed:

$$\begin{bmatrix} V_{ds} & V_{qs} \end{bmatrix}^T = \left|P(\theta_s)\right| \cdot \begin{bmatrix} V_\alpha & V_\beta \end{bmatrix}^T; \quad P(\theta_s) = \begin{bmatrix} \cos\theta_s & \sin\theta_s \\ -\sin\theta_s & \cos\theta_s \end{bmatrix} \quad (1.60)$$

The α, β, x, y equations (in stator coordinates) of the 6 phase IM are thus!

$$\begin{aligned} &V_\alpha = R_s i_\alpha + L_s \frac{di_\alpha}{dt} + L_m \frac{di_{\alpha r}}{dt}; \quad V_\beta = R_s i_\beta + L_s \frac{di_\beta}{dt} + L_m \frac{di_{\beta r}}{dt} \\ &V_x = R_s i_x + L_{ls} \frac{di_x}{dt}; \quad V_y = R_s i_y + L_{ls} \frac{di_y}{dt} \end{aligned} \quad (1.61)$$

$$\begin{aligned} 0 &= R_r i_{\alpha r} + L_r \frac{di_{\alpha r}}{dt} + L_m \frac{di_{\alpha r}}{dt} + \omega_r L_r i_{\beta r} + \omega_r L_m i_{\beta s} \\ 0 &= R_r i_{\beta r} + L_r \frac{di_{\beta r}}{dt} + L_m \frac{di_{\beta r}}{dt} - \omega_r L_r i_{\alpha r} - \omega_r L_m i_{\alpha s} \end{aligned} \quad (1.62)$$

ω_r is the electrical rotor speed and R_s, L_s, R_r, L_r, L_{ls}, L_m are the typical parameters of the orthogonal (dq) model of cage rotor IM.

The transformation (1.58)–(1.60) is orthogonal (direct power conservation) and thus thetorque is:

$$T_e = p \cdot L_m \cdot \left(i_{\beta r} \cdot i_\alpha - i_{\alpha r} \cdot i_\beta \right) \tag{1.63}$$

(The factor m/2 is missing; m—number of phases)

It should be noted that the V_x, V_y equations are not related to torque production, so, in a first attempt, i_x and i_y may be regulated to zero. They may be used, however, to introduce current harmonics for other dedicated purposes, especially for operation under some faulty phases [16–20], to reduce torque pulsations and/or reduce copper losses. So far the zero components have been neglected as two or a single null point connection of phases was considered. However if, for example, no null points are considered, their equations have to be added (and used for postfault operation) [16]:

$$V_{0+} = R_s i_{0+} + L_{ls} \frac{di_{0+}}{dt}; \quad V_{0-} = R_s i_{0-} + L_{ls} \frac{di_{0-}}{dt} \tag{1.64}$$

Equations (1.64) are similar to those of x, y subspace as they are not related to net torque production (L_{ls} is the stator leakage inductance).

The same equations as above may be used after open phase faults but the decoupling of α-β, x-y and $0_+ - 0-$ subspaces is lost. For example, if $i_{c2} = 0$ (phase C_2 open), the α, x and 0+, which do not contain i_{c2} term, remain unchanged (as in matrix $|T_{6phase}|$) but

$$\begin{aligned} i_\beta &= \frac{1}{\sqrt{3}} \left[\frac{\sqrt{3}}{2} \left(i_{b1} - i_{c1} \right) + \frac{1}{2} \left(i_{a2} + i_{b2} \right) \right] \\ i_y &= \frac{1}{\sqrt{3}} \left[\frac{\sqrt{3}}{2} \left(i_{c1} - i_{b1} \right) + \frac{1}{2} \left(i_{a2} + i_{b2} \right) \right]; \quad i_{0-} = \frac{1}{\sqrt{3}} \left(i_{a2} + i_{b2} \right) \end{aligned} \tag{1.65}$$

For two isolated neutrals: $i_{0+} = i_{0-} = 0$ and for $i_{c2} = 0$, $i_{a2} = -\, i_{b2}$ and thus $i_y = -i_\beta$, $i_{0-} = 0$.

For a single neutral point ($i_{c2} = 0$):

$$i_{0-} = \frac{1}{\sqrt{3}} \left(i_{a2} + i_{b2} \right) = i_\beta + i_y; \quad i_{0+} = -i_{0-} \tag{1.66}$$

Such simplifications are very useful in fault-tolerant control.

A similar approach has been applied to add a number of phases to IMs [18] or PMSMs [19]. For a 9 phase IM the generalized Park transformation for the stator is considering the $2\pi/9$ phase shift between adjacent phases:

$$P_{gphase} = \begin{matrix} 0 \\ d_1 \\ q_1 \\ d_3 \\ q_3 \\ d_5 \\ q_5 \\ d_7 \\ q_7 \end{matrix} \begin{bmatrix} \frac{1}{\sqrt{2}} & \frac{1}{\sqrt{2}} & \cdots\cdots & \frac{1}{\sqrt{2}} \\ \cos\theta & \cos(\theta-2\pi/9) & \cdots\cdots & \cos(\theta-8\pi/9) \\ -\sin\theta & \sin(\theta-2\pi/9) & \cdots\cdots & \sin(\theta-8\pi/9) \\ \cos 3\theta & \cos 3(\theta-2\pi/9) & \cdots\cdots & \cos 3(\theta-8\pi/9) \\ -\sin 3\theta & -\sin 3(\theta-2\pi/9) & \cdots\cdots & -\sin 3(\theta-8\pi/9) \\ \cos 5\theta & \cos 5(\theta-2\pi/9) & \cdots\cdots & \cos 5(\theta-8\pi/9) \\ -\sin 5\theta & -\sin 5(\theta-2\pi/9) & \cdots\cdots & -\sin 5(\theta-8\pi/9) \\ \cos 7\theta & \cos 7(\theta-2\pi/9) & \cdots\cdots & \cos 7(\theta-8\pi/9) \\ -\sin 7\theta & -\sin 7(\theta-2\pi/9) & \cdots\cdots & -\sin 7(\theta-8\pi/9) \end{bmatrix} \cdot \sqrt{\frac{2}{9}} \quad (1.67)$$

with θ—the angle of rotor flux.

In this case, 0, d_r, q_r row vectors are related to harmonics subspaces 9i (i = 0, 1, 2, 3,…) and, respectively, γ, 9i ± γ (i = 1, 2, 3,…).

Stator equations of the kth dq sequence (k = 1, 3, 5, 7) for a 9 phase IM are:

$$\begin{aligned} V_{sdk} &= R_s i_{sdk} - k\omega\Psi_{sdk} + \frac{d\Psi_{sdk}}{dt}; \quad V_{s0} = R_s i_{s0} + \frac{d\Psi_{s0}}{dt} \\ V_{sqk} &= R_s i_{sqk} + k\omega\Psi_{sqk} + \frac{d\Psi_{sqk}}{dt}; \quad \Psi_{s0} = L_{ls} i_{s0} \\ V_{\gamma qk} &= R_{rk} i_{rqk} + k(\omega - p_1\omega_m)\Psi_{sdk} + \frac{d\Psi_{rqk}}{dt}; \\ V_{\gamma dk} &= R_{rk} i_{rdk} - k(\omega - p_1\omega_m)\Psi_{rqk} + \frac{d\Psi_{rdk}}{dt}; \end{aligned} \quad (1.68)$$

The rotor resistance R_{rk} refers to the kth harmonic (k=1, 3, 5, 7).

$$\begin{aligned} \Psi_{sd,q,k} &= (L_{ls} + L_{mk}) i_{sd,qk} + L_{mk} i_{rd,qk} \\ \Psi_{rd,q,k} &= (L_{lr} + L_{mk}) i_{rd,qk} + L_{mk} i_{sd,qk} \end{aligned} \quad (1.69)$$

With torque T_e:

$$T_e = p_1 \left[\sum_{k1,3,5,7} L_{mk} \left(i_{sdk} i_{rqk} - i_{sqk} i_{rdk} \right) \right] \quad (1.70)$$

The similarity with the dq model of 3 phase IMs is notable but so are the differences. The harmonic dq subspaces are used under faulty operation to reduce copper losses and torque pulsations.

1.13 SUMMARY

- Electric machines undergo transients when connected to (disconnected from) the power source, for load variations, and during exposure to steep front voltage pulses from atmospheric sources, or from PWM converter for energy and motion control.
- The investigation of machine transients or of high-frequency behavior needs dedicated modeling that is both precise enough and practical in terms of CPU time.
- High-frequency models are treated separately as they include the electric machine–distributed stray capacitor network.
- Most transients, especially for energy and motion control, refer to low frequency and they are treated by circuit models among which the dq0 model and the space phasor (complex variable) model have gained widespread acceptance and are introduced here intuitively.

  The machine circuit parameters for such models are either calculated in the design stage or measured in special tests.
- The dq0 (orthogonal axis) model's success is explained by its merit of exhibiting constant (rotor position independent) self and mutual inductances in contrast to the phase-coordinate model. The former will be applied in the following chapters dedicated to a.c. machine transients.
- The dq0 physical model concept is introduced based on commutator orthogonal windings with all brushes aligned to dq axes that are solidary with the machine part that has magnetic (and winding) anisotropy. In this way motion-induced voltages are avoided in the anisotropic part of the machine, and the dq0 model shows constant inductances. The dq0 model is applicable to 2,3 or more phase a.c. machines.
- The zero-sequence current, which occurs for three-phase a.c. machine modeling, besides the physical dq model, completes the equivalence. It is zero for star connection or for symmetric (balanced) operation. In the dq0 model, stator and rotor variables have the same frequency $\omega_1 - \omega_b$, under steady state in contrast to real a.c. machines.
- For the modeling of 6-,9-,1two-phase a.c. machines, 2(3,4) *dq* axes pairs plus 1(2,3) zero-sequence currents are required for full equivalence in terms of power, losses, flux linkages, and torques. The zero sequences act only on the stator (rotor) leakage inductances and resistances.
- For the synchronous machine dq0 model, the brush speed (*dq* axes speed) $\omega_b = \omega_r$ (rotor speed) because the rotor of the SM is d.c. (or PM) excited or of reluctance type ($L_{dm} \neq L_{qm}$). In this way, for steady state, it is d.c. in the dq0 model, that is, the ideal case for control design.
- For the induction machine dq0 model, the brush speed (*dq* axes speed) is indifferent, but $\omega_b = 0, \omega_r, \omega_1$ are most used. To get d.c. under steady state, $\omega_b = \omega_1$, that is, synchronous coordinates.
- For wound rotor IMs with PWM converter frequency control in the rotor, the *dq* axes may also be attached to the rotor ($\omega_b - \omega_r$) Stator coordinates ($\omega_b = 0$) may be used for IMs to investigate softstarter or stator–inverter-fed behavior.

- Magnetic crosscoupling saturation may be elegantly (though not locally) accounted for, in the dq0 model, by introducing transient inductances for single but distinct magnetization curves along axes d and q, with magnetization inductances, L_{dm}, L_{qm}, dependent only on the resultant magnetization current, I_m.
- Skin (frequency) influence on leakage inductances and resistances in the rotor of IMs and SMs may be handled in the dq0 model by adding additional fictitious constant parameter circuits in parallel. Three circuits in parallel suffice, in practice, for all SMs and IMs.
- The space phasor (complex variable) model is assembled from the dq model $\bar{V}_s = V_d + jV_q$, etc. It provides an abbreviated form of the equations and alludes to physical intuitional interpretations related to the traveling field concept. The zero-sequence current is, again, additional.
- The equivalence of a.c. machine stator or rotor symmetric three-phase windings to the dq0 model is based here on the mmf equivalence, but it conserves power and torque to a 3/2 ratio. Reactive power of a.c. machines may also be calculated through the dq0 model, even with d.c. under steady state.
- The d.c. brush machine corresponds to a simplified case of the dq0 model by its own nature, due to stator and rotor magnetic field axes orthogonality, because of the placement of brushes in the neutral axis.
- As the voltage, current, flux linkage, power, torque equivalence between three-phase a.c. machines and the dq0 model has been introduced here, only the parameter (inductances, resistances) correspondence is needed; this will be done in Chapters 9 and 10.
- We should point out that the main assumptions on which the dq0 model–m phase a.c. machines are based are
    - o Nonsymmetric windings or magnetic anisotropies present either in the rotor or in the stator, with dq axes attached to this part of the machine.
    - o For distributed symmetric a.c. windings on the stator with sinusoidal armature flux density in the airgap, the salient pole rotor case is handled in this case by replacement with a thin fictitious (superconducting) flux barrier in axis d but for constant airgap.
    - o For nonoverlapping (concentrated) windings on the stator, but no windings on the rotor: the rotor may be anisotropic magnetically or may have PMs but the stator self and mutual inductances should vary sinusoidally with the rotor position. Sinusoidal inductance distributions are crucial for the dq0 model because of the *dq* axes coupling by motion emfs (sinusoidal in time).
    - o For all other cases the phase-coordinate models should be used from start (nonsymmetric three-phase windings on stator and rotor, etc.) as shown in subsequent chapters.
    - o Linear a.c. machines (LIMs and LSMs) follow the same pattern in modeling for transients but dynamic end effects (due to open magnetic circuit at stator (mover) ends, in the presence of Gauss flux law as constraint) make the modeling more complicated for high-speed applications and for a small number of poles on the primary part.
    - o Multiple dq0 models are feasible in multiphase (5, 6, 7, 9…) a.c. machines.

1.14 PROPOSED PROBLEMS

1.1 Does a separately excited d.c. brush motor qualify as a dq0 model? Demonstrate why and write the two voltage equations.
Hint: Check Section 1.4, add the excitation winding in place of PMs, observing zero motion emf in the latter.

1.2 Based on the fact that the dq0 model of a distributed a.c. three-phase SM with salient poles rotor ($L_{dm} > L_{qm}$) is strictly valid only if the latter has constant airgap, draw the pertinent d (or q) axis flux barrier rotor for $2\,p_1 = 2$ poles; for $L_{dm} < Lqm$ place PMs in the flux barrier.
Hints: Place a diametrical flux barrier in axis d for $L_{dm} > L_{qm}$ and in axis q for PM-filled barrier case and keep the rotor cylindrical.

1.3 Strip the three-phase SM from the cage and d.c. excitation windings on the rotor and put PM d-axis poles instead; simplify the machine dq0 model equations and rewrite the flux–current relationship as well as the torque simplified equations.
Hints: Check Equations 1.12 through 1.20 and simplify them; replace $L_{dm}\, I_F$ by Ψ_{PM}.

1.4 Figure 1.11 shows a PM salient rotor pole ($L_{dm} < L_{qm}$) two-phase SM machine.
The stator windings do not have the same number of turns but use the same copper weight. Can this machine be treated for transients with the dq0 model (is the zero-sequence current nonzero)? In what coordinates (ω_b = ?), if one winding in the stator is eliminated, can the machine still be treated by the dq model? Why is this so?
Hint: Two orthogonal stator windings with the same copper weight are basically symmetric.

1.5 Figure 1.11b shows the case of a two-pole two-phase IM whose stator windings are fully symmetric and so is the rotor cage. Can the dq0 model be applied to it? In what coordinates (ω_b = ?)?
Hint: Two-phase symmetrical IM resembles the dq model without the zero-sequence component.

1.6 A six-slot/four-pole three-phase PMSM with tooth-wound coils as in Figure 1.12 has the PM flux linkage in the stator coils sinusoidal with the rotor position. Can the dq model be applied to this nonsalient pole (constant magnetic airgap) machine? In what coordinates (ω_b =?)?

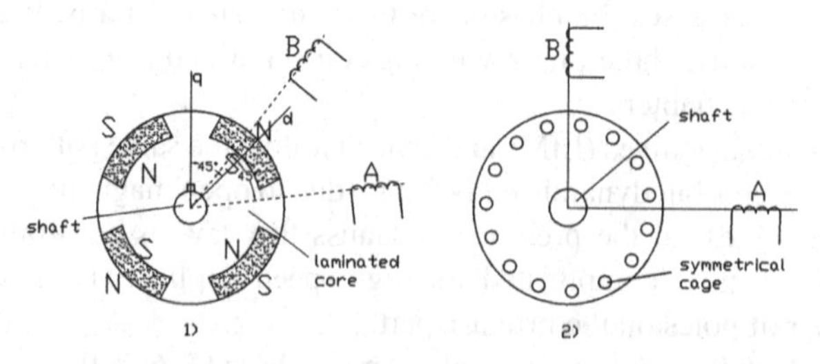

FIGURE 1.11 Salient pole PM rotor two-phase SM (a) and cage rotor two-phase IM (b).

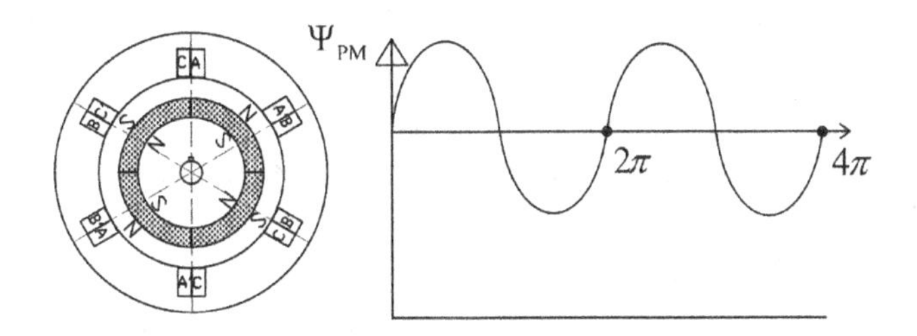

FIGURE 1.12 Six-slot/four-pole PMSM with sinusoidal PM flux linkage.

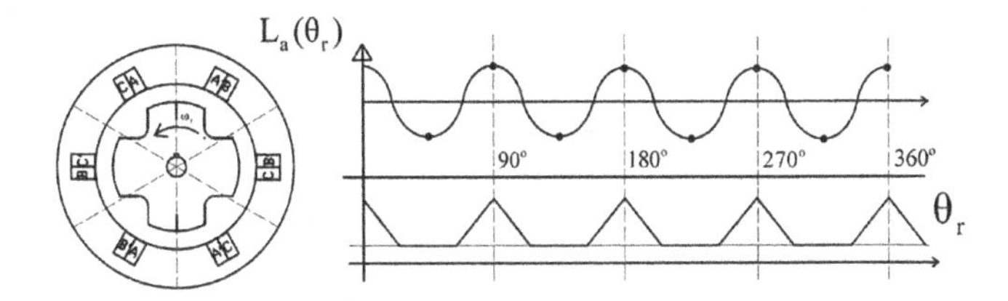

FIGURE 1.13 Six-slot/four-pole three-phase switched reluctance machine.

Hints: The stator winding mmf has a $2p_1 = 4$ harmonic that makes this machine synchronous; the other stator mmf harmonics are added to stator leakage inductance as they do not produce nonzero average torque.

1.7 A three-phase six-slot/four-pole switched reluctance machine (Figure 1.13) has zero mutual inductances but the self-inductances vary as

$$L_{A,B,C} = L_{se} + L_0 \cos\left(4\theta_r - (i-1)\frac{2\pi}{3}\right)$$

Can the dq0 model be used in this case? In what coordinates (ω_b = ?)? Will it cause a problem if the actual current in the stator phases is sinusoidal in terms of torque production? With sinusoidal current will it be torque pulsations?

For the case of linear ramp and flat portion variation of $L_{A,B,C}$ with rotor position (Figure 1.13), may the dq0 model still be used? If not, why?

Hints: Notice that with sinusoidal inductances, the machine becomes a synchronous one with salient poles.

REFERENCES

1. R.N. Park, The reaction theory of synchronous machines: A generalized method of analysis, *AIEE Trans.* 48, 1929, 716–730.
2. W.V. Lion, *Transient Analysis of Alternating Current Machinery: An Application of the Method of Symmetrical Components*, MIT, Cambridge, MA, 1954.

3. I. Racz and K.P. Kovacs, *Transient Regimes of AC Machines*, Springer Verlag, 1995 (the original edition in German, 1959).
4. J. Stepina, *Complex equations for electric machines at transient conditions*, Proceedings of ICEM-1990, Cambridge, MA, vol. 1, pp. 43–47.
5. I. Boldea and S.A. Nasar, *Electric Machines Dynamics*, MacMillan Publishing Company, New York, 1986.
6. D.W. Novotny and T.A. Lipo, *Vector Control and Dynamics of AC Machines*, OUP, Oxford, U.K., 1996.
7. N. Bianchi, *Finite Element Analysis of Electric Machines*, CRC Press, Taylor & Francis Group, New York, 2005.
8. M.A. Arjona and D.C. MacDonald, A new lumped steady-state model derived from FEA, *IEEE Trans*. EC-14, 1999, 1–7.
9. V. Ostovic, *Dynamics of Saturated Electric Machines*, Springer Verlag, New York, 1989.
10. S. Yamamura, *Spiral Vector Theory of AC Circuits and Machines*, Clarendon Press, Oxford, U.K., 1992.
11. I. Boldea and S.A. Nasar, Unified treatment of core losses and saturation in orthogonal axis model of electric machines, *Proc. IEE*, 134(6), 1987, 355–363.
12. E. Levi, Saturation modeling in dq model of salient pole synchronous machines, *IEEE Trans*. EC-14, 1999, 44–50.
13. B. Mirafzal, G. Skibinski, R. Tallam, D. Schlegel, and R. Lukaszewski, Universal induction motor model for low to high frequency response characteristics, *Proceedings of IEEE-IAS*, Tampa, FL, 2006.
14. G. Vidmar, D. Miljavec, "A universal high-frequency three-phase electric-motor model suitable for delta- and star-winding connections", *IEEE Trans*, PE–30(8), 2015, 4365–4376.
15. I. Gonzales-Prieto, M. J. Duran, J.J. Aciego, C. Martin, F. Barrero, "Model predictive control of 6-phase IM drives using virtual voltage vectors", *IEEE Trans.*, IE-65(1), 2018, 27–37.
16. H. S. Che, M. J. Duran, E. Levi, M. Jones, W.-P. Hew, N. Abd. Rahim, "Postfault operation of an asymmetrical six–phase induction machine with single and two isolated neutral points", *IEEE Trans.*, PE-29(10), 2014, 5406–5416.
17. I. Gonzales-Prieto, M. J. Duran, F. J. Barrero, "Fault tolerant control of 6-phase IM drives with variable current injection", *IEEE Trans.*, PE-32(10), 2017, 7894–7903.
18. Z. Liu, Z. Zheng, Y. Li, "Enhancing fault-tolerant ability of a nine-phase IM drive system using fuzzy logic current controllers", *IEEE Trans.*, EC-32(2), 2017, 759–769.
19. S. Sadeghi, L. Guo, H. A. Toliyat, L. Parsa, "Wide operational speed range of 5–phase PM machines by using different stator winding configurations", *IEEE Trans.*, IE-59(6), 2012, 2621–2631.
20. S. Kallio, M. Andriollo, A. Tortella, J. Karttunen, "Decoupled dq model of double star interior permanent magnet synchronous motors", *IEEE Trans.*, IE-60(6), 2013, 2486–2494.

2 Transients and Control Principles of Brush–Commutator DC Machines

2.1 INTRODUCTION

The d.c. brush–commutator—with PM stator—motor is still favorite for low-power levels in many applications such as small fans, robotics, and automotive auxiliaries. The lower cost of PWM converters for unidirectional motion (in fact most) applications makes the d.c. brush PM small motors even more attractive.

On the other hand, d.c. excited brush–commutator machines are still used for traction in urban and interurban transportation in numerous countries. The d.c. generator mode is seldom used, despite its resistive only (small, in other words) voltage regulation, which is an advantage in standalone applications.

At low speeds (100 rpm or so) and with, say, 1 MW reversible drives in metallurgy, with a 3/1 starting to rated torque, the d.c. brush motor drive (with a dual converter (rectifier) for four-quadrant operation) is still cost/performance-wise very competitive. The slotless rotor configuration of PM d.c. brush motor provides for the fastest torque response to date in servodrives.

Finally, the a.c. brush series (universal) motor is in use today, with voltage control only, for some home and construction tool applications [1–4].

So, again, though the brush–commutator machines have been considered a dying breed, they seem to die hard. This is one reason why their transients are treated in this chapter.

The second reason is that they represent the simplest second (third)-order system that can serve as a practical introduction to a.c. machine transients. The latter, when flux orientation control is used, undergo a forceful reduction in the model order to come closer to the d.c. brush machine, which provides naturally decoupled flux (excitation flux) and torque control.

2.2 ORTHOGONAL (DQ) MODEL OF DC BRUSH MACHINES WITH SEPARATE EXCITATION

In Section 1.4, we have derived the d.c. PM brush machine dq model. Here, by leaving only one winding on the stator (along axis d, the field winding) and one along axis q in the rotor (the armature winding), we obtain the separately excited d.c. brush–commutator machines (Figure 2.1).

DOI: 10.1201/9781003216018-2

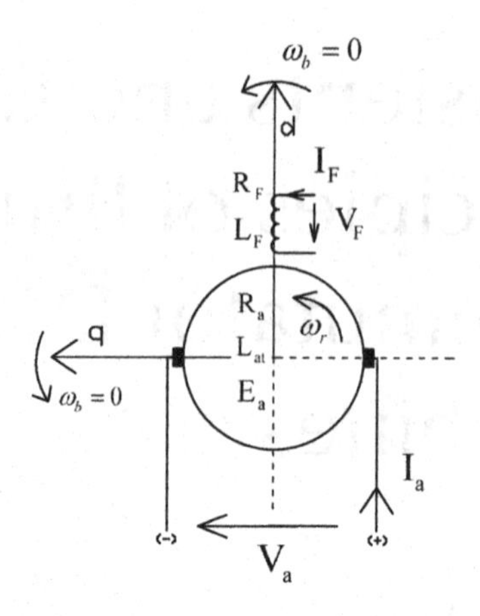

FIGURE 2.1 dq model of separately excited d.c. brush machine.

The d.c. excitation circuit in Figure 2.1 is separately supplied from a d.c. source, but by simple mathematical constraints, it may be shunted or connected in a series to the brushes. The commutation poles, if any, are lumped into the rotor armature winding.

Shunt excitation: $V_F = V_a$
Series excitation: $I_F = I_a$; $V_F + V_a = V_{source}$

The field-circuit equation is straightforward as there is no motion-induced voltage in it ($\omega_b = 0$), and the orthogonality of the winding axes precludes any pulsational-induced voltage from armature winding:

$$R_F I_F - V_F = -L_{Ft} \frac{dI_F}{dt} \tag{2.1}$$

On the rotor, there is a motion-induced voltage and a pulsational-induced voltage:

$$R_a I_a - V_a = -L_{at} \frac{dI_a}{dt} - \omega_r \Psi_{dr}; \quad \Psi_{dr} = L_{dm} I_F \tag{2.2}$$

Ψ_{dr} is produced by a fictitious armature rotor winding placed in axis d_r. As this does not exist, its flux linkage is produced solely by the field winding from the stator. In other words, L_{dm} is reduced to the rotor, but L_F, R_F, and V_F are not.

The emf E_r is [5]:

$$E_r = k_\Phi n \Phi_p; \quad k_\Phi = \frac{p_1 N}{a}; \quad \omega_r = (2\pi n) p_1 \tag{2.3}$$

where

n is the speed in rps,
N is the total number of conductors in rotor slots,
$2a$ is the current path count, and
p_1 is the pole pairs.

If magnetic saturation is considered, the transient inductances in Equations 2.1 and 2.2, L_{Ft}, L_{at}, are

$$L_{Ft} = L_F + \frac{\partial L_F}{\partial i_F} i_F \le L_F$$
$$L_{at} = L_a + \frac{\partial L_a}{\partial i_a} i_a \le L_a \tag{2.4}$$

The inductance L_{dm} (I_F) is the main inductance as it occurs in the motion-induced voltage.

The electromagnetic torque is simply calculated from P_e:

$$P_e = T_e 2\pi n = \omega_r \Psi_{dr} I_a = \omega_r L_{dm} I_F I_a \tag{2.5}$$

It is also possible that L_{dm} depends not only on I_F but also on I_a, due to crosscoupling magnetic saturation (Chapter 1).

These are, however, cases typical for high overloading (such as traction, metallurgical, hand tool applications). In general, the L_F (I_F), L_{Ft} (I_F), L_{at} (I_a), and L_{dm} (I_F) functions are sufficient and may be obtained through adequate tests.

Adding the motion equations:

$$\frac{J}{p_1}\frac{d\omega_r}{dt} = T_e - T_{load}; \quad \frac{d\theta_r}{dt} = \frac{\omega_r}{p_1} \tag{2.6}$$

we obtain a fourth-order system (Equations 2.1, 2.2, and 2.6) with some parameters (inductances) dependent on some variables but independent of rotor position, as expected.

Denoting d/dt by s (Laplace operator), they lead to the structural diagram in Figure 2.2. The structural diagram depicts products of variables ($I_F I_a$, $I_F \omega_r$) and actual variable inductances. A nonlinear system is thus obtained.

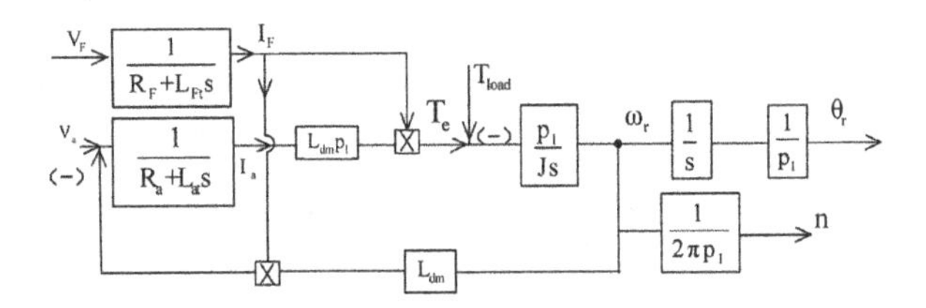

FIGURE 2.2 Structural diagram of separately excited d.c. brush machine.

To simplify our dealings with transients, we introduce the following three types of them:

- Electromagnetic transients (n = constant)
- Electromechanical transients (both electromagnetic and mechanical variables vary)
- Mechanical transients (only mechanical variables vary)

2.3 ELECTROMAGNETIC (FAST) TRANSIENTS

Let us consider a separately excited d.c. brush machine operating as a generator at a constant speed (ω_r = const).

From Equations 2.1 through 2.3, the load equations are added:

$$\begin{aligned} V_F &= R_F I_F + L_{Ft}\frac{di_F}{dt} \\ V_a &= R_a I_a + L_a\frac{di_a}{dt} + \omega_r L_{dm} I_F \\ V_a &= -R_{load} - L_{load}\frac{di_a}{dt} \end{aligned} \quad (2.7)$$

Or, in Laplace form ($d/dt \rightarrow s$):

$$\frac{\tilde{I}_F}{\tilde{V}_F} = \frac{1}{R_F + sL_{Ft}}; \quad \tilde{I} = \frac{\tilde{V}_a}{R_a + sL_{at}} - \frac{\omega_r L_{dm}\tilde{V}_F}{(R_F + sL_{Ft})(R_a + sL_{at})} \quad (2.8)$$

$$\frac{\tilde{V}_a}{\tilde{V}_F} = \frac{\omega_r L_{dm}(R_{load} + sL_{load})}{(R_F + sL_{Ft})(R_{load} + sL_{load} + R_a + sL_{at})} \quad (2.9)$$

Besides R_{load} and L_{load}, an emf E_a may be added to simulate a d.c. brush motor load.

It is evident that the field current circuit introduces a delay in the output voltage $\tilde{V}_a$ by Equation 2.9. Also, a second-order system has been obtained from Equation 2.9, which is easy to handle, with constant inductances in the machine by the linear system routines.

Example 2.1: Sudden V_F Increase and Short-Circuit Transients

Let us consider a d.c. generator with separate excitation with the following data: $R_a = 0.1\ \Omega$, $R_F = 1\ \Omega$, $L_a = 0.5$ mH, $L_F = 0.5$ H, $V_{an} = 200$ V, $I_{an} = -100$ A, $I_{Fn} = 5$ A, at $n = 1500$ rpm, for resistive load R_{load}.

Calculate the output voltage transfer function (2.9) and the $V_a(t)$ and $I_a(t)$ after a 20% step increase in V_F. Also calculate the sudden short-circuit current variation.

Solution:

The load resistance R_{load} is

$$R_{load} = \frac{-V_{an}}{I_{an}} = \frac{-200}{(-100)} = 2\Omega \tag{2.10}$$

For Equation 2.9, we still need $\omega_r L_{dm}$:

$$\begin{aligned} V_{an} &= R_a I_{an} + \omega_r L_{dm} I_{Fn}; \quad 200 = 0.1 \times (-100) + \omega_r L_{dm} \times 5 \\ \omega_r L_{dm} &= 42\ \Omega \end{aligned} \tag{2.11}$$

So Equation 2.9 becomes

$$\frac{\tilde{V}_a}{\tilde{V}_F} = \frac{42 \cdot 2}{(1+0.5s)(2+0.1+0.005s)}$$

For a 20% increase in V_F, we have

$$\tilde{V}_F = I_{Fn} \cdot R_F \frac{0.2}{s} = \frac{5 \times 2 \times 0.2}{s} = \frac{2}{s}$$

so

$$\tilde{V}_a = \frac{42 \cdot 2 \cdot 2}{s(1+0.5s)(2.1+0.005s)}$$

or

$$\Delta v_a(t) = 2\left(20 - 20.9e^{-2t} + 0.95e^{-420t}\right)(\mathrm{V})$$

The load current variation $\Delta i_a(t)$ is

$$\Delta i_a(t) = -\frac{\Delta v_a(t)}{R_{load}} \tag{2.12}$$

As we can see, the output voltage and current variation (response) at a 20% increase in field-circuit supply voltage are stable (attenuated) and nonperiodic. This is a special merit of the d.c. generator as a controlled power source. However, as the field-circuit time constant $T_F = L_F/R_F$ is large, the response is slow. A large voltage-ceiling in V_F is required if a quick response is needed in the output voltage when V_F varies.

This is very similar to the case when the magnetization flux (current) varies in a.c.machines. This is how the flux orientation control of a.c.machines has come into play.

The sudden short circuit of the same d.c. generator is also an electromagnetic transient ($V_a = 0$, $I_F = I_{Fn}$):

$$\tilde{I}_{sc} = -\frac{\omega_r L_{dm} I_{Fn}}{R_a + sL_a} \tag{2.13}$$

(Continued)

Example 2.1: (Continued)

with the solution

$$I_{sc}(t) = -\frac{\omega_r L_{dm} I_{Fn}}{R_a} + Ae^{-\frac{t}{\tau_e}}; \quad \tau_e = L_a/R_a \tag{2.14}$$

Also, at $t = 0$ $I_{sc}(0) = -I_n$:

$$I_{sc}(t) = -2100 + 2000e^{-0.005t} \tag{2.15}$$

So the short-circuit transient is very fast because the armature electric time constant is small: $\tau_e = 5$ ×; it is even less than 1 ms for slotless rotor windings. So the sudden short circuit of the d.c. brush machine (with separate or PM excitation) is very fast and dangerous to the machine, because the final current is only R_a resistance limited. Such an event has to be avoided by means of fast protection in all d.c. brush motor drives.

2.4 ELECTROMECHANICAL TRANSIENTS

Most transients are electromechanical (both electrical and mechanical variables vary). To approach the problem gradually, let us first tackle constant excitation (or PM) flux transients.

2.4.1 Constant Excitation (PM) Flux, Ψ_{DR}

From Equations 2.1 through 2.6, we are now left with two equations, if speed control is targeted:

$$\begin{gathered} V_a = R_a I_a + L_{at}\frac{di_a}{dt} + \omega_r \Psi_{dm}; \quad \Psi_{dm} = L_{dm} I_F = \Psi_{PM} \\ \frac{J}{p_1}\frac{d\omega_r}{dt} = p_1 \Psi_{dm} I_a - T_{load} - B\omega_r \end{gathered} \tag{2.16}$$

In Laplace form, Equations 2.16 becomes

$$\begin{gathered} \tilde{V}_a = (R_a + sL_a)\tilde{i}_a + \tilde{\omega}_r \Psi_{dm} \\ s\tilde{\omega}_r = \frac{p_1^2}{J}\Psi_{dm}\tilde{i}_a - \left(\tilde{T}_{load} + B\tilde{\omega}_r\right)\frac{p_1}{J} \end{gathered} \tag{2.17}$$

We may now extract the open-loop transfer functions for two cases:
– constant load torque $T_{load} = \text{const}\left(\tilde{T}_{load} = 0\right)$

$$\tilde{i}_a = \frac{\tilde{V}_a(Js + Bp_1)}{(R_a + sL_a)(Js + Bp_1) + p_1^2\Psi_{dm}^2} = G_i(s)\tilde{V}_a; \quad \tilde{\omega}_r = -\frac{p_1 \Psi_{dm}\tilde{i}_a}{\frac{Js}{p_1} + B} = G_\omega(s)\tilde{i}_a \tag{2.18}$$

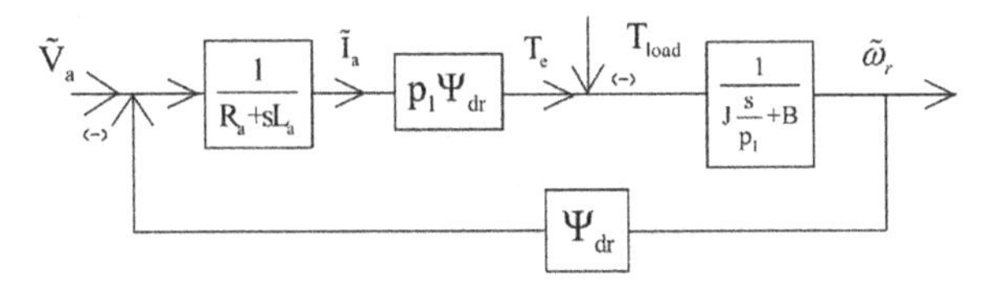

FIGURE 2.3 The structural diagram of a d.c. brush PM machine.

– constant voltage $V_a = \text{const}\left(\tilde{V}_a = 0\right)$

$$\begin{aligned}\tilde{i}_a &= \frac{\tilde{T}_{load} p_1 \Psi_{dm}}{\left(R_a + sL_a\right)\left(Js + Bp_1\right) + p_1^2 \Psi_{dm}^2} = G_{it}\left(s\right)\tilde{T}_{load}; \\ \tilde{\omega}_r &= -\frac{\left(R_a + sL_a\right)\tilde{i}_a}{\Psi_{dm}} = G_{\omega t}\left(s\right)\tilde{i}_a\end{aligned} \tag{2.19}$$

Figure 2.3 illustrates the structural diagram corresponding to Equations 2.17.

A second-order system has been obtained. As analysis through linear systems routines is straightforward, the eigenvalues $s_{1,2}$ from Equation 2.17 are obtained simply and, for $B = 0$ (zero friction torque):

$$s_{1,2} = \frac{-1 \pm \sqrt{1 - 4\tau_e / \tau_{em}}}{2\tau_e}; B = 0; \tau_{em} = \frac{JR_a}{\left(p_1 \Psi_{dm}\right)^2}; \tau_e = \frac{L_a}{R_a} \tag{2.20}$$

While $\tau_e = \tau_a$ has already been defined as an electrical time constant, τ_{em} is the electromechanical time constant. With $\tau_{em} > 4\tau_e$, the eigenvalues are real and negative, so the machine response is attenuated and nonperiodical. For $\tau_{em} < 4\tau_e$ (small inertia applications), the response is still attenuated but periodic. Again, the PM (constant flux) d.c. brush machine braves ideal transient behavior due to its inner feedback (speed, emf) loop (Figure 2.3).

Example 2.2: DC Brush PM Motor Transients

A PM d.c. brush motor has the following data: $P_n = 50$ W, $V_n = 12$ Vd.c., $\eta_n = 0.9$, $R_a = 0.12\ \Omega$, $\tau_e = 2$ ms, $p_1 = 1$, and $n_n = 1500$ rpm.

Calculate $\Psi_{dr} = \Psi_{PM}$ (PM flux linkage), rated electromagnetic torque, speed and current transients for sudden reduction of V_a from 12 to 10 Vd.c., where the machine inertia $J = 2 \times 10^{-4}$kg · m² and $J' = 1 \times 10^{-3}$kg · m², and for constant load torque.

Solution:

First, from Equation 2.16, with d/dt = 0

$$V_n = R_a I_n + \omega_{rn} \Psi_{PM}; \quad \omega_{rn} = 2\pi p_1 n_n$$

(Continued)

Example 2.2: (Continued)

with

$$I_n = \frac{P_n}{\eta_n V_n} = \frac{50}{0.9\times 12} = 4.63\ \text{A}$$

So

$$\Psi_{PM} = \frac{12 - 0.12\times 4.63}{2\pi(1500/60)} = 0.0729\ \text{Wb}$$

$$T_{en} = p_1 \Psi_{PM} I_n = 1\cdot 0.0729\times 4.63 = 0.3375\ \text{N m} = T_{load}$$

From Equation 2.17, we can eliminate $\tilde{i}_a$ and switch back s to d/d*t* to obtain

$$\tau_{em}\tau_e \frac{d^2\omega_r}{dt^2} + \tau_{em}\frac{d\omega_r}{dx} + \omega_r = \frac{V_a(t)}{\Psi_{PM}} - T_{load}\frac{R_a}{p_1\Psi_{PM}^2} \tag{2.21}$$

With $\tau_e = 2\times 10^{-3}$s; τ_{em} from Equation 2.20 is:

$$\tau_{em}\frac{JR_a}{(p_1\Psi_d)^2} = \begin{cases} \nearrow \dfrac{0.0002\times 0.12}{0.0729^2} = 4.510\times 10^{-3} & s < 4\tau_e \\ \searrow \dfrac{0.001\times 0.12}{0.0729^2} = 22.58\times 10^{-3} & s > 4\tau_e \end{cases}$$

So the eigenvalues are (Equation 2.20)

$$\underline{\gamma}_{1,2} = \frac{-1 \pm \sqrt{1 - 4\tau_e/\tau_{em}}}{2\tau_e} = \begin{matrix} \nearrow \dfrac{-1 \pm j0.877}{4\times 10^{-3}} \\ \searrow \dfrac{-1 \pm 0.803}{4\times 10^{-3}} \end{matrix}$$

The initial value of speed is the same for both inertia values: $n_n = 1500$rpm ($(\omega_r)_{t0}$ = $2\pi 1500/60 = 157$rad/s). Also, the final value of speed ω_{rf} is the same as the load torque value holds:

$$(\omega_r)_{t=\infty} = \frac{(V_a)_{t=\infty}}{\Psi_{PM}} - \frac{T_{em}R_a}{p_1\Psi_{PM}^2} = \frac{10}{0.0729} - \frac{0.3375\times 0.12}{1\times 0.0724^2} = 129.55\ \text{rad/s} \tag{2.22}$$

In addition,

$$\left(\frac{d\omega_r}{dt}\right)_{t=0} = 0 \tag{2.23}$$

The solution of Equation 2.21 for speed is now straightforward:

$$\omega_r(t) = (\omega_r)_{t=0} + Ae^{-250t}\cos(219.25t + \varphi) \tag{2.24}$$

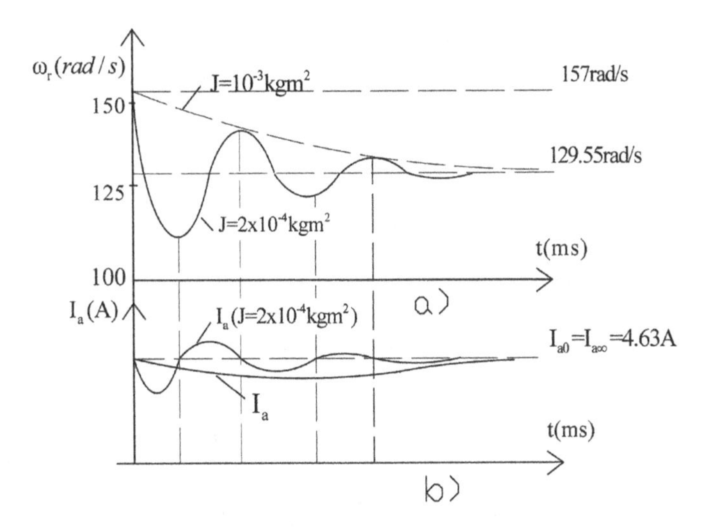

FIGURE 2.4 (a) Speed and (b) current response of a d.c. brush PM motor to sudden voltage reduction from 12 to 10 Vd.c., for constant torque.

$$\omega_r(t) = (\omega_r)_{t=0} + A_1 e^{-49.25t} + A_2 e^{-450t} \tag{2.25}$$

Based on boundary conditions (2.22) and (2.23), A, φ in Equation 2.24 and A_1 and A_2 in Equation 2.25 are easily found.

In any case, for the small inertia, the speed response is periodic and attenuated, while for the large inertia, it is nonperiodic and attenuated. Fast response open loop for low inertia implies speed response overshooting and attenuated oscillations (Figure 2.4).

The current response is obtained from Equation 2.17

$$i_a(t) = \left(\frac{J}{p_1}\frac{d\omega_r}{dt} + T_{load}\right) / (p_1 \Psi_{PM}) \tag{2.26}$$

Note: A similar problem could be solved for constant voltage but with a step increase (decrease) in load torque.

The only difference is that $(di_a/dt)_{t=0} = 0$ instead of $(d\omega_r/dt)_{t=0} = 0$, as it was for constant load torque.

2.4.2 Variable Flux Transients

To extend the speed range above the rated (base) speed at rated rotor voltage, flux weakening is required. So d.c. excitation is required. Let us consider here separate excitation.

The transients model, ready for numerical solution, is collected from Equations 2.1, 2.2, and 2.6:

$$\frac{dI_F}{dt} = \frac{V_F - R_F I_F}{L_{Ft}}; \quad \frac{di_a}{dt} = \frac{V_a - R_a I_a - \omega_r L_{dm} I_F}{L_{at}} \tag{2.27}$$

$$\frac{d\omega_r}{dt} = \frac{p_1}{J}\left(p_1 L_{dm} I_F I_a - T_{load} - B\omega_r\right); \quad \frac{d\theta_r}{dt} = \frac{\omega_r}{p_1} \tag{2.28}$$

As we now have products of variables, *small deviation theory* is used to linearize the system:

$$\begin{aligned} &V_F = V_{F0} + \Delta V_F; \quad V_a = V_0 + \Delta V_a; \quad T_{load} = T_{L0} + \Delta T_L \\ &I_F = I_{F0} + \Delta I_F; \quad I_a = I_0 + \Delta I_a, \quad \omega_r = \omega_{r0} + \Delta\omega_r; \quad \theta_r = \theta_{r0} + \Delta\theta_r \end{aligned} \tag{2.29}$$

For initial conditions ($t = 0$), with d/dt in Equations 2.27 and 2.28 as zero:

$$\begin{aligned} &V_{F0} = I_{F0} R_F; \quad V_0 = R_a I_{a0} + \omega_r L_{dm} I_{F0} \\ &\theta_{r0} = \theta_0; \quad T_{L0} + B\omega_{r0} = p_1 L_{dm} I_{F0} I_{a0} \end{aligned} \tag{2.30}$$

For small deviations, Equations 2.27 and 2.28 with Equation 2.29 in matrix Laplace form are

$$\begin{vmatrix} \Delta V_F \\ \Delta V_a \\ \Delta T_L \\ 0 \end{vmatrix} = \begin{vmatrix} R_F + sL_{Ft} & 0 & 0 & 0 \\ \omega_{r0} L_{dm} & R_a + sL_{at} & L_{dm} I_{F0} & 0 \\ p_1 L_{dm} I_0 & p_1 L_{dm} I_{F0} & -\frac{Js}{p_1} - B & 0 \\ 0 & 0 & -1/p_1 & 0 \end{vmatrix} \cdot \begin{vmatrix} \Delta I_F \\ \Delta I_a \\ \Delta\omega_r \\ \Delta\theta_r \end{vmatrix} \tag{2.31}$$

Various transfer functions between inputs ΔV_F, ΔV_a, and ΔT_L and output (variable) deviations ΔI_F, ΔI_a, $\Delta\omega_r$, and $\Delta\theta_r$ may be extracted from Equation 2.31. It is however evident that the eigenvalues are now obtained from the equation:

$$\Delta(s) = \left(R_F + sL_{Ft}\right)\left[\left(R_a + sL_a\right)\left(\frac{Js}{p_1} + B\right) + p_1^2 L_{dm}^2 I_{F0}^2\right] = 0 \tag{2.32}$$

So the field-circuit produces a separate (decoupled) negative real eigenvalue $\gamma_3 = -R_F/L_{Ft}$, while the other two, $\gamma_{1,2}$, are the same as for the constant flux transients (2.20), but calculated for the initial field current I_{F0} (flux).

The magnetic decoupling between the field circuit and the armature circuit (due to the orthogonal placing of the two windings) leads to this independent large time constant of the system.

Flux transients are slow, so flux weakening (reducing I_F) is accompanied by slower torque (I_a) current and torque transients.

This situation is very similar to flux weakening in flux orientation control of a.c. machines where the decoupling of field current and torque current components in ac-phase current is done mathematically, online, through DSPs.

This similarity has led to the introduction of the flux orientation (vector) control of a.c. machines that revolutionized motor/generator control through power electronics.

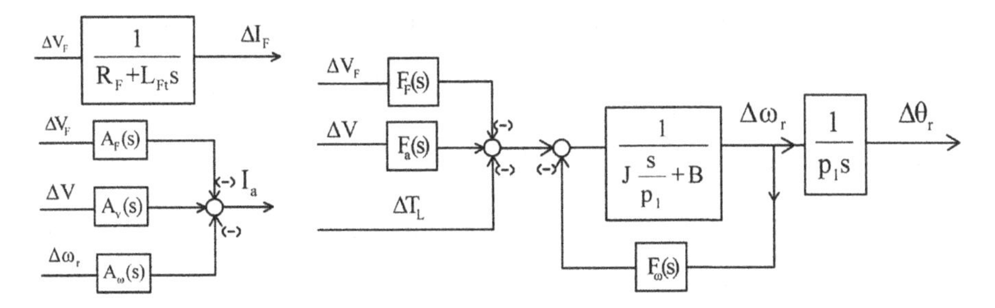

FIGURE 2.5 Structural diagram for the separately excited d.c. brush motor.

A rather elaborated structural diagram can be extracted from Equation 2.31 (Figure 2.5).

2.4.3 DC Brush Series Motor Transients

The generated scheme of the d.c. brush series machine is shown in Figure 2.6a. It has only one voltage equation and one motion equation (for speed control):

$$\begin{aligned} V_a &= (R_a + R_{Fs})I_a + (L_{at} + L_{Fst}(i_a))\frac{di_a}{dt} + \omega_r L_{dm}(i_a)I_a \\ T_e &= p_1 L_{dm}(i_a)I_a I_a \\ \frac{J}{p_1}\frac{d\omega_r}{dt} &= T_e - T_{load} \end{aligned} \tag{2.33}$$

$L_{dm}(i_a)$ reveals the fact that magnetic saturation is almost unavoidable, because the torque current i_a is also variable as the field current is (or it is proportional to it).

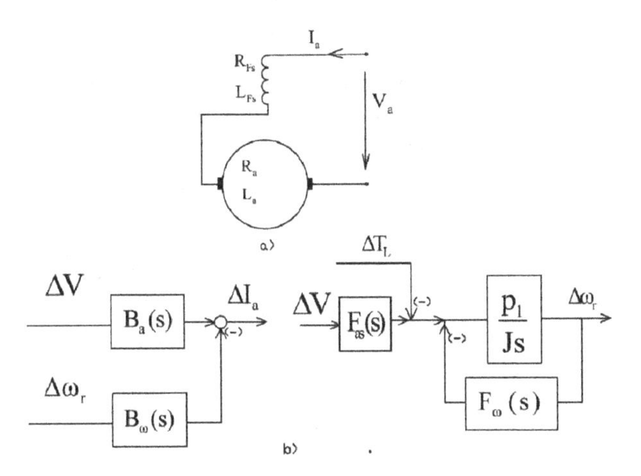

FIGURE 2.6 DC brush series motor: (a) general scheme and (b) structural diagrams.

Small deviation theory is required for linearization and then for control design:

$$V_a = V_0 + \Delta V;\quad \omega_r = \omega_{r0} + \Delta\omega_r;\quad I_a = I_0 + \Delta I_a;\quad T_L = T_{L0} + \Delta T_L \tag{2.34}$$

Also for the initial situation at (d/dt = 0):

$$V_0 = \left(R_a + R_{Fs}\right) I_0 + \omega_{r0} L_{dm} I_0;\quad p_1 L_{dm} I_0^2 = T_{L0} \tag{2.35}$$

From Equations 2.33 through 2.35 the small deviation model arises:

$$\begin{vmatrix} \Delta V \\ \Delta T_L \end{vmatrix} = \begin{vmatrix} R_a + R_{Fs} + \omega_{r0} L_{dm} + s\left(L_{at} + L_{Fst}\right) & L_{dm} I_0 \\ 2 p_1 L_{dm} I_0 & -\dfrac{Js}{p_1} \end{vmatrix} \cdot \begin{vmatrix} \Delta I_a \\ \Delta\omega_r \end{vmatrix} \tag{2.36}$$

The eigenvalues are extracted from the characteristic equation of Equation 2.36:

$$\Delta(s) = \left(R_a + R_{Fs} + \omega_{r0} L_{dm} + s\left(L_{at} + L_{Fst}\right)\right) Js + 2\left(p_1 L_{dm} I_0\right) = 0 \tag{2.37}$$

Comparing with $\Delta(s)$ of separate excitation d.c. brush machines in Equation 2.32, the equivalent electrical time constant is

$$\tau_{es} = \left(L_{at} + L_{Fst}\right) / \left(R_a + R_{Fs} + \omega_{ro} L_{dm}\right) \tag{2.38}$$

So the equivalent electrical time constant, τ_{es}, decreases with speed only if magnetic saturation decreases with speed (because I_a decreases with speed). In any case, it appears that $\tau_{es} < \tau_e = L_a/R_a$, so a quicker response is expected.

The structural diagrams of Equation 2.36 are shown in Figure 2.6b

2.5 BASIC CLOSED-LOOP CONTROL OF DC BRUSH PM MOTOR

Based on Equations 2.18 and 2.19 we may introduce speed or torque basic close-loop control to be explored more fully in electric drives. Here, we only illustrate speed control d.c. brush PM motor transfer functions (Figure 2.7) based on Equations 2.18 and 2.19.

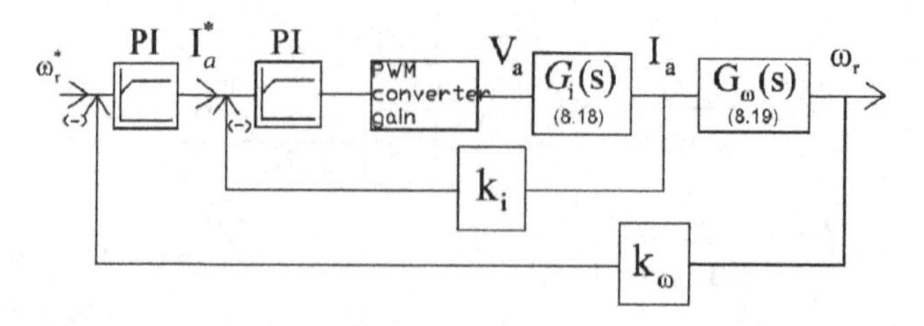

FIGURE 2.7 Cascaded basic close speed control with inner current loop for a d.c. brush PM motor.

K_i and K_ω are the gains of the current and speed sensors. There are two close loops, one fast (current loop) and one slower (speed loop). Voltage change in PWM converter (d.c.–d.c. converter, in general) is fast because of the large switching frequency and thus the converter is modeled as a constant gain. This is acceptable as long as the current is continuous. Discontinuous current leads to sluggish (chartic) speed response and hence should be avoided.

2.6 D.C.–D.C. CONVERTER-FED D.C. BRUSH PM MOTOR

To accomplish speed control, the average voltage, V_{av}, has to be changed in a d.c.–d.c. converter. Here, only the single quadrant d.c.–d.c. converter is considered (Figure 2.8a). Let us consider the case of discontinuous current to explore its consequences. Figure 2.8b shows the source IGBT current, I_g, terminal voltage, $V_a(t)$, and motor discontinuous current, $i_a(t)$.

The reverse of the switching period ($1/T_s$) is larger than 250 Hz; in many cases, it is in the kilohertz range. A constant frequency, (T_s), pulse width modulation is used to modify the average voltage, V_{av}, applied to the motor. Speed pulsations are considered negligible here due to a much larger electromechanical time constant, T_{em}.

The voltage equation for the IGBT "on" and for the diode D "on," respectively, is

$$R_a I_a + L_a \frac{di_a}{dt} + \omega_r \Psi_{PM} = V_0; \quad 0 \le t < T_{on} \tag{2.39}$$

$$R_a I_a + L_a \frac{di_a}{dt} + \omega_r \Psi_{PM} = 0; \quad T_{on} \le t < T_s \tag{2.40}$$

$$T_e = p_1 \Psi_{PM} i_a$$

For a discontinuous current ($\lambda < 1$), the solutions of i_a in Equations 2.39 and 2.40 are straightforward:

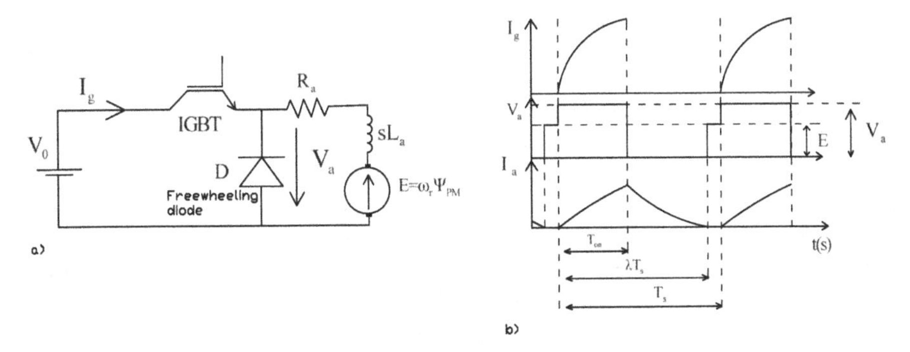

FIGURE 2.8 Basic d.c.–d.c. converter-driven d.c. brush PM motor scheme (a) and current and voltage waveforms (b).

$$I_a(t) = \frac{1}{R_a}(V_0 - \omega_r \Psi_{PM}) + C_1 e^{-t/\tau_e}; 0 \le t < T_{on}$$
$$= -\frac{\omega_r \Psi_{PM}}{R_a} + C_2 e^{-(t-\lambda T_s)/\tau_e}; T_{on} \le t < \lambda T_s \quad (2.41)$$
$$I_a(t) = 0; \lambda T_s \le t < T_s; \tau_e = L_a/R_a$$

The continuity of current at $t = 0$, T_{on}, and T_s yields the three unknowns C_1, C_2, and λ:

$$C_1 = -\frac{1}{R_a}(V_0 - \omega_r \Psi_{PM}); C_2 = \omega_r \Psi_{PM}/R_a$$

$$\lambda T_s = T_{on} + T_e \ln\left[\frac{V_a}{\omega_r \Psi_{PM}}\left(1 - e^{-T_{on}/\tau_e}\right) + e^{-T_{on}/\tau_e}\right] \quad (2.42)$$

for the continuous current $\lambda = 1$ in Equation 2.42. Then, from the same continuity conditions we find, at $t = 0$, T_{on}, and T_s, new conditions for C_1 and C_2:

$$C_2 = C_1 + V_0/R_a; C_1 e^{-T_{on}/\tau_e} + V_a/R_a = C_2 e^{-(T_s - T_{on})/\tau_e} \quad (2.43)$$

The average values of the armature voltage, V_{av}, for the two cases is

$$V_{av} = \frac{T_{on}}{T_s} V_0 = \alpha V_0; \quad \text{for} \quad \lambda = 1 \text{ (continuous current)}$$
$$V_{av} = \frac{T_{on}}{T_s} V_0 + (1 - \lambda)\omega_r \Psi_{PM}; \quad \text{(discontinuous current)} \quad (2.44)$$

The average motor current, I_{av}, and torque, T_{av}, are

$$I_{av} = \frac{1}{T}\int_0^T I_a(t)\,dt$$
$$T_{av} = p_1 \Psi_{PM} I_{av} \quad (2.45)$$

So the average gain of the d.c.–d.c. converter is larger for a discontinuous current. To offset the armature effect of the discontinuous current on the dynamic behavior of the motor, the latter situation is first detected and then the modulation index (gain), α^*, is increased artificially by $\Delta\alpha$ from a lookup table so as to maintain the average voltage that would exist for the continuous current at $\alpha^* + \Delta\alpha = \alpha_a$. For control purposes, the d.c.–d.c. converter can then be approximated by a sample followed by a zero-order hold, with α_a gain. Controlled rectifiers and 2,4 quadrant buck/boost d.c.–d.c. converters are also used to control d.c. brush motors, but this subject is not discussed here [5,6].

2.7 PARAMETERS FROM TEST DATA/LAB 2.1

In the preceding sections, the parameters involved in the transient equations of d.c. brush machines were explained. In this section, we briefly discuss how these parameters can be determined from experiments. Machines of medium (traction) or large (metallurgy) powers carry large armature currents. In some cases, measures are taken by compensating windings to cancel the armature reaction field, but magnetic saturation due to field current, I_F, still exists and varies with I_F. On the other hand, low-power machines do not have interpoles and compensating windings and, in such cases, large armature currents may influence the excitation field, and thus the level of magnetic saturation. PM d.c. brush motors with surface PM stator poles, with armature windings in slots or "in air," do not exhibit notable or variable magnetic saturation.

Only one basic test for parameter estimation is used here: current decay standstill tests. Standstill tests do not imply the coupling of another machine, and the energy consumption for them is small. Standstill tests include step and frequency response tests. We present here only step response (current decay) tests.

Let us consider the machine shown in Figure 2.9a that is supplied through a d.c.–d.c. converter at an average initial current, I_0, in the armature circuit (Figure 2.9b). Then the IGBT is opened and the current freewheels through the diode until it reaches, say, $0.01i_0$, after time t_{da}. The freewheeling diode voltage, V_d, and the current, I_a, are acquired through a computer interface.

The machine voltage equation is

$$R_a i_a + L_a \frac{di_a}{dt} + V_d(t) = 0; \quad (i_a)_{t=0} = I_0 \tag{2.46}$$

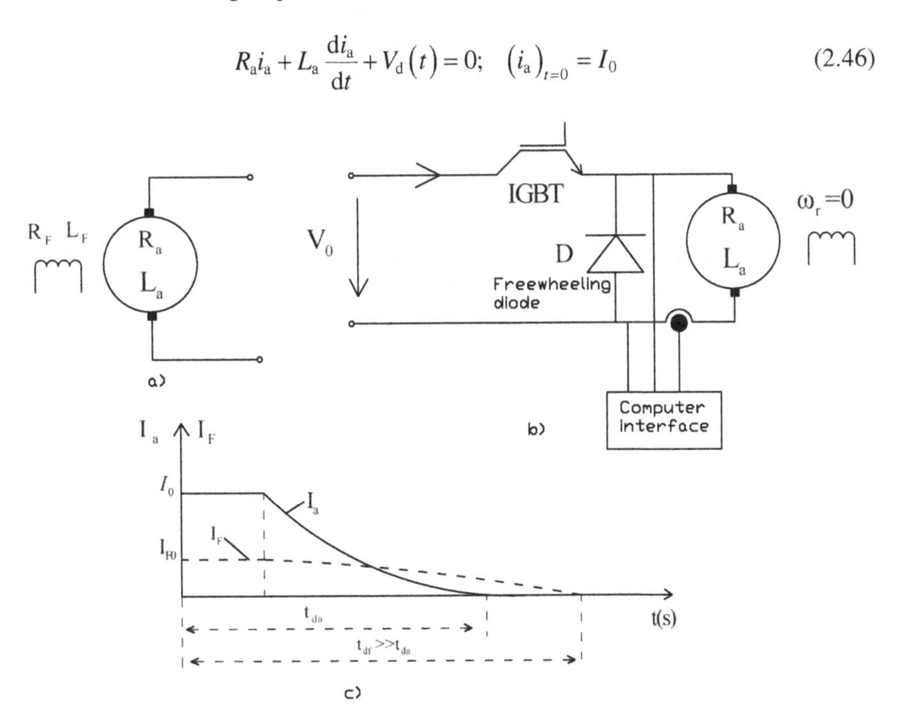

FIGURE 2.9 (a) Separately excited d.c. brush machine, (b) fed at standstill from a d.c.–d.c. converter, and (c) current decay after IGBT turn off.

Considering the final current as zero (in reality, $0.01i_0$), the integration of Equation 2.46 leads to

$$L_a i_0 = \int_0^{t_{da}} V_d(t)\,dt + \int_0^{t_{da}} R_a i_a(t)\,dt \quad (2.47)$$

Now, if we have already measured the armature resistance, R_a, by using the initial diode voltage, V_{d0}, and initial current, i_0, i.e., $R_a = V_{d0}/i_0$, Equation 2.47 yields the armature inductance, L_a, that corresponds to i_0. A few different values of i_0 may be chosen to check the eventual variation of L_a with i_0 and to adopt an average value from measurements. The same test may be performed for the field circuit, with an open armature circuit, to obtain $L_F(I_{F0})$.

Additionally, the current decay test in one axis (d or q) may be done in the presence of a d.c. constant current in the other axis, to check the so-called cross-coupling saturation effect. The inclusion of the freewheeling diode voltage, V_d, is important, especially in high-rated current or low-voltage machines.

Note: To determine the $L_{dm}i_F$ or Ψ_{PM} excitation flux in emf, E, a no-load motor test may be run, with known i_a, V_a, and ω_r and neglected or known brush voltage drop, ΔV_{brush}:

$$E = \omega_r L_{dm} i_F \approx V_a - R_a I_a - \Delta V_{brush} \quad (2.48)$$

The inertia, J, can be found from a free deceleration test with $I_F = 0$, and measured $\omega_r(t)$:

$$\frac{J}{p_1}\frac{d\omega_r}{dt} = -p_{mec}\frac{p_1}{\omega_r} \quad (2.49)$$

with p_{mec} segregated from the motor no-load tests at constant speed but gradually smaller voltage, V_a, and field current, I_F. For a d.c. brush PM machine, this method does not work because core losses may not be separated from mechanical losses. In this case, J may be measured by the pendulum method. Complete testing of d.c. brush machines is described in standards such as those of NEMA, IEEE, and IEC.

2.8 SUMMARY

- Single d.c. excitation, or PM, brush machines with one stator and one wound rotor with fixed brushes (or fixed magnetic fields, or stator coordinates) fit into the simplified dq model.
- Single d.c. excitation may be separate, shunt, or series type, but the structural diagram of the d.c. brush machine shows two electrical time constants and variable product nonlinearities.
- The field circuit is decoupled from the armature circuit.
- The order of the separately excited d.c. brush machine set of equations is four, with I_F, I_a, ω_r, and θ_r as variables and V_a, V_F, and T_{load} as inputs.

- For electromagnetic (fast) transients, the speed may be considered constant.
- A second-order system is obtained for a d.c. generator, at a constant speed, supplying an R_L and L_L load.
  The excitation circuit introduces its large time constant to delay the V_a response to V_F variations, but the voltage regulation is very small ($R_a I_a$).
- Sudden short circuit at terminals is an electromagnetic (fast) transient, which is also dangerous above 5% of rated speed, at rated field current, I_{Fn} (or PM).
- For constant flux linkage (or PM) electromechanical transients, the order of the d.c. brush machine system is two, with i_a and ω_r as variables and V_a and T_{load} as inputs.
- With constant parameters (inductances), the second-order system is linear, and thus easy to investigate.
- The two eigenvalues always have negative real parts, so the response is always stable, but it may be oscillatory if $4\tau_e > \tau_{em}$, where τ_e is the electrical constant and τ_{em} is the electromechanical time constant.
- For variable flux electromechanical transients, one more large real and negative eigenvalue is added, $\gamma_3 = -L_F/R_F$, to the case of constant flux transients.
  This is why, for fast response, it is desirable to keep the excitation flux constant.
- The d.c. brush series motor electromechanical transients can be described, after linearization, as a second-order system, but with an electric time constant that decreases with speed ω_r.
- When d.c.–d.c. converter fed, the d.c. brush machine perceives a variable average voltage, V_{av}, and may operate under a continuous or discontinuous current mode. The discontinuous mode has to be avoided to escape sluggish control, especially at low speeds.
- Close-loop control of d.c. brush motor for variable speed is only introduced here along with the d.c. current decay standstill tests for parameter identification.

2.9 PROPOSED PROBLEMS

2.1 A separately excited d.c. generator running at constant speed supplies a load with $R_l = 1\Omega$, $L_l = 1$H. The armature resistance is $R_a = 0.1\Omega$, and $L_a = 0$. The field circuit, characterized by $R_F = 50\Omega$ and $L_F = 5$H, is suddenly connected to a 120 V d.c. source. For the given speed, $E/i_F = \omega_r L_{dm} = 40$V/A, determine the buildup of the armature current.
Hint: Check Example 2.1.

2.2 A d.c. brush PM motor having an armature resistance, $R_a = 0.12\Omega$ and $L_a = 0$, $2p_1 = 2$, is started on a load, $T_L = 0.2 + 10^{-3} \times \omega_r$, from rest, by connecting it to a 12 V d.c. source. The torque is constant, $p_1\Psi_{PM} = 0.073$N m /A, and the inertia, $J = 2 \times 10^{-4}$kg m^2.
Calculate the speed, $\omega_r(t)$, armature current, $i_a(t)$, and torque, $T_e(t)$, during the starting process.
Hint: Use Equation 2.21, with $\tau_e = 0$, and consider $V_a = 12$V d.c. = constant.

2.3 The motor in Problem 2.2 operates at 1500 rpm at steady state. Subsequently, the load torque is decreased stepwise by 20%. Calculate the steady-state armature current and then the speed, ω_r, and torque, T_e, during a stepwise torque decrease.

Hint: Check Example 2.2 and notice the new boundary condition $(d\, i_a/d\, t)_{t=0} = 0$.

2.4 A series d.c. brush motor has the data $R_a = 2R_F = 1\ \Omega$, $L_a = 10^{-2}$ H, $L_{Fs} = 0.5$ H, $p_1 = 2$ and operates at $n_n = 1500$ rpm from $V_{d.c.} = 500\ V$ and at $I_a = 100$ A. All but rotor and stator windings losses are neglected. After calculating the rated emf, E, L_{dm} $(\omega_r L_{dm} I_{an} = E)$, and the steady state, T_e, determine the machine eigenvalues $\gamma_{1,2}$ for $n_n = 1500$ rpm and for $n = 750$ rpm. Calculate the current and speed transients at constant voltage $(\Delta V = 0)$ for a 20% increase in the load torque at 1500 rpm.

Hint: Check Section 2.4 and Equations 2.35 through 2.37.

2.5 A d.c. brush PM machine with $R_a = 1\ \Omega$, $L_a/R_a = 5 \times 10^{-3}$ s, $p_1 = 1$, and emf = 0.05 V/rad/s operates at $\omega_r = 120$ rad/s. The motor is fed from a d.c.–d.c. converter with the switching period $T_s = 10^{-3}$ s and $V_a = 12$ V d.c.. Assuming constant speed and instantaneous d.c.–d.c. converter commutation, determine the armature current profile.

Hint: Check Section 2.6.

REFERENCES

1. P.C. Sen, *Thyristor DC Drives*, John Wiley & Sons, New York, 1980.
2. T. Kenjo and S. Nagamori, *Permanent Magnet and Brushless DC Motors*, Chapter 7, Clarendon Press, Oxford, U.K., 1985.
3. I. Boldea and S.A. Nasar, *Electric Machines Dynamics*, Chapter 3, MacMillan Publishing Company, New York, 1986.
4. H.A. Toliyat and G.B. Kliman (eds.), *Handbook of Electric Motors*, 2nd edn., Chapter 6, Marcel Dekker, New York, 2004.
5. I. Boldea and S.A. Nasar, *Electric Drives*, 2nd edn., CRC Press, Taylor & Francis Group, New York, 2005, 3rd edn., 2016.
6. C.M. Ong, *Dynamic Simulation of Electric Machinery*, Chapter 8, Prentice Hall, Englewood Cliffs, NJ, 1998.

3 Synchronous Machine Transients and Control Principles

3.1 INTRODUCTION

In the companion book (on steady state), we investigated the steady state of synchronous machines, when the currents and voltages have constant amplitude, and the frequency is constant and equal to the electric speed, $\omega_1 = \omega_{ro}$. Also, the voltage power angle, $\boldsymbol{\delta}_v$, and the torque were constant.

During transient processes such as connection (or disconnection) from (to) the power grid, or when the load varies or the SM is fed through PWM state converters for energy or motion control at variable speed, all or most electric (amplitude of voltages and currents) or mechanical (power angle, $\boldsymbol{\delta}_v$, torque, T_e, speed, ω_r) variables vary in time.

During transients, the steady-state model of the SM is not operational. On the other hand, the dq (complex variable) model—described in Chapter 1—is quite suitable for the modeling of transients of a.c. machines, especially for synchronous machines.

The phase coordinate model is introduced here first and then the parameter equivalence with the dq model is worked out. The modeling of transients is treated in general for the dq model of SM, and then typical transients are dealt with. Transients can be classified as

- Electromagnetic transients
- Electromechanical transients: small deviation and large deviation theories
- Electromagnetic and electromechanical transients for controlled flux
- Variable speed SMs, for modern drives and generator control

Modeling of transients for a split-phase capacitor PMSM (or reluctance SM) is described in detail. Also, rectangular current-controlled PMSM and switched reluctance motor modeling is illustrated in detail.

Finally, standstill current decay and frequency response tests for the SM parameter estimation are described extensively, as they are already a part of recent IEEE standards.

DOI: 10.1201/9781003216018-3

3.2 PHASE INDUCTANCES OF SMS

We start with the salient pole SM (Figure 3.1a and b). The distributed a.c. windings are considered to have inductances that vary sinusoidally with the rotor position (Figure 3.1c).

The flux barrier in Figure 3.1b, provides constant airgap, which, for a sinusoidal ideal winding distribution, produces sinusoidal airgap flux density. This is a physical justification for the main condition: airgap flux density sinusoidal distribution for the equivalence of dq model and the actual machine.

Note: The case of winding-less rotor in a surface PMs (constant airgap) PMSM with nonoverlapping coil a.c. stator windings (q < 0.5 slots/pole/phase) and

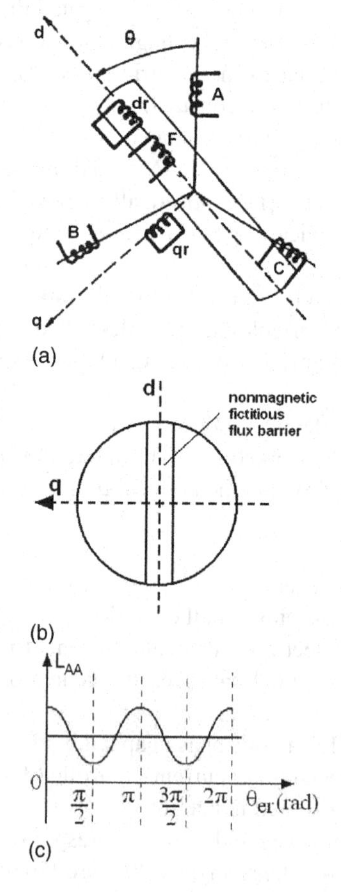

FIGURE 3.1 The phase circuits and inductances of SMs: (a) phase circuits, (b) two-pole salient rotor with constant airgap and one flux barrier, and (c) phase A self-inductance vs. rotor position, θ_{er} (electric angle).

sinusoidal emf can also be treated by the dq0 model. In this case, the synchronous inductances, $L_d = L_q = L_s$, are constant.

The phase A inductance (Figure 3.1c) is expressed as

$$L^{\theta_{er}}_{AA,BB,CC} = L_{sl} + L_0 + L_2 \cos\left(2\theta_{er} + (i-1)\frac{2\pi}{3}\right); \quad i = 1,2,3 \tag{3.1}$$

For $\theta_{er} = 0$, and π, 2π, L_{AA} is maximum, and L_{sl} is the leakage inductance. Also, the mutual stator inductances, L_{AB}, $_{BC}$, $_{CA}$, are

$$L^{\theta_{er}}_{BC,CA,AB} = M_0 + L_2 \cos\left(2\theta_{er} + (i-1)\frac{2\pi}{3}\right); \quad i = 1,2,3 \tag{3.2}$$

For a symmetric three-phase distributed winding ($q \geq 2$), the constant component of mutual inductance, M_0, is

$$M_0 = L_0 \cos\left(\frac{2\pi}{3}\right) = -\frac{L_0}{2} \tag{3.3}$$

The stator/rotor mutual inductances are straightforward, because of the admitted sinusoidal distribution of windings and the constant airgap:

$$L_{A,B,C,F} = M_F \cos\left(\theta_{er} + (i-1)\frac{2\pi}{3}\right); \quad i = 1,2,3 \tag{3.4}$$

$$L_{A,B,C,dr} = M_{dr} \cos\left(\theta_{er} + (i-1)\frac{2\pi}{3}\right) \tag{3.5}$$

$$L_{A,B,C,qr} = -M_{qr} \sin\left(\theta_{er} + (i-1)\frac{2\pi}{3}\right) \tag{3.6}$$

L_{sl}, L_0, L_2, M_F can be determined using standstill tests when a single phase (A) is a.c. fed at a low frequency (to neglect the core losses) and when we measure V_{A0}, V_{B0}, and V_{F0} for two distinct rotor positions, $\theta_{er} = 0$ (axis d), and $\theta_{er} = \frac{\pi}{2}(\text{axis } q)$:

$$I_{A0} \approx \frac{V_{A0}}{\omega L_{AA}}, L_{AB} = \frac{V_{B0}}{\omega I_{A0}}, L_{AF} = \frac{V_{Fo}}{\omega I_{A0}} \tag{3.7}$$

It is true that these measurements do not necessarily reflect the actual magnetic saturation in the machine at the rated speed and load, but they serve to enforce the above assumptions.

Moreover, FEM may be used to calculate the above-mentioned inductance coefficients. Determination of M_{dr} and M_{qr} require more elaborated tests, explained in paragraph 3.19 of this chapter.

3.3 PHASE COORDINATE MODEL

In a matrix form, the stator phase coordinate circuit model for the three-phase SM is

$$\left|I_{\mathrm{ABCF}drqr}\right| \times \left|R_{\mathrm{ABCF}drqr}\right| - \left|V_{\mathrm{ABCF}drqr}\right| = -\frac{\mathrm{d}}{\mathrm{d}t}\left\{\left|L_{\mathrm{ABCF}drqr}^{(\theta_{\mathrm{er}})}\right| \times \left|I_{\mathrm{ABCF}drqr}\right|\right\} \tag{3.8}$$

with

$$\left|V_{\mathrm{ABCF}drqr}\right| = \left|V_{\mathrm{A}}, V_{\mathrm{B}}, V_{\mathrm{C}}, V_{\mathrm{F}}, 0, 0\right|^{\mathrm{T}} \tag{3.9}$$

$$\left|I_{\mathrm{ABCF}drqr}\right| = \left|I_{\mathrm{A}}, I_{\mathrm{B}}, I_{\mathrm{C}}, I_{\mathrm{F}}^{\mathrm{r}}, I_{dr}^{\mathrm{r}}, I_{qr}^{\mathrm{r}}\right|^{\mathrm{T}} \tag{3.10}$$

$$\left|R_{\mathrm{ABCF}drqr}\right| = \mathrm{Diag}\left|R_{\mathrm{s}}, R_{\mathrm{s}}, R_{\mathrm{s}}, R_{\mathrm{F}}^{\mathrm{r}}, R_{dr}^{\mathrm{r}}, R_{qr}^{\mathrm{r}}\right| \tag{3.11}$$

$$\left|L_{\mathrm{ABCF}drqr}^{(\theta_{\mathrm{er}})}\right| = \begin{pmatrix} L_{\mathrm{AA}}(\theta_{\mathrm{er}}) & L_{\mathrm{AB}}(\theta_{\mathrm{er}}) & L_{\mathrm{CA}}(\theta_{\mathrm{er}}) & L_{\mathrm{AF}}(\theta_{\mathrm{er}}) & L_{\mathrm{A}dr}(\theta_{\mathrm{er}}) & L_{\mathrm{A}qr}(\theta_{\mathrm{er}}) \\ L_{\mathrm{AB}}(\theta_{\mathrm{er}}) & L_{\mathrm{BB}}(\theta_{\mathrm{er}}) & L_{\mathrm{BC}}(\theta_{\mathrm{er}}) & L_{\mathrm{BF}}(\theta_{\mathrm{er}}) & L_{\mathrm{B}dr}(\theta_{\mathrm{er}}) & L_{\mathrm{B}qr}(\theta_{\mathrm{er}}) \\ L_{\mathrm{CA}}(\theta_{\mathrm{er}}) & L_{\mathrm{BC}}(\theta_{\mathrm{er}}) & L_{\mathrm{CC}}(\theta_{\mathrm{er}}) & L_{\mathrm{CF}}(\theta_{\mathrm{er}}) & L_{\mathrm{C}dr}(\theta_{\mathrm{er}}) & L_{\mathrm{C}qr}(\theta_{\mathrm{er}}) \\ L_{\mathrm{AF}}(\theta_{\mathrm{er}}) & L_{\mathrm{BF}}(\theta_{\mathrm{er}}) & L_{\mathrm{CF}}(\theta_{\mathrm{er}}) & L_{\mathrm{F}}^{\mathrm{r}} & L_{\mathrm{F}dr}^{\mathrm{r}} & 0 \\ L_{\mathrm{A}dr}(\theta_{\mathrm{er}}) & L_{\mathrm{B}dr}(\theta_{\mathrm{er}}) & L_{\mathrm{C}dr}(\theta_{\mathrm{er}}) & L_{\mathrm{F}dr}^{\mathrm{r}} & L_{dr}^{\mathrm{r}} & 0 \\ L_{\mathrm{A}qr}(\theta_{\mathrm{er}}) & L_{\mathrm{B}qr}(\theta_{\mathrm{er}}) & L_{\mathrm{C}qr}(\theta_{\mathrm{er}}) & 0 & 0 & L_{qr}^{\mathrm{r}} \end{pmatrix} \tag{3.12}$$

Only the rotor self-inductances, $L_{\mathrm{F}}^{\mathrm{r}}$, L_{dr}^{r}, L_{qr}^{r}, and the mutual inductance, L_{Fdr}^{r}, are independent of the rotor position in Equation 3.12. For the nonsalient pole rotor SMs with d.c. electromagnetic or surface PM excitation, $L_2 = 0$ and, thus, L_{AA}, L_{BB}, L_{CC}, and $L_{\mathrm{BC,CA,AB}}$ are also constant. Still, the stator/rotor mutual inductances are dependent on the rotor position. (The PM excitation may be seen as a constant, i_{F}, d.c. excitation circuit.)

Multiplying Equation 3.8 by $[I]^{\mathrm{T}}$, we obtain

$$[I]^{\mathrm{T}} \times [V] = [I]^{\mathrm{T}}[I][I] + \frac{\mathrm{d}}{\mathrm{d}t}\left(\frac{1}{2}[I]^{\mathrm{T}}\left|L(\theta_{\mathrm{er}})\right|[I]\right) + \frac{1}{2}[I]^{\mathrm{T}} \times \left|\frac{\partial L(\theta_{\mathrm{er}})}{\partial \theta_{\mathrm{er}}}\right| \times [I] \times \frac{\mathrm{d}\theta_{\mathrm{er}}}{\mathrm{d}t} \tag{3.13}$$

In the absence of iron losses, the last term in Equation 3.13 is the electromagnetic power, P_e

$$P_e = T_e \times \frac{d\theta_{er}}{p_1\, dt} = \frac{1}{2}\left[I\right]^T \left|\frac{\partial L(\theta_{er})}{\partial \theta_{er}}\right| \left[I\right] \times \frac{d\theta_{er}}{dt} \tag{3.14}$$

So, the electromagnetic torque is

$$T_e = \frac{p_1}{2}\left[I\right]^T \left|\frac{\partial L(\theta_{er})}{\partial \theta_{er}}\right| \left[I\right] \tag{3.15}$$

The motion equations are then added:

$$\frac{J}{p_1}\frac{d\omega_r}{dt} = T_e - T_{load};\quad \frac{d\theta_{er}}{dt} = \omega_r \tag{3.16}$$

The SM model described above exhibits an eighth order, with six currents, ω_r, θ_{er} as variables and four voltages, and load torque, T_{load}, as inputs. Such a high-order system with products of variables and variable coefficients (position-dependent inductances) can be solved only numerically, requiring large CPU time. It is also very difficult to apply to a system control design.

The phase coordinate model may be used in special cases, such as for uniform airgap SMs without a rotor cage, the so-called brushless d.c. PM motors supplied with rectangular bipolar d.c. currents through PWM converters in variable frequency (speed) drives.

3.4 dq0 MODEL TO THREE-PHASE SM PARAMETERS RELATIONSHIPS

The dq0 model for the SM, in rotor coordinates, has been introduced in Chapter 1, together with the voltage, current, and flux equivalence relationships:

$$\frac{d\psi_d}{dt} = V_d - R_s I_d + \omega_r \psi_q;\quad \frac{d\psi_q}{dt} = V_q - R_s I_q - \omega_r \psi_d \tag{3.17}$$

$$\frac{d\psi_F^r}{dt} = V_F^r - R_F^r I_F^r;\quad \frac{d\psi_{dr}^r}{dt} = -R_{dr}^r I_{dr}^r;\quad \frac{d\psi_{qr}^r}{dt} = -R_{qr}^r I_{qr}^r \tag{3.18}$$

$$\frac{d\Psi_0}{dt} = V_0 - R_s I_0;\ \frac{d\omega_r}{dt} = \frac{p_1}{J}\left(\frac{3p_1}{2}\left(\Psi_d i_q - \Psi_q i_d\right) - T_{load}\right) \tag{3.19}$$

and

$$\begin{vmatrix} i_d \\ i_q \\ i_0 \end{vmatrix} = \left|S_{dq0}(\theta_{er})\right| \begin{vmatrix} I_A \\ I_B \\ I_C \end{vmatrix}; \quad \theta_{er} = \int \omega_r dt + \theta_0 \tag{3.20}$$

$$S_{dq0}(\theta_{er}) = \frac{2}{3} \begin{vmatrix} \cos(-\theta_{er}) & \cos\left(-\theta_{er} + \frac{2\pi}{3}\right) & \cos\left(-\theta_{er} - \frac{2\pi}{3}\right) \\ \sin(-\theta_{er}) & \sin\left(-\theta_{er} + \frac{2\pi}{3}\right) & \sin\left(-\theta_{er} - \frac{2\pi}{3}\right) \\ \frac{1}{2} & \frac{1}{2} & \frac{1}{2} \end{vmatrix} \tag{3.21}$$

Equivalence 3.20 is also valid for voltages V_d, V_q, and V_0 and for flux linkages ψ_d, ψ_q, and ψ_0:

$$\begin{vmatrix} \psi_d \\ \psi_q \\ 0 \end{vmatrix} = \left|S_{dq0}(\theta_{er})\right| \begin{vmatrix} \psi_A \\ \psi_B \\ \psi_C \end{vmatrix} \tag{3.22}$$

The rotor variables of the SM need not be changed for the dq0 model because the rotor windings (F, d_r, q_r) are orthogonal by nature.

From Equation 3.13 and inductance matrix 3.12, the stator flux linkages, ψ_A, ψ_B, ψ_C are

$$\begin{aligned} \psi_A &= L_{AA}I_A + L_{AB}I_B + L_{CA}I_C + L^r_{AF}I^r_F + L^r_{Adr}I^r_{dr} + L^r_{Aqr}I^r_{qr} \\ \psi_B &= L_{AB}I_A + L_{BB}I_B + L_{CB}I_C + L^r_{BF}I^r_F + L^r_{Bdr}I^r_{dr} + L^r_{Bqr}I^r_{qr} \\ \psi_C &= L_{CA}I_A + L_{CB}I_B + L_{CC}I_C + L^r_{CF}I^r_F + L^r_{Cdr}I^r_{dr} + L^r_{Cqr}I^r_{qr} \end{aligned} \tag{3.23}$$

In Equation 3.23, the inverse Park transformation is used, in order to eliminate stator currents:

$$\begin{vmatrix} I_A \\ I_B \\ I_C \end{vmatrix} = \frac{3}{2}\left[S_{dq0}\right]^T \begin{vmatrix} i_d \\ i_q \\ i_0 \end{vmatrix} \tag{3.24}$$

Finally, we obtain

$$\begin{aligned} \psi_d &= L_{sl}I_d + \frac{3}{2}(L_0 + L_2)I_d + \frac{3}{2}M_F I^r_F + \frac{3}{2}M_{dr}I^r_{dr} \\ \psi_q &= L_{sl}I_q + \frac{3}{2}(L_0 - L_2)I_d + \frac{3}{2}M_{qr}I^r_{qr} \end{aligned} \tag{3.25}$$

But for the dq0 model with the rotor reduced to the stator (Chapter 1), ψ_d, and ψ_q are

$$\begin{aligned}\psi_d &= L_{sl}I_d + L_{dm}\left(I_d + I_F + I_{dr}\right)\\ \psi_q &= L_{sl}I_q + L_{qm}\left(I_q + I_{qr}\right)\end{aligned} \tag{3.26}$$

The parameter equivalence between Equations 3.25 and 3.26 is thus straightforward:

$$L_{dm} = \frac{3}{2}\left(L_0 + L_2\right);\quad L_{qm} = \frac{3}{2}\left(L_0 - L_2\right) \tag{3.27}$$

$$I_F = I_F^r K_F;\quad K_F = \frac{3}{2}\frac{M_F}{L_{dm}};\quad I_{dr} = I_{dr}^r K_{dr};\quad K_{dr} = \frac{3}{2}\frac{M_{dr}}{L_{dm}} \tag{3.28}$$

$$I_{qr} = I_{qr}^r K_{qr};\quad K_{qr} = \frac{3}{2}\frac{M_{qr}}{L_{qm}} \tag{3.29}$$

From the conservation of rotor power and losses

$$V_F = \frac{V_F^r}{K_F};\quad R_F = R_F^r\frac{1}{K_F^2};\quad R_{dr} = R_{dr}^r\frac{1}{K_{dr}^2};\quad R_{qr} = R_{qr}^r\frac{1}{K_{qr}^2} \tag{3.30}$$

Also, from the conservation of the leakage magnetic field energy

$$L_{Fl} = L_{Fl}^r\frac{1}{K_F^2};\quad L_{drl} = L_{drl}^r\frac{1}{K_{dr}^2};\quad L_{qrl} = L_{qrl}^r\frac{1}{K_{qr}^2};\quad L_{Fdr} = L_{Fdr}^r\frac{1}{K_F K_{dr}} \tag{3.31}$$

The stator resistance, R_s, and phase leakage inductance, L_{sl}, remain in the dq0 model with their values in the real three-phase machine. L_{Fl}, L_{drl}, L_{qrl}, and L_{Fdr} are rotor leakage inductances reduced to the stator. Now L_{dm}, and L_{qm} are the cyclic magnetization inductances derived in the companion book, Vol. 1, for the steady and state.

For the PMSM without a rotor cage, the dq0 model greatly simplifies to

$$\frac{d\psi_d}{dt} = V_d - R_s I_d + \omega_r\psi_q;\quad \psi_d = \psi_{PM} + L_d I_d \tag{3.32}$$

$$\frac{d\psi_q}{dt} = V_q - R_s I_q - \omega_r\psi_d;\quad \psi_q = L_q I_q;\quad L_d < L_q \tag{3.33}$$

The torque Equation 3.19 is

$$T_e = \frac{3}{2}p_1\left(\psi_{PM} + \left(L_d - L_q\right)I_d\right)I_q \tag{3.34}$$

For the three-phase reluctance synchronous motor, $\psi_{PM} = 0$ and $L_d >> L_q$.

3.5 STRUCTURAL DIAGRAM OF THE SM dq0 MODEL

The dq0 model of the SM Equations 3.17 through 3.19 and the rotor flux/current relationships

$$\begin{aligned} \psi_F &= L_{Fl} I_F + \psi_{dm} \\ \psi_{dr} &= L_{drl} I_{dr} + \psi_{dm} \\ \psi_{qr} &= L_{qrl} I_{qr} + \psi_{qm} \end{aligned} \tag{3.35}$$

may be put under the Laplace form $\left(\frac{d}{dt} \to s\right)$ as

$$\begin{aligned} &(V_F - s\psi_{dm}) \times \frac{1}{R_F(1+s\tau'_F)} = I_F \\ &V_d - R_s(1+s\tau'_s) I_d + \omega_r \psi_q = s\psi_{dm} \\ &I_d + I_F = \frac{\psi_{dm}(1+s\tau_{dr})}{L_{dm}(1+s\tau'_{dr})} \\ &V_q - R_s(1+s\tau'_s) I_q - \omega_r \psi_d = s\psi_{qm} \\ &\frac{\psi_{qm}(1+s\tau_{qr})}{L_{qm}(1+s\tau'_{qr})} = i_q \\ &\tau'_s = L_{sl}/R_s; \quad \tau_{dr} = (L_{drl} + L_{dm})/R_{dr}; \quad \tau_{qr} = (L_{qrl} + L_{qm})/R_{qr} \\ &\tau'_{dr} = L_{drl}/R_{dr}; \quad \tau'_{qr} = L_{qrl}/R_{qr}; \quad \tau'_F = L_{Fl}/R_F \end{aligned} \tag{3.36}$$

The above-mentioned equations are illustrated in the form of a structural diagram in Figure 3.2.

We may identify small (milliseconds to tens of milliseconds) time constants, τ'_s, τ'_F, τ'_{dr}, and τ'_{qr} and two large (hundreds of milliseconds) time constants τ_{dr}, and τ_{qr}.

For steady state, at $\omega_b = \omega_r$ (rotor coordinates), $s = 0$ and thus the structural diagram looses the influence of all time constants. In the absence of d, q axes rotor cages, τ_{dr}, τ'_{dr}, τ_{qr}, and τ'_{qr} are eliminated from the structural diagram all together.

Now if the machine has PMs or a variable reluctance rotor and no cage on the rotor, the structural diagram simplifies notably (Figure 3.3).

Once the motion emfs, $\omega_r\psi_d$, and $\omega_r\psi_q$, are added (in axis *d*) or subtracted (in axis *q*) to the respective voltages, V_d, V_q, only the stator leakage (small) time constants remain active.

This was how rectangular bipolar current control in PMSMs originated. The same equations (Equation 3.36) may be arranged into the well-known equivalent circuits

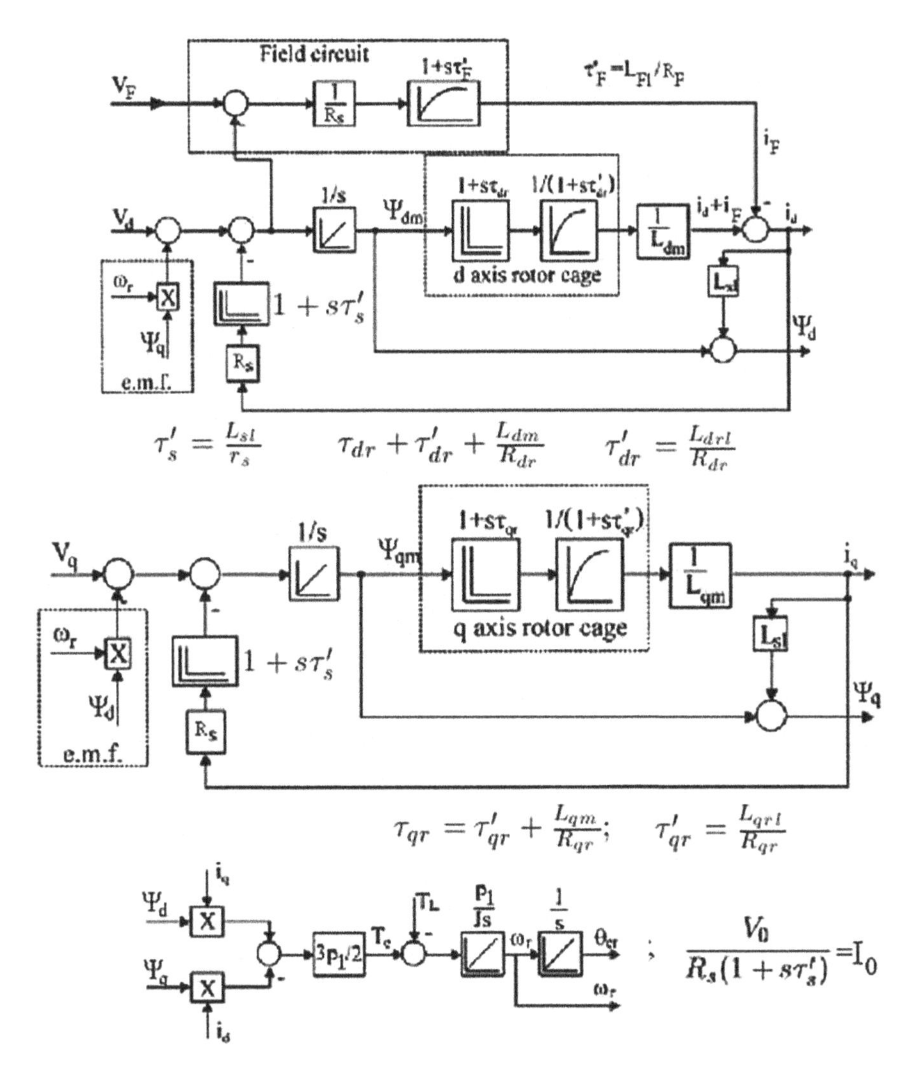

FIGURE 3.2 Structural diagram of the dq0 model of SMs: (a) axis d, (b) axis q, and (c) motion equations.

for transients (Figure 3.4). For the PM machine, the d.c. excitation circuit is replaced by a constant current source, $i_{F0} = \psi_{PM}/L_{dm}$.

Again, steady state is obtained for $s = 0$; iron losses are not yet considered. They do not have much influence other than during the first 3–5 ms into the transients [1]. However, in large iron core loss machines, they might be included as resistances in parallel with the motion emfs, $\omega_r\psi_d$ and $\omega_r\psi_d$. At zero speed, $\omega_r\psi_d = 0$ and $\omega_r\psi_q = 0$, and the equivalent circuits become very instrumental for the parameter estimation

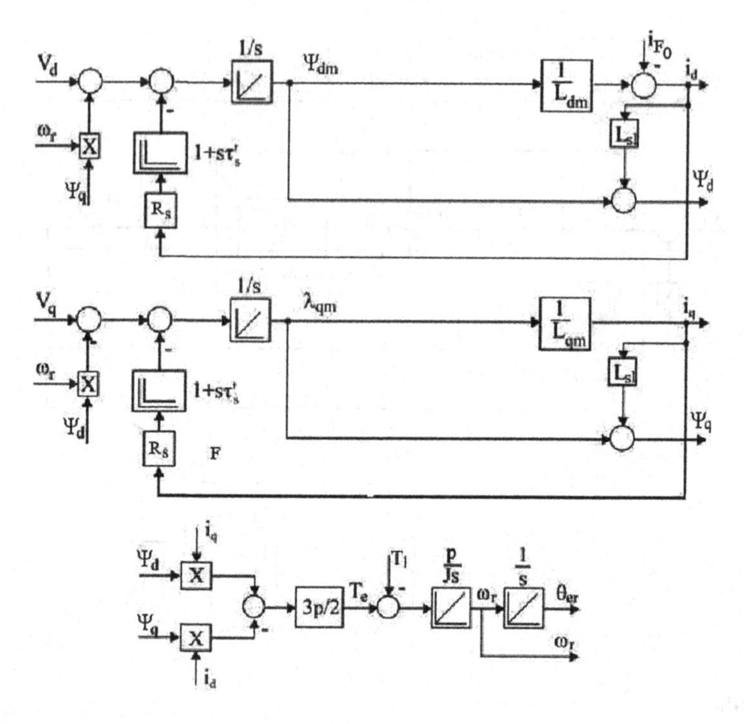

FIGURE 3.3 Structural diagram for the PM and reluctance SMs (without damper cage): (a) axis *d*, (b) axis *q*, and (c) motion equations.

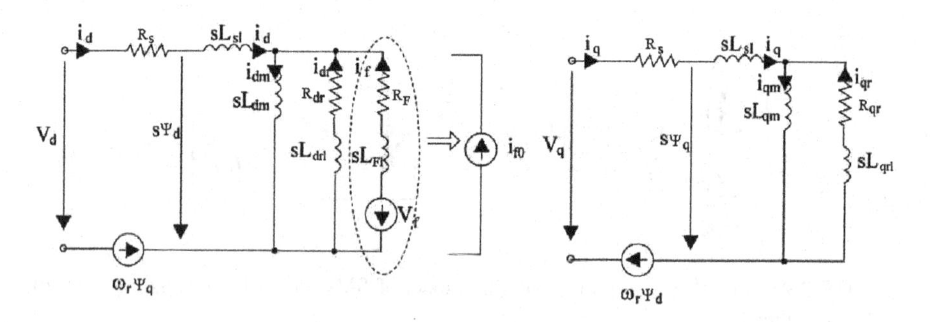

FIGURE 3.4 SM dq model equivalent circuits for transients.

from the standstill current decay or frequency response tests described in paragraph 3.19 of this chapter.

3.6 PU dq0 MODEL OF SMS

The pu (per unit) system is used to norm the system of equations by expressing voltages, currents, flux linkages, resistances, inductances, torque, and power in relative units.

Such a denotation limits all numerical values of variables to, say, 30, at most; it also brings more generality to the results and leads the way toward global standardization.

To build a pu system, base voltages are required. There is more than one way of doing this.

One method that is widely accepted is described here:

- $V_{no}\sqrt{2}$—base voltage (peak value of rated phase voltage)
- $I_{no}\sqrt{2}$—base current (peak value of rated phase current)
- $X_{no} = V_{no}/I_{no} = \omega_{10}L_{no}$—base inductance (reactance at base frequency, ω_{10})
- $\psi_{no} = V_{no}\sqrt{2}/\omega_{10}$—base flux linkage
- $P_{no} = 3V_{no}I_{no}$—base power
- $T_{no} = P_{no} \times p_1/\omega_{10}$—base torque
- $H = \dfrac{J\omega_{10}^2}{2p_1^2P_{no}}$—inertia in seconds

Now if time is measured in seconds, then $\dfrac{d}{dt} \rightarrow \dfrac{1}{\omega_{10}}\dfrac{d}{dt}$ in the original model (Equations 3.17 through 3.19):

$$\begin{gathered}
\frac{1}{\omega_{10}}\frac{d\psi_d}{dt} = V_d - r_s i_d + \omega_r\psi_q; \quad \frac{1}{\omega_{10}}\frac{d\psi_{qr}}{dt} = V_q - r_s i_q - \omega_r\psi_d \\
\frac{1}{\omega_{10}}\frac{d\psi_F}{dt} = V_F - r_F i_F; \quad \frac{1}{\omega_{10}}\frac{d\psi_{dr}}{dt} = -r_{dr}i_{dr}; \quad \frac{1}{\omega_{10}}\frac{d\psi_{qr}}{dt} = -r_{qr}i_{qr} \\
\frac{d\omega_r}{dt} = \frac{1}{2H}\left(\psi_d i_d - \psi_q i_d - t_{shaft}\right); \quad t_e = \psi_d i_q - \psi_q i_d; \quad \frac{1}{\omega_{10}}\frac{d\theta_{er}}{dt} = \omega_r
\end{gathered} \tag{3.37}$$

The flux/current relationships in the pu system remain practically the same as for the actual dq variables, but the difference is that lower case letters are often used (Equation 3.38):

$$\begin{aligned}
\psi_d &= l_{sl}i_d + l_{dm}\left(i_d + i_{dr} + i_F\right) \\
\psi_q &= l_{sl}i_q + l_{qm}\left(i_q + i_{qr}\right) \\
\psi_F &= l_{Fl}i_F + l_{dm}\left(i_d + i_{dr} + i_F\right) + l_{Fdrl}\left(i_{dr} + i_F\right) \\
\psi_{dr} &= l_{drl}i_d r + l_{dm}\left(i_d + i_{dr} + i_F\right) + l_{Fdrl}\left(i_{dr} + i_F\right) \\
\psi_{qr} &= l_{qrl}i_q + l_{qm}\left(i_q + i_{qr}\right)
\end{aligned} \tag{3.38}$$

The mutual inductance, l_{Fdrl}, allows for an additional leakage flux coupling between the d axis cage and the field circuit on the rotor.

Example 3.1: PU Parameters in Ohms

A large synchronous motor operated at the power grid has the design data: P_n = 1800 kW, efficiency $\eta_n = 0.983$, cos $\varphi_n = 0.9$, $f_1 = 50$ Hz, $V_{nl} = 10.0$ KV, $n_1 = 100$ rpm, $l_{dm} = 0.6$(pu), $l_{qm} = 0.4$(pu), $l_{sl} = 0.1$(pu), $l_{drl} = 0.11$(pu), $r_{dr} = 0.05$(pu), l_{qrl} = 0.025(pu), $r_s = 0.01$(pu), $l_{Fl} = 0.17$(pu), $r_F = 0.016$(pu), and $l_{Fdrl} = -0.037$(pu).

Let us consider that the field current, i_F, for the case in point (rated power) is i_F = 2.5 (pu) and $H = 2$ s. Calculate the rated current, number of pole pairs, p_1, base

(Continued)

torque, base reactances, all resistances and inductances in ohms, inertia J (in kg m^2), i_F in amperes, and V_F in volts. Also, calculate the time constants, τ_F', τ_s', τ_{dr}, τ_{dr}', τ_{qr}, τ_{qr}', in seconds.

Solution:

According to the efficiency, η_1, and power factor definitions, the rated current I_n is simply

$$I_n = \frac{P_n}{\sqrt{3}V_{nl}\eta_n \cos\varphi_n} = \frac{1800\times 10^3}{10\sqrt{3}\times 10^3\times 0.9\times 0.983} = 1{,}176 \text{ A} \quad (3.39)$$

The number of poles

$$2p_1 = \frac{2f_1}{n_1} = \frac{2\times 50}{(100/60)} = 60$$

The base torque

$$T_{no} = P_{no}\times p_1/\omega_{10} = 1{,}800\times 10^3\times 30/(2\pi 50) = 171{,}970 \text{ Nm} = 171.970 \text{ kNm}$$

The base reactance

$$X_n = \frac{V_{nl}}{I_n\sqrt{3}} = \frac{10\times 10^3}{\sqrt{3}\times 1176} = 4.915\ \Omega$$

The actual inertia

$$J = 2H\left(\frac{p_1}{\omega_{10}}\right)^2 P_{n0} = 2\times 2\times\left(\frac{30}{314}\right)^2\times 1{,}800{,}000 = 65.72\times 10^3 \text{ kg m}^2$$

The field current i_F in amperes

$$i_F(A) = i_F(\text{pu})\times I_n = 2.5\times 1176 = 2940 \text{ A}$$

The field circuit voltage V_F (reduced to the stator)

$$V_F(V) = R_F(\Omega)\, i_F(\text{A}) = r_F(\text{pu})X_n I_F(\text{A}) = 0.016\times 4.915\times 2940 = 231.2 \text{ V}$$

All resistances and reactances may be calculated as

$$R_s(\Omega) = r_s(\text{pu})\times X_n(\Omega) = 0.01\times 4.915 = 0.04915\ \Omega$$
$$X_{sl}(\Omega) = l_{sl}(\text{pu})\times X_n(\Omega) = 0.1\times 4.915 = 0.4915\ \Omega$$

The time constants in seconds are

$$\tau_F' = \frac{l_{Fl}}{r_F}\frac{1}{\omega_{10}} = \frac{0.17}{0.016}\times\frac{1}{314} = 3.38\times10^{-2}\ \text{s}$$

$$\tau_s' = \frac{l_{sl}}{r_s}\frac{1}{\omega_{10}} = \frac{0.1}{0.01}\times\frac{1}{314} = 3.185\times10^{-2}\ \text{s}$$

$$\tau_{dr} = \frac{l_{drl}+l_{dm}}{r_{dr}}\frac{1}{\omega_{10}} = \frac{0.11+0.6}{0.05}\times\frac{1}{314} = 4.522\times10^{-2}\ \text{s}$$

$$\tau_{dr}' = \frac{l_{drl}}{r_{dr}}\frac{1}{\omega_{10}} = \frac{0.1}{0.05}\times\frac{1}{314} = 6.369\times10^{-3}\ \text{s}$$

$$\tau_{qr} = \frac{l_{qm}+l_{qrl}}{r_{qr}}\frac{1}{\omega_{10}} = \frac{0.4+0.025}{0.03}\times\frac{1}{314} = 4.51\times10^{-2}\ \text{s}$$

$$\tau_{qr}' = \frac{l_{qrl}}{r_{qr}}\frac{1}{\omega_{10}} = \frac{0.025}{0.03}\times\frac{1}{314} = 2.654\times10^{-3}\ \text{s}$$

3.7 BALANCED STEADY STATE VIA THE dq0 MODEL

A balanced steady state means, for the ideal grid-connected SM, symmetric sinusoidal phase voltages and currents:

$$\begin{aligned} V_{A,B,C} &= V_0\sqrt{2}\times\cos\left(\omega_1 t - \left(i-1\frac{2\pi}{3}\right)\right); \quad i=1,2,3 \\ I_{A,B,C} &= I_0\sqrt{2}\times\cos\left(\omega_1 t - (i-1)\frac{2\pi}{3} - \varphi_1\right); \quad i=1,2,3 \end{aligned} \tag{3.40}$$

Applying the Park transformation (Equation 3.21) in rotor coordinates ($\theta_{er} = \omega_1 t + \theta_0$) to *A, B, C* voltages and currents in Equation 3.40 yields

$$\begin{aligned} V_{d0} &= V_0\sqrt{2}\cos(\theta_0); \qquad I_{d0} = I_0\sqrt{2}\cos(\theta_0 - \varphi_1) \\ V_{q0} &= -V_0\sqrt{2}\sin(\theta_0); \qquad I_{q0} = -I_0\sqrt{2}\sin(\theta_0 - \varphi_1) \end{aligned} \tag{3.41}$$

A positive φ_1 means a lagging power factor (motor association of voltage/current signs). As the stator voltages and currents in the dq model, for the balanced steady state, are d.c., the field current is d.c. and d/dt = 0 and, thus, $i_{dr} = i_{qr} = 0$ (no damper cage currents):

$$V_{F0} = R_F I_{F0}; \quad i_{dr0} = i_{qr0} = 0 \tag{3.42}$$

The torque, T_e, (Equation 3.34) is simplified to

$$T_e = \frac{3}{2}p_1\left(\Psi_d i_q - \Psi_q i_d\right) = \frac{3}{2}p_1\left(L_{dm}i_{F0} + \left(L_d - L_q\right)i_{d0}\right)i_{q0} \tag{3.43}$$

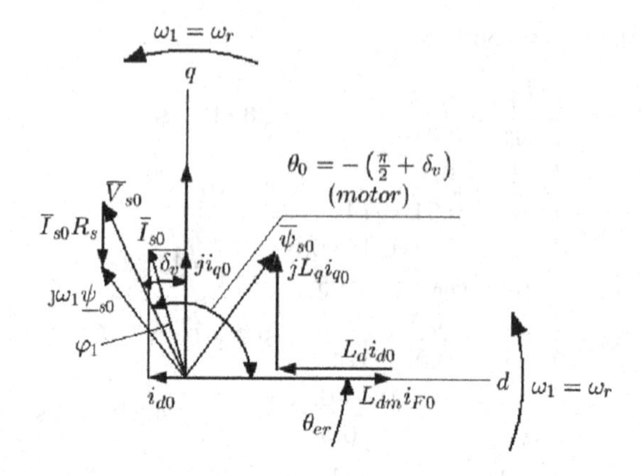

FIGURE 3.5 Space phasor diagram of MS for the balanced steady state.

The dq stator equations for the steady state ($d/dt = 0$) may be written in a space phasor form as

$$\overline{V}_{s0} = R_s \overline{i}_{s0} + j\omega_r \overline{\psi}_{s0};\quad \overline{\psi}_{s0} = \psi_{d0} + j\psi_{q0};\quad \overline{i}_{s0} = i_{d0} + j i_{q0}$$
$$\psi_{d0} = L_{dm} i_{F0} + L_d i_d;\quad \psi_{q0} = L_{qr} i_{q0} \tag{3.44}$$

So we can represent Equation 3.44 in a space phasor (vector) diagram as shown in Figure 3.5.

The space phasor (vector) diagram reproduces all phase shift angles of the phasor diagram of the SM described the in Chapter 6, of companion book, Vol.1. Here, the angles for the phase phasors are "space" angles whereas those for the phase phasors described in Chapter 6, (Vol.1) were "time" angles.

The relationships between θ_0 in the space phasor diagram (dq model) and the voltage power angle, $\boldsymbol{\delta}_v$, (Chapter 6, Vol.1) is

$$\theta_0 = -\left(\frac{\pi}{2} + \delta_v\right);\quad \delta_v > 0 \tag{3.45}$$

for motor operation and

$$\theta_0 = -\left(\frac{\pi}{2} + \delta_v\right);\quad \delta_v < 0 \tag{3.46}$$

for generator operation mode.

Example 3.2: Balanced Steady-State Operation with dq0 Model

The SM in Example 3.1 operates as a motor at an unity power factor and at $\boldsymbol{\delta}_v = 30°$.

Calculate

a. The emf, E_s (no-load voltage)
b. V_{do}, i_{d0}, V_{q0}, i_{q0}, V_{s0}, and i_{s0}
c. Steady-state short-circuit current and braking torque

To solve the problem, we make use of the space phasor diagram in Figure 3.5 where $\varphi_1 = 0$ and $\delta_v = 30°\left(\frac{\pi}{6}\right)$.

From Equations 3.41 and 3.45

$$V_{d0} = V_0\sqrt{2}\times\cos\left(-\frac{\pi}{2}-\delta_v\right) = -V_0\sqrt{2}\sin\delta_v = -5780\sqrt{2}\frac{1}{2} = -4074\text{V}$$
$$I_{d0} = I_0\sqrt{2}\times\cos\left(-\frac{\pi}{2}-\delta_v\right) = -I_0\sqrt{2}\sin\delta_v \quad (3.47)$$

$$V_{q0} = -V_0\sqrt{2}\times\sin\left(-\frac{\pi}{2}-\delta_v\right) = -V_0\sqrt{2}\cos\delta_v = 5780\sqrt{2}\frac{\sqrt{3}}{2} = 7091.77\text{V}$$
$$I_{q0} = -I_0\sqrt{2}\times\sin\left(-\frac{\pi}{2}-\delta_v\right) = I_0\sqrt{2}\cos\delta_v \quad (3.48)$$

Note: For the generator, φ_1 in Equation 3.45 would have been $\varphi_1 = \pi$. From Example 3.1, $V_0 = \frac{V_{nl}}{\sqrt{3}} = \frac{10\times10^3}{\sqrt{3}} = 5780$ V

The space phasor diagram (Figure 3.5) with $\varphi_1 = 0$ provides

$$V_{do} = -\omega_1\psi_{q0} + R_s i_{d0} = -\omega_1 L_q i_{q0} + R_s i_{d0}$$
$$V_{qo} = \omega_1\psi_{d0} + R_s i_{q0} = \omega_1\left(L_d i_{d0} + L_{dm} i_{F0}\right) + R_s i_{q0} \quad (3.49)$$

From Equations 3.47 in 3.53, we can compute the two remaining unknowns, I_0 and i_F, if the machine inductances (reactances) are known (Example 3.1).

$$X_d = \omega_1 L_d = \left(l_{dm} + l_{sl}\right)X_n = \left(0.6+0.1\right)4.915 = 3.44\ \Omega$$
$$X_q = \omega_1 L_q = \left(l_{qm} + l_{sl}\right)X_n = \left(0.4+0.1\right)4.915 = 2.4575\ \Omega$$
$$\omega_1 L_{dm} = l_{dm}X_n = 0.6\times4.915 = 2.95\ \Omega$$
$$R_s = 0.04915\ \Omega$$

A second-order equation has to be solved for this purpose. But to simplify the solution we may neglect the effect of the stator resistance ($r_s = 0.01$(p.u.)) here.

In this case

$$i_{q0} = \frac{-V_{d0}}{X_q} = \frac{V_0\sqrt{2}\sin\delta_v}{X_q} = \frac{5780\times\sqrt{2}\times0.5}{2.4575} = 1658\text{ A}$$

$$i_{d0} = -i_{q0}\times\tan\delta_v = -1658\times\frac{1}{\sqrt{3}} = -958\text{ A}$$

(Continued)

Example 3.2: (Continued)

The stator current phasor is $i_{s0} = I_0\sqrt{2} = i_{q0}/\cos\delta_v = 1658/0.867 = 1912$ A (peak phase value; 1356.2 A (RMS phase value)).

$$i_{F0} \approx \frac{\left(V_{q0} - X_d i_{d0}\right)}{X_{dm}} = \frac{5780\sqrt{2}\times\left(\frac{\sqrt{3}}{2}\right) - 3.44(-958)}{2.95} = 3506.8 \text{ A}$$

The no-load voltage is $E_0 = X_{dm} i_{F0} = 3506.80 \times 2.95 = 10{,}345$ V (peak phase value; 7336 V (RMS, phase value)) For the short circuit, we just put $V_{d0} = V_{q0} = 0$ in the steady-state Equation 3.49 and i_{dsc} and i_{qsc}:

$$\begin{aligned} 0 &= -X_q i_{qsc} + R_s i_{dsc} \\ 0 &= X_d i_{dsc} + X_{dm} i_F + R_s i_{qsc} \end{aligned} \tag{3.50}$$

We retain the stator resistance in Equation 3.50 for generality. In the kilowatt range, for example, PMSMs, neglecting R_s would lead to ignoring the braking torque during shortcircuit, with small I_d, I_q, *and* I_{qm} (which would become 0.5 pu). This would mean ignoring 30%–60% rated torque braking.

The solution of Equation 3.50 is straightforward:

$$i_{qsc} = i_{dsc} \times \frac{R_s}{X_q}; \quad i_{dsc} = \frac{-X_{dm} i_{F0}}{X_d + R_s^2/X_q} \tag{3.51}$$

And the torque, from Equation 3.43, or from stator copper losses, is

$$T_{esc3} = -\frac{3}{2} R_s \left(i_{dsc}^2 + i_{qsc}^2\right) \frac{p_1}{\omega_1} \tag{3.52}$$

$$\begin{aligned} i_{dsc} &= -\frac{10345}{3.44 + 0.04915^2/2.4575} = -3006.4 \text{ A} \\ i_{qsc} &= -3006.4 \times \frac{0.04915}{2.4575} = -60.128 \text{ A} \end{aligned}$$

From Equation 3.52, the braking torque is $T_{esc3} = -\frac{3}{2} 0.04915\left(3006.4^2 + 60.128^2\right)\frac{30}{314} = -63{,}666$ N m = – 63.666 kN m; in p.u. terms (see Example 3.1; with $T_{en} =$ 171.90 kNm) $t_{esc3} = T_{esc3}/T_{en} = -63.666/171.90 = -0.37$ (p.u.).

It should be noticed that even at the short circuit, the SM works as a generator, so $T_{esc3} < 0$, as both i_{dsc} and i_{qsc} are negative in the dq model.

3.8 LAPLACE PARAMETERS FOR ELECTROMAGNETIC TRANSIENTS

From the equivalent circuits for transients (Figure 3.4), the flux current relationships in Laplace terms may be extracted, after elimination of the rotor cage currents

$$\begin{aligned}\psi_d(s) &= L_d(s)i_d(s) + G(s)V_F(s) \\ \psi_q(s) &= L_q(s)i_q(s)\end{aligned} \tag{3.53}$$

with

$$\begin{aligned}L_d(s) &= L_d\frac{(1+s\tau_d')(1+s\tau_d'')}{(1+s\tau_{d0}')(1+s\tau_{d0}'')} \\ L_q(s) &= L_q\frac{(1+s\tau_q'')}{(1+s\tau_{q0}'')} \\ G(s) &= \frac{L_{dm}}{R_F}\frac{1+s\tau_{dr}'}{(1+s\tau_{d0}')(1+s\tau_{d0}'')}\end{aligned} \tag{3.54}$$

where $L_d(s)$, and $L_q(s)$, G(s) are the so-alled operational (Laplace) parameters for SMs. Their form is the same in pu, where only $l_d(s)$, $l_q(s)$, and $g(s)$ would appear in Equation 3.54 instead of $L_d(s)$, $L_q(s)$ and $G(s)$, because the time constants are still measured in seconds:

$$\begin{aligned}\tau_d' &= \left(L_{\text{Fl}} + L_{\text{F}drl} + \frac{L_{dm}L_{\text{sl}}}{L_{dm}+L_{\text{sl}}}\right)\frac{1}{R_{\text{F}}} \\ \tau_{d0}' &= \left(L_{dm} + L_{\text{F}drl} + L_{\text{Fl}}\right)\frac{1}{R_{\text{F}}} \\ \tau_d'' &= \frac{1}{R_{dr}}\left(L_{drl} + \frac{L_{dm}L_{\text{F}drl}L_{\text{Fl}} + L_{dm}L_{\text{sl}}L_{\text{Fl}} + L_{\text{sl}}L_{\text{F}drl}L_{\text{Fl}}}{L_{dm}L_{\text{Fl}} + L_{\text{Fl}}L_{\text{sl}} + L_{dm}L_{\text{Fl}} + L_{\text{se}}L_{Fdrl} + L_{\text{sl}}L_{dm}}\right) \\ \tau_{d0}'' &= \frac{1}{R_{dr}}\left(L_{drl} + \frac{L_{\text{Fl}}\left(L_{dm}+L_{\text{F}drl}\right)}{L_{\text{Fl}} + L_{dm} + L_{Fdrl}}\right) \\ \tau_{q0}'' &= \frac{1}{R_{qr}}\left(L_{qrl} + L_{qm}\right); \quad \tau_{dr}' = L_{drl}/R_{dr} \\ \tau_q'' &= \frac{1}{R_{qr}}\left(L_{qrl} + \frac{L_{qm}L_{sl}}{L_{qm}+L_{sl}}\right)\end{aligned} \tag{3.55}$$

It should be noticed that the Laplace parameters do not contain any information related to machine speed; this is because the dq0 model of SM is used in rotor coordinates. The initial and final values of $L_d(s)$ and $L_q(s)$ corresponding to *subtransient* L_d'' and L_q'', and *synchronous inductances* L_d and L_q are

$$L_d'' = \lim_{s\to\infty(t\to 0)} L_d(s) = L_d\frac{T_d'T_d''}{T_{d0}'T_{d0}''} < L_d \tag{3.56}$$

$$L_q'' = \lim_{s\to\infty(t\to 0)} L_q(s) = L_q \frac{T_q''}{T_{q0}''} < L_q$$
$$L_d = \lim_{s\to 0(t\to\infty)} L_d(s)$$
$$L_q = \lim_{s\to 0(t\to\infty)} L_q(s)$$

In the absence of damper cage along the *d axis*, the so-called transient inductance axis, L_d', is defined as

$$L_d'' < L_d' = \lim_{s\to\infty, T_d''=T_{d0}''} L_d(s) = L_d \frac{T_d'}{T_{d0}'} < L_d \quad (3.57)$$

L_d'' and L_q'' reflect the SM's initial reaction to transients, based on the flux conservation law. Additional (transient) currents occur in the rotor to conserve the initial value of flux that "looses" the "support" of initial stator currents, which change quickly but not instantaneously. It is expected that after a sudden short circuit at SM terminals, the currents in the stator and rotor change dramatically. Indeed, the sudden short-circuit current is large in SMs with a strong (low resistance) rotor cage. For PMSMs without any rotor cage, only synchronous inductances exist, and thus the transients at a sudden short circuit are lower and slower.

As even with the dq0 model the study of transients in general is faced with difficult mathematical hurdles, the transients are approximated according to three categories:

1. Electromagnetic transients
2. Electromechanical transients
3. Mechanical transients

3.9 ELECTROMAGNETIC TRANSIENTS AT CONSTANT SPEED

During fast transients, the speed can be approximated as constant. These are considered electromagnetic transients. The stator voltage build-up in a synchronous generator at no load is such a typical transient, but the sudden three-phase short circuit is the most important one recognized by industry.

The investigation of transients for constant machine inductances (magnetic saturation level does not vary) in time can be undertaken using Equation 3.17 with d/dt = s, and with the Laplace definition of parameters Equations 3.53 and 3.54

$$\begin{aligned} V_d(s) &= R_s i_d(s) + s\left[L_d(s) i_d(s) + G(s) V_F(s)\right] - \omega_r L_q(s) I_q(s) \\ V_q(s) &= R_s i_q(s) + s L_q(s) i_q(s) + \omega_r\left[L_d(s) i_d(s) + G(s) V_F(s)\right] \end{aligned} \quad (3.58)$$

We stress that Equations 3.58 are written in Laplace form, and thus deal only with the deviations of variables $i_d(s)$ and $i_q(s)$ and of inputs $V_d(s)$, $V_q(s)$ and $V_F(s)$ with respect to their initial values (at $t = 0$). The rotor speed, ω_r, is considered constant.

Example 3.3: Voltage Build-Up

Let us consider the case of an SM, at no load and at speed ω_r, whose full-field circuit voltage is applied suddenly. Calculate the stator voltage components and phase voltage build-up expressions in pu for l_{dm} = 1.2 pu, l_{Fl} = 0.2 pu, V_{F0} = 0.005833 pu, r_F = 0.01 pu, and ω_r = 1 pu (ω_{10} = 377 rad/s)

Solution:

For a voltage build-up, the field circuit pu voltage step up $V_{F0}(s)$ in the Laplace form is

$$V_F(s) = \frac{V_{F0}}{s}\omega_{10} \tag{3.59}$$

It should be noticed that in pu terms, s is replaced by s/ω_{10} (in our case ω_{10} = 377 rad/s (60 Hz)). The rated speed is inferred as ω_r = 1 pu.

For no load, $i_d(s) = 0$, $i_q(s) = 0$ and thus, from Equation 3.58

$$\begin{aligned} V_d(s) &= g(s)\frac{s}{\omega_{10}}V_F(s) \\ V_q(s) &= \omega_r g(s) V_F(s) \end{aligned} \tag{3.60}$$

But from Equation 3.54, with l_{dm}/r_F instead of L_{dm}/R_F, in the absence of rotor cage $\tau_d'' = 0$, $\tau_{dr}' = 0$, the stator/field winding transfer function $g(s)$ is

$$g(s) = \frac{l_{dm}}{r_F}\frac{1}{\left(1 + (l_{dm} + l_{Fl})\frac{s}{r_F\omega_{10}}\right)} \tag{3.61}$$

With Equations 3.59 and 3.61, Equations 3.60 have these straightforward solutions:

$$\begin{aligned} V_d(t) &= \frac{V_{F0} l_{dm}}{l_{dm} + l_{Fl}} e^{-\frac{t}{\tau_{d0}'}}; \quad \tau_{d0}' = \frac{l_{dm} + l_{fl}}{r_F\omega_{10}} \\ V_q(t) &= \omega_r \frac{l_{dm} V_{F0}}{r_F}\left(1 - e^{-t/\tau_{d0}'}\right) \end{aligned} \tag{3.62}$$

With V_{F0} = 0.005833 pu, r_F = 0.01 pu, l_{dm} = 1.2 pu, l_{Fl} = 0.2 pu, ω_r = 1 pu, and ω_{10} = 377 rad/s, Equations 3.62 become

$$\begin{aligned} V_d(t) &= 0.05e^{-2.7\times t}; \quad \text{in pu} \\ V_q(t) &= 0.70\left(1 - e^{-2.7\times t}\right); \quad \text{in pu} \end{aligned}$$

The stator phase voltage, $V_A(t)$, is obtained through the invterse Park transformation:

$$V_A(t) = V_d(t)\cos(\omega_{10}t) - V_q\sin(\omega_{10}t) \tag{3.63}$$

For $\omega_r \neq 1$ pu in Equation 3.63, it would be ($\omega_r \times \omega_{10}t$) instead of ($\omega_{10}t$), to reflect the actual frequency of the stator voltage which is equal to the actual electric speed, ω_r, in radians per second.

3.10 SUDDEN THREE-PHASE SHORT CIRCUIT FROM A GENERATOR AT NO LOAD/LAB

For a generator at no load (at steady state), we have (from (3.44))

$$I_{d0} = I_{q0} = 0, \quad V_{d0} = 0, \quad \left(\delta_{V0} = 0\right), \quad V_{q0} = \omega_r L_{dm} i_{F0}, \quad \theta_0 = -\pi/2 \tag{3.64}$$

Also

$$V_{F0} = R_F i_{F0} \quad \psi_{q0} = 0, \quad \psi_{d0} = L_{dm} i_{F0}$$

As already $V_{d0} = 0$, to simulate a sudden short circuit, a step voltage, $-V_{q0}$, is applied to the q axis of Equation 3.58:

$$\begin{bmatrix} 0 \\ -\dfrac{V_{q0}}{s} \end{bmatrix} = \begin{bmatrix} R_s + sL_d(s) & -L_q(s)\omega_r \\ L_d(s)\omega_r & R_s + sL_q(s) \end{bmatrix} \begin{vmatrix} i_d(s) \\ i_q(s) \end{vmatrix} \tag{3.65}$$

Note that we are now operating with actual (and not pu) variables (we switch back and forth to make the reader get used to both systems).

Solving for $i_d(s)$ and $i_q(s)$ in Equation 3.65, after neglecting terms containing R_s^2 and using the approximation

$$\frac{R_s}{2}\left(\frac{1}{L_d(s)} + \frac{1}{L_q(s)}\right) \approx \frac{R_s}{2}\left(\frac{1}{L''_d} + \frac{1}{L''_q}\right) = 1/\tau_a \tag{3.66}$$

we obtain

$$i_d(s) = -\frac{V_{q0}\omega_r^2}{s\left(s^2 + \dfrac{2s}{\tau_a} + \omega_r^2\right)} \frac{1}{L_d(s)}; \quad i_q(s) = -\frac{V_{q0}\omega_r}{\left(s^2 + \dfrac{2s}{\tau_a} + \omega_r^2\right)} \frac{1}{L_q(s)} \tag{3.67}$$

And finally, after considerable analytical work, with $L_d(s)$ and $L_q(s)$ of Equation 3.54,

$$i_d(t) = -\frac{V_{q0}}{\omega_r}\left[\frac{1}{L_d} + \left(\frac{1}{L'_d} - \frac{1}{L_d}\right)e^{-t/\tau'_d} + \left(\frac{1}{L''_d} - \frac{1}{L'_d}\right)e^{-t/\tau''_d} - \frac{1}{L''_d}e^{-t/\tau_a}\cos(\omega_r t)\right] \tag{3.68}$$

$$i_q(t) \approx -\frac{V_{q0}}{\omega_r L''_q}\sin(\omega_r t)$$

The sign (−) in $i_q(t)$ is related to the fact that under short circuit the machine operates as a generator, and dq0 model equations are given for the motor mode. The phase A current is obtained again using the inverse Park transformation (Equation 3.18):

$$\begin{aligned} i_A(t) &= i_d\cos(\omega_r t+\gamma_0) - i_q\sin(\omega_r t+\gamma_0) \\ &= -\frac{V_{q0}}{\omega_r}\left(\left(\frac{1}{L_d}+\left(\frac{1}{L_d'}-\frac{1}{L_d}\right)e^{-t/\tau_d'}+\left(\frac{1}{L_d''}-\frac{1}{L_d'}\right)e^{-t/\tau_d''}\right)\cos(\omega_r t+\gamma_0)\right. \\ &\quad \left. -\frac{1}{2}\left(\frac{1}{L_d''}+\frac{1}{L_q''}\right)e^{-t/\tau_a}-\frac{1}{2}\left(\frac{1}{L_d''}-\frac{1}{L_q''}\right)e^{-t/\tau_a}\cos(2\omega_r t+\gamma)\right) \end{aligned} \tag{3.69}$$

Ignoring stator resistance leads to the elimination of the periodic components of the frequency, ω_r, in $i_d(t)$ and $i_q(t)$, while in $i_A(t)$, only the frequency term, ω_r, remains with two time constants, τ_d' and τ_d''. The field current transients are related to the stator current by

$$\begin{aligned} i_F(s) &= -sG(s)i_d(s) \\ i_F(s) &= i_{F0}+i_{F0}\frac{L_d-L_d'}{L_d}\left[e^{-t/\tau_d'}-\left(1-\frac{\tau_{dr}'}{\tau_d''}\right)e^{-t/\tau_d''}-\frac{\tau_{dr}'}{\tau_d''}e^{-t/\tau_a}\cos(\omega_r t)\right] \end{aligned} \tag{3.70}$$

The peak short-circuit current, i_{Amax}, can be obtained from Equation 3.69 for $\partial i_A/\partial t = 0$, but approximately it is

$$i_{Amax} \approx \frac{V_{q0}}{\omega_r L_d''}\times 1.8 = \frac{V_0\sqrt{2}}{\omega_r L_d''}\times 1.8 = I_{sc3}\times\frac{L_d}{L_d''}\times 1.8\times\sqrt{2} \approx (8-20)I_{sc3} \tag{3.71}$$

I_{sc3} is the RMS phase current for the steady-state short circuit.

The flux linkages $\psi_d(s)$ and $\psi_q(s)$ are

$$\psi_d(s) = L_d(s)i_d(s);\quad \psi_q(s) = L_q(s)i_q(s) \tag{3.72}$$

and finally for $(1/\tau_a \ll \omega_r)$

$$\Psi_d(t) \approx \Psi_{d0}+\Psi_d(t) = \frac{V_{q0}}{\omega_r}e^{-t/\tau_a}\cos(\omega_r t)+L_{dm}i_{F0} \tag{3.73}$$

$$\Psi_q(t) \approx 0+\Psi_q(t) = -\frac{V_{q0}}{\omega_r}e^{-t/\tau_a}\sin(\omega_r t)$$

From this approximation, it appears that flux linkages vary only if $R_s \neq 0$, and their variation is fast.

The electromagnetic torque during shortcircuit is

$$T_{esc}(t) = \frac{3}{2}p_1\left[\psi_d(t)i_q(t)-\psi_q(t)i_d(t)\right] \tag{3.74}$$

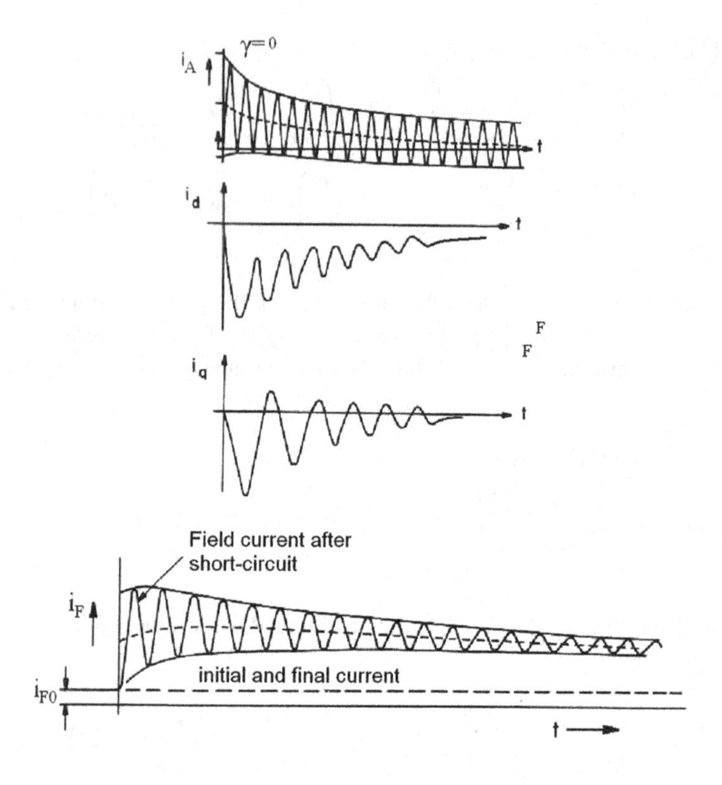

FIGURE 3.6 Sudden three-phase short-circuit currents in SMs.

Note: It is evident that during the sudden shortcircuit, the magnetic saturation level decreases continuously from no-load saturated to steady-state nonsaturated magnetic conditions. Despite the situation, the complicated mathematical description ignores this, just to obtain analytical solutions valuable for intuitional interpretation. A qualitative view of $i_d(t)$, $i_q(t)$, $i_F(t)$, $i_A(t)$ is shown in Figure 3.6.

Typical values of transient parameters in pu and time constants (in seconds) are given in Table 3.1.

The sudden short circuit may serve to identify SM time constants in axis *d*, while the peak short-circuit current is useful in designing the stator end connections mechanically, against largest electrodynamic forces at i_{Amax}.

3.11 ASYNCHRONOUS RUNNING OF SMS AT A GIVEN SPEED

Synchronous motors with d.c. excitation and starting/damper cage windings, when operated at a constant frequency and voltage power grid, are often started in the asynchronous mode, with the excitation winding connected first to a resistor, $R_x \approx 10R_F$. Then, when the speed stabilizes at a certain value, $\omega_r = \omega_1(1 - S)$, below the synchronous speed, $\omega_1 = \omega_{r0}$, the d.c. field circuit is commutated to the d.c. source; after a few oscillations, the SM eventually synchronizes. This is an electromechanical transient that is

TABLE 3.1
Typical SG Parameter Values

Parameter	Two-Pole Turbogenerator	Hydrogenerators
l_d (pu)	0.9–1.5	0.6–1.5
l_q (pu)	0.85–1.45	0.4–1.0
$l_d'(\text{pu})$	0.12–0.2	0.2–0.5
$l_d''(\text{pu})$	0.07–0.14	0.13–0.35
l_{Fdrl} (pu)	−0.05 to +0.05	−0.05 to +0.05
l_0 (pu)	0.02–0.08	0.02–0.2
l_{sl} (pu)	0.07–0.14	0.15–0.2
r_s (pu)	0.0015–0.005	0.002–0.02
$\tau_{d0}'(s)$	2.8–6.2	1.5–9.5
τ_d (s)	0.35–0.9	0.5–3.3
$\tau_d''(s)$	0.02–0.05	0.01–0.05
$\tau_{d0}''(s)$	0.02–0.15	0.01–0.15
$\tau_q''(s)$	0.015–0.04	0.02–0.06
$\tau_{q0}''(s)$	0.04–0.08	0.05–0.09
$l_q''(\text{pu})$	0.2	0.45

discussed later in the chapter. Here, we treat the average asynchronous torque at various slips. The dq voltages, V_d and V_q, can be expressed in relation to the stator voltages:

$$V_{A,B,C}(t) = V_0\sqrt{2}\cos(\omega_1 t + \delta_v) \tag{3.75}$$

$$\begin{aligned} V_d + jV_q &= \frac{2}{3}\left(V_A(t) + V_B(t)e^{j\frac{2\pi}{3}} + V_C(t)e^{-j\frac{2\pi}{3}}\right)e^{-j\omega_r t} \\ &= V_0\sqrt{2}\left[\cos((\omega_1 - \omega_r)t + \delta_v) - j\sin((\omega_1 - \omega_r)t + \delta_v)\right] \end{aligned} \tag{3.76}$$

$$\omega_1 - \omega_r = S\omega_1; \quad \omega_r = \omega_1(1 - S) \tag{3.77}$$

We may now introduce the complex expression of V_d and V_q separately:

$$\underline{V}_d(jS\omega_1) = V_0\sqrt{2}e^{j(S\omega_1 t + \delta_v)}; \quad \underline{V}_q = jV_0\sqrt{2}e^{j(S\omega_1 t + \delta_v)} \tag{3.78}$$

With $s = jS\omega_1$ in the Laplace form, the stator equations with $V_F(jS\omega_1) = 0$ but R_F replaced by $(R_F + R_x)$ in $L_d(jS\omega_1)$, become

$$\begin{aligned} \underline{V}_d(jS\omega_1) &= (R_s + jS\omega_1 L_d(jS\omega_1))\underline{I}_d(jS\omega_1) - \omega_1(1 - S)L_q(jS\omega_1)\underline{I}_q(jS\omega_1) \\ \underline{V}_q(jS\omega_1) &= (R_s + jS\omega_1 L_q(jS\omega_1))\underline{I}_q(jS\omega_1) + \omega_1(1 - S)L_d(js\omega_1)\underline{I}_d(jS\omega_1) \end{aligned} \tag{3.79}$$

From Equation 3.54, the complex inductances, $L_d(jS\omega_1)$ and $L_q(jS\omega_1)$, are

$$L_d(jS\omega_1) = \underline{L}_d = \frac{(1 + jS\omega_1\tau_d')(1 + jS\omega_1\tau_d'')}{(1 + jS\omega_1\tau_{d0}')(1 + jS\omega_1\tau_{d0}'')} L_d; \quad R_F \to R_F + R_x \tag{3.80}$$

$$L_q(jS\omega_1) = \underline{L}_q = \frac{(1 + jS\omega_1\tau_q'')}{(1 + jS\omega_1\tau_{q0}')} L_q \tag{3.81}$$

The a.c. field current, $I_F(jS\omega_1)$, is Equation 3.70:

$$\underline{I}_F(jS\omega_1) = -jS\omega_1 G(jS\omega_1)\underline{I}_d(jS\omega_1) \tag{3.82}$$

The average torque T_e is simply

$$T_{eav} = \frac{3}{2} p_1 \operatorname{Re}\left(\underline{\psi}_d(jS\omega_1)\underline{I}_q^*(jS\omega_1) - \underline{\psi}_q(jS\omega_1)\underline{I}_d^*(jS\omega_1)\right) \tag{3.83}$$

with

$$\underline{\psi}_d(jS\omega_1) = \underline{L}_d\underline{I}_d; \quad \underline{\psi}_q(jS\omega_1) = \underline{L}_q\underline{I}_q \tag{3.84}$$

We may thus calculate $I_d(jS\omega_1)$, $I_q(jS\omega_1)$ and $I_F(jS\omega_1)$, T_{eav} from Equations 3.79 through 3.84. For zero stator resistance, ($R_s = 0$), from Equation 3.79

$$\underline{I}_d \approx \frac{\underline{V}_d}{j\omega_1 \underline{L}_d}; \underline{I}_q \approx \frac{\underline{V}_q}{j\omega_1 \underline{L}_q} \tag{3.85}$$

So the average torque is

$$T_{eav} = \frac{3}{2} p_1 \frac{\left(V_0\sqrt{2}\right)^2}{\omega_1} \operatorname{Re}\left[\frac{1}{j\omega_1 \underline{L}_d^*(jS\omega_1)} + \frac{1}{j\omega_1 \underline{L}_q^*(jS\omega_1)}\right] \tag{3.86}$$

Now, the currents, $i_d(t)$, and $i_q(t)$, are finally

$$\begin{aligned} i_d(t) &= \operatorname{Re}\left[\underline{I}_d e^{j(S\omega_1 t + \delta_v)}\right] \\ i_q(t) &= \operatorname{Re}\left[\underline{I}_q e^{j(S\omega_1 t + \delta_v)}\right] \\ \psi_d(t) &= \operatorname{Re}\left[\underline{\psi}_d e^{j(S\omega_1 t + \delta_v)}\right] \\ \psi_q(t) &= \operatorname{Re}\left[\underline{\psi}_q e^{j(S\omega_1 t + \delta_v)}\right] \end{aligned} \tag{3.87}$$

and

$$i_A(t) = i_d(t)\cos(\omega_1(1-S)t) - i_q(t)\sin(\omega_1(1-S)t) \tag{3.88}$$

In this interpretation, $i_d(t)$ and $i_q(t)$ will experience only slip frequency, while the stator current, $i_A(t)$, will have a fundamental frequency, ω_1, and, if $R_s \neq 0$, the additional frequency will be $\omega_1(1-2S) = \omega_1'$.

The asynchronous torque will experience pulsations at a frequency of $2S\omega_1$ that may be as high as 50% of the rated synchronous torque. These are mainly due to the asymmetry of the rotor windings (and due to magnetic anisotropy along the d and q axes). The instantaneous torque in the general torque formula is

$$T_e = \frac{3}{2}p_1\left(\psi_d(t)i_q(t) - \psi_q(t)i_q(t)\right) \tag{3.89}$$

The average asynchronous torque versus slip, S, for a SM in the per unit system is as follows: $V_0\sqrt{2} = 1$ pu, $l_{sl} = 0.15$ pu, $l_{dm} = 1.0$ pu, $l_{Fl} = 0.3$ pu, $l_{qm} = 0.6$ pu, $l_{qrl} = 0.12$ pu, $L_{drl} = 0.2$ pu, $r_s = 0.012$ pu, $r_{dr} = 0.03$ pu, $r_{qr} = 0.04$ pu, $r_F = 0.03$ pu ($r_F + r_x = 10r_F$), as shown in Figure 3.7.

It should be noted that when $R_x = 10R_F$ is connected to the field circuit, more asynchronous torque is obtained (Figure 3.7b).

Example 3.4: DC Field Current (or PM) Rotor-Induced Asynchronous Stator Losses

Nonzero d.c. excitation (or PMs) on the rotor produces additional (asynchronous) losses in the stator windings, at speed frequency $\omega_1' = \omega_1(1-S)$. It is required to derive the expression of this torque and calculate the speed where its maximum occurs for the SM with data above in the text.

Solution:

In rotor coordinates, the dq currents, I_d' and I_q', are d.c. and the stator windings can be considered shortcircuited (because the power grid internal impedance is zero: infinite-power grid). Also, the rotor currents are $I_{dr} = I_{qr} = 0$ and $I_F = I_{F0}$: $V_F(s) \rightarrow V_{F0} = R_F i_{F0}$. So, with $V_d = V_q = 0$, $s \rightarrow 0$; $V_F \rightarrow R_F i_{F0}$ Equations 3.79 degenerate into steady-state Equation 3.90:

$$\begin{aligned}
0 &= R_s I_d' - \omega_1(1-S)L_q I_q' \\
0 &= (1-S)\omega_1\left(L_d I_d' + L_{dm} I_{F0}\right) + R_s I_q'
\end{aligned} \tag{3.90}$$

The solutions of Equation 3.90 are straightforward:

$$\begin{aligned}
I_d' &= \frac{-L_{dm}L_q I_{F0}\omega_1^2(1-S)^2}{R_s^2 + (1-S)^2\omega_1^2 L_d L_q} \\
I_q' &= \frac{-L_{dm}I_{F0}\omega_1(1-S)R_s}{R_s^2 + (1-S)^2\omega_1^2 L_d L_q}
\end{aligned} \tag{3.91}$$

(Continued)

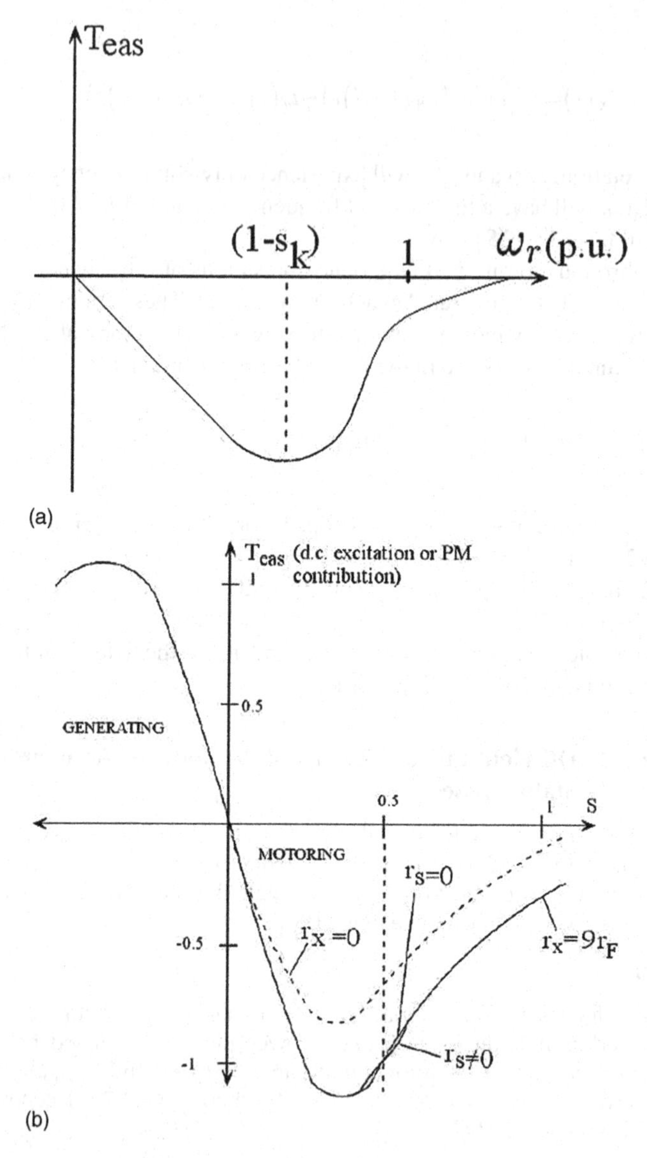

FIGURE 3.7 (a) Asynchronous average torque in pu vs. slip, S, of an SM and (b) d.c. (or PM) rotor field asynchronous torque.

The torque corresponds to the stator winding losses, W'_{co}:

$$W'_{co} = \frac{3}{2} R_s \left(I_d'^2 + I_q'^2 \right) = -T_{edc} \frac{\omega_1 (1-S)}{p_1} \tag{3.92}$$

$$T_{edc} = -\frac{3}{2} p_1 R_s \left(L_{dm} I_{F0} \right) 2\omega_1 (1-S) \frac{\left[R_s^2 + L_q^2 \omega_1^2 (1-S)^2 \right]}{\left[R_s^2 + L_d L_q \omega_1^2 (1-S)^2 \right]^2} \tag{3.93}$$

The maximum torque occurs at S_k':

$$S_k' \approx 1 - \sqrt{\frac{2L_dL_q\omega_1^2 - R_s^2}{2L_q^2\omega_1^2 + L_dL_q\omega_1^2}} \tag{3.94}$$

for the above data given in pu

$$S_k' \approx 1 - \sqrt{\frac{2l_dl_q - r_s^2}{2l_q^2 + l_dl_q}} \approx 1 - \sqrt{\frac{2\times1\times0.6 - 0.012^2}{2\times0.6^2 + 1\times0.6}} \approx 0.28 \tag{3.95}$$

So at 72% of rated speed, the torque, T_{edc}, Figure 3.7b, is maximum. For $l_d = l_q$ and $r_s = 0$, S_k' is $\left(1-\sqrt{2/3}\right) = 0.1875$. For PMSM, $L_{dm}i_{F0} = \psi_{PM}$, which is the PM flux linkage in the dq model.

This torque is dependent on stator resistance, and thus it is important in low-power machines (PMSMs). It should be noticed that the operating mode is, in fact, as a generator in short circuit at speed $\omega_r = \omega_1(1 - S)$. The results from Example 3.3 can be used to explain this case. This point, though redundant, is reiterated here for the benefit of the reader. When self-starting of a PMSM at power grid takes place, this torque may decay or even hamper the process of self-synchronization, especially for a large torque load. Also, in case of a PWM converter supply failure, this torque continuously brakes the work machine and in some applications, such as car steering by a wire, it becomes an additional design constraint.

3.12 REDUCED-ORDER DQ0 MODELS FOR ELECTROMECHANICAL TRANSIENTS

For power systems stability and control investigation, where many synchronous generators (SGs) work in parallel, in the point of common connection (interest: PCC), the modeling of SGs has to be detailed, while, for those at a distance, simplified models may be used, to save computing time.

A few such approximations have gained wide acceptance and are thus presented here:

3.12.1 Neglecting Fast Stator Electrical Transients

Neglecting the pulsational stator voltages in Equation 3.17, we obtain

$$\left(\frac{d\psi_d}{dt}\right)_{\omega_1=\omega_{ro}} = \left(\frac{d\psi_q}{dt}\right)_{\omega_1=\omega_{ro}} \tag{3.96}$$

Now we are left with two options: to consider that the speed does not vary, or, if the speed varies ($\omega_r \neq \omega_1$), the machine inductances become dependent on rotor position.

In essence, the approximation is better if in the motion-induced voltages we use ω_1 instead of ω_r, [2], but keep the speed varying (not much) through the motion equation:

$$V_d = R_s I_d - \omega_1 \psi_q; \quad V_q = R_s I_q + \omega_1 \psi_q$$

$$\frac{d\psi_F}{dt} = V_F - R_F I_F, \quad \frac{d\psi_{dr}}{dt} = -R_{dr} I_{dr}, \quad \frac{d\psi_{qr}}{dt} = -R_{qr} I_{qr} \tag{3.97}$$

$$\frac{d\omega_r}{dt} = \frac{p_1}{J}\left[p_1\left(\psi_d I_q - \psi_q I_d\right) - T_{load}\right]; \quad \frac{d\theta_{er}}{dt} = \omega_r \tag{3.98}$$

To conclude, neglecting stator transients means disregarding the attenuated components of the frequency, ω_1 and $2\omega_1$, in the transient currents and torque.

3.12.2 Neglecting Stator and Rotor Cage Transients

Now, in addition, the rotor cage transients are neglected:

$$\frac{d\psi_{dr}}{dt} = \frac{d\psi_{qr}}{dt} = 0; \quad I_{dr} = I_{qr} = 0 \tag{3.99}$$

So the dq model looses two more orders:

$$V_d = R_s I_d - \omega_r \psi_q; \quad V_q = R_s I_q + \omega_r \psi_d \tag{3.100}$$

$$\frac{d\psi_F}{dt} = V_F - R_F I_F \tag{3.101}$$

The motion Equations 3.98 still hold. Only the field current transients are considered, and here the transient inductance, L_d', comes into play.

3.12.3 Simplified (Third-Order) dq Model Adaptation for SM Voltage Control

This model starts with the third-order model mentioned above (Equations 3.97 through 3.101) and, in order to prepare the model for field voltage regulation, the field current, I_F, is eliminated and thus a new transient emf, e_q', (in pu) is defined for the generator mode:

$$e_q' = \omega_r \frac{l_{dm}}{l_F} \psi_F; \quad l_F = l_{dm} + l_{Fl} \tag{3.102}$$

The stator equation along the q axis (Equation 3.100) becomes

$$V_q = -r_s i_q - x_d' i_d - e_q'; \quad x_d' = \omega_r \left(l_d - \frac{l_{dm}^2}{l_F} \right) \tag{3.103}$$

Now Equation 3.101 may be rearranged as

$$\dot{e}_q' = \frac{V_F - e_q' + i_d\left(x_d - x_d'\right)}{\tau_{d0}'}; \quad \tau_{d0}' = \frac{l_F}{r_F\omega_{10}} \tag{3.104}$$

The motion Equations 3.98 hold and the initial value of the transient e_q' is

$$\left(e_q'\right)_{t=0} = \omega_{r0}\frac{l_{dm}}{l_F}\left[l_F\cdot\left(i_F\right)_{t=0} + l_{dm}\cdot\left(i_d\right)_{t=0}\right] \tag{3.105}$$

For more on the subject, the reader can refer to [3].

Example 3.5: Biaxial Excitation Generator for Automobiles

Let us consider a synchronous machine with a distributed three-phase a.c. winding placed in uniform slots, with a rotor with high magnetic saliency. This rotor is made of flux barriers (in the q axis) filled with weak (B_r = 0.6—0.8 [T], remanent flux density) PMs and with a d.c. excitation winding along the d axis (Figure 3.8)—BEGA [4]. Obtain the dq model for this configuration and draw the space phasor diagram for $i_d = 0$ and $\psi_q = 0$ and discuss the results.

Solution:

The stator dq equations in the space phasor form are as in Equation 3.36, but the expressions of the flux linkages, ψ_d and ψ_q, are slightly different:

$$\begin{aligned} \bar{V}_s &= R_s \underline{i}_s + j\omega_r \underline{\psi}_s + \frac{d\underline{\psi}_s}{dt} \\ \bar{\psi}_s &= \psi_d + j\psi_q; \quad \psi_d = L_{dm} i_F + L_d i_d; \quad L_d > L_q \\ \psi_q &= L_q i_q - \psi_{PMq}; \quad \psi_F = L_{Fl} i_F + L_{dm}\left(i_F + i_d\right) \\ \frac{d\psi_F}{dt} &= V_F - R_F i_F \end{aligned} \tag{3.106}$$

The torque is

$$T_e = \frac{3}{2}p_1\left(\psi_d i_q - \psi_q i_d\right) = \frac{3}{2}p_1\left[L_{dm} i_F i_q + \psi_{PMq} i_d + \left(L_d - L_q\right) i_d i_q\right]$$

Hence the torque has three terms, suggesting that more torque may be produced per given winding loss or per stator current. However, the torque is limited by magnetic saturation for a given cooling system and given admissible temperatures. To limit the stator current for maximum torque, it appears that by setting i_d = 0 and $L_q i_q - \psi_{PM} = 0$, the torque remains with the first term:

$$\left(T_e\right)_{i_d=0, \psi_q=0} = \frac{3}{2}L_{dm} i_F i_q; \quad i_q = \frac{\psi_{PMq}}{L_q} = \text{constant} \tag{3.107}$$

(Continued)

Example 3.8: (Continued)

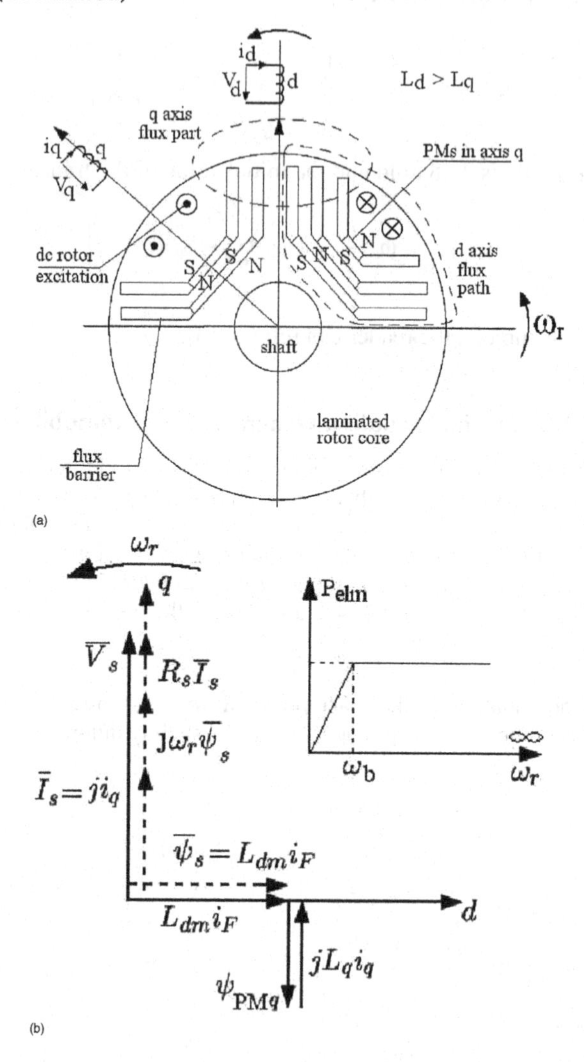

FIGURE 3.8 BEGA (a) rotor cross-section and (b) vector diagram for motoring at i_d=0, Ψ_q=0.

But, for the steady state, if $i_d = 0$ and $\psi_q = 0$, the stator voltage sole current equation becomes

$$\overline{I_s} = 0 + ji_q; \quad V_s = -R_s i_s + \omega_r L_{dm} i_F \tag{3.108}$$

Hence the power factor is implicitly unity, and Equation 3.108 shows that for the stator, only the resistive voltage drop occurs. This is exactly the situation in a separately excited d.c. brush machine, which has minimum (resistive) voltage regulation. The situation is brought about by controlling $i_d = 0$ and $i_q = \psi_{PM}/L_q =$ const. By reducing i_F in these conditions, the speed may be increased theoretically to infinity at constant power, if mechanical losses are neglected. A wide but constant power speed range with the least overrating of SM and its PWM converter is thus obtained. This is typical in starters/alternators and for most autonomous synchronous generators.

This is just one example of the many innovative SM configurations that can bring extraordinary control for new, challenging, applications at small and medium powers. It may be rightfully argued that the rotor is not very rugged mechanically, but for every merit, most often, there is also a drawback.

3.13 SMALL-DEVIATION ELECTROMECHANICAL TRANSIENTS (IN PU)

The small deviation theory investigates small transients around an initial steady-state point (situation). From the space phasor diagram (Figure 3.5) and Example 3.2, we continue with the steady-state conditions here (for motoring $\delta_v > 0$):

$$\begin{aligned} V_d &= -V\sqrt{2}\sin(\delta_v) = -\omega_r L_q I_q + R_s i_d \\ T_{e0} &= \frac{3}{2} p_1 \left(L_{dm} i_{F0} + (L_d - L_q) i_{d0}\right) i_{q0} \\ V_q &= V\sqrt{2}\cos(\delta_v) = \omega_r \left(L_d i_d + L_{dm} i_F\right) + R_s i_q \end{aligned} \tag{3.109}$$

with $V_d = V_{d0}$, $V_q = V_{q0}$, $i_d = i_{d0}$, $i_q = i_{q0}$, $\delta_v = \delta_{v0}$ (initial voltage power angle), and $T_{e0} = T_{L0}$.

For large transients, we need to use the general Equation 3.62, but for small deviations (from Equation 3.109):

$$\begin{aligned} &V_d = V_{d0} + \Delta V_d; \quad V_q = V_{q0} + \Delta V_q; \quad \delta_v = \delta_{v0} + \Delta\delta_v \\ &\Delta V_d = -\Delta V\sqrt{2}\sin\delta_{v0} - V_0\sqrt{2}\cos\delta_{v0}\Delta\delta_v \\ &\Delta V_q = \Delta V\sqrt{2}\cos\delta_{v0} - V_0\sqrt{2}\sin\delta_{v0}\Delta\delta_v \end{aligned} \tag{3.110}$$

Also

$$\begin{aligned} &V_F = V_{F0} + \Delta V_F; \quad T_L = T_{L0} + \Delta T_L; \quad T_e = T_{e0} + \Delta T_e \\ &\omega_r = \omega_{r0} + \Delta\omega_r; \quad \omega_1 = \omega_{r0} + \Delta\omega_1 \end{aligned} \tag{3.111}$$

The flux/current relationships are also linearized:

$$\begin{aligned} &\Delta\psi_d = L_{sl}\Delta i_d + L_{dm}\Delta i_{dm}; \quad \Delta i_{dm} = \Delta i_d + \Delta i_F + \Delta i_{dr} \\ &\Delta\psi_q = L_{sl}\Delta i_q + L_{qm}\Delta i_{qm}; \quad \Delta i_{qm} = \Delta i_q + \Delta i_{qr} \\ &\Delta\psi_F \approx L_{Fl}\Delta i_F + L_{dm}\Delta i_{dm}; \quad \Delta\psi_{dr} = L_{drl}\Delta i_{dr} + L_{dm}\Delta i_{dm} \\ &\Delta\psi_{qr} = L_{qrl}\Delta i_{qr} + L_{qm}\Delta i_{qm} \end{aligned} \tag{3.112}$$

Now, the dq0 model of Equations 3.17 through 3.19 is linearized to obtain

$$\frac{d\Delta\psi_d}{dt} = \Delta V_d - R_s\Delta i_d + \omega_{r0}\Delta\psi_q + \Delta\omega_r\psi_{q0}$$
$$\frac{d\Delta\psi_q}{dt} = \Delta V_q - R_s\Delta i_q - \omega_{r0}\Delta\psi_d - \Delta\omega_r\psi_{d0}$$
$$\frac{d\Delta\psi_F}{dt} = \Delta V_F - R_F\Delta i_F; \quad \frac{d\Delta\psi_{dr}}{dt} = -R_{dr}\Delta i_{dr}$$
$$\frac{d\Delta\psi_{qr}}{dt} = -R_{qr}\Delta i_{qr}$$
$$\Delta T_e = \frac{3}{2}p_1\left(\Delta\psi_d i_q + \psi_{d0}\Delta i_q - \Delta\psi_q i_{d0} - \psi_{q0}\Delta i_d\right) \tag{3.113}$$

with $\psi_{d0} = L_d i_{d0} + L_{dm} i_{F0}; \psi_{q0} = L_q i_{q0}; \ i_{dm0} = i_{d0} + i_{F0}; \ i_{qm0} = i_{q0}$

$$\frac{d}{dt}\Delta\delta_v = \Delta\omega_1 - \Delta\omega_r; \quad \frac{J}{p_1}\frac{d}{dt}\Delta\omega_r = \Delta T_e - \Delta T_L \tag{3.114}$$

The variation of stator frequency, $\Delta\omega_1$, which is in fact that of the power supply, is introduced here as an additional variable, but it may be considered zero ($\Delta\omega_1 = 0$) to simplify the computational process. The above model can be arranged into a matrix format

$$\left[L\right]\left[\Delta\dot{X}\right] = |R||\Delta X| + B|\Delta U| \tag{3.115}$$

with

$$\Delta U = \left[\Delta V, \Delta V, \Delta V_F, 0, 0, \Delta T_L, \Delta\omega_1\right]^T \tag{3.116}$$

$$\Delta X = \left[\Delta i_d, \Delta i_q, \Delta i_F, \Delta i_{dr}, \Delta i_{qr}, \Delta\omega_r, \Delta\delta_v\right]^T \tag{3.117}$$

$$\left[L\right] = \begin{bmatrix} L_{sl}+L_{dm} & 0 & L_{dm} & L_{dm} & 0 & 0 & 0 \\ 0 & L_{sl}+L_{qm} & 0 & 0 & L_{qm} & 0 & 0 \\ L_{dm} & 0 & L_{Fl}+L_{dm} & L_{dm} & 0 & 0 & 0 \\ L_{dm} & 0 & L_{dm} & L_{drl}+L_{dm} & 0 & 0 & 0 \\ 0 & L_{qm} & 0 & 0 & L_{qrl}+L_{qm} & 0 & 0 \\ 0 & 0 & 0 & 0 & 0 & 1 & 0 \\ 0 & 0 & 0 & 0 & 0 & 0 & 1 \end{bmatrix} \tag{3.118}$$

$$|B| = \begin{vmatrix} -\sqrt{2}\sin(\delta_{v0}) \\ \sqrt{2}\cos(\delta_{v0}) \\ 1 \\ 0 \\ 0 \\ -1 \\ 1 \end{vmatrix} \tag{3.119}$$

$$|R| = [\|C + D\|] \tag{3.120}$$

$$|C| = \begin{vmatrix} -R_s & \omega_{r0}(L_{sl} + L_{qm}) & 0 \\ -\omega_{r0}(L_{sl} + L_{dm}) & -R_s & -\omega_{r0}L_{dm} \\ 0 & 0 & -R_F \\ 0 & 0 & 0 \\ 0 & 0 & 0 \\ \frac{3}{2}p_1[(L_{sl} + L_{dm})i_{q0} - \psi_{q0}] & -\frac{3}{2}p_1[(L_{sl} + L_{qm})i_{d0} + \psi_{d0}] & -\frac{3}{2}p_1L_{dm}i_{q0} \\ 0 & 0 & 0 \end{vmatrix} \tag{3.121}$$

$$|D| = \begin{vmatrix} 0 & \omega_{r0}L_{qm} & \psi_{q0} & -V_0\sqrt{2}\cos(\delta_{v0}) \\ -\omega_{r0}L_{dm} & 0 & -\psi_{dr0} & -V_0\sqrt{2}\sin(\delta_{v0}) \\ 0 & 0 & 0 & 0 \\ -R_{dr} & 0 & 0 & 0 \\ 0 & -R_{qr} & 0 & 0 \\ \frac{3}{2}p_1L_{dm}i_{q0} & -\frac{3}{2}L_{dm}p_1i_{d0} & 0 & 0 \\ 0 & 0 & -1 & 0 \end{vmatrix} \tag{3.122}$$

The derivation of the matrices of $[L]_{7\times7}$ and $[R]_{7\times7}$ from Equations 3.113 and 3.114 is straightforward. If the current deviations are replaced by flux deviations, $\Delta\psi_d$, $\Delta\psi_q$, $\Delta\psi_F$, $\Delta\psi_{dr}$, $\Delta\psi_{qr}$, the matrix $[L]$ becomes a 6 × 1 simple, single column matrix, which is easy to solve numerically.

Once the terms of Equation 3.115 are known, with $d/dt \rightarrow s$ the linear system theory can be used to investigate the small deviation transients. One particular transfer function, $\Delta\omega_r(s)$, is of special interest for the motor mode:

$$\frac{J}{p_1}s^2\Delta\delta_v = -A_{\delta_v}(s)\Delta\delta_v + A_{\omega_1}(s)\Delta\omega_1 - A_F(s)\Delta V_F + A_V(s)\Delta V + \Delta T_L \tag{3.123}$$

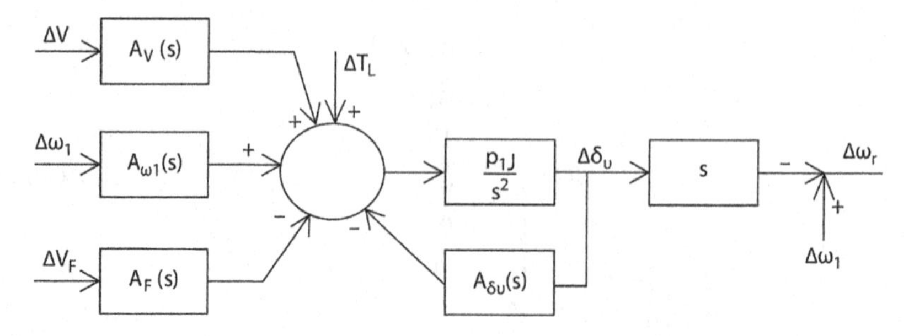

FIGURE 3.9 Structural diagram of SM for small deviation transients.

The coefficients, $A_{\delta v}(s)$, $A_{\omega 1}(s)$, $A_F(s)$, and $A_V(s)$ also stem from the above equations. Equation 3.123 leads to the structural diagram shown in Figure 3.9, which reveals the complexity involved in control for a multiple input system.

Example 3.6

Investigation of Response to Forced Torque Pulsations by Small Deviation Theory

Solution:

The forced mechanical torque pulsations write:

$$\Delta T_L = \sum \Delta T_{Lv} \cos(\omega_v t + r_v) \tag{3.124}$$

In a first approximation, all other inputs are zero ($\Delta V = 0$, $\Delta\omega_1 = 0$, $\Delta V_F = 0$).

Equation 3.123 degenerates into

$$\frac{J}{p_1} s^2(\Delta\delta_v) = -A_{\delta v}(s)\Delta\delta_v + \Delta T_L; \quad s\Delta\delta_v = -\Delta\omega_r \tag{3.125}$$

We now divide the first right-hand term of Equation 3.125 into two terms, by separating its real and imaginary parts (with $s = j\omega_\upsilon$):

$$\left[\frac{J}{p_1}\omega_v^2 + A_{\delta vi}(\omega_v) j\omega_v + A_{\delta vr}\right]\underline{\Delta\delta_v} = \Delta\underline{T}_{Lv} \tag{3.126}$$

Equation 3.126 is written in complex number terms as it refers to one load torque harmonic pulsation. The fact that the coefficients $A_{\delta vi}$ and $A_{\delta vr}$ vary with the torque pulsation frequency allows us to investigate with better precision the transient response of SM driven by a prime mover or by driving a load machine with mechanical torque pulsations.

3.14 LARGE-DEVIATION ELECTROMECHANICAL TRANSIENTS

In transients with large variations of variables, the complete dq model of SM has to be used. Typical examples are the asynchronous starting and self-synchronization of

SMs with d.c. rotor or PM excitation (or the SG synchronization to the power grid). In both cases, the rotor is provided with a cage on the rotor.

3.14.1 Asynchronous Starting and Self—Synchronization of DC—Excited SMs/Lab 3.2

The stator voltages are considered symmetric:

$$V_{\mathrm{A,B,C,F}} = V\sqrt{2}\cos\left(\omega_1 t - \left(i - 1\frac{2\pi}{3}\right)\right); \quad i = 1,2,3 \tag{3.127}$$

Using the Park transformation with $\theta = \int(\omega_r dt + \theta_0)$, we obtain, in rotor coordinates,

$$\begin{aligned} V_d(t) &= V\sqrt{2}\cos(\omega_1 t - \theta); \quad \frac{d\theta}{dt} = \omega_r \\ V_q(t) &= -V\sqrt{2}\sin(\omega_1 t - \theta) \end{aligned} \tag{3.128}$$

As expected, the rotor speed varies continuously during asynchronous motor acceleration. The dq model in pu of Equations 3.6 and 3.38 holds.

For the d.c.-excited SM, the field circuit is connected to a resistance R_x and thus $V_F = -R_x i_F$ in the dq model (Figure 3.10).

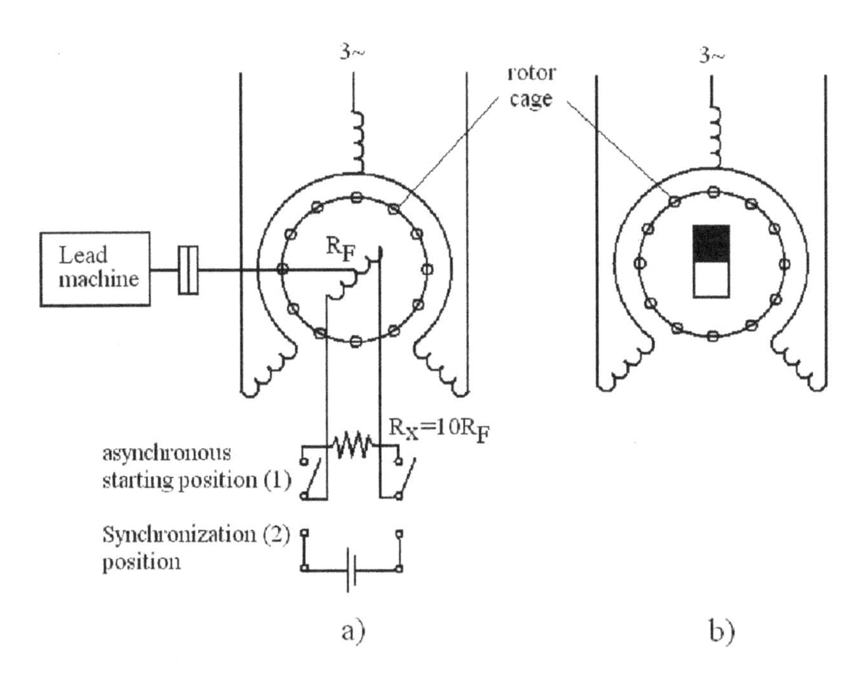

FIGURE 3.10 SM self-synchronization with (a) d.c. excitation and (b) PM rotor.

The angle θ_0 refers to the initial (at start) phase of the stator voltages. After the acceleration process settles at a speed $\omega_{ra} < \omega_1$ ($\omega_{ra}/\omega_1 \approx 0.95$—$0.98$), at a certain chosen moment, the field circuit is disconnected from the resistance, R_x, and is connected to the d.c. excitation source. A transient process occurs, where both the asynchronous and synchronous torques interact and the SM eventually synchronizes. The success of the synchronization depends on the load torque, and mainly on the initial voltage power angle, $\boldsymbol{\delta}_v$, when the d.c. voltage is connected to the rotor. A zero value of $\delta_v = -\theta_0 - \frac{\pi}{2} = 0$, is considered optimal because $\omega_r < \omega_1$, and thus inevitably the rotor trails the stator field and the synchronous torque is motoring from the start, toward a more probable synchronization. As the rotor slip (S) is small, the frequency of field current is small ($f_F = Sf_1$), and thus its circuit is almost resistive. The zero crossing of the slip frequency field current corresponds to zero emf. This means, however, that there is maximum stator flux in the latter, which means a zero power angle when the d.c. voltage is applied to the rotor. This may serve as a means for repetitious starting, necessary in many applications.

The asynchronous starting and synchronization in (pu) for the large SM in Example 3.1 are shown in Figure 3.11 (for torque and speed).

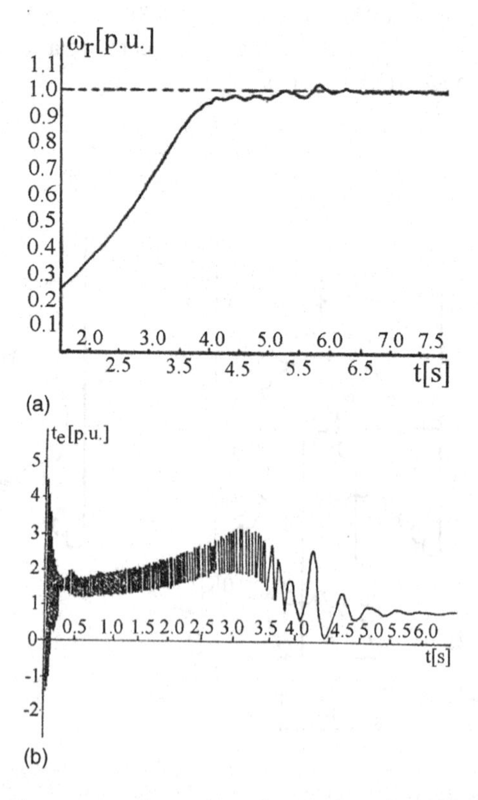

FIGURE 3.11 Asynchronous starting and synchronization of a 1800 KW, 100 rpm, SM: (a) Electromagnetic torque transients and (b) speed transients.

At $t = 4.15$ s, the d.c. voltage is connected to the rotor and the SM synchronizes at 6 s under a load of 0.8 pu. Notable torque pulsations are visible and the peak transient torque goes over 4.5 pu during starting.

3.14.2 Asynchronous Self-Starting of PMSMs to Power Grid

For the asynchronous self-starting of PMSM, the complete dq model is considered. But the latter gets simplified because the field circuit is eliminated and $L_{dm}i_F = \psi_{PM}$. The dq voltages in Equation 3.128 still hold. The dq model, Equations 3.32 through 3.34, is added here for completeness:

$$\begin{aligned}
&\frac{d\psi_d}{dt} = V_d - R_s i_d + \omega_r \psi_q; \quad \psi_d = L_{sl} i_d + L_{dm}\left(i_d + i_{dr}\right) + \psi_{PMd} \\
&\frac{d\psi_q}{dt} = V_q - R_s i_q - \omega_r \psi_d; \quad \psi_q = L_{sl} i_q + L_{dm}\left(i_q + i_{qr}\right) \\
&\frac{d\psi_{dr}}{dt} = -R_{dr} i_{dr}; \quad \frac{d\psi_{qr}}{dt} = -R_{qr} i_{qr}; \quad \psi_{dr} = L_{drl} i_{dr} + \psi_{PMd} + L_{dm} i_d \\
&\psi_{qr} = L_{qrl} i_{qr} + L_{qm}\left(i_q + i_{qr}\right); \quad T_e = \frac{3}{2} p_1 \left(\psi_d i_q - \psi_q i_d\right)
\end{aligned} \tag{3.129}$$

$$\frac{J}{p_1}\frac{d\omega_r}{dt} = T_e - T_L; \quad \frac{d\theta}{dt} = \omega_r \tag{3.130}$$

3.14.3 Line-to-Line and Line-to-Neutral Faults

Line-to-line and line-to-neutral faults (Figure 3.12a and b) are kind of extreme transients in the sense that their short-circuit current peaks are very large for PMSMs without a rotor cage. They have to be known for the proper design of the power source (for example, a PWM converter) protection system.

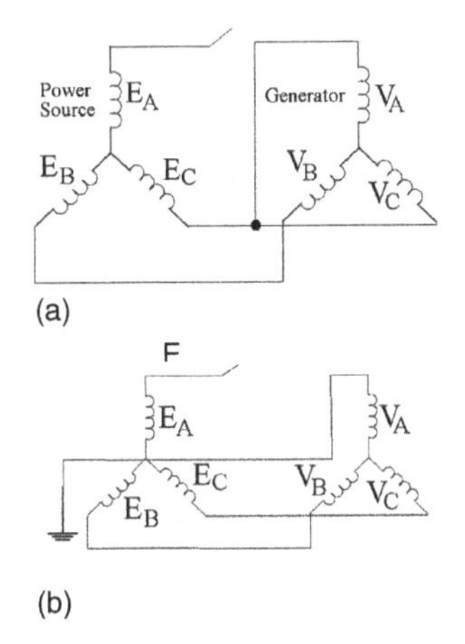

FIGURE 3.12 (a) Line-to-line and (b) line-to-neutral faults.

We suppose that the power source (grid) is of infinite power and thus its voltages are

$$E_{A,B,C}(t) = V\sqrt{2}\cos\left(\omega_1 t - (i-1)\frac{2\pi}{3}\right); \quad i = 1,2,3 \tag{3.131}$$

From Figure 3.12a

$$\begin{aligned} &V_B - V_C = E_B - E_C \\ &\quad V_A = V_C; \quad I_A + I_B + I_C = 0, \quad \text{so} \quad V_A + V_B + V_C = 0 \end{aligned} \tag{3.132}$$

Consequently,

$$V_C(t) = V_A(t) = \frac{1}{3}(E_C - E_B); \quad V_B(t) = -2V_C(t) \tag{3.133}$$

for the line-to-line shortcircuit and

$$\begin{aligned} &V_B - V_C = E_B - E_C \\ &V_C - V_A = E_C; \quad V_A + V_B + V_C = 0 \end{aligned} \tag{3.134}$$

for the line-to-neutral fault (Figure 3.12b).

Hence

$$V_A = -\frac{(E_C + E_B)}{3}; \quad V_B = V_A + E_B; \quad V_C = V_A + E_C \tag{3.135}$$

Practically, $V_A(t)$, $V_B(t)$, and $V_C(t)$ can be expressed as functions of time. With the Park transformation

$$V_d + jV_q = \frac{2}{3}\left(V_A(t) + V_B(t)e^{j\frac{2\pi}{3}} + V_C(t)e^{-j\frac{2\pi}{3}}\right)e^{-j\theta} \tag{3.136}$$

and $d\theta/dt = \omega_r$, all we need is ready to apply the dq model and find all variables during such severe faults.

3.15 TRANSIENTS FOR CONTROLLED FLUX AND SINUSOIDAL CURRENT PMSMS

Sinusoidal current control, at a constant d axis (ψ_d) or total stator flux (ψ_s), called flux orientation or vector control, presupposes voltage-source PWM converters in variable speed electric drives.

In such cases, no damper cage is placed on the rotor to reduce current ripples (because of large L_d and L_q values) and to avoid additional rotor losses. *Note*: There is a scalar control (*V/f* or *i&f*) where the frequency is ramped up and the amplitude of the voltage rises proportionally with frequency:

$$V_1 = V_0 + kf_1 \tag{3.137}$$

In such cases, stabilizing loops are added to preserve synchronism dynamically, and thus a damper cage is beneficial on the rotor. The complete dq model for transients as described earlier in this chapter is to be used then.

3.15.1 Constant d-Axis (ψ_d) Flux Transients in Cageless SMs

PM action corresponds to a fictitious constant field current rotor winding. Taking up the dq model in Equations 3.17 through 3.19 and simplifying it for the cageless rotor with constant field current and constant i_{d0}, and $\psi_{d0} = \text{const}$:

$$V_d = R_s i_{d0} - \omega_r L_q i_q; \quad \psi_{d0} = L_d i_{d0} + \psi_{PMd}$$
$$L_q \frac{di_q}{dt} = V_q - R_s i_q - \omega_r \psi_{d0} \tag{3.138}$$

$$\frac{J_1}{p_1}\frac{d\omega_r}{dt} = \left(T_e - T_L - B\omega_r\right)$$
$$T_e = \frac{3}{2}p_1\left(\psi_{d0} i_q - L_q i_q i_{d0}\right) = \frac{3}{2}p_1\left(\psi_{d0} - L_q i_{d0}\right) i_q \tag{3.139}$$

For the steady state $d/dt = 0$

$$V_{d0} = R_s i_{d0} - \omega_r L_q i_{q0}; \quad V_{d0}^2 + V_{q0}^2 = V_s^2$$
$$V_{q0} = R_s i_{q0} + \omega_r \psi_{d0} \tag{3.140}$$

Now, neglecting R_s Equations 3.140 become

$$V_s^2 = \omega_r^2\left(L_q^2 i_{q0}^2 + \psi_{d0}^2\right) \tag{3.141}$$

Replacing i_{q0} with T_e (from Equation 3.139), Equation 3141 becomes

$$V_s = \omega_r \sqrt{L_q^2 \frac{4}{9}\frac{T_e^2}{p_1^2\left(\psi_{d0} - L_q i_{d0}\right)^2} + \psi_{d0}^2} \tag{3.142}$$

The no-load ideal speed $(\omega_{r0})_{Te=0}$ is

$$\omega_{r0} = \frac{V_s}{\psi_{d0}} = \frac{V_s}{\psi_{PMd} + L_d i_{d0}} \tag{3.143}$$

For the infinite ideal no-load speed

$$i_{d0} = -\frac{\psi_{PMd}}{L_d} \tag{3.144}$$

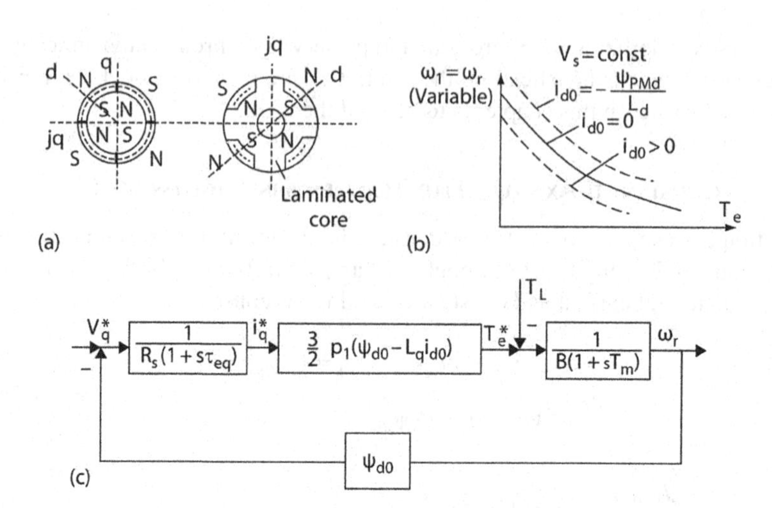

FIGURE 3.13 (a) PMSMs and (b) their speed/torque, for constant ψ_{d0} and (c) voltage structural diagram.

This equality has been proved to be a key design condition where a wide speed range but constant electromagnetic power is required. It goes without saying that the speed, ω_r, has to vary in tact with the stator frequency, $\omega_1 = \omega_{r0}$.

The above equations allow us to calculate the speed versus torque for a given value of voltage, V_s (Figure 3.13).

The model for transients where only one electrical time constant τ_{eq} occurs is

$$i_q\left(1+s\tau_{eq}\right)=\left(V_q-\omega_r\Psi_{d0}\right)/R_s;\tau_{eq}=L_q/R_s \tag{3.145}$$

$$\omega_r\left(1+s\tau_m\right)=\frac{1}{B}\left[\frac{3}{2}p_1\left(\Psi_{d0}-L_qi_{d0}\right)i_q-T_L\right] \tag{3.146}$$

$$\tau_m=\frac{J_1}{p_1B} \tag{3.147}$$

τ_m is the mechanical time constant related to the friction torque component, proportional to speed. The structural diagram built after Equations 3.145 and 3.146 Figure 3.13b is similar to that of the PM d.c. brush rotor (Chapter 2).

The V_d voltage is calculated a posteriori with the ω_r and i_q variables determined from Equations 3.145 through 3.147. As expected, the evolution of transients is now much simpler.

We may eliminate i_q from Equations 3.145 and 3.146:

$$\widetilde{\omega_r}\left[s^2\tau_m\tau_{eq}R_sB+s\left(\tau_m+\tau_{eq}\right)R_sB+R_sB+\frac{3}{2}p_1\left(\Psi_{d0}-L_qi_{d0}\right)\Psi_{d0}\right]$$
$$=\frac{3}{2}p_1\left(\Psi_{d0}-L_qi_{d0}\right)\widetilde{V_q}-\widetilde{T_L}R_s\left(1+s\tau_{eq}\right) \tag{3.148}$$

$$i_q = \left(V_q - \widetilde{\omega_r}\Psi_{d0}\right)/R_s/\left(1+s\tau_{eq}\right) \tag{3.149}$$

A second-order system similar to that for the d.c. brush PM was obtained.

Similar transients are expected for $\tilde{V}_q$ and $\tilde{T}_L$ inputs. The voltage, V_d, is just calculated a posteriori from Equation 3.120:

$$\tilde{V}_d = R_s i_{d0} - \tilde{\omega}_r L_q i_q \tag{3.150}$$

Note: All the above simplifications are valid for constant i_{d0} and PM excitation.

Example 3.7

Consider a PMSM with the following data: $\psi_{PMd} = 1$ Wb, $R_s = 1\ \Omega$, $L_d = 0.05$ H, $L_q = 0.10$ H, $p_1 = 2$, $B = 0.01$ Nms, and $\tau_m = 0.3$ s.

Calculate the electrical time constant, τ_{eq}, inertia, J, the current, i_{d0}, for infinite zero-torque speed and the eigenvalues for the speed response.

Solution:

From Equations 3.145 through 3.147

$$\tau_{eq} = L_q / R_s = 0.1/0.1 = 0.1s$$

$$J_1 = \tau_m p_1 B = 0.3\times 2\times 0.01 = 6\times 10^{-3}\left[kg\,m^2\right]$$

The value of i_{d0} for infinite zero-torque speed (Equation 3.144) is

$$i_{d0} = -\frac{-\psi_{PMd}}{L_d} = -\frac{1}{0.05} = -20\text{A}$$

The eigenvalues for the speed response (Equation 3.148) correspond to the characteristic equation:

$$s^2\left(0.3\times 0.1\times 1\times 0.01\right) + s\left(0.3\times 0.1\right)\times 1\times 0.01 + 1\times 0.01 + \frac{3}{2}\times 2\left(0 - 0.1\times 20\right)\times 0 = 0$$

$$0 = 10^{-2} + s^2\times 3\times 10^{-4} + s\times 4\times 10^{-3};$$

$$\begin{aligned} s_{12} &= \frac{-2\times 10^{-3} \pm \sqrt{4\times 10^{-6} - 10^{-2}\times 3\times 10^{-4}}}{3\times 10^{-4}} \\ &= -3.33,\text{respectively},-10.0 \end{aligned} \tag{3.151}$$

Consequently, the speed response should be stable and aperiodic. For $i_d = i_{d0} = -\psi_{PM}/L_q(\psi_{d0} = 0)$, Equation 3.149 gets simplified to

$$\left(\tilde{i}_q\right)_{\Psi_{d0}=0} = \frac{\tilde{V}_q}{R_s\left(1+s\tau_{eq}\right)} \tag{3.152}$$

3.15.2 Vector Control of PMSMs at Constant ψ_{D0} (I_{D0} = const)

On the basis of Equations 3.138 and 3.139, we can introduce the vector current control of PMSMs (Figure 3.14).

The reference value of i_d is considered zero up to base speed, ω_b, and then is made negative for flux weakening above ω_b. Then, the reference torque, T_e^*, is "delivered" by the speed regulator. From the torque expression, with T_e^*, and i_{d0}^* known, the reference torque current, i_q^*, is calculated. Then, with i_d^* and i_q^* known, using the Park transformation from rotor to stator coordinates (θ_{er}), the reference a.c. phase currents, $i_a^*(t)$, $i_b^*(t)$, and $i_c^*(t)$, are calculated and then regulated via a.c. current regulators of various configurations.

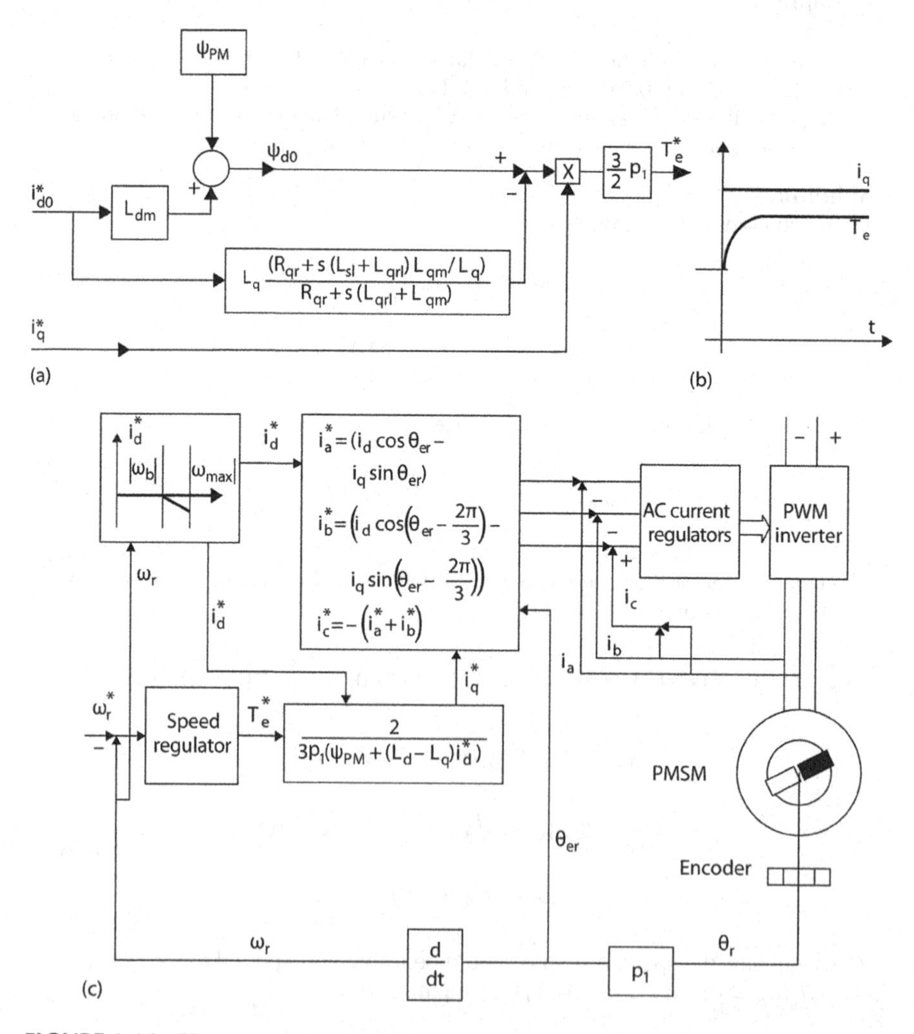

FIGURE 3.14 Vector control of PMSM with constant i_{d0}. (a) PMSM structural diagram at i_{d0}, (b) torque response to sudden i_q^* increase, and (c) basic vector control scheme.

The current regulators provide the PWM the means to produce the required a.c. voltages that supply the motor through the PWM inverter. This is how variable speed with high performance is obtained.

3.15.3 Constant Stator Flux Transients in Cageless SMs at cos $\varphi_1 = 1$

Constant stator (total) flux conditions, $|\psi_s| = |\psi_d + j\psi_q|$, may be maintained when the machine is loaded, only for d.c. excitation on the rotor. Also, for this case, operation at unity power factor is typical. Representative applications use large SMs supplied from a PWM (two or three, multi, level) voltage source converter in variable speed drives (up to 50 MW, 60 kV gas compressor drive units).

Again, we start with the stator equation, but for the unity power factor and constant stator flux

$$V_d + jV_q = V_s = R_s i_s + \omega_r \psi_s$$
$$\psi_d = L_d i_d + L_{dm} i_F; \quad \psi_q = L_q i_q \tag{3.153}$$

$$\left(L_{Fl} + L_{dm}\right)\frac{di_F}{dt} + L_{dm}\frac{di_d}{dt} = V_F - R_F i_F \tag{3.154}$$

$$\psi_d^2 + \psi_q^2 = \psi_s^2 \tag{3.155}$$

The motion equation is

$$\frac{J}{p_1}\frac{d\omega_r}{dt} = T_e - T_L; \quad T_e = \frac{3}{2} p_1 \psi_s i_s \tag{3.156}$$

The space phasor diagram is shown in Figure 3.15b.

The voltage power angle, $\boldsymbol{\delta}_v$, is equal to the current angle, $\boldsymbol{\delta}_i$, and flux angle, $\boldsymbol{\delta}_{\psi s}$, due to the unity power factor. The conditions for the unity power factor are

$$\sin\left(\delta_v\right) = \frac{L_q i_q}{\psi_s} = -\frac{i_d}{i_s}; \quad i_d < 0 \tag{3.157}$$

And thus

$$\tan\delta_v = L_q i_s / \psi_s \tag{3.158}$$

$$L_{dm} i_F = \psi_s \cos\delta_v - L_d i_d; \quad i_d < 0 \tag{3.159}$$

So, for a given stator flux, ψ_s^*, and torque, T_e^*, from Equation 3.156

$$i_s^* = \frac{2}{3}\frac{T_e^*}{p_1 \psi_s^*} \tag{3.160}$$

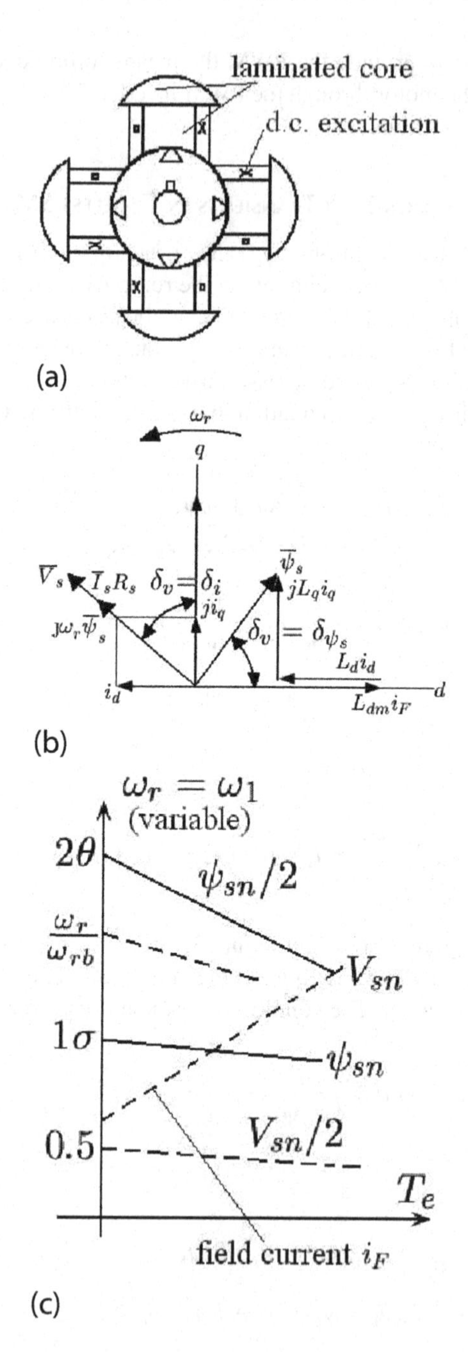

FIGURE 3.15 (a) DC-excited cageless rotor SM, (b) its space phasor diagram at cos φ_1, and (c) speed/torque curves.

From Equation 3.158 δ_v^* is calculated, and then

$$i_d = -i_s^* \sin\delta_v^*; \quad i_q = i_s^* \cos\delta_v^* \tag{3.161}$$

Finally, from Equation 3.159, I_F^* is calculated. This sequential approach is useful for achieving variable speed drive control. It is worth noting that for the steady state, Equation 3.153 can be written as

$$V_s = R_s \frac{2}{3}\frac{T_e^*}{p_1\psi_s^*} + \omega_r\psi_s^* \tag{3.162}$$

Equation 3.162 suggests a linear speed/torque curve. This is similar to the constant excitation flux d.c. brush machines. It is implicit that the frequency, ω_1, varies in tact with ω_r.

The speed may be controlled by

- Voltage control, V_s, up to base speed, ω_{rb}
- Flux weakening (ψ_s decreases) above ω_{rb} (Figure 3.15c)

As seen in Figure 3.15, the speed decreases a little with torque, denoting stable behavior while the excitation (field) current, to maintain the unity power factor, increases with torque, as expected.

Example 3.8

A constant ψ_s, $\cos\varphi_s = 1$, variable-speed SM at 1800 kW, where $V_{nl} = 4.2$ kV, $f_0 =$ 60 Hz, $2p_1 = 60$ with d.c. excitation and a cageless rotor $x_d = 0.6$ pu, $x_{dm} = 0.5$ pu, $x_q = 0.4$ pu, $R_s = 0.01$ pu, operates at the unity power factor from 10% to 100% rated current.
Calculate

a. Rated current, I_n, with efficiency $\eta_r = 0.985$, unity power factor, rated speed, and the required field current, i_{Fn}
b. No-load ideal speed for full voltage and 50% voltage for i_{Fn} and for full voltage and $i_{Fn}/2$ (magnetic saturation is neglected)
c. ω_r (T_e) and i_F (T_e), for $V_{nl}/2$ and for 25%, 50%, and 200% I_n

Solution:

a. The rated current, I_n, is calculated from the definition of efficiency:

$$I_n = \frac{P_n}{\eta_n\sqrt{3}V_{nl}\cos\varphi_n} = \frac{1800\times10^3}{0.985\times\sqrt{3}\times4.2\times10^3\times1.0} = 251.50 \text{ A}$$
$$I_{sn} = I_n\sqrt{2} = 251.50\sqrt{2} = 354.62 \text{ A}$$

(Continued)

Example 3.8: (Continued)

The voltage Equation 3.153 yields:

$$V_{sn} = \frac{V_{nl}}{\sqrt{3}}\sqrt{2} = R_s i_{sn} + 2\pi f_b \psi_{sn}$$

$$\psi_{sn} = \frac{4.2\times10^3\sqrt{2/3} - 0.01\dfrac{4.2\times10^3}{251.50\sqrt{3}}\times 251.50\sqrt{2}}{2\pi 60} = 8.9938 \text{ Wb}$$

The electromagnetic torque is

$$T_{en} = \frac{P_n}{\dfrac{\omega_{rb}}{p_1}} = \frac{1800\times10^3\times 30}{2\pi 60} = 143.312\times10^3 \text{ N m}$$

Tan δ_v Equation 3.158 yields

$$\tan\delta_{V_n} = L_q\frac{i_{sn}}{\psi_{sn}} = \frac{0.4}{\psi_{sn}}\frac{V_{nl}}{\sqrt{3}}\frac{I_n\sqrt{2}}{I_n\omega_{rb}} = \frac{0.4\times 4200\sqrt{2/3}}{8.9938\times 120\times\pi} = 0.40;\quad \delta_{V_n} = 22^\circ \quad \text{(A)}$$

So the rated voltage (flux, current) power angle is 22°, which is quite a realistic value.

Now from Equation 3.161

$$i_{dn} = -I_{sn}\sin\delta_{V_n} = -354.62\times\sin 22^\circ = -132.84 \text{ A} \quad \text{(B)}$$

So the required field current, i_{Fn}, Equation 3.159 is

$$i_F = \frac{\psi_s\cos\delta_v - L_d i_d}{L_{dm}} = \frac{8.9938\times 0.927}{0.5\times\dfrac{4200}{\sqrt{3}}25/0.5\times 120\pi} - \frac{0.6}{0.5}(-132.84) = 650.876 + 159.41 = 810.284 \text{ A} \quad \text{(C)}$$

Note that the field current is reduced to the stator.

b. The no-load ideal speed (Equation 3.162) ω_{r0} is

$$\omega_{r0} = \frac{V_s}{\psi_s} = \frac{4200\times\sqrt{2/3}}{8.9438} = 380.609 \text{ rad/s}$$

This is slightly larger than the rated speed, $\omega_{rb} = 2\pi 60 = 376.8$ rad/s. For half the voltage, $(V_{sn}/2)$, the ideal no-load speed is halved to 190.304 rad/s. On the other hand, for $i_{Fn}/2$, $\psi_s = \psi_{sn}/2$ and thus the no-load ideal speed is doubled, 2 × 380.99 rad/s.

c. To calculate the field current for $i_s = (25\%, 50\%)I_{sn}$, and unity power factor and full flux, we just have to repeat the above process, starting with tan δ_v, then i_d, then i_F from Equations A through C and the torque, $T_e = \frac{3}{2}p_1\psi_{sn}i_s$ from Equation 3.160.

3.15.4 Vector Control of SMs with Constant Flux (ψ_s) and cos φ_s = 1

A basic vector control scheme for constant ψ_s and unity power factor is shown in Figure 3.16

- The reference stator flux, ψ_s^*, is set constant up to base speed (when full voltage is attained), and then decreased inverse proportionally to speed.
- The reference torque, T_e^*, is the output of the speed regulator and it has also to be limited (reduced) above base speed.
- On the basis of known ψ_s^*, and T_e^* on line, i_s^*, δ_v^*, and i_F^* are calculated as in Equations 3.157 through 3.161; then V_s is calculated as in Equation 3.153 and the voltage angle $\theta_{vs} = \left(\frac{\pi}{2} + \delta_v + \theta_{er}\right)$ in Equation 3.24 is used for the Park transformation to yield $V_A^*(t)$, $V_B^*(t)$, and $V_C^*(t)$, which are then open-loop PWM-"fabricated" in the inverter.
- Simultaneously, the reference field current, i_F^*, (after reduction to the rotor via k_F) is close-loop regulated through a d.c.–d.c. (or a.c.–d.c.) PWM low rating converter.
- For an easy correction of parameter mismatch, the power factor angle φ_s may be measured and the reference current, i_F^*, is corrected with the output of an additional PI regulator of φ_s (set to zero, $\varphi_s^* = 0$).

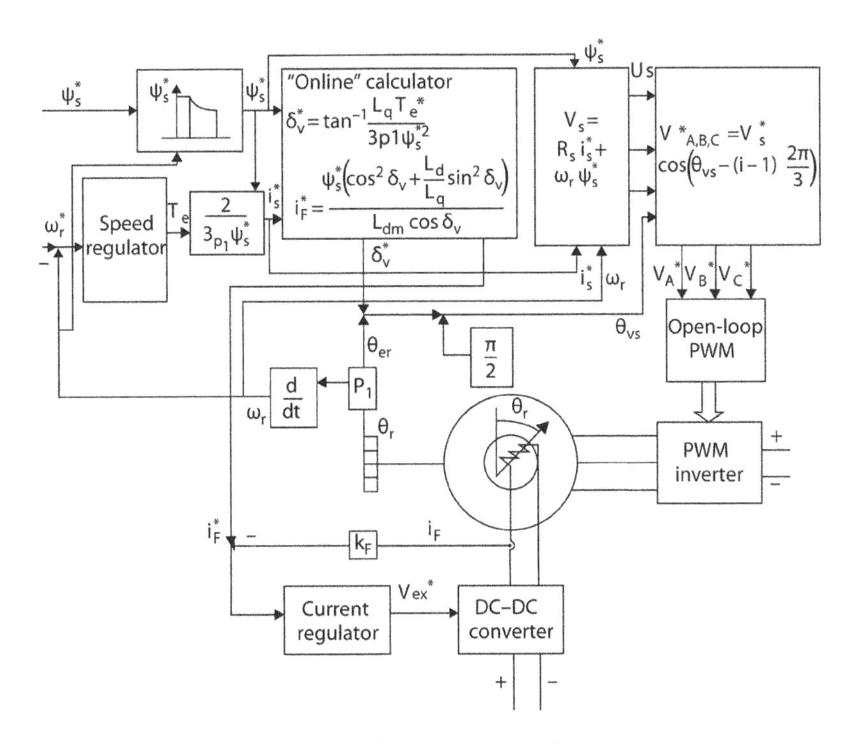

FIGURE 3.16 Basic vector control of SMs with constant flux, ψ_s, and cos φ_s = 1.

The absence of stator current regulators means that other safe current protection means have to be used. Figure 3.16 shows an example of the usefulness of transients modeling for the design of variable speed motor/generator control (more on electric drives can be found in [5]).

3.16 TRANSIENTS FOR CONTROLLED FLUX AND RECTANGULAR CURRENT SMS

Rectangular current control of SMs is used in two extreme cases:

- Low-power cageless rotor PMSM with proximity Hall sensors and PWM voltage source inverters for variable speed and for lower costs (BLDC motor) (Figure 3.17a through d)
- High-power cage rotor d.c. excitation SMs with proximity Hall sensors and current source inverters for lower costs.

As both are widely used in industry, we tackle the SM transients for both, which may serve as a foundation for further studies in electric drives.

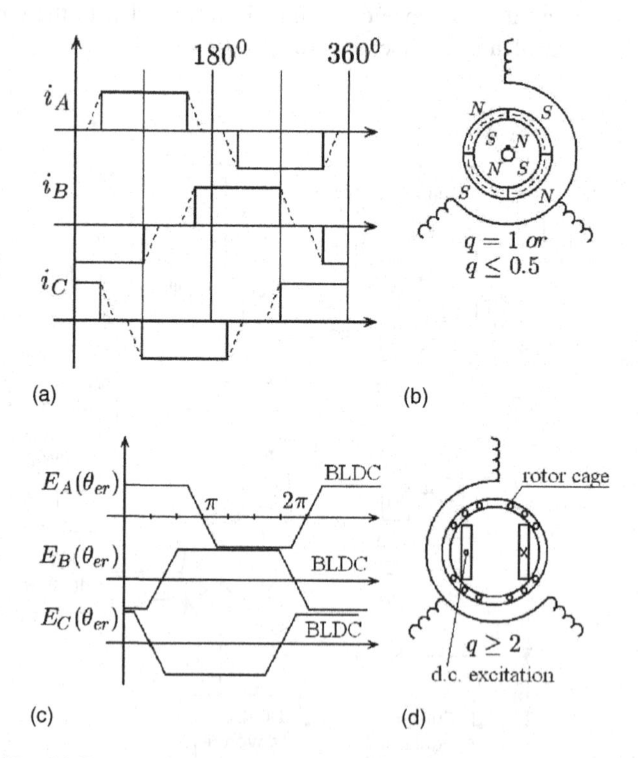

FIGURE 3.17 (a) Rectangular currents, (b) surface PM rotor (BLDC) motor, (c) emfs for BLDC PM motor, and (d) d.c.-excited cage rotor SM.

3.16.1 Model of Brushless DC-PM Motor Transients

The PMSM with a cageless rotor, with trapezoidal emf—with $q = 1$ or $q < 0.5$ slots/pole/phase stator windings, controlled with ideal rectangular (in fact trapezoidal) a.c. currents, intact with rotor position, by three proximity Hall sensors (120° apart), and supplied by a PWM voltage source inverter, is called the brushless d.c. motor (or BLDC-PM motor).

Most BLDC-PM motors have (in general) surface PMs on the rotor and thus $L_d = L_q = L_s$, independent of the rotor position. But the PM-produced emfs in the stator phase are rather trapezoidal (ideally rectangular and up to 180° wide, both positive and negative), Figure 3.17c.

The emfs may be decomposed into odd harmonics as

$$\begin{aligned} E_A(t) &= \sum_{\nu=1,3,5,\ldots} \omega_r \Psi_{PM\nu} \cos(\nu\theta_{er}) \\ E_B(t) &= \sum_{\nu=1,3,5,\ldots} \omega_r \Psi_{PM\nu} \cos\left(\nu\left(\theta_{er} - \frac{2\pi}{3}\right)\right) \\ E_C(t) &= \sum_{\nu=1,3,5,\ldots} \omega_r \Psi_{PM\nu} \cos\left(\nu\left(\theta_{er} + \frac{2\pi}{3}\right)\right) \end{aligned} \tag{3.163}$$

For practical purposes, the 1st, 3rd, 5th harmonics suffice.

As surface PM pole rotors are considered, the cyclic inductance, L_s, is

$$L_s = L_{sl} + \frac{4}{3}L_g; \quad L_{AB} = -L_g/3 \tag{3.164}$$

So the phase coordinate model is straightforward:

$$\begin{vmatrix} L_s & 0 & 0 \\ 0 & L_s & 0 \\ 0 & 0 & L_s \end{vmatrix} \frac{d}{dt}\begin{vmatrix} I_A \\ I_B \\ I_C \end{vmatrix} = \begin{vmatrix} V_A(t) \\ V_B(t) \\ V_C(t) \end{vmatrix} - \begin{vmatrix} R_s & 0 & 0 \\ 0 & R_s & 0 \\ 0 & 0 & R_s \end{vmatrix}\begin{vmatrix} i_A \\ i_B \\ i_C \end{vmatrix} - \begin{bmatrix} E_A(t) \\ E_B(t) \\ E_C(t) \end{bmatrix} \tag{3.165}$$

The torque is

$$T_e = \frac{E_A(t)i_A(t) + E_B(t)i_B(t) + E_C(t)i_C(t)}{(\omega_r/p_1)} \tag{3.166}$$

$$\frac{J}{p_1}\frac{d\omega_r}{dt} = T_e - T_{load}; \quad \frac{d\theta_{er}}{dt} = \omega_r \tag{3.167}$$

If the voltage waveforms are applied in tact to rotor position (θ_{er}) by a PWM converter and with phase commutation every 60° (electrical), the machine behaves much like a d.c. machine.

While Equations 3.163 through 3.167 constitute the complete model, for steady state, ignoring the commutation of phases, the 120° wide, smooth current blocks lead to a two-phase operation of the machine (Figure 3.17):

$$V_{dc} = V_A - V_B = 2L_s\left(\frac{di_a}{dt}\right)_{=0} + 2R_s i_{A0} + E_A - E_B \tag{3.168}$$

Consider one voltage pulse, $V_A - V_B = V_{dc}$, and constant (180° wide) emf "blocks," $E_A - E_B = 2E$:

$$V_{dc} = 2R_s i_{A0} + 2E \tag{3.169}$$

From Equation 3.169, the electromagnetic torque is

$$T_e = \frac{2Ei_{A0}}{\omega_r/p_1} \tag{3.170}$$

$$E = \omega_r \psi_{PM} \tag{3.171}$$

Hence

$$V_{dc} = 2R_s i_{A0} + 2\omega_r \psi_{PM} \tag{3.172}$$

$$V_{dc} = R_s T_e/(\psi_{PM} p_1) + 2\omega_r \psi_{PM} \tag{3.173}$$

But this is again similar to the d.c. brush PM machine with a known linear speed/torque curve. As the phase currents are "in phase" with the phase emfs, during commutation of phases (from AB, AC, CB to BA), the current in phase B will go to zero and the current in phase C will take its place. The process evolves with the participation of the capacitance, C_{dc} (Figure 3.18).

To keep the current waveform flat, voltage chopping is required and possible up to the base speed where a full V_{dc} voltage is applied. The fact that the position sensor is required only to produce six signals (shifted by $\pi/3$ electric radians) leads to low-cost proximity Hall sensors. For more on brushless d.c.-PM motor control, the reader can refer to [6].

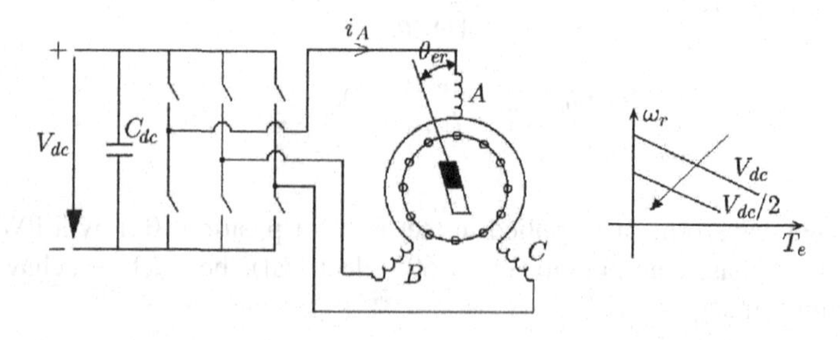

FIGURE 3.18 Brushless d.c. PM motor with a voltage source PWM inverter.

3.16.2 DC-Excited Cage Rotor SM Model for Rectangular Current Control

As here the SM has a cage rotor, with a d.c. excitation rotor circuit, the complete dq model may be used to simulate any transients, including those with rectangular current control. It is supplied by a current source inverter (Figure 3.19a).

However, to simplify the treatment, we may assume that for the commutation of phases (for "rectangular" current control), load (emf) commutation is used. This means leading power factor for the fundamental component of rectangular current, or machine over-excitation. Moreover, the commutation inductance is $L_c \approx \left(L_d'' + L_q''\right)/2$, where L_d'' and L_q'' are subtransient inductances as defined in previous sections, and which are smaller as the rotor cage gets stronger.

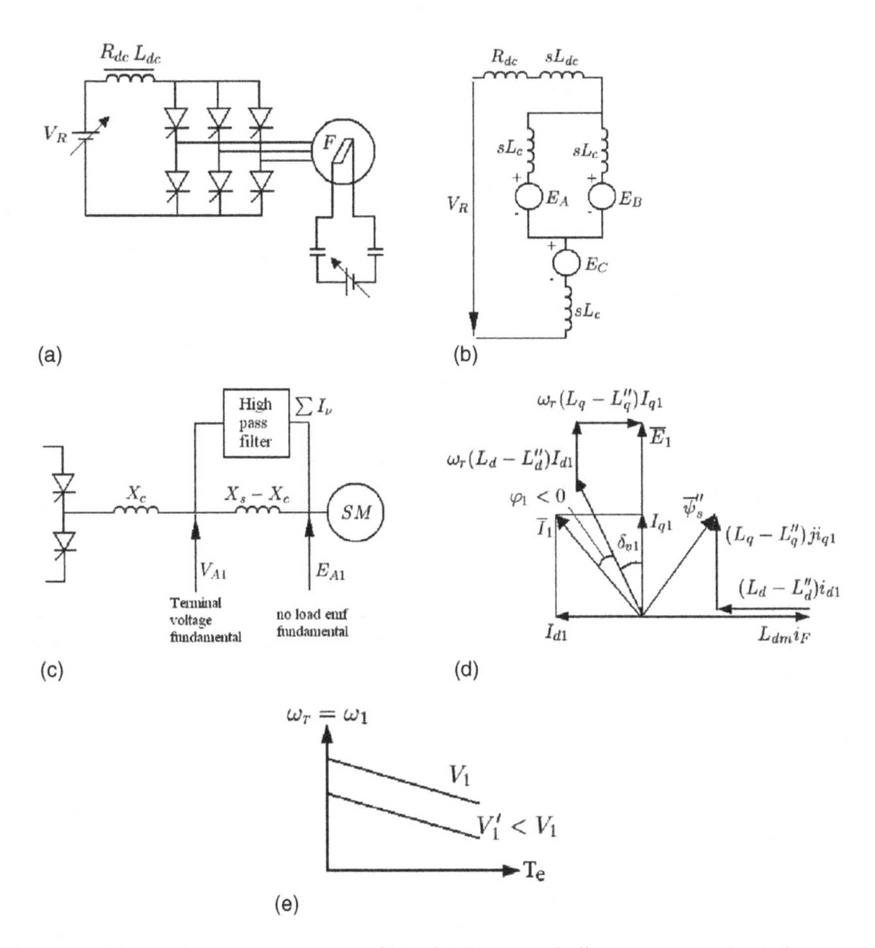

FIGURE 3.19 DC-excited cage rotor SM with "rectangular" current control: (a) the current source inverter plus SM, (b) equivalent circuit during "load" commutation of phases (from AC′ to BC′), (c) steady-state equivalent circuit, (d) vector diagram for the fundamental components, and (e) speed (torque) linear dependence, with position control and leading power factor, $\varphi_i = -(8–12)°$.

The SM may be thus modeled by L_c for the phase commutation (transients) and by $L_s - L_c$ for the steady state (Figure 3.19b).

The commutation process is more subtle but the in essence, the smaller L_c, the larger the load (phase) current that can be commutated within the 15–30° (electrical) available for it. We may treat the steady state approximately for the fundamental component of current by assuming a constant flux behind subtransient reactance (inductance), ψ''_s, and a leading power factor angle, Figure 3.19:

$$\begin{aligned} &V_{s1} \approx R_s i_{s1} \cos\varphi_1 + \omega_r \psi''_s \\ &\psi_{d1} = L_{dm} i_F + \left(L_d - L''_d\right) i_{d1}; \quad \psi''_s = \sqrt{\psi_{d1}^2 + \psi_{q1}^2} \\ &\psi_{q1} = \left(L_q - L''_q\right) i_{q1}; \quad T_e = \frac{3}{2} p_1 \psi''_s i_{s1} \cos\varphi_1 \end{aligned} \tag{3.174}$$

From the vector diagram in Figure 3.19d

$$\begin{aligned} \tan\left(\delta_{v1} - \varphi_1\right) &= -I_d / I_q \\ \tan\delta_{v1} &= \psi_{d1} / \psi_{q1} \end{aligned} \tag{3.175}$$

For given ψ_s^* and T_e^*, and $\varphi_1^* < 0$, we may calculate using Equations 3.174 and 3.175, as for the case of $\cos\varphi_1 = 1$, i_{s1}^*, δ_{v1}^*, I_{d1}^*, I_{q1}^*, and I_F^*. As expected, i_F^* increases with torque, as for $\cos\varphi_1 = 1$, but this time for larger i_F^* values.

The voltage Equation 3.174 again leads to a stable (linear) speed/torque curve. For more on the current source inverter SM drives, the reader is directed to [5].

3.17 SWITCHED RELUCTANCE MACHINE MODELING FOR TRANSIENTS

SRMs are doubly salient, singly (stator) excited electric machines with passive rotors [7,8]. Their nonoverlapping (tooth wound) coils (phases) are turned on sequentially (in relation to rotor position) to produce torque, through d.c. voltage pulses that produce unipolar phase currents. Three or more phase SRMs may be started from any position but single-phase SRMs need a self-starting artifact, a parking PM, stepped rotor airgap, an energy drain cage additional winding, short-circuit coil (shaded poles) on rotor, or a zone of easy saturation (with slots) on a part of rotor poles.

Most SRMs lack mutual flux between phases, which makes them more fault tolerant, at the price of loosing an important torque component.

A typical three-phase 6/4 SRM is shown in Figure 3.20a; its phase inductances vary with the rotor position as shown in Figure 3.20b. The simplified flux/current/position family of curves in Figure 3.20c reveals the potential source of torque production.

The flux/current/position curves obtained by FEM or by tests may be approximated by various methods to analytical expressions. For the linear dependencies shown in Figure 3.19c:

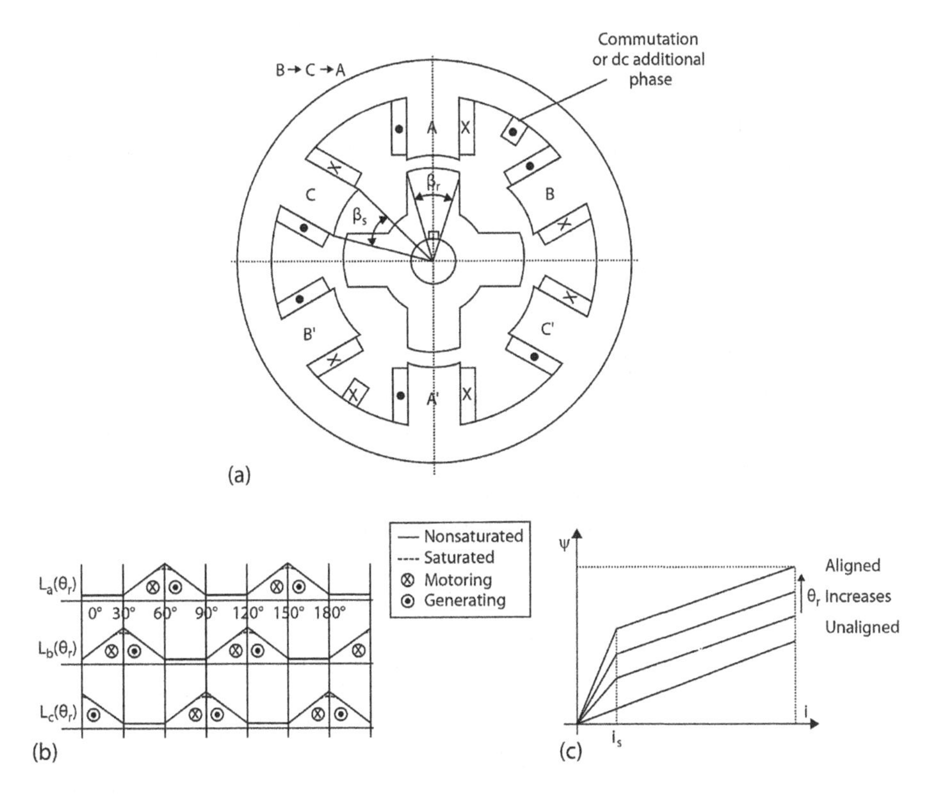

FIGURE 3.20 (a) Three-phase 6/4 SRM, (b) inductances, and (c) simplified flux/current/position curves.

$$\psi \approx \left(L_u + \frac{K_s(\theta_r - \theta_0)}{i_s}\right) i; \quad \text{for } i \le i_s$$
$$\psi \approx L_u i + K_s(\theta_r - \theta_0); \quad \text{for } i \ge i_s. \tag{3.176}$$

θ_r varies only from the unaligned to the aligned position. As there is no coupling between phases and the machine has double saliency, phase coordinates are to be used:

$$V_{A,B,C} = R_s i_{A,B,C} + \frac{d_s \psi_{A,B,C}(i,\theta_r)}{dt} \tag{3.177}$$

$$T_{e,A,B,C} = \left(\frac{\partial W_{m\ \text{coenergy}}}{\partial \theta_r}\right)_{i=\text{const}} \tag{3.178}$$

We may write Equation 3.177 as

$$V_{A,B,C} = R_s i_{A,B,C} + \frac{\partial \psi_{A,B,C}(i,\theta_r)}{\partial i}\frac{di}{dt} + \frac{\partial \psi_{A,B,C}(i,\theta_r)}{\partial \theta_r}\frac{d\theta_r}{dt} \tag{3.179}$$

$$J\frac{d\Omega_r}{dt} = T_e - T_{load} - B\Omega_r; \quad \frac{d\theta_r}{dt} = \Omega_r \tag{3.180}$$

$E_i = \dfrac{\partial\psi_{A,B,C}(i,\theta_r)}{\partial\theta_r}$ looks like an emf but it is a pseudo-emf because the torque expression:

$$T_e = \frac{1}{2}\Sigma\frac{E_i i_i}{\Omega_r} = \frac{1}{2}\Sigma i_i^2\frac{\partial L_i}{\partial\theta_r}; \quad \Omega_r = \frac{d\theta_r}{dt} \tag{3.181}$$

is valid only for $L_i(\theta_r) = \psi_i/L_i$, that is, for the linear case (no magnetic saturation). E_i changes sign with the rotor position (it is ⊕ for a rising inductance slope and ⊖ for a decaying slope) and thus with positive current, positive (motoring) and negative torques are obtained.

There are N_r (rotor pole number) energy cycles per phase per revolution. So, the SRM operates like an SM with N_r pole pairs. The average torque T_{eav} is

$$T_{eav} = mN_r(T_{eav})_{cycle} \tag{3.182}$$

where m is the number of stator phases.

A small-signal model of SRM per phase may be developed for the linear case and,

$$\begin{aligned} i &= i_0 + \Delta i; \quad \Omega_r = \Omega_{r0} + \Delta\Omega_r \\ V &= V_0 + \Delta V; \quad T_L = T_{L0} + \Delta T_L \end{aligned} \tag{3.183}$$

by using Equations 3.177 through 3.180:

$$\left(s + \frac{1}{\tau_e}\right)\Delta i + \frac{K_b}{L_{av}}\Delta\Omega_r = \frac{\Delta V}{L_{av}} \tag{3.184}$$

$$-\frac{1}{J}K_b\Delta i + \left(s + \frac{1}{\tau_m}\right)\Delta\Omega_r = -\frac{\Delta T_L}{J} \tag{3.185}$$

with

$$R_e = R_s + \frac{\partial L}{\partial\theta_r}\omega_{r0}; \tau_e = \frac{L_{av}}{R_e}; L_{av} \approx \frac{L_{max} + L_{min}}{2};$$

$$K_b = \frac{dL}{d\theta_r}i_0; \Delta E = K_b\Delta\Omega_r; E \approx K_b\Omega_r; \tau_m = \frac{J}{B} \tag{3.186}$$

Equations 3.184 and 3.185 lead to the linearized structural diagram of Figure 3.21a with its reduced form ready for control, shown in Figure 3.21b.

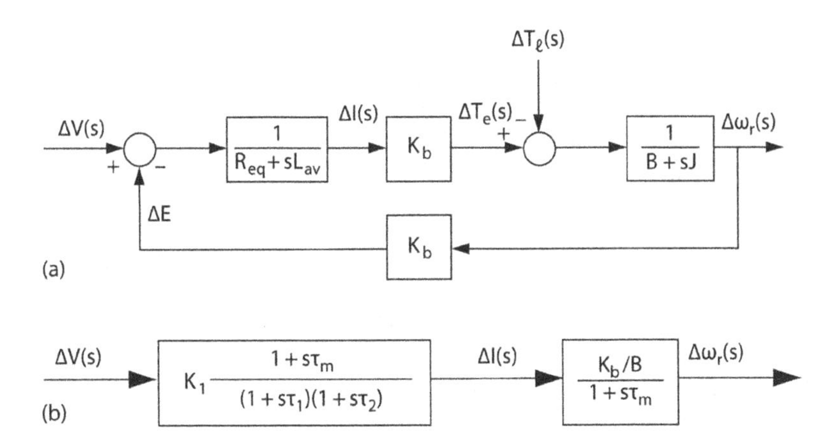

FIGURE 3.21 SRM: (a) structural diagram and (b) its reduced form.

In deriving the small-signal model, L was considered constant as L_{av}, where its derivative with the rotor position was still calculated. The structural diagram is very similar to that of the d.c. brush series machine where K_b varies with the initial current, i_0, and $\partial L/\partial\theta_r$ (with rotor position) and R_e varies with $dL/d\theta_r$ and Ω_r. $\tau_m = J/B$ is the mechanical time constant. The eigenvalues of the linear model, τ_1 and τ_2, dependent on i_0, Ω_{r0}, and $\partial L/\partial\theta_r$ can be extracted from Equations 3.184 and 3.185 and both of them have a negative real part, so that the response is stable; however, it may be periodic or aperiodic as for d.c. the brush series machines.

One major difference with respect to the d.c. brush series machine is that this one is motoring for $\partial L/\partial\theta_r > 0$ and generating for $\partial L/\partial\theta_r < 0$. Numerous applications for SRMs have been proposed, but only a few have reached the markets. Their ruggedness to temperature and chemically aggressive environments makes them favorites for niche applications. For more on SRMs and drives, reference [9].

Example 3.9

A three-phase 6/4 SRM has the phase inductance (Figure 3.20b) with $L_{min} = 2mH$ and $L_{max} = 10$ mH and the phase resistance $R_s = 1.0\ \Omega$ is operated at $n = 9000$ rpm and controlled with an ideal constant current, I_0, over the entire $\theta_{dwell} = 30°$ that corresponds to the stator and rotor pole spans, with an average voltage, $V_0 = 96V_{dc}$. Calculate:

a. The constant current, I_0 max., phase flux linkage and average torque, $T_{e0} = T_{L0}$, at $\omega_r = \omega_{r0} = 2\pi\ 9000/60 = 300\ \pi$rad/s
b. For a constant load torque and 10% increase in average voltage V_0 ($\Delta V = +0.1V_0$), calculate the eigenvalues and current and speed transients using the small-signal approach. $\tau_m = J/B = 0.1$ s, $B = 3.33 \times 10^{(-4)}$ N m s

Solution:

a. We take Equation 3.180 and integrate it over time, t_{on}, corresponding to 30°:

(Continued)

Example 3.9: (Continued)

$$t_{on} = \frac{\pi/6}{2\pi n} = \frac{\pi/6}{2\pi 9000/60} = 0.555\times10^{-3}\ s$$

and obtain

$$V_0 t_{0n} = R_s I_0 t_{0n} + \left(L_{max} - L_{min}\right) I_0$$

$$I_0 = \frac{96\times0.555\times10^{-3}}{(10-2)\times10^{-3} + 1\times0.555\times10^{-3}} = 6.2048\ A$$

The maximum flux linkage is $\psi_{max} = L_{max} I_0 = 10 \times 10^{-3} \times 6.2048 = 0.062048$ Wb.

The average torque may be calculated from the power balance per cycle, considering that there is ideally a single phase working at any given time, $T_{eav}\ 2\pi n = (V_{dc0} - R_s I_0) I_0$.

$$T_{eav0} = \frac{(96-1\times6.2048)\times6.2048}{2\pi(9000/60)} = 0.591\,Nm$$

The load torque is equal to the motor torque for the initialization of transients:

$$T_{eav0} = T_{L0} + B\omega_{ra} = T_{L0} + 3.33\times10^{-4}\times2\pi150; T_{L0} = 0.366\ Nm$$

b. The inductance varies from L_{min} to L_{max}, linearly from $\theta_r = 0$ to $\theta_r = \pi/6$, so

$$L(\theta_r) = L_{min} + \left(L_{max} - L_{min}\right)\frac{\theta_r}{\pi/6}[H]; 0 \le \theta_r \le \pi/6$$

Consequently

$$K_b = \frac{\partial L}{\partial \theta_r} i_0 = \left[\left(L_{max} - L_{min}\right)\frac{6}{\pi}\right] i_0 = \left[(10-2)\frac{6}{\pi}\right]10^{-3}\times6.2048$$
$$= 0.09485\ \text{Wb}$$

From Equation 3.186

$$R_e = R_s + \frac{\partial L}{\partial \theta_r}\omega_{r0} = 1.0 + 15.2866\times10^{-3}\times300\pi = 15.4\ \Omega$$

The inertia, J, is

$$J = \tau_m B = 0.33\times10^{-4}\ \text{kg m}^2$$

$$\tau_e = L_{av}/R_e = \frac{(2+10)\times10^{-3}}{2\times15.4} = 0.3895\times10^{-3}\ \text{s}$$

Hence the Equations 3.184 and 3.185 become (for $\Delta V = +3.6\text{V}$, $\Delta T_L = 0$)

$$\left(s+\frac{1}{0.3895\times10^{-3}}\right)\Delta i(s)+\frac{0.09485}{6\times10^{-3}}\Delta\Omega_r(s)-\frac{3.6}{s\times6\times10^{-3}}$$

$$-\frac{0.09485}{0.33\times10^{-4}}\Delta i(s)+\left(s+\frac{1}{0.1}\right)\Delta\Omega_r(s)=0$$

The eigenvalues are obtained from the characteristic equation:

$$(s+2567)(s+10)+\frac{0.09485^2}{0.333\times10^{-4}\cdot6\times10^{-3}}=s^2+2557.7+4.756\times10^4=0$$

$$s_{1,2}=\frac{-2577.7\pm\sqrt{2577.7^2-4\times4.756\times10^4}}{2\cdot2539.8}=-18.6,\text{ respectively}-2558.4$$

Now, the solution for speed transients is straightforward:

$$\omega_r(t)=\omega_{r0}+\Delta\omega_r=\omega_{r0}+A_1e^{-18.6t}+A_2e^{-2558.4t}+\left(\omega_{rfinal}-\omega_{r0}\right)$$

The final speed value, for constant torque (constant current, i_0, and thus same flux variation), is straightforward from the voltage equation integrated over the new time, t'_{on}, which corresponds to a $\pi/6$ rotor angle rotation:

$$(V_0+\Delta V)t'_{on}=R_sI_0t'_{on}+\left(L_{max}-L_{min}\right)I_0$$

$$t'_{on}=\frac{(10-2)10^{-3}\times6.2048}{96+9.6-1\times6.2048}=0.5\times10^{-3}\text{ s}$$

So, the speed, $\omega_{r\,final}$, is

$$\omega_{r\,final}=\omega_{r0}\frac{t_{on}}{t'_{on}}=300\pi\times\frac{0.555\times10^{-3}}{0.5\times10^{-3}}=1046\text{ rad/s}$$

Also at $t = 0$, $\omega_r = \omega_{r0} = 942$ rad/s and $(d\omega_r/dt)_{t-0} = 0$. Thus, both constants, A_1 and A_2, can be found:

$$A_1+A_2=\omega_{r0}-\omega_{r\,final}=942-1046=-104.0$$

$$-18.6A_1-2558.4A_2=0$$

$$A_1=-104.6\ A_2=0.76$$

The current transients start and end at $I_0 = 6.2048$ A and follow Equation 3.185 with $\Delta T_L = 0$:

$$i=i_0+\Delta i=i_0+\left(\Delta\omega_r(t)\right)\times\frac{J}{\tau_mK_b}+\frac{J}{K_b}\times\frac{d\left(\Delta\omega_r(t)\right)}{dt}$$

A graphical representation of speed $\Delta\omega_r(t)$ and current $\Delta i(t)$ variations (in %) is shown qualitatively in Figure 3.22.

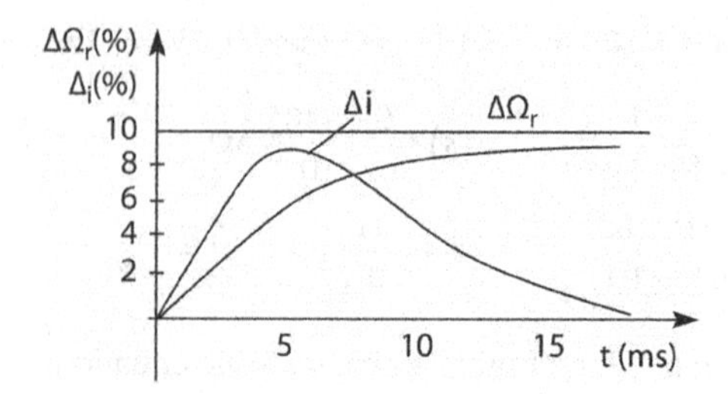

FIGURE 3.22 Speed, $\Delta\omega_r$, and current, Δ_i, small transients at a step rise of 10% voltage for a constant load torque.

Note: The response is stable as expected, but it is also aperiodic, because the inertia and mechanical time constant, τ_m, are large with respect to the equivalent electrical time constant, τ_e.

3.18 SPLIT-PHASE CAGE ROTOR SM TRANSIENTS

A split-phase cage rotor SM (Figure 3.23) is used for small power single-phase grid-connected (constant frequency (speed), constant voltage) domestic applications, as better efficiency can be obtained than with split-phase IMs.

For the four-pole machine shown in Figure 3.23, the geometrical axes that correspond to the electrical orthogonal axes are $\pi/4$ apart. PMs may or may not be placed in the rotor flux barriers. Strong PMs are preferable for the steady state but they produce a large braking torque during self-starting, (Example 3.4), which hampers self-starting and synchronization unless a larger starting capacitor is used.

Self-starting transients are essential for split-phase SMs, even for synchronous reluctance machines, ($L_d < L_q$), with or without weak PMs in the *d*-axis (in our case).

The machine has magnetic saliency on the rotor, so if the dq model is to be used, it can be tied to the rotor. But, in this case, the stator main and auxiliary windings should be symmetric (equivalent) in the sense that when the auxiliary winding is

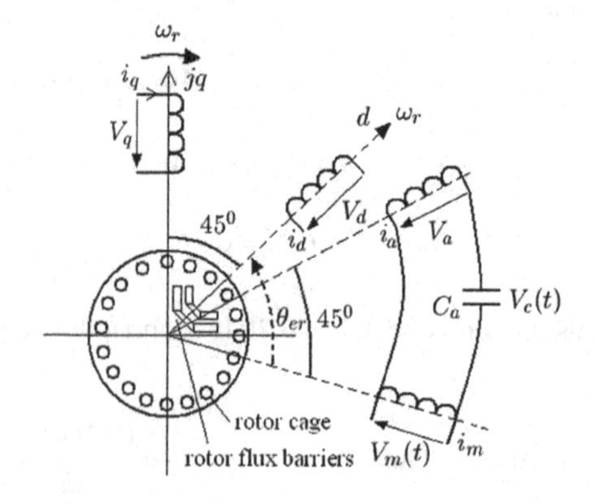

FIGURE 3.23 Split-phase cage rotor SM.

reduced to the main winding its resistance, R_a', and leakage inductance, L_{al}', should be equal to R_m and L_{ml}, respectively,

$$R_m = R_a\left(\frac{W_m K_{W_m}}{W_a K_{W_a}}\right)^2 = R_s; \quad a = \frac{W_a K_{W_a}}{W_m K_{W_m}}$$
$$L_{ml} = L_{al}\left(\frac{W_m K_{W_m}}{W_a K_{W_a}}\right)^2 = L_{sl} \tag{3.187}$$

This equivalence is met using the same copper weight in the main and auxiliary windings. If this condition is not met, phase coordinates should be used from the start.

Let us suppose that Equations 3.187 hold. In this case, the dq model may be used here directly:

$$\frac{d\Psi_d}{dt} = -i_d R_s + V_d + \omega_r \Psi_q$$

$$\frac{d\Psi_q}{dt} = -i_q R_s + V_q + \omega_r \Psi_d$$

$$\frac{d\Psi_{dr}}{dt} = -i_{dr} R_{rd}; \frac{d\Psi_{qr}}{dt} = -i_{qr} R_{rq};$$

$$T_e = p_1\left(\Psi_d i_q - \Psi_q i_d\right) \tag{3.188}$$

$$\Psi_d = L_{sl} i_d + L_{dm}\left(i_d + i_{dr}\right) + \Psi_{PM}$$

$$\Psi_q = L_{sl} i_q + L_{qm}\left(i_q + i_{qr}\right)$$

$$\Psi_{dr} = L_{drl} i_{dr} + L_{dm}\left(i_d + i_{dr}\right) + \Psi_{PM}$$

$$\Psi_{qr} = L_{qrl} i_{qr} + L_{qm}\left(i_q + i_{qr}\right)$$

with $L_d < L_q$

$$\frac{J}{p_1}\frac{d\omega_r}{dt} = T_e - T_{load}; \quad \frac{d\theta_{er}}{dt} = \omega_r \tag{3.189}$$

and θ_{er} is the rotor position in electrical degrees.

The equivalence, in terms of voltages and stator currents, between the real machine and the dq model is

$$
\begin{aligned}
V_{\mathrm{m}}(t) &= V\sqrt{2}\cos(\omega_1 t+\gamma)\\
V_{\mathrm{a}}'(t) &= \frac{V_{\mathrm{m}}(t)-V_{\mathrm{c}}(t)}{a}; \quad \frac{\mathrm{d}V_{\mathrm{c}}}{\mathrm{d}t}=\frac{i_a}{C_{\mathrm{a}}}\\
V_d &= V_{\mathrm{m}}(t)\cos\theta_{\mathrm{er}}+V_{\mathrm{a}}'(t)\sin\theta_{\mathrm{er}}\\
V_q &= -V_{\mathrm{m}}(t)\sin\theta_{\mathrm{er}}+V_{\mathrm{a}}'(t)\cos\theta_{\mathrm{er}}\\
I_{\mathrm{m}} &= I_d\cos\theta_{\mathrm{er}}-I_q\sin\theta_{\mathrm{er}}\\
I_{\mathrm{a}}' &= I_d\sin\theta_{\mathrm{er}}+I_q\cos\theta_{\mathrm{er}}\\
I_{\mathrm{a}} &= I_{\mathrm{a}}'/a
\end{aligned}
\tag{3.190}
$$

Magnetic saturation may be included in a simplified manner only in q axis by the $L_{qm}(i_{qm})$ function, $i_{qm} = i_q + i_{qr}$. It is all that is needed if the state variables are ψ_d, ψ_q, ψ_{dr}, ψ_{qr}, V_c, ω_r, and θ_{er}.

3.19 STANDSTILL TESTING FOR SM PARAMETERS/LAB 3.3

By SM parameters we mean

- d and q axes magnetization curve families: $\psi_{dm}(i_{dm}, i_{qm})$, and $\psi_{qm}(i_{dm}, i_{qm})$
- Transient and subtransient inductances and time constants: l_d, l_d', l_d'', and l_q'' (in pu), and τ_{d0}', τ_{d0}'', τ_{q0}'', τ_d', τ_d'', τ_q'', and τ_{dr} in seconds
- Stator leakage inductance and resistance: R_s and L_{sl}
- Field circuit leakage inductance and resistance: R_F and L_{Fl}
- Rotor to stator field winding reduction factor: $K_{iF} = i_f / i_F^*$
- Rotor inertia: H (in seconds)

A few remarks are in order:

- The above-mentioned complete set of parameters is typical for single cage (per axis) and field circuit rotor SMs. For the skin effect affected SMs (large power, or solid rotor), one more cage is added along the d axis and two more such circuits along the q axis, in order to correctly model the machine for transients in the 0.001–100 Hz frequency spectrum.
- PMSMs, or d.c.-excited SMs without a cage on the rotor, lack all sub (or sub-sub) transient inductances and time constants. So the parameter identification is significantly simpler.
- Though there are standard tests for parameter identification in the running machine, ($\omega_r \neq 0$), such as sudden short-circuit current waveforms processing, only standstill tests are introduced here, because they are recent, comprehensive, require less hardware and testing time, and make full use of the dq model.

3.19.1 SATURATED STEADY-STATE PARAMETERS, L_{dm} AND L_{qm}, FROM CURRENT DECAY TESTS AT STANDSTILL

In what follows, we adopt the concept of unique magnetization curves, $\psi_{dm}^*(i_m)$ and $\psi_{qm}^*(i_m)$, as presented in [1,9,10] and the total magnetization current, $i_m = \sqrt{i_{dm}^2 + i_{qm}^2}$

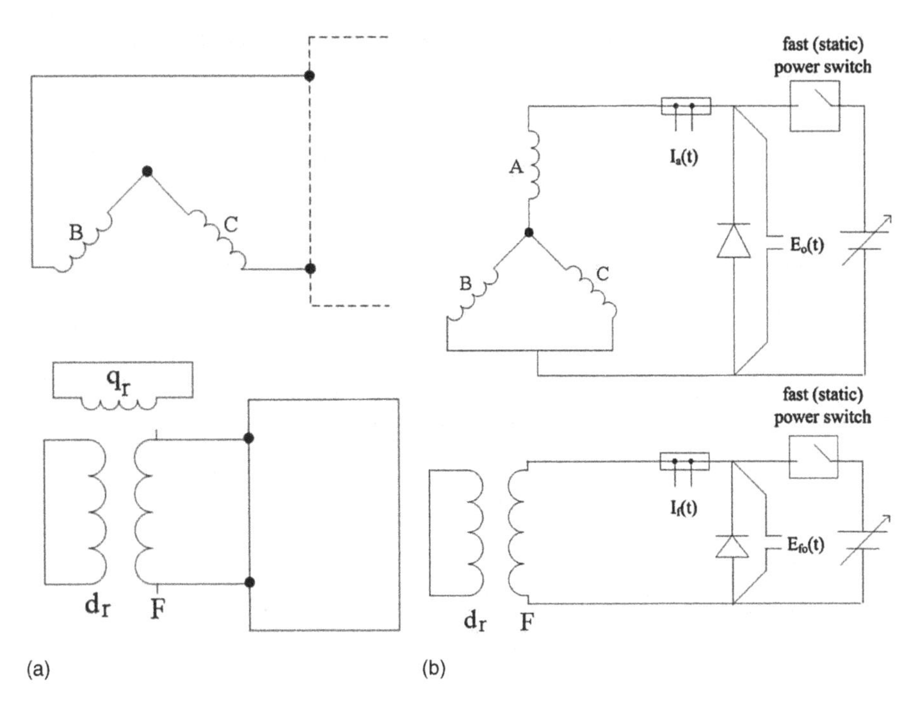

FIGURE 3.24 Current decay tests in the (a) *d* axis and (b) *q* axis.

(Chapter 1, par. 1.7). Current decay tests at standstill are done in the *d,q* axes and in the field circuit, from various initial values of currents: i_{d0}, i_{q0}, and i_{F0}. The test arrangements for current (flux) decay tests for the axes d,q are shown in Figure 3.24. The placing of rotor in the *d* axis ($L_d > L_q$) may be achieved easily when the d.c. source is connected to the arrangement in d axis (Figure 3.24a).

Note: In PM rotors with flux barriers, finding the location of the rotor in the *d* axis and the polarity of the PM needs special care (as $L_d < L_q$).

After a certain d.c. current is installed in the stator (or in the field winding), the stator is disconnected and the stator current decays through the free wheeling diode (as done for the d.c. brush machine; see Chapter 2). At zero speed ($\omega_r = 0$), the dq model of stator equations is

$$i_d R_s - V_d = -\frac{d\psi_d}{dt}; \quad i_q R_s - V_q = -\frac{d\psi_q}{dt} \tag{3.191}$$

but with the Park transformation (θ_{er}) for the *d* axis

$$i_d = \frac{2}{3}\left(i_A + i_B \cos\frac{2\pi}{3} + i_C \cos\left(-\frac{2\pi}{3}\right)\right) = i_A \tag{3.192}$$

$$i_q = \frac{2}{3}\left(0 + i_B \sin\frac{2\pi}{3} + i_C \sin\left(-\frac{2\pi}{3}\right)\right) = 0; \quad (i_B = i_C) \tag{3.193}$$

The flux in the q axis is zero (for Figure 3.24a). After shortcircuiting the stator over the freewheeling diode:

$$R_s\int_0^\infty i_d(t)\,dt+\frac{2}{3}\int_0^\infty V_{diode}(t)=(\psi_d)_{initial}-(\psi_d)_{final} \tag{3.194}$$

The final flux linkage is solely produced by the field current, i_{F0}, if any:

$$\psi_{d\,final}=L_{dm}i_{F0} \tag{3.195}$$

Current decay tests may be run on the field circuit from some initial value of i_{F0} with the stator open, but this time the stator voltage $V_{ABC}(t)=V_{diode}(t)=3/2V_d(t)$ is recorded:

$$\psi_{d\,initial}(i_{F0})=\int_0^\infty \frac{2}{3}V_{ABC}(t)\,dt;\quad i_A=0 \tag{3.196}$$

By running the d axis stator (for zero i_F) test and then the field circuit (stator open) current decay test, we can infer from Equations 3.194 and 3.196:

$$(\psi_{initial})_{i_F=0}=(L_{dm}+L_{sl})i_{d0}=\psi_{dm}(i_{d0})+L_{sl}i_{d0} \tag{3.197}$$

$\psi_{dm}(i_{F0})$ and $\psi_{dm}(i_{d0})$, with L_{sl} given, are thus calculated. For equal values of fluxes, we can calculate the field current reduction ratio:

$$K_F=\frac{i_{d0}}{i_{F0}} \tag{3.198}$$

We now return to Equation 3.194, with $(\psi_d)_{final}$ from the field current decay curve as

$$\psi_{d\,initial}(i_{dm0}=i_{d0}+i_{F0})=L_{sl}i_{d0}+\psi_{dm}(i_{dm0}) \tag{3.199}$$

But $i_{dm0}=i_{m0}$ in our case because $i_{q0}=0$, and thus the test yields the $\psi_{dm}(i_{dm})$ saturation curve.

For the q axis (Figure 3.24b)

$$i_d=0;\quad i_A=0;\quad i_B=-i_C; \tag{3.200}$$

Also

$$\begin{aligned} i_q&=\frac{2}{3}\left(i_B\sin\frac{2\pi}{3}+i_C\sin\left(-\frac{2\pi}{3}\right)\right)=\frac{2}{\sqrt{3}}i_B\\ V_q&=\frac{2}{3}(V_B-V_C)\sin\frac{2\pi}{3}=\frac{V_B-V_C}{\sqrt{3}} \end{aligned} \tag{3.201}$$

As done for the d axis, the current decay test yields:

$$\psi_q\left(i_{\mathrm{m}0}\right)=\psi_{q\text{ initial}}\left(i_{q0},i_{\mathrm{F}0}\right)=\frac{2}{\sqrt{3}}\int i_{\mathrm{B}}R_{\mathrm{s}}\mathrm{d}t+\frac{1}{\sqrt{3}}\int_0^{\infty}V_{\mathrm{diode}}\mathrm{d}t \tag{3.202}$$

The final flux in the q axis is zero even if $i_{\mathrm{F}0}\neq 0$ in the field circuit, to secure a desired level of magnetic saturation. Now the initial magnetization current $i_{\mathrm{m}0}$ is

$$i_{\mathrm{m}0}=\sqrt{i_{q0}^2+i_{\mathrm{F}0}^2} \tag{3.203}$$

The tests can be performed with zero or nonzero d.c. field current (to check the cross-coupling magnetic saturation importance). If the field current shows transients during the q axis decay test, this is a sign of a significant cross-coupling saturation effect.

Note: Similar tests may be run in any rotor position with Equations 3.172 through 3.179 used to calculate simultaneously the $\psi_{dm}(i_{\mathrm{m}})$ and $\psi_{qm}(i_{\mathrm{m}})$ curves but, in this case, with nonzero $i_{\mathrm{F}0}$, the magnetization current, $i_{\mathrm{m}0}$, is

$$i_{\mathrm{m}0}=\sqrt{\left(i_{d0}+i_{\mathrm{F}0}\right)^2+i_{q0}^2} \tag{3.204}$$

The machine magnetization inductances, $L_{dm}(i_{\mathrm{m}0})$ and $L_{qm}(i_{\mathrm{m}0})$, are

$$L_{dm}^*\left(i_{\mathrm{m}0}\right)=\frac{\psi_{dm}^*\left(i_{\mathrm{m}0}\right)}{i_{\mathrm{m}}};\quad L_{qm}^*\left(i_{\mathrm{m}0}\right)=\frac{\psi_{qm}^*\left(i_{\mathrm{m}0}\right)}{i_{\mathrm{m}0}} \tag{3.205}$$

These inductances pertain to the unique d and q axes magnetization curves [9]. Typically, such $\psi_{dm}^*\left(i_{\mathrm{m}}\right)$ curves for a 3 KW SM are shown in Figure 3.25 where $\psi_{dm}^*\left(i_{\mathrm{F}0}\right)$ was also calculated from a no-load running generator test, to verify the results obtained.

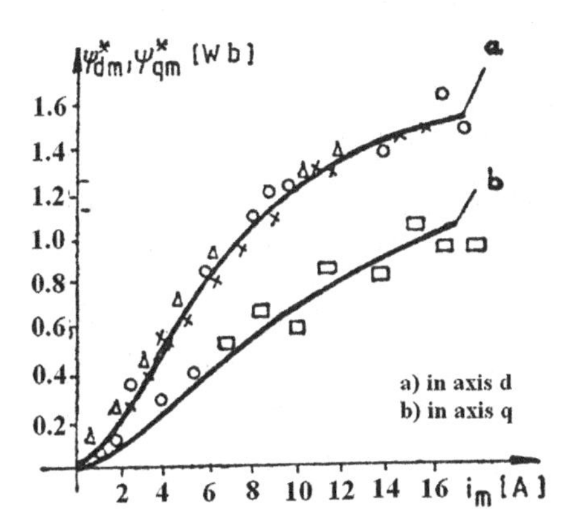

FIGURE 3.25 Current decay tests in (a) axis d and (b) axis q.

Note: The d.c. current decay tests transients may be used also to identify all operational parameters of $L_d(s)$, $L_q(s)$, and $G(s)$ using curve fitting methods [11,12].

3.19.2 Single Frequency Test for Subtransient Inductances, L_d'' and L_q''

The subtransient inductances, L_d'' and L_q'' (or reactances, in pu, x_d'' and x_q''), refer to fast transients. At standstill, if we supply the machine, at frequency ω_1, from a single-phase transformer as shown in Figure 3.24a and b (without the free-wheeling diode), and measure voltages E_{ABC}, I_A, P_d and E_{BC}, I_B, P_q we obtain

$$Z_d'' = \frac{2}{3}\frac{E_{ABC}}{I_A}; \quad R_d'' = \frac{2}{3}\frac{P_d}{I^2} \tag{3.206}$$

$$Z_q'' = \frac{1}{2}\frac{E_{BC}}{I_B}; \quad R_q'' = \frac{P_q}{2I_B^2} \tag{3.207}$$

$$\begin{aligned} X_d'' &= \sqrt{\left(Z_d''\right)^2 - \left(R_d''\right)^2}; \quad L_d'' = X_d''/\omega_1 \\ X_q'' &= \sqrt{\left(Z_q''\right)^2 - \left(R_q''\right)^2}; \quad L_q'' = X_d''/\omega_1 \end{aligned} \tag{3.208}$$

The negative sequence parameters, R_-, X_- can be adapted as

$$R_- \approx \left(R_d'' + R_q''\right)/2 \quad \text{or} \quad R_- = \sqrt{R_d'' R_q''} \tag{3.209}$$

$$X_- \approx \left(X_d'' + X_q''\right)/2 \quad \text{or} \quad X_- = \sqrt{X_d'' X_q''} \tag{3.210}$$

Note: In reality, for the negative sequence, the rotor experiences $2f_1$ frequency (and not f_1 frequency). So we may increase the standstill frequency to $2f_1$ and repeat the tests to check the relative significance influence of the frequency (skin) effects in the rotor.

3.19.3 Standstill Frequency Response Tests

The standstill frequency response tests (SSFRS) refer to arrangements in Figure 3.24 (the freewheeling diode is eliminated), where the machine is fed through almost sinusoidal voltages of variable frequency, from 0.001 (0.01) to 100 Hz for SGs at the power grid. The machine voltage, current, and their phase angle, for the *d* and *q* axes tests, and closed or open field windings, are done for each frequency; (8–10) values per decade are acquired. The tests are done at 10% of rated current, as the duration of tests is lengthy and machine overheating has to be avoided.

The current/voltage dq–ABC relationships remain the same as for current decay tests.

But the d–q relationships based on Equation 3.158, and $\omega_r = 0$ and $V_F(s) = 0$, are

$$\underline{Z}_d(j\omega) = \frac{\underline{V}_d(j\omega)}{\underline{I}_d(j\omega)} = R_s + j\omega L_d(j\omega); \quad \underline{V}_d(j\omega) = \frac{2}{3}\underline{V}_{ABC}$$
$$I_d(i\omega) = \underline{I}_A \tag{3.211}$$

$$\underline{Z}_q(j\omega) = \frac{\underline{V}_q(j\omega)}{\underline{I}_q(j\omega)} = R_s + j\omega L_q(j\omega); \quad \underline{V}_q(j\omega) = \frac{(\underline{V}_B - \underline{V}_C)}{\sqrt{3}}$$
$$I_q(j\omega) = \underline{I}_B\frac{2}{\sqrt{3}} \tag{3.212}$$

where $L_d(j\omega)$ and $L_q(j\omega)$ are of the form in Equation 3.58:

$$L_{d,q}(j\omega) = L_{d,q}(0)\frac{\left(1 + j\omega\tau_{d,q}'''\right)\left(1 + j\omega\tau_{d,q}''\right)\left(1 + j\omega\tau_{d,q}'\right)}{\left(1 + j\omega\tau_{d,q0}'''\right)\left(1 + j\omega\tau_{d,q0}''\right)\left(1 + j\omega\tau_{d,q0}'\right)} \tag{3.213}$$

In Equation 3.213, three-rotor circuits have been considered along each axis (sufficient for large power SGs with a solid rotor).

Also, from

$$-j\omega G(j\omega) = \frac{I_F(j\omega)}{\underline{I}_d(j\omega)} \tag{3.214}$$

with $G(j\omega)$ as in Equation 3.54 for short-circuited field winding and acquired $i_F(j\omega)$:

$$G(j\omega) = \frac{L_{dm}(0)}{R_F}\frac{(1 + p\tau_{dr1})(1 + p\tau_{dr2})}{\left(1 + j\omega\tau_{d0}'''\right)\left(1 + j\omega\tau_{d0}''\right)\left(1 + j\omega\tau_{d0}'\right)} \tag{3.215}$$

Typical experimental results are shown in Figure 3.26.

Once $L_d(j\omega)$, $L_q(j\omega)$, and $j\omega G(j\omega)$ are obtained experimentally, finding the time constants in their expressions Equations 3.213 through 3.215 become a problem for curve fitting [12], with some methods requiring the calculation of gradients and others avoiding them, such as pattern search (IEEE-standard 115-1995). Straightforward analytical expressions of SSFR parameters are given in [13]. For the phenomenology of multiple rotor circuits models, an intuitive method to identify the time constants based on finding the phase response extremes' frequencies is presented in [14]. For a complete description of SM testing, see the IEEE standard 115-1995, while for PMSM consult [15].

3.20 LINEAR SYNCHRONOUS MOTOR TRANSIENTS

As LSM [16–19] may be built with active guideway, and d.c. excitation, or superconducting d.c. excitation, or PM excitation on board of the mover (Figure 3.27a through c).

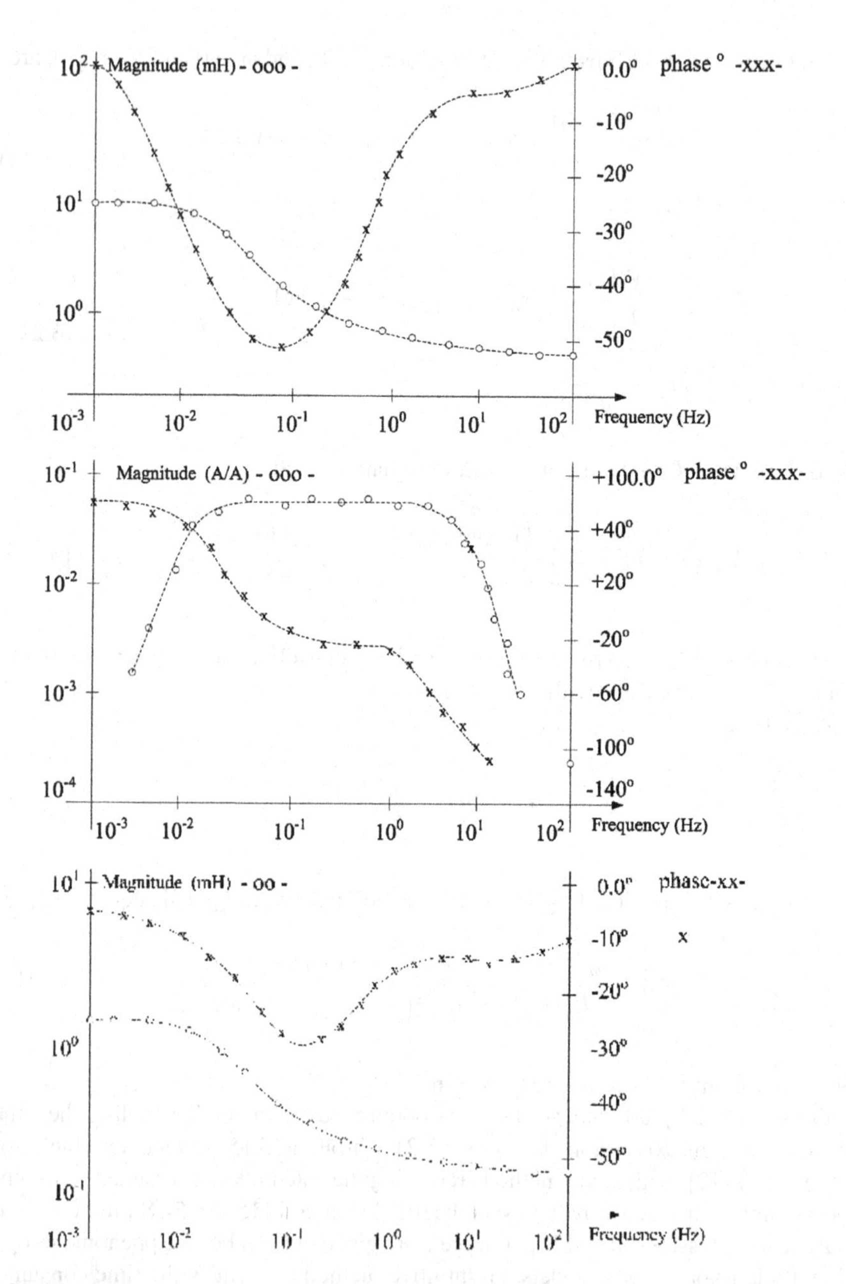

FIGURE 3.26 Typical (a) $L_d(j\omega)$, (b) $j\omega G(j\omega)$, and (c) $L_q(j\omega)$.

For limited travel, the PMs are placed along the track and the three-phase winding is placed on board of mover (Figure 3.28a). Finally, it is possible to have both PMs (or d.c. excitation) and the a.c. windings on the mover, with a salient homopolar pole core structure along the track (Figure 3.28b).

But whatever the configuration, if the number of poles exceeds $2p_1 \geq 6$, the end effects due to the openness of the magnetic circuit along the direction of motion are

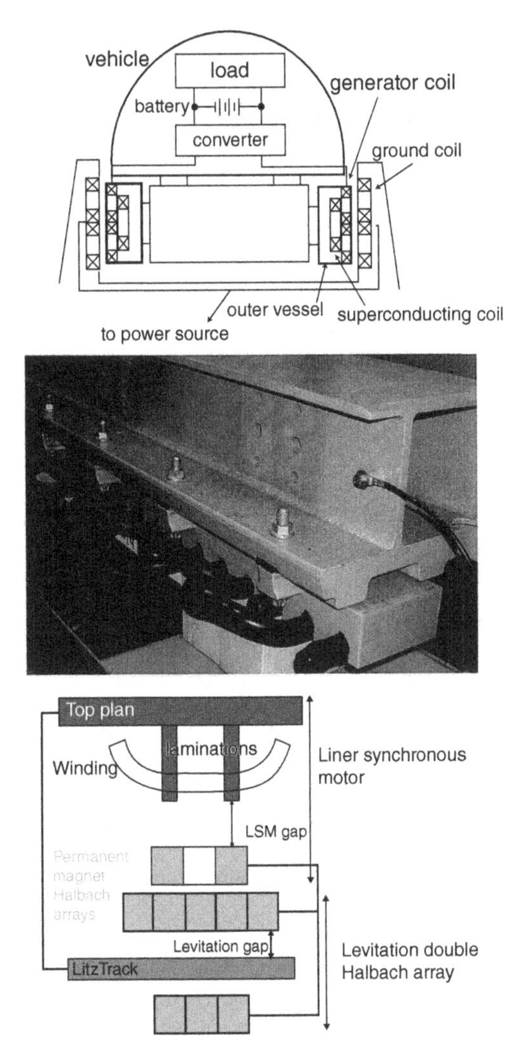

FIGURE 3.27 Active guideway LSMs with (a) d.c. excitation, (b) d.c. superconducting excitation, and (c) PM excitation.

mild; the dq model typical for rotary SMs applies here approximately but with three differences in variables:

- Synchronous (rotor) speed, ω_r rad/s, turns into linear synchronous (field) speed, $U_s = 2\tau f_1$(m/s); U_m is the mover actual speed in m/s.
- Electromagnetic torque, T_e, (Nm) is replaced by thrust $F_e(N)$.
- Rotor angle variable, θ_r, (rad) changes into the linear variable, $x(m)$; the angle θ_{er} in the inductance expression changes to $x\pi/\tau$.

The electromagnetic power, P_{elm}, changes from $P_{elm} = T_e\omega_r/p_1$ to $P_{elm} = F_e \cdot 2\tau f_1$. Consequently, the dq model of LSMs is

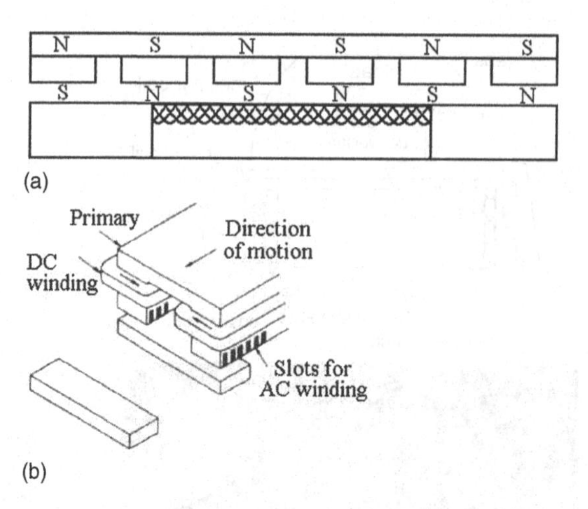

FIGURE 3.28 LSM with (a) PM guideway and (b) d.c. excitation and a.c. winding on the mover.

$$\begin{aligned}
&\bar{V}_s = R_s\bar{I}_s + jU_m\bar{\psi}_s\frac{\pi}{\tau} + \frac{d\bar{\psi}_s}{dt}; \quad V_F = R_F i_F + \frac{d\psi_F}{dt} \\
&\bar{\psi}_s = \psi_d + j\psi_q \\
&\psi_d = L_{dm}(g)i_F + L_d(g)i_d; \quad \psi_F = L_F(g)i_F + L_{dm}(g)i_d; \quad \psi_q = L_q(g)i_q
\end{aligned} \tag{3.216}$$

g-the magnetic (or mechanical) airgap

$$F_e = \frac{3}{2}\frac{\pi}{\tau}(\psi_d i_q - \psi_q i_d); \quad M_m dU_m/dt = F_e - F_{load}; \quad dx/dt = U_m$$

$$\begin{vmatrix} V_d \\ V_q \\ V_0 \end{vmatrix} = \frac{2}{3}\begin{vmatrix} \cos\left(-\frac{\pi}{\tau}x\right) & \cos\left(-\frac{\pi}{\tau}x + \frac{2\pi}{3}\right) & \cos\left(-\frac{\pi}{\tau}x - \frac{2\pi}{3}\right) \\ \sin\left(-\frac{\pi}{\tau}x\right) & \sin\left(-\frac{\pi}{\tau}x + \frac{2\pi}{3}\right) & \sin\left(-\frac{\pi}{\tau}x - \frac{2\pi}{3}\right) \\ \frac{1}{2} & \frac{1}{2} & \frac{1}{2} \end{vmatrix} \tag{3.217}$$

The sign $\left(-\frac{\pi}{\tau}x\right)$ in Equation 3.217 is valid for the active guideway LSM, but it turns to $\left(+\frac{\pi}{\tau}x\right)$ for the passive guideway LSM, where the armature (a.c.) winding is on board of the mover. This is so because the coordinate system of the dq model is attached to the d.c.-excited or passive (magnetically anisotropic) part of the LSM.

Also, the dq model is valid for distributed ($q \geq 2$) a.c. armature windings with uniform slots, and for fractionary ($q \leq 0.5$) tooth wound coil windings where the PM-produced emf varies sinusoidally with the mover position, x, as for rotary SMs.

Linear switched reluctance or stepper motors require, in general, phase coordinate models as do their rotary counterparts.

The LSM may be controlled, as rotary machines, for constant ψ_{d0} or constant ψ_s (unity power factor is possible only with d.c. excitation LSM). In addition, LSM may be controlled for integrated propulsion and suspension control (for MAGLEVs). In the latter case, the stator (or airgap flux) is controlled in order to dynamically keep the airgap, g, constant (within ±20%–25%).

The normal force, F_n, which produces suspension is

$$F_{\mathrm{na}} \approx \frac{3}{2}\left(i_d \frac{\partial \psi_{\mathrm{d}}}{\partial g} + i_q \frac{\partial \psi_q}{\partial g} \right) \tag{3.218}$$

The state feedback or variable structure control has been proved adequate for the active suspension control [20].

Note: For the d.c. excitation LSM, suspension control is performed through the field current, while i_d and i_q control is done through propulsion control, which is much faster than the former. For more on LSM control, see [16–20].

3.21 SUMMARY

- By SM transients, we mean the slow to fast variation of SM voltages, currents, flux magnitudes or frequency (speed) in time.
- The two reaction steady-state models of SM are not applicable to SM transients.
- The phase coordinate model of SM shows variable self- and mutual inductances with the rotor position, and thus is computer-time prohibitive in its solving of transients. It is, consequently, not practical for the fast computation of transients and for control design.
- The dq0 model of SM in rotor coordinates is quite suitable in modeling transients as its inductances are independent of the rotor position. It is also widely used for SM control in modern, variable speed, drives.
- The dq0 model applies, in general, only to distributed ($q \geq 2$) a.c. windings and to constant airgap configurations. To create magnetic saliency, flux barriers along the d axis are provided.
- The dq0 model may also be applied (with caution) to $q \leq 0.5$ (slots/pole/phase) PMSMs (with tooth-wound coils) where the emf is sinusoidal in time and the rotor is winding-less.
- For specially shaped rotor/stator poles, even for a switched reluctance (or stepper) machine ($q \leq 0.5$) with double magnetic saliency, but with inductances having sinusoidal variation of the rotor position (and no rotor windings), the dq0 model can be applied with caution.
- The dq0 model of SM with a single-cage rotor and d.c. excitation represents an eighth-order system.
- The pu dq0 model equations use $\frac{1}{\omega_{10}} \mathrm{d}/\mathrm{d}t$ instead of d/dt and have all parameters (resistances and inductances) in p.u., but the inertia, H, and various time constants are in seconds.

- As expected, in rotor coordinates, the dq0 model variables for the balanced steady state of the SM are all d.c..
- For the symmetric stator, the space phasor form of equations is not only feasible, but also useful.
- For an ideal no-load generator in the dq model, not only $i_{d0} = i_{q0} = 0$ but also $V_{d0} = 0$ and $V_{q0} = V_0\sqrt{2}$ (V_0 is the RMS phase voltage).
- For a steady-state three-phase short circuit, the machine is not saturated and the current is in general smaller than $3I_{rated}$, even for a cageless rotor PMSM. Due to stator resistances, there is a sizable short-circuit braking torque, which may be a design constraint for some fault-tolerant applications.
- Transients at constant speed are purely electromagnetic; for constant SM parameters (inductances), their operational parameters are defined as $L_d(s)$ and $L_q(s)$, and $sG(s)$, if the SM changes its inductances along the d and q axes during transients. The initial values, L_d'' and L_q'', are called subtransient inductances, and L_d', the transient inductance, corresponds to later transients when there are no longer any rotor cage effects. Finally, at steady state, the SM shows L_d and L_q. The operational parameters also contain a few time constants, depending on the number of the d, q rotor axes shorted circuits used to simulate frequency (skin) effects.
- The three-phase sudden short-circuit current waveform may be processed so that the d axis parameters, L_d'', L_d', T_d'', T_d', T_{d0}'', and T_{d0}', can be determined by curve fitting methods.
- Asynchronous running can also be treated in Laplace formulation with $s = jS\omega_1$, (S is the slip), to calculate asynchronous torque versus speed; this information is useful in estimating the asynchronous starting capabilities of SMs at power grid.
- Simplified dq0 models include neglecting stator and/or also rotor transients, when the SM model with d.c. excitation retains a third order.
- For electromechanical transients, the speed also varies and the dq model equations are linearized. They have ΔV, $\Delta\omega_1$, ΔV_F, and ΔT_L as inputs and Δi_d, Δi_q, Δi_{dr}, Δi_{qr}, Δi_F, $\Delta\omega_r$, and $\Delta\boldsymbol{\delta}_v$ as variables.
- Speed response to load torque pulsations can easily be calculated by the small deviation theory.
- Large deviation transients make use of the full dq model in rotor coordinates. Asynchronous starting and self-synchronization and line-to-line and line-to-neutral faults constitute large transients. The results are useful for refined machine and protection design.
- Constant d axis flux, ψ_{d0}^*, or stator flux, ψ_s^*, control are used in variable speed (and frequency) control.
- The dq model reveals for constant ψ_{d0}^* frequency (speed) control—applicable especially to PMSMs—a decreasing speed with torque characteristic. The ideal no-load speed is $\omega_{r0} = V_s/\psi_{d0}$; $\psi_{d0} = \psi_{PM} + L_d i_{d0}$. For $i_{d0} = -\psi_{PM}/L_d$, $\omega_{r0} \rightarrow \infty$; it is ideal for constant power wide-speed range, needed for many applications.
- For constant stator flux, ψ_s^*, (and unity power factor), the speed/torque curve is a descending straight line as in separately excited d.c. motors. The field

current, i_F, for cos $\varphi_1 = 1$, increases also with torque, as expected. The case is typical for the voltage-source PWM inverter-fed large SM drives. No damper cage is provided on the rotor.

- Two basic vector control drives schemes for constant ψ_d^0 or Ψ_S are introduced.
- For rectangular current, two-phase conducting, controlled PMSMs (brushless d.c. PM motors) and cage rotor d.c.-excited rotor SM (supplied from current source inverters for variable speed), the basic transient equations and speed/torque curves are derived. Again, linear descending speed versus torque curves are obtained. These two situations correspond to the two typical variable speed SM drives described in some detail.
- Switched reluctance motors have double saliency, and thus the phase coordinate model is used. It is not a traveling field machine but the frequency of current pulses, f_1, into a stator phase is directly related to the number of rotor salient poles, N_r, and speed, n: $f_1 = nN_r$.
- A small deviation model of SRM reveals a transient behavior very similar to the d.c. brush series motor. Due to its ruggedness, the SRM is suitable in thermally or chemically aggressive environments. Stepper motors without or with PMs (hybrid) operate as an SRM but are supplied open loop (by frequency ramping), so as not to loose steps and are applied in harsh environments.
- Split-phase cage rotor SMs with magnetic rotor saliency ($L_d > L_q$) or with PMs ($L_d \leq L_q$) may be used for home appliances, to increase efficiency and reduce motor size for direct power grid operation. The dq model in rotor coordinates holds only if the two orthogonal stator windings (main and auxiliary), are equivalent (made essentially of same copper weight). For auxiliary-start-only-windings, this is not the case, in general, and thus the phase coordinates are required.
- Testing for SM parameters is essential for control design. Standstill flux *dq* current/decay tests and frequency response (SSFRs) are described in detail to exemplify the practicality of dq model; for complete SM testing, see the IEEE standard 115-1995.
- Linear progressive motion synchronous motors are very similar to rotary ones in terms of topology and modeling for transients (see [16,17]), with some end effects to be considered at high speed or, for $2p_1 = 2,4$ poles, motor applications.
- Linear oscillatory motion single-phase PMSMs are being used for small refrigerator compressor drives. They work at resonance (electrical frequency, f_e, is equal to the mechanical spring proper frequency, f_m (f_e=f_m)); see [16–20].

3.22 PROPOSED PROBLEMS

3.1 Draw the space phasor diagrams for SM motoring and generating at unity power factor.
Hint: Check Section 3.7, Figure 3.5, and Equations 3.45 and 3.46.

3.2 For the large SM in Example 3.1, calculate the inductances l_d'', l_d', l_q'' and τ_{d0}'', τ_{d0}', τ_d'', τ_d', τ_q'', and τ_{q0}''.

Hint: Check Equation 3.55.

3.3 For the MS in Example 3.1, calculate peak value of $i_d(t)$, $i_F(t)$, $i_A(t)$ during sudden three-phase short circuit from no load. Consider that $V_{q0} = 1.2(V_{nl}/\sqrt{3})\sqrt{2}$ and $i_{F0} = 2I_n$.
Hint: See Equations 3.68 through 3.70.

3.4 The machine in Example 3.5 with the parameters, $\psi_{PMd} = 1.0$ Wb, $L_d = 0.05$ H, $L_{dm} = 0.9\,L_d$, $L_q = 0.1$ H, $R_s = 1\ \Omega$, $p_1 = 2$, operates at the unity power factor at $V_{s0} = V_0\sqrt{2} = 18\sqrt{2}V$. Calculate the corresponding speed, ω_{rb}, rated stator current, torque, and electromagnetic power. For $3\omega_{rb}$, calculate the torque and electromagnetic power.
Hint: Check Example 3.5.

3.5 Find the eigenvalues of SM for the small deviation dq model. The initial situation corresponds to the following data: $L_d = L_q = 0.1\ H$, $L_{sl} = 0.1L_d = L_{Fl} = L_{drl} = L_{qrl}$, $\delta_{v0} = 30°$; $\omega_{r0} = 314$ rad/s, $R_s = 1\ \Omega = R_F$; $R_{dr} = R_{qr} = 3R_s$, $V_{s0} = V_0\sqrt{2} = 220\sqrt{2}$ V, $i_{F0} = 2i_{d0}$; $J = 1$ kg m^2, $p_1 = 2$.
Hint: Calculate first i_{d0}, i_{F0}, i_{q0}, ψ_{d0}, and ψ_{q0}.

3.6 A PMSM with $\psi_{PMd} = 2$ Wb, $R_s = 2\ \Omega$, $L_d = L_q = 0.2$ H, $p_1 = 4$, $B = 5 \times 10^{-3}$ Nms, and $\tau_m = 0.4$s is supplied from $V_s = V_0\sqrt{2} = 120\sqrt{2}$ V at $n = 1500$ rpm.
Calculate
a) The current, i_q, torque, efficiency, and power factor, for $i_{d0} = 0$.
b) For constant speed and $i_{d0} = 0$, at 10% increase in the load torque, calculate the speed and current, i_q, transients.

3.7 A cageless rotor large SM with d.c. excitation is controlled by a PWM voltage source inverter.
The motor data are as follows $P_n = 5$ MW, $V_{s0} = 4200\sqrt{2/3}$ V, $n_b = 10$ rpm, $f_b = 5$ Hz, $\eta = 0.985$, $\cos\varphi_1 = 1$, (only copper losses are considered), $x_d = x_q = 0.65$ pu, $X_{dm} = 0.55$ pu.
Calculate
a) The voltage power angle $\boldsymbol{\delta}_v$, i_{d0}, i_{q0}, and i_{F0} if $E_o = 1.2V_{s0}$ at n_b and P_n
b) Same deliverables as in (a) but for $n_{max} = 1.6\,n_b$
Hint: Follow closely Example 3.6.

3.8 Draw the speed/torque curve for the rectangular current control of brushless d.c. motor with $V_{dc} = 42\ V$, $R_s = 1\ \Omega$, and $\psi_{PM} = 0.1$ Wb (the emf is considered flat and 180° wide for both polarities)
Hint: Use Equations 3.169 through 3.173.

3.9 A cage d.c.-excited rotor, large SM has the following data: $V_{s1} = 4200\sqrt{2/3}$ V, $I_{s1} = 1000$ A, $f_b = 60$ Hz, $n_b = 1800$ rpm has the data $x_d = 1.3$, $x_q = 0.6$ (pu). $L''_d = 0.3\,L_d$, $L''_d = 0.3\,L_q$, $r_s = 0.01$(pu), $\varphi_1 = -10$ (leading power factor) operates with rectangular current control from a current-source inverter at I_{s1}.
Calculate
a) Torque, T_{em}, voltage power angle $\boldsymbol{\delta}_{v1}$, I_{d1}, I_{q1}, and I_F
b) No-load ideal speed
c) For same voltage at $n_{max} = 2n_b$, $\varphi_1 = -10°$, I_{s1}, calculate again $\boldsymbol{\delta}_{v1}$, I_{d1}, I_{q1}, I_F
Hint: Check Section 3.10 and Equations 3.174 and 3.175.

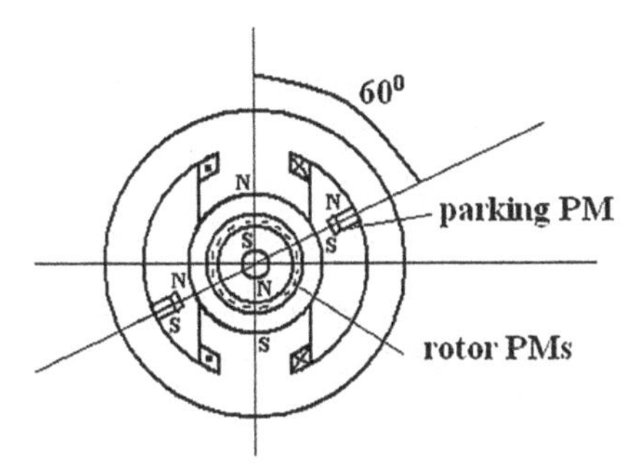

FIGURE 3.29 Single-phase PMSM.

3.10 A cageless surface PM two-pole rotor single-phase SM (Figure 3.29) has a parking PM which provides a $\theta_{r\ initial} = 60°$ away from the stator pole alignment of rotor poles, for self-starting. The emf varies sinusoidally with rotor position, θ_r, and the main PM cogging torque (at zero) varies sinusoidally with $2\theta_r$. Write the phase coordinate model of the machine. Write a MATLAB®-Simulink® code to simulate the machine transients. Include pertinent data for a small (100 W, 3600 rpm) motor and apply the code for the voltages $V(t) = V\sqrt{2}\cos(\omega_r t + \gamma)$, and $V = V_0 + K_r\omega_r$, where ω_r is the current speed. Vary γ and discuss the results.
Hint: Use phase coordinates, with constant stator inductances and sinusoidal emf.

3.11 A small excursion single-sided linear PMSM carrier with active guideway is controlled at zero i_d and develops, thrust $F_{xn} = 1$ kN, for a pole pitch $\tau = 0.06$ m at $U_m = 1.56$ m/s, $2p_1 = 8$, thrust density is 2.5 N/cm^2; and the airgap $g = 1$ mm; surface PMs, $h_{PM} = 3.0$ mm, thick.
The PM span is equal to the pole pitch and the PM airgap flux density $B_{gPM} = B_r \times h_{PM}/(h_{PM} + g) = 0.8$ T, $L_{sl} = 0.35L_{dm}$, ($L_{dm} = L_{qm}$), number of phase turns $W_sK_{Ws} = 750$ turns/phase.
Calculate (for steady state)

a. PM flux linkage, $\psi_{PMd} = W_sK_{W_s}\frac{2}{\pi}B_{gPM}\tau l_i$, where l_i is the stack length to be calculated from the thrust density
b. the magnetization inductance

$$L_m = \frac{6\mu_0\left(W_sK_{W_s}\right)^2}{\pi^2 p\left(g + h_{PM}\right)}\tau l_i \tag{3.219}$$

c. With $\psi_{PMd} = L_m i_{F0}$, calculate i_{F0} corresponding to the PMs' excitation
d. Stator current, i_q, for rated thrust F_{xn}, and $i_d = 0$
e. Values of $\psi_d = \psi_{PM}$, ψ_q, f_1, ω_1 (stator frequency), and V_s (stator voltage) for zero resistance losses
f. Normal attraction force, F_{na}, both from Equation 3.218 and $F_{na} \approx \frac{B_{gPM}^2}{2\mu_0}2p_1\tau l_i$, and the ratio F_{na}/F_{xn}.

Hints: Check Section 3.20 (selective results $W_s K_{Ws} = 750$, $f_1 = 13$ Hz, $V_s \approx 130\sqrt{2}$ V (peak phase voltage)), $F_n \approx 10$ KN (10/1 normal (suspension) to propulsion force).

REFERENCES

1. I. Boldea and S.A. Nasar, Unified treatment of core losses and magnetic saturation in the orthogonal axis model of electric machines, *IEE Proc. B*, 134(6), 1987, 355–365.
2. P.C. Krause, F. Mazari, T.L. Skvarenina, and D.W. Olive, The theory of neglecting stator transients, *IEEE Trans.*, PAS-93, 1976, 729–737.
3. J. Machowski, J.W. Bialek, and J.R. Bumby, *Power System Dynamics and Stability*, John Wiley & Sons, New York, 1997.
4. I. Boldea and V. Coroban, BEGA—Vector control for wide constant power speed range at unity power factor, *Record of OPTIM-2006*, Brasov, Romania, 2006.
5. I. Boldea and S.A. Nasar, *Electric Drives*, 2nd Edition, Chapters 11 and 13, CRC Press, Taylor & Francis Group, New York, 2005.
6. R. Krishnan, *Electric Motor Drives*, Prentice Hall, Upper Saddle River, NJ, 2001.
7. T.J. Miller, *Switched Reluctance Motors and Their Control*, Clarendon Press, Oxford, 1993.
8. R. Krishnan, *Switched Reluctance Motor Drives*, CRC Press, Boca Raton, FL, 2001.
9. I. Boldea, *Electric Generators Handbook*, Vol. 1, Synchronous Generators, Chapter 5, CRC Press, Taylor & Francis Group, New York, 2005.
10. M. Namba, J. Hosoda, S. Dri, and M. Udo, Development for measurement of operating parameters of SG and control system, *IEEE Trans.*, PAS-200(2), 1981, 618–628.
11. A. Keyhani, S.I. Moon, A. Tumageanian, and T. Leskan, *Maximum likelihood estimation of synchronous machine parameters from flux decay data, Proceedings of ICEM-1992*, Manchester, U.K., Vol. 1, pp. 34–38, 1992.
12. P.L. Dandeno and H.K. Karmaker, Experience with standstill frequency response (SSFR) testing of salient pole synchronous machines, *IEEE Trans.*, EC-14(4), 1999, 1209–1217.
13. S.D. Umans, I.A. Malick, and G.L. Wlilson, Modeling of solid iron rotor turbogenerators, Part 1&2, *IEEE Trans.*, PAS-97(1), 1978, 269–298.
14. A. Watson, A systematic method to the determination of SM parameters from results of frequency response tests, *IEEE Trans.*, EC-15(4), 2000, 218–223.
15. D. Iles-Klumpner, I. Boldea et al., Experimental characterization of IPMSM with tooth-wound coils, *Record of EPE-PEMC 2006*, Porto Rose, Slovenia, 2006.
16. I. Boldea and S.A. Nasar, *Linear Motion Electromagnetic Systems*, John Wiley & Sons, New York, 1985.
17. I. Boldea and S.A. Nasar, *Linear Electric Actuators and Generators*, Cambridge University Press, Cambridge, UK, 1997.
18. J. Gieras, *Linear Synchronous Drives*, CRC Press, Boca Raton, FL, 1994.
19. I. Boldea and S.A. Nasar, *Linear Motion Electromagnetic Devices*, Taylor & Francis Group, New York, 2001.
20. I. Boldea, *Linear Electric Machines, Drives, Drives and MAGLEVs Handbook*, CRC Press, Boca Raton, FL, 2013.

4 Induction Machines Transients and Control Principles

An induction machine (IM) is built with 1-, 2-, 3- (5 or 6-) phase a.c. primary stator windings and a cage or wound rotor. In the latter case, the rotor winding is of a three-phase a.c. type, and is connected through copper sliprings and brushes to an external source (PWM converter) that provides variable voltage and frequency. The so-called wound rotor or doubly fed induction machine is thus obtained. Also linear induction motors (LIMS) with aluminum-slab-on-iron or ladder secondary, or with two (3)-phase a.c. secondary winding on board of mover have also been built for niche applications.

All the above-mentioned classes of IMs that are discussed in this chapter should be modeled for transients.

The three-phase cage-secondary (rotor) IMs require a separate discussion. We start with the phase variable model of an IM with three-phase windings both on the stator and on the rotor (the rotor cage, when healthy, is equivalent to a symmetric three-phase winding).

4.1 THREE-PHASE VARIABLE MODEL

Let us consider a symmetric three-phase winding stator and rotor IM (Figure 4.1).

The machine equations (Figure 4.1) for the phase variables (coordinates) may be directly written in a matrix form:

$$\left[I_{\text{ABCabc}}\right]\left[R_{\text{ABCabc}}\right]-\left[V_{\text{ABCabc}}\right]=-\frac{\mathrm{d}}{\mathrm{d}t}\left[\Psi_{\text{ABCabc}}^{(\theta_{\text{er}},t)}\right] \tag{4.1}$$

with

$$\left[\Psi_{\text{ABCabc}}^{(\theta_{\text{er}},t)}\right]=\left[L_{\text{ABCabc}}^{(\theta_{\text{er}})}\right]\left[I_{\text{ABCabc}}\right] \tag{4.2}$$

The resistance matrix is diagonal:

$$\left[R_{\text{ABCabc}}\right]=\text{Diag}\left[R_{\text{s}}\quad R_{\text{s}}\quad R_{\text{s}}\quad R_{\text{r}}\quad R_{\text{r}}\quad R_{\text{r}}\right] \tag{4.3}$$

but the inductance matrix is 6 × 6 and the stator/rotor mutual inductances vary with the electrical rotor position. The stator-only and rotor-only inductances do not vary

DOI: 10.1201/9781003216018-4

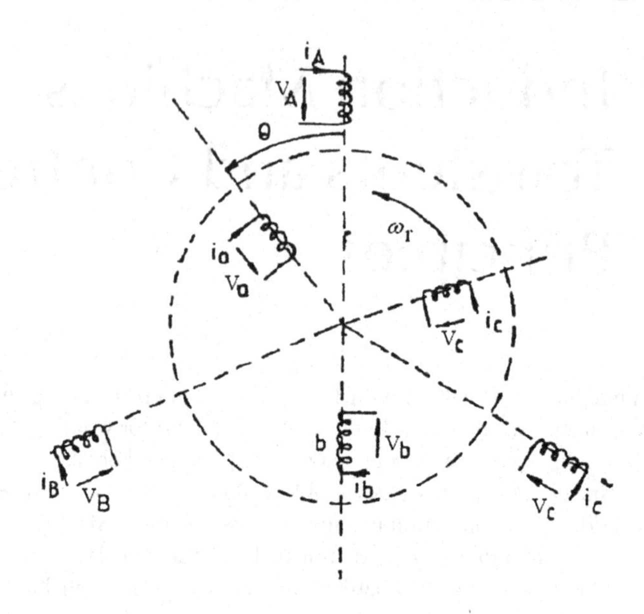

FIGURE 4.1 Three-phase IM.

with the rotor position (slot openings are neglected). For distributed a.c. windings ($q \geq 2$), the variation of the mutual inductances with the electrical rotor position is rather sinusoidal. So, the inductance matrix, $L_{\text{ABCabc}}^{(\theta_{er})}$, is straightforward:

$$L_{\text{ABCabc}}^{(\theta_{er})} = \begin{array}{c} \text{A} \\ \text{B} \\ \text{C} \\ \text{a} \\ \text{b} \\ \text{c} \end{array} \begin{bmatrix} L_{sl}+L_{os} & -L_{os}/2 & -L_{os}/2 & L_{sr}\cos\theta_{er} & L_{sr}\cos(\theta_{er}+2\pi/3) & L_{sr}\cos(\theta_{er}-2\pi/3) \\ -L_{os}/2 & L_{sl}+L_{os} & -L_{os}/2 & L_{sr}\cos(\theta_{er}-2\pi/3) & L_{sr}\cos\theta_{er} & L_{sr}\cos(\theta_{er}+2\pi/3) \\ -L_{os}/2 & -L_{os}/2 & L_{sl}+L_{os} & L_{sr}\cos(\theta_{er}+2\pi/3) & L_{sr}\cos(\theta_{er}-2\pi/3) & L_{sr}\cos\theta_{er} \\ L_{sr}\cos\theta_{er} & L_{sr}\cos(\theta_{er}-2\pi/3) & L_{sr}\cos(\theta_{er}+2\pi/3) & L_{rl}+L_{or} & -L_{or}/2 & -L_{or}/2 \\ L_{sr}\cos(\theta_{er}+2\pi/3) & L_{sr}\cos\theta_{er} & L_{sr}\cos(\theta_{er}-2\pi/3) & -L_{or}/2 & L_{rl}+L_{or} & -L_{or}/2 \\ L_{sr}\cos(\theta_{er}-2\pi/3) & L_{sr}\cos(\theta_{er}+2\pi/3) & L_{sr}\cos\theta_{er} & -L_{or}/2 & -L_{or}/2 & L_{rl}+L_{or} \end{bmatrix} \quad (4.4)$$

with

$$\frac{L_{or}}{L_{sr}} = \frac{L_{sr}}{L_{os}}; \quad L_{os}\cos 2\pi/3 = -L_{os}/2; \quad L_{or}\cos 2\pi/3 = -L_{or}/2 \quad (4.5)$$

From Equations 4.1 and 4.2, and after multiplication with $[I_{ABCabc}]^{T,}$ we obtain

$$\left[V_{ABCabc}\right]\left[I_{ABCabc}\right]^{T} = \underbrace{\left[I_{ABCabc}\right]\left[I_{ABCabc}\right]^{T}\left[R_{ABCabc}\right]}_{\text{winding losses}} + \underbrace{\frac{d}{dt}\left\{\frac{\left[I_{ABCabc}\right]\left[L_{ABCabc}^{(\theta_{er})}\right]\left[I_{ABCabc}\right]^{T}}{2}\right\}}_{\frac{\partial W_{mag}}{\partial t}} + \underbrace{\frac{1}{2}\left[I_{ABCabc}\right]\left[\frac{\partial L_{ABCabc}^{(\theta_{er})}}{\partial \theta_{er}}\right]\left[I_{ABCabc}\right]^{T}\frac{d\theta_{er}}{dt}}_{P_{elm}=\frac{T_e}{p_1}\frac{d\theta_{er}}{dt}} \quad (4.6)$$

The instantaneous torque, T_e, "springs" from the electromagnetic power, P_{elm}:

$$T_e = \frac{P_{elm}}{p_1\frac{d\theta_{er}}{dt}} = \frac{p_1}{2}\left[I_{ABCabc}\right]\left[\frac{\partial L_{ABCabc}^{(\theta_{er})}}{\partial \theta_{er}}\right]\left[I_{ABCabc}\right] \quad (4.7)$$

The motion equations complete the phase variable model:

$$\frac{J}{p_1}\frac{d\omega_r}{dt} = T_e - T_{load}; \quad \frac{d\theta_{er}}{dt} = \omega_r; \quad \omega_r = 2\pi p_1 n \quad (4.8)$$

An eighth-order nonlinear system with variable coefficients (through $[L^{(\theta_{er})}{}_{ABCabc}]$) has been obtained. It is evident that such a system is difficult to handle as it requires a large CPU time in digital simulations. The dq model, in its space phasor form, as derived in Chapter 1, is the key to a practical approach to IM transients.

4.2 DQ (SPACE PHASOR) MODEL OF IMS

The dq model of an IM with single rotor circuits per the d, q axes—from Chapter 1 (Figure 4.2)—is given here in short:

$$\begin{aligned} i_d R_s - V_d &= -\frac{\partial \Psi_d}{\partial t} + \omega_b \Psi_q; \quad i_q R_s - V_q = -\frac{\partial \Psi_q}{\partial t} - \omega_b \Psi_d \\ i_{dr} R_r - V_{dr} &= -\frac{\partial \Psi_{dr}}{dt} + (\omega_b - \omega_r)\Psi_{qr}; \quad i_{qr} R_r - V_{qr} = -\frac{\partial \Psi_{qr}}{dt} - (\omega_b - \omega_r)\Psi_{dr} \end{aligned} \quad (4.9)$$

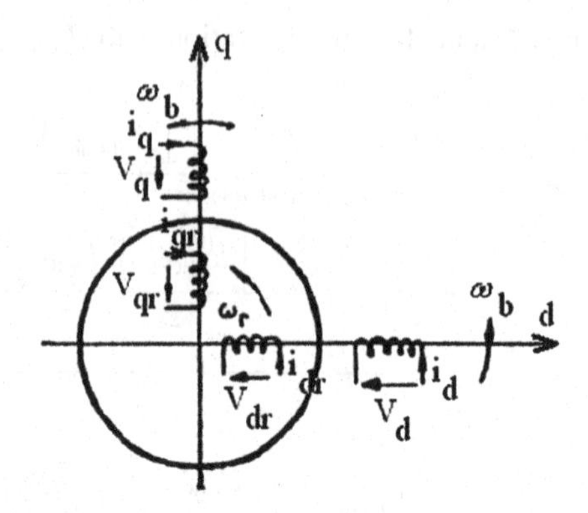

FIGURE 4.2 dq model of IM.

$$T_e = \frac{3}{2} p_1 \left(\Psi_d i_q - \Psi_q i_d \right); \quad \frac{J}{p_1} \frac{d\omega_r}{dt} = T_e - T_{load}; \quad \frac{d\theta_b}{dt} = \omega_b \tag{4.10}$$

$$P_s = \frac{3}{2} \left(V_d i_d + V_q i_q \right); \quad Q_s = \frac{3}{2} \left(V_d i_q - V_q i_d \right) \tag{4.11}$$

P_s, Q_s are the active and reactive input powers.

As described by Equations 4.10 and 4.11, the torque and power equivalence of the dq model with the three-phase IM has been included. In a space phasor notation

$$\begin{aligned} \overline{V}_s &= V_d + jV_q; \quad \overline{V}_r = V_{dr} + jV_{qr} \\ \overline{\psi}_s &= \psi_d + j\psi_q; \quad \overline{\psi}_r = \psi_{dr} + j\psi_{qr} \end{aligned} \tag{4.12}$$

$$\begin{aligned} \overline{i}_r &= i_{dr} + ji_{qr} \\ \overline{i}_s R_s - \overline{V}_s &= -\frac{\partial \overline{\psi}_s}{\partial t} - j\omega_b \overline{\psi}_s \\ \overline{i}_r' R_r - \overline{V}_r' &= -\frac{\partial \overline{\psi}_r'}{\partial t} - j\left(\omega_b - \omega_r\right) \overline{\psi}_r' \end{aligned} \tag{4.13}$$

with

$$\begin{aligned} T_e &= \frac{3}{2} p_1 \text{Real}\left[j\overline{\psi}_s \ \overline{i}_s^* \right] \\ \overline{\psi}_s &= L_{sl} \overline{i}_s + L_m \left(\overline{i}_s + \overline{i}_r' \right) \\ \overline{\psi}_r' &= L_{rl}' \overline{i}_r' + L_m \left(\overline{i}_s + \overline{i}_r' \right) \end{aligned} \tag{4.14}$$

It is evident that in the space phasor (dq) model, the rotor has been reduced to the stator (because we directly add $\overline{i}_s$ and $\overline{i}_r'$).

4.3 THREE-PHASE IM–DQ MODEL RELATIONSHIPS

The space phasor transformation of currents (Chapter 1) is

$$\begin{aligned}\overline{i_s} &= \frac{2}{3}\left[i_A(t)+i_B(t)e^{j\frac{2\pi}{3}}+i_C(t)e^{-j\frac{2\pi}{3}}\right]e^{-j\theta_b}\\ \overline{i_r} &= \frac{2}{3}\left[i_a(t)+i_b(t)e^{j\frac{2\pi}{3}}+i_c(t)e^{-j\frac{2\pi}{3}}\right]e^{-j(\theta_b-\theta_{er})};\quad \frac{d\theta_{er}}{dt}=\omega_r\end{aligned} \tag{4.15}$$

$$i_{0s}=\frac{1}{3}(i_A+i_B+i_C);\quad i_{0r}=\frac{1}{3}(i_a+i_b+i_c) \tag{4.16}$$

The same transforms, Equations 4.15 and 4.16, hold for flux linkages, $\overline{\psi_s}$ and $\overline{\psi_r}'$. If we apply them, and then use the inductance matrix $[L^{(\theta er)}{}_{ABCabc}]$, Equation 4.4, we finally obtain

$$\begin{aligned}\overline{\psi_s} &= L_{sl}\overline{i_s}+\frac{3}{2}L_{0s}\left(\overline{i_s}+\overline{i_r'}\right);\quad \overline{i_r'}=\overline{i_r}\cdot L_{sr}/L_{0s}\\ \overline{\psi_r'} &= L_{rl}'\overline{i_r'}+\frac{3}{2}L_{0s}\left(\overline{i_s}+\overline{i_r'}\right);\quad \overline{\psi_r'}=\overline{\psi_r}\cdot L_{sr}/L_{0r}=\overline{\psi_r}\cdot L_{0s}/L_{sr}\\ \overline{i_r'} &= \overline{i_r}\cdot L_{0r}/L_{sr};\quad L_{r1}'=L_{rl}\left(L_{sr}/L_{0r}\right)^2;\quad R_r'=R_r\left(L_{sr}/L_{0r}\right)^2\\ \psi_{0s} &= L_{sl}i_{0s};\quad \psi_{0r}=L_{rl}i_{0r};\quad V_{0s}=R_{0s}i_{0s}+L_{sl}\frac{di_{0s}}{dt};\quad V_{0r}'=R_r'i_{0r}'+L_{rl}'\frac{di_{0r}'}{dt}\end{aligned} \tag{4.17}$$

Notice again that $L_{0r}/L_{sr} = L_{sr}/L_{0s}$.

In Equations 4.17 the rotor has been already reduced to the stator. Comparing Equation 4.14 with Equation 4.17 we see that

$$L_m = \frac{3}{2}L_{0s} \tag{4.18}$$

One can notice that L_m is the cyclic magnetization inductance of IM, as defined in Chapter 5, Vol. 1 (companion book). As the leakage inductances, L_{sl}, L_{rl}' remain the same through the three-phase IM–dq model transition, so do the phase resistances, R_s, R_r'. The winding losses are

$$P_{cos}=\frac{3}{2}R_s\left|\overline{i_s}\right|^2;\quad P_{cor}=\frac{3}{2}R_r'\left|\overline{i_r}'\right|^2 \tag{4.19}$$

As expected, the resistance and inductance relationships between the three-phase IM and its dq (space phasor) model are very simple, because of the "blessing" of the sinus and cosinus in the inductance expressions. Note that the zero sequences (V_{0s}, i_{0s}, V_{0r}', i_{0r}') in Equation 4.17 complete the IM–dq model equivalences.

4.4 MAGNETIC SATURATION AND SKIN EFFECTS IN THE dq MODEL

In Chapter 5, Vol. 1 (companion book) we introduced the unique magnetization curve of the IM as

$$\psi_{\mathrm{m}}\left(i_{\mathrm{m}}\right) = L_{\mathrm{m}}\left(i_{\mathrm{m}}\right) i_{\mathrm{m}}; \quad \overline{i_{\mathrm{m}}} = \overline{i_{\mathrm{s}}} + \overline{i_{\mathrm{r}}}' \tag{4.20}$$

where i_{m} is the total (magnetization) current:

$$i_{\mathrm{m}} = \sqrt{i_{dm}^2 + i_{qm}^2}; \quad i_{dm} = i_d + i'_{dr}; \quad i_{qm} = i_q + i'_{qr} \tag{4.21}$$

The stator and rotor space phasors, $\overline{\psi_{\mathrm{s}}}, \overline{\psi_{\mathrm{r}}}'$, may be written with an airgap (main) flux space phasor, ψ_{m}, as the common term:

$$\overline{\psi_{\mathrm{m}}} = L_{\mathrm{m}} \overline{i_{\mathrm{m}}}; \quad \overline{\psi_{\mathrm{s}}} = L_{\mathrm{sl}} \overline{i_{\mathrm{s}}} + \overline{\psi_{\mathrm{m}}}; \quad \overline{\psi_{\mathrm{r}}}' = L'_{\mathrm{rl}} \overline{i_{\mathrm{r}}}' + \overline{\psi_{\mathrm{m}}} \tag{4.22}$$

The time derivative of the main flux, $\overline{\psi_{\mathrm{m}}}$, is straightforward:

$$\begin{aligned} \frac{\mathrm{d}\overline{\psi_m}}{\mathrm{d}t} &= L_{\mathrm{m}} \frac{\mathrm{d}\overline{i_{\mathrm{m}}}}{\mathrm{d}t} + \frac{\partial L_{\mathrm{m}}}{\partial i_{\mathrm{m}}} i_{\mathrm{m}} \frac{\mathrm{d}\overline{i_{\mathrm{m}}}}{\mathrm{d}t} = \left(L_{\mathrm{m}} + \frac{\partial L_{\mathrm{m}}}{\partial i_{\mathrm{m}}} i_{\mathrm{m}} \right) \frac{\mathrm{d}\overline{i_{\mathrm{m}}}}{\mathrm{d}t} \\ &= L_{\mathrm{mt}} \frac{\mathrm{d}\overline{i_{\mathrm{m}}}}{\mathrm{d}t}; \quad L_{\mathrm{mt}} = \frac{\mathrm{d}\psi_{\mathrm{m}}}{\mathrm{d}i_{\mathrm{m}}} \end{aligned} \tag{4.23}$$

Consequently, the transient magnetization inductance, L_{mt}, occurs in the main flux time derivative, but in the flux Ψ_{m}, Equation 4.22, the normal inductance, L_{m}, holds. Both L_{mt} and L_{m} depend on the magnetization current.

Note: If only flux variables are used, with currents as dummy variables, L_{mt} is not required. Unfortunately, in IM control, often currents are controlled, and thus L_{mt} occurs inevitably.

The rotor skin effect can be simulated by multiple fictitious cages arranged in parallel. Let us consider two cages in the rotor:

$$\begin{aligned} &\overline{V_{\mathrm{s}}} = R_{\mathrm{s}} \overline{i_{\mathrm{s}}} + L_{\mathrm{sl}} \frac{\mathrm{d}\overline{i_{\mathrm{s}}}}{\mathrm{d}t} + \frac{\mathrm{d}\overline{\Psi_{\mathrm{m}}}}{\mathrm{d}t} + j\omega_{\mathrm{b}}\left(L_{\mathrm{sl}} \overline{i_{\mathrm{s}}} + \overline{\Psi_{\mathrm{m}}}\right) \\ &0 = R'_{r1} \overline{i'_{r1}} + L'_{rl1} \frac{\mathrm{d}\overline{i_{r1}}}{\mathrm{d}t} + \frac{\mathrm{d}\overline{\Psi_{\mathrm{m}}}}{\mathrm{d}t} + j\left(\omega_{\mathrm{b}} - \omega_{\mathrm{r}}\right)\left(L'_{rl1} \overline{i'_{r1}} + \overline{\Psi_{\mathrm{m}}}\right) \\ &0 = R'_{r2} \overline{i'_{r2}} + L'_{rl2} \frac{\mathrm{d}\overline{i_{r2}}}{\mathrm{d}t} + \frac{\mathrm{d}\overline{\Psi_{\mathrm{m}}}}{\mathrm{d}t} + j\left(\omega_{\mathrm{b}} - \omega_{\mathrm{r}}\right)\left(L'_{rl2} \overline{i'_{r2}} + \overline{\Psi_{\mathrm{m}}}\right) \\ &\frac{\mathrm{d}\overline{\Psi_{\mathrm{m}}}}{\mathrm{d}t} = L_{\mathrm{mt}}\left(i_{\mathrm{m}}\right) \frac{\mathrm{d}\overline{i_{\mathrm{m}}}}{\mathrm{d}t}; \quad \overline{i_{\mathrm{m}}} = \overline{i_{\mathrm{s}}} + \overline{i'_{r1}} + \overline{i'_{r2}}; \quad \overline{\Psi_{\mathrm{m}}} = L_{\mathrm{m}}\left(i_{\mathrm{m}}\right) \overline{i_{\mathrm{m}}} \\ &T_{\mathrm{e}} = \frac{3}{2} p_1 \operatorname{Re}\left(j \overline{\Psi_{\mathrm{s}}} \overline{i_{\mathrm{s}}}^*\right) = p_1 L_{\mathrm{m}} \operatorname{Re}\left[j \overline{i_{\mathrm{s}}} \left(\overline{i'_{r1}} + \overline{i'_{r2}}\right)\right]; \quad L_{mt} = \frac{d\Psi_m}{di_m} \end{aligned} \tag{4.24}$$

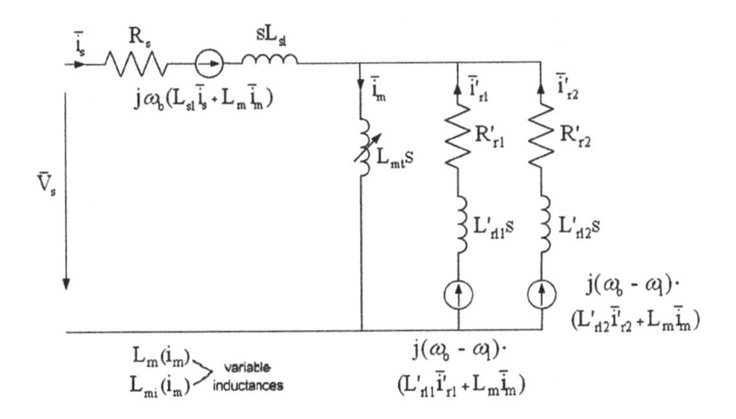

FIGURE 4.3 Space phasor equivalent circuit of IM with saturation and skin effect.

With $d\overline{\Psi_m}/dt$ as the common voltage, Equations 4.24 suggests a circuit equivalent to that shown in Figure 4.3, though in reality there may be a leakage flux coupling between the two virtual or real cages. Its influence is lumped into the fictitious cage parameters. There are motion emfs in the stator and in the rotor as ω_b occurs generally. For $\omega_b = \omega_r$ (rotor coordinates), the rotor emf disappears while for $\omega_b = 0$ (stator coordinates) the emf disappears in the stator. For synchronous coordinates ($\omega_b = \omega_1$), the emf appears both in the stator and in the rotor.

Note: The fictitious cage parameters may be found from the standstill frequency response tests (described later in this chapter), as done for the SM.

4.5 SPACE PHASOR MODEL STEADY STATE: CAGE AND WOUND ROTOR IMS

For steady state, let us consider sinusoidal voltages, constant speed, and load torque:

$$V_{A,B,C}^{(t)} = V\sqrt{2}\cos\left(\omega_1 t - (i-1)\frac{2\pi}{3}\right); \quad i = 1,2,3 \tag{4.25}$$

From the space phasor transformation (Equation 4.15), with $\theta_b = \omega_1 t$, for synchronous coordinates ($\omega_b = \omega_1$):

$$\overline{V_s} = \frac{2}{3}\left[V_A(t) + V_B(t)e^{j\frac{2\pi}{3}} + V_C(t)e^{-j\frac{2\pi}{3}}\right]e^{-j\omega_1 t} \tag{4.26}$$

Making use of Equation 4.25 in Equation 4.16 yields

$$\overline{V_s} = V\sqrt{2} = V_d + jV_q \Rightarrow V_d = V\sqrt{2}; \quad V_q = 0 \tag{4.27}$$

The voltage space phasor is a d.c. quantity and so are the currents and flux linkages of the space phasor model under steady-state and synchronous coordinates, both in the stator and in the rotor. With d/dt = 0 (in general, $\frac{\mathrm{d}}{\mathrm{d}t} = j\left(\omega_1 - \omega_b\right)$), Equations 4.13 becomes

$$\overline{V_{s0}} = R_s\overline{i_{s0}} + j\omega_1\overline{\Psi_{s0}};\quad \overline{\Psi_{s0}} = L_{sl}\overline{i_{s0}} + \overline{\Psi_{m0}};\quad \overline{\Psi_{m0}} = L_m\left(\overline{i_{s0}} + \overline{i'_{r0}}\right) \tag{4.28}$$

$$\overline{V'_{r0}} = R'_r\overline{i'_{r0}} + jS\omega_1\overline{\Psi'_{r0}};\quad \overline{\Psi'_{r0}} = L'_{rl}\overline{i'_{r0}} + \overline{\Psi_{m0}};\quad S = \left(\omega_1 - \omega_r\right)/\omega_1 \tag{4.29}$$

$$T_e = \frac{3}{2}p_1\,\mathrm{Re}\left(j\overline{\Psi}_{s0}\overline{i}^*_{s0}\right) \tag{4.30}$$

where S is the slip.

The space phasor equivalent circuit for the steady state (as suggested by Equations 4.28 and 4.29) is shown in Figure 4.4. It is similar to that of one phase in terms of the phasors (at frequency ω_1) described in Chapter 5, Vol.1 (companion book).

It should be noticed that the rotor voltage, $\overline{V'_{r0}}$, is also d.c. (in synchronous coordinates). But the phase (position) of $\overline{V'_{r0}}$ with respect to $\overline{V_{s0}}$ is variable with respect to the load in the motor or in the generator operation. Let us now draw the space phasor diagrams of the IM, first for the cage rotor and motor and generator operation modes (Figure 4.5a and b).

We should notice that at steady state, for the cage rotor IM, the rotor flux and current space phasors, $\overline{\Psi}'_{r0}, \overline{i'_{r0}}$, are orthogonal. The same is true even for transients, but with a constant rotor flux, $\Psi_r = \Psi_{r0}$ = const. This is the so-called rotor flux orientation (vector) control principle. The slip is positive for motoring ($S >$ 0) and negative for generating ($S < 0$), but in both cases the stator flux space phasor amplitude is largest, $\Psi_{s0} > \Psi_{m0} > \Psi_{r0}$, a clear sign that the machine magnetization arises from the power supply in both cases (from an external source, anyway). For the wound rotor (doubly fed) induction machine, we have to define first the

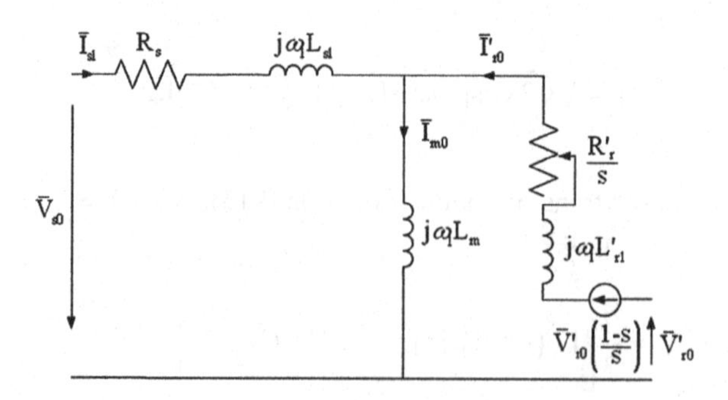

FIGURE 4.4 The space phasor circuit of the IM under a steady state.

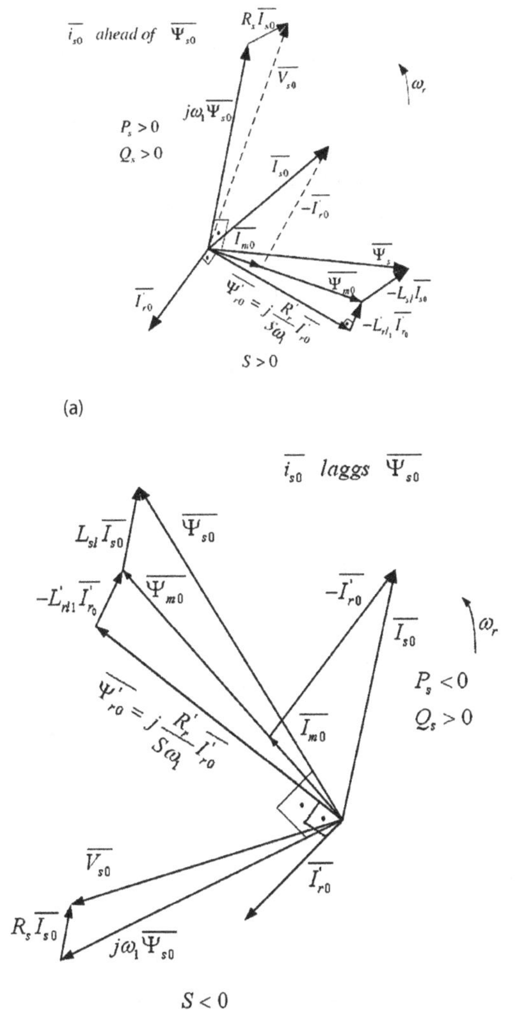

FIGURE 4.5 Space phasor diagram of cage rotor IM: (a) motor mode and (b) generator mode.

no-load ideal speed, ω_{r0} (slip S_0), which corresponds to zero-rotor current in Equation 4.14:

$$\overline{V'_{r0}} = jS_0\omega_1\overline{\Psi'_{r0}}; \quad \omega_{r0} = \omega_1\left(1 - S_0\right) \tag{4.31}$$

So the ideal no-load speed may occur for positive (subsynchronous) or negative (hypersynchronous) S_0:

$$\begin{aligned} &\omega_{r0} < \omega_1 \quad \text{for } S_0 > 0 \\ &\omega_{r0} > \omega_1 \quad \text{for } S_0 < 0 \end{aligned} \tag{4.32}$$

The sign and relative value of S_0 depends on the amplitude and phase of $\overline{V'_{r0}}$ (with respect to $\overline{\Psi}'_{r0}$ or $\overline{V}_{s0}$). Both motoring and generating is possible around S_0. It is feasible to magnetize the machine from the rotor or from the stator sources or from both. The active and reactive powers of the stator (Equation 4.18) and the rotor are

$$P_s = \frac{3}{2}\mathrm{Re}\left(\overline{V}_{s0}^*\overline{i}_{s0}^*\right) = \frac{3}{2}\left(V_{d0}i_{d0} + V_{q0}i_{q0}\right)$$

$$Q_s = \frac{3}{2}Im\left(\overline{V}_{s0}\overline{i}_{s0}^*\right) = -\frac{3}{2}\left(V_{d0}i_{q0} - V_{q0}i_{d0}\right)$$

$$P_r^r = \frac{3}{2}\mathrm{Re}\left(\overline{V}'_{r0}\overline{i}'^*_{r0}\right) = \frac{3}{2}\left(V'_{dr0}i'_{dr0} + V'_{qr0}i'_{qr0}\right)$$

$$\left(Q'_r\right)^r_{s\omega 1} = \frac{3}{2}Im\left(\overline{V}'_{r0}\overline{i}'^*_{r0}\right) = -\frac{3}{2}\left(V'_{dr0}i'_{qr0} - V'_{qr0}i'_{dr0}\right)$$

$Q_r'^r$ is considered at the slip frequency ($S\omega_1$) in synchronous coordinates; in other words, $Q_r'^r/S$ is the reactive power of the rotor source seen at the stator frequency. So the reactive power produced at the $S\omega_1$ frequency is "amplified" $\frac{1}{S}$ times when observed at the ω_1 frequency; this is based on the fact that the corresponding magnetic energy is conserved.

Let us consider the case of $S < 0$ (supersynchronous) operation at the unity rotor power factor in the motor and generator modes (Figure 4.6) based (still) on Equations 4.28 through 4.30.

The unity rotor power factor means a minimum kVA of the rotor power source; it also means that the machine magnetization is done from the stator $\left(\Psi_{s0} > \Psi'_{r0}\right)$. For the unity power factor in the stator, the magnetization arises from the rotor power source, and thus $\Psi'_{r0} > \Psi_{s0}$.

Example 4.1: Space-Phasor Steady State of the Cage Rotor IM

Let us consider a three-phase cage rotor IM with the following data: $P_n = 1.5$ kW, $\eta_n = 0.9$, $\cos\varphi_n = 0.9$, $f_1 = 60$ Hz, $V_{s0} = 120\sqrt{(2)}$, $r_s = 0.04$(pu), $r_r' = 0.03$(pu), $l_{sl} = l_{rl}' = 0.1$(pu), $2p_1 = 4$ poles, $l_m = 3$ (pu), and operating at a frequency, $f_1 = 60$ Hz, and, slip S = 0.03. Calculate R_s, i_n, L_{sl}, L_{rl}', L_m in Henry, $\overline{I}_{s0}$, $\overline{I}'_{r0}$, $\overline{\Psi}'_{r0}$, $\overline{\Psi}_{s0}$, and T_e (torque), and speed, ω_r (and n in rps).

Solution:

The pu norm reactance, X_n, is (as for the SM)

(Continued)

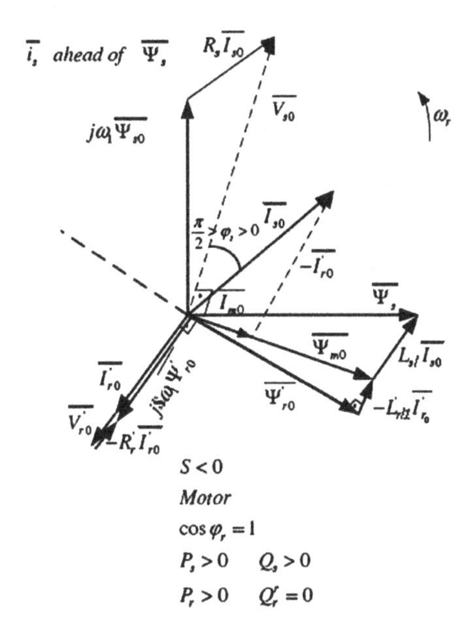

(a)

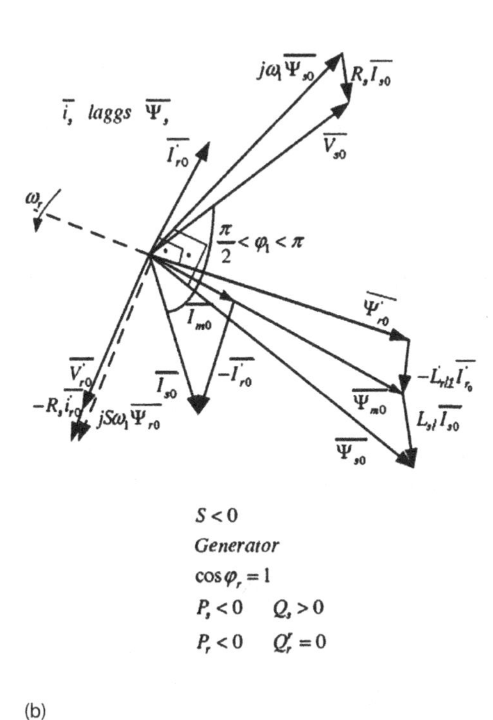

(b)

FIGURE 4.6 Space phasor diagram of doubly fed IM at $S < 0$: (a) motoring and (b) generating, at the unity rotor power factor.

Example 4.1: (Continued)

$$X_n = \frac{V_{nph}\sqrt{2}}{I_{nph}\sqrt{2}};\quad I_{nph} = \frac{P_n}{3V_{nph}\eta_n\cos\phi_n} = \frac{1500}{3\times120\times0.9\times0.9} = 5.144\text{ A}$$

$$X_n = \frac{120\sqrt{2}}{5.144\sqrt{2}} = 23.328\,\Omega;\quad R_s = r_s \times X_n = 0.04\times23.328 = 0.933\,\Omega,$$

$$R_r' = 0.03\times23.328 = 0.6998\,\Omega,\ L_{sl} = L_{rl}' = 0.1\times\frac{X_n}{2\pi f_1}$$

$$= \frac{0.1\times23.328}{2\pi\times60} = 6.19\text{ mH}$$

$$L_m = 3\times\frac{X_n}{2\pi f_1} = 0.1857\text{H}$$

From Equations 4.28 and 4.29

$$120\sqrt{2} = \left[0.933 + j2\pi60(0.00619+0.1857)\right]\overline{i_{s0}} + j2\pi60\times0.1857\overline{I_{r0}'}$$
$$0 = \left[0.6998 + j0.03\times2\pi60(0.00619+0.1857)\right]\overline{I_{r0}'} + j\times0.03$$
$$\times2\pi60\times0.1857\times\overline{I_{s0}}$$

Both $\overline{I_{s0}}$ and $\overline{I_{r0}'}$ can be easily calculated from the above equations:

$$\overline{I_{r0}'} = -6.5596 + j1.593$$
$$\overline{I_{s0}} = 6.393 - j3.362$$
$$\overline{V_{s0}} = 120\sqrt{2}$$

So, the flux linkages, $\bar{\Psi}_{s0}$ and $\bar{\Psi}_{r0}'$, are

$$\bar{\Psi}_{s0} = \frac{\bar{V}_{s0} - R_s\bar{i}_{s0}}{j\omega_1} = \frac{120\sqrt{2} - 0.933(6.393 - j3.362)}{j2\pi60}$$
$$\bar{\Psi}_{r0}' = \frac{-R_r'\bar{I}_{r0}'}{jS\omega_1} = \frac{-0.6998\times(-6.5596 + j1.159)}{j\times0.03\times2\pi\times60}$$
$$\bar{\Psi}_{s0} = -j0.4332 + 0.008$$
$$\bar{\Psi}_{r0}' = -j0.40608 - 0.0717$$

It is evident that $\Psi_{s0} > \Psi_{r0}'$ (in amplitude) and is leading (motoring).
The torque, T_e, is

$$T_e = \frac{3}{2}p_1\,\text{Re}\left(j\bar{\Psi}_{s0}\bar{i}_{s0}^*\right) = \frac{3}{2}\times2\times\text{Re}\left[(0.008 - j0.4332)(6.343 + j3.362)\right]$$
$$= 4.522\text{ Nm}$$

Example 4.2: Doubly Fed IM Steady State

Let us consider a doubly fed IM with the following parameters: $R_s = R_r' = 0.018\ \Omega$, $X_{sl} = X_{rl} = 0.18\ \Omega$, $X_m = 14.4\ \Omega$ at $f_n = 50$ Hz, $2p_1 = 4$, $V_{sn\ line} = 6000$ V, $I_{1n/phase} = 1204$ A (RMS), and having a star connection. The ratio of rotor to stator turns is $K_{rs} = 4/1$, while the slip is S = −0.25. For cos $\varphi_s = 1$ (stator unity power factor), calculate the rotor active and reactive powers, P_r^r and Q_r^r and the total active delivered power by the generator $P_{gen} = P_s + P_r^r$.

Solution:

With Equation 4.28 at the unity stator power factor, $V_{s0} = R_s i_{s0} + j\omega_1\bar{\Psi}_{s0}$, and i_{s0} is negative because the generator mode is considered here.

$$\frac{6000}{\sqrt{2}} \times \sqrt{2} = 0.018 \times (-1204) + j2\pi \times 50\overline{\Psi_{s0}}$$

$$\Psi_{s0} = -j15.64\text{Wb}$$

But,

$$\begin{aligned}\bar{\Psi}'_{r0} &= \bar{\Psi}_{s0}\frac{L_{rl}+L_m}{L_m} - L_{sc}i_{s0} = -j15.64 \times \frac{(14.4+0.18)}{14.4} - \frac{0.36}{314} \times (-1204)\\ &= -j15.8355 + 1.380\end{aligned}$$

Hence, the rotor current is

$$\begin{aligned}I'_{r0} &= \frac{\bar{\Psi}'_{r0} - L_m i_s}{L_m + L'_{rl}} = \frac{-j15.835 + 1.380 - 14.4 \times (-1024)/314}{(14.58/314)}\\ &= \frac{-j15.835 + 56.49}{4.6433 \times 10^{-2}} = -j341.03 + 1219.29\\ I'_{r0} &= 1266 \text{ A} > I_{s0}\end{aligned}$$

because the machine is magnetized from the rotor.

The rotor voltage is obtained from Equation 4.19:

$$\begin{aligned}\bar{V}'_{r0} &= R'_r\bar{I}'_{r0} + jS\omega_1\bar{\Psi}'_{r0}\\ &= 0.018 \times (-j341.28 + 1219) + j(-0.25)2\pi 50 \times (-j15.8255 + 1.380)\\ &= -1221 - j114.468\end{aligned}$$

The stator active power, P_s (cos $\varphi_s = 1$) is

$$\begin{aligned}P_s &= \frac{3}{2}V_{s0}i_{s0} = \frac{3}{2} \times \frac{6000\sqrt{2}}{\sqrt{3}}(-1024) = -8.8316 \times 10^6 \text{ W} = -8.8316 \text{ MW}\\ S_r^r &= P_r^r + jQ_r^r = \frac{3}{2}\left(\bar{V}'_{r0}\bar{I}'^{*}_{r0}\right)\\ &= \frac{3}{2}(-1221 - j114.246)(1219.29 + j341.03)\\ &= -2.174 \text{ MW} + j0.55566 \text{ MVAR}\end{aligned}$$

(Continued)

Example 4.2: (Continued)

So, the total delivered active power of the doubly fed IG, P_{gen}, is

$$P_{gen} = P_s + P_r^r = (-88316 - 2174)\text{ MW} \approx -11\text{ MW}$$

The rotor reactive power to the stator would be

$$Q_r = Q_r^r / |S| = \frac{0.55566}{0.25}\text{ MVAR} = 2.2226419\text{ MVAR}$$

But for the design of the rotor side converter Q_r^r and P_r^r are paramount. As $K_{rs} = L_{0r}/L_{sr} = 4/1$, it means that the actual rotor voltage V_{r0} is four times larger than V_{r0}' while the current I_{r0} is four times smaller than I_{r0}'. The DFIG is dominant in modern wind generation systems at variable speeds.

4.6 ELECTROMAGNETIC TRANSIENTS

As for synchronous machines, there are fast transients in cage IMs during which the motor speed may be considered constant. Such transients are called electromagnetic transients.

For fast transients, it seems appropriate to use rotor coordinates ($\omega_b = \omega_r$) in the dq (space phasor) model. Let us consider the IM with a dual rotor cage (to account for the rotor skin effect). There are no motion emfs in the rotor and, for pu equations, d/dt is replaced by s/ω_{10}, as for the SMs:

$$\begin{aligned}
&\overline{V}_s = \overline{i}_s r_s + \frac{s}{\omega_{10}}\overline{\Psi}_s + j\omega_{r0}\overline{\Psi}_s; \quad \overline{\Psi}_s = l_{sl}\overline{i}_{sl} + l_m\left(\overline{i}_s + \overline{i}_{r1}' + \overline{i}_{r2}'\right)\\
&0 = \overline{i}_{r1}' r_{r1}' + \frac{s}{\omega_{10}}\overline{\Psi}_{r1}'; \quad \overline{\Psi}_{r1}' = i_{rl1}'\overline{i}_{r2}' + l_m\left(\overline{i}_s + \overline{i}_{r1}' + \overline{i}_{r2}'\right)\\
&0 = \overline{i}_{r2}' r_{r2}' + \frac{s}{\omega_{10}}\overline{\Psi}_{r2}'; \quad \overline{\Psi}_{r2}' = i_{rl2}'\overline{i}_{r2}' + l_m\left(\overline{i}_s + \overline{i}_{r1}' + \overline{i}_{r2}'\right)\\
&T_e = \text{Re}\left[jl_m\left(\overline{i}_s \times \left(\overline{i}_{r1}'^{*} + \overline{i}_{r2}'^{*}\right)\right)\right]
\end{aligned} \tag{4.33}$$

where ω_{10} is the norm angular frequency in radians per second.

We may eliminate i_{r1}' and i_{r2}' from Equations 4.33 and obtain the stator flux, $\overline{\Psi}_s$, in the Laplace form:

$$\begin{aligned}
&\overline{i}_s = l(s)\overline{i}_s(s)\\
&l(s) = (l_{sl} + l_{ml})\frac{(1+s\tau'')(1+s\tau')}{(1+s\tau_0'')(1+s\tau_0')}
\end{aligned} \tag{4.34}$$

The operational inductance, $l(s)$, is similar to that of an SM with two rotor circuits in parallel, with the time constants in seconds:

$$\begin{aligned}
\tau'' &= \frac{1}{r'_{r1}\omega_{10}}\left(l'_{rl1} + \frac{l_m l'_{rl2} l_{sl}}{l_m l'_{rl2} + l_m l_{sl} + l'_{rl2} l_{sl}}\right) \\
\tau''_0 &= \frac{1}{r'_{r1}\omega_{10}}\left(l'_{rl1} + \frac{l'_{rl2} l_m}{l'_{rl2} + l_m}\right) \\
\tau' &= \frac{1}{r'_{r2}\omega_{10}}\left(l'_{rl2} + \frac{l_m l_{sl}}{l_m + l_{sl}}\right) \\
\tau'_0 &= \frac{1}{r'_{r2}\omega_{10}}\left(l'_{rl2} + l_m\right)
\end{aligned} \tag{4.35}$$

Again, subtransient, l'', transient, l', and synchronous, l_s, inductances are defined:

$$\begin{aligned}
l'' &= \lim_{\substack{s\to\infty \\ (t\to 0)}} = \left(l_{sl} + l_m\right)\frac{\tau'\tau''}{\tau'_0\tau''_0} \\
l' &= \lim_{\substack{s\to\infty \\ \tau''=\tau''_0=0}} = \left(l_{sl} + l_m\right)\frac{\tau'}{\tau'_0} \\
l_s &= \lim_{\substack{s\to 0 \\ (t\to\infty)}} \left(l(1)\right) = l_{sl} + l_m
\end{aligned} \tag{4.36}$$

4.7 THREE-PHASE SUDDEN SHORT CIRCUIT/LAB 4.1

A three-phase sudden short circuit may occur accidentally at large IMs terminals and produce significant faults in the local power grid.

Alternatively, as for the SM, the sudden short circuit may be produced right after disconnection from the power grid (less than 1 ms later) and then, with the stator current acquired, machine parameters can be identified by curve fitting. We maintain P.U. variables as this case is of primary interest to power system analysis.

To simplify the mathematics, let us consider that the machine runs at ideal no load ($\omega_r = \omega_{r0} = \omega_1$) when the sudden short circuit occurs.

So, the initial current, i_{s0}, is

$$\bar{i}_{s0} = \frac{\bar{V}_{s0}}{r_s + j\omega_1 l_s} \tag{4.37}$$

Let us consider the machine phase voltages as

$$V_{ABC} = V\sqrt{2}\cos\left(\omega_1 t + \varphi_0 - (i-1)\frac{2\pi}{3}\right); \quad i = 1,2,3 \tag{4.38}$$

$$\overline{v}_{s0}=\frac{2}{3}\frac{1}{V\sqrt{2}}\left(V_A+V_Be^{j\frac{2\pi}{3}}+V_Ce^{-j\frac{2\pi}{3}}\right)e^{-j\omega_1 t}=v_{s0}e^{j\psi_0} \tag{4.39}$$

To short-circuit the machine, $\overline{v}_s=-\overline{v}_{s0}$ has to be applied. But the machine voltage equation (from Equations 4.33 and 4.34) is

$$\overline{i}_s'(s)=\frac{\overline{v}_s(s)}{r_s+\left(\frac{s}{\omega_{10}}+j\omega_1\right)\cdot l(s)} \tag{4.40}$$

with

$$\overline{v}_s(s)=-\frac{v_{s0}e^{j\Psi_0}}{\left(\frac{s}{\omega_{10}}\right)} \tag{4.41}$$

From Equations 4.40 and 4.41

$$\overline{i}_s'(s)=\frac{-v_{s0}e^{j\Psi_0}}{\frac{s}{\omega_{10}}\left(\frac{r_s}{l(s)}+\frac{s}{\omega_{10}}+j\omega_1\right)l(s)} \tag{4.42}$$

Approximating

$$\tau_a=\frac{l(s)}{\omega_{10}r_s}\approx\frac{l''}{\omega_{10}r_s} \tag{4.43}$$

yields

$$i_s'(s)\approx-v_{s0}e^{j\Psi_0}\left[\frac{\omega_{10}}{\left(\frac{s}{\omega_{10}}+\frac{1}{\tau_a\omega_{10}}+j\omega_1\right)\left(s+\frac{1}{\tau'}\right)}\left(\frac{1}{l'}-\frac{1}{l_s}\right)\right.$$
$$\left.+\frac{1}{\frac{s}{\omega_{10}}\left(\frac{s}{\omega_1}+\frac{1}{\tau_a\omega_{10}}+j\omega_1\right)}\frac{1}{l_s}+\frac{\omega_{10}}{\left(\frac{s}{\omega_{10}}+\frac{1}{\tau_a\omega_{10}}+j\omega_1\right)\left(s+\frac{1}{\tau''}\right)}\left(\frac{1}{l''}-\frac{1}{l_s'}\right)\right] \tag{4.44}$$

Finally, the resultant current space vector, $\bar{i}_s(t)$, is

$$\bar{i}_s(t) = \bar{i}_{s0} + \bar{i}_s'(t) = -\frac{v_{s0}e^{j\Psi_0}}{\omega_1}\left[\left(e^{-\left(\frac{1}{\tau_a\omega_{10}}+j\omega_1\right)t\omega_{10}} - e^{-\frac{t}{\tau'}}\right)\left(\frac{1}{l'}-\frac{1}{l_s}\right)\right.$$
$$\left. + \left(e^{-\left(\frac{1}{\tau_a\omega_{10}}+j\omega_1\right)t\omega_{10}} - e^{-\frac{t}{\tau''}}\right)\left(\frac{1}{l''}-\frac{1}{l'}\right) + \left(e^{-\left(\frac{1}{\tau_a\omega_{10}}+j\omega_1\right)t\omega_{10}} - 1\right)\frac{1}{l_s}\right] \quad (4.45)$$

The phase A current, $i_A(t)$, becomes

$$i_A(t) = \mathrm{Re}\left(\bar{i}_s(t)e^{j\omega_1\omega_{10}t}\right) = \frac{v_{s0}}{\omega_1 l''}e^{-\frac{t}{\tau_a}}\sin\Psi_0 - \left[\left(\frac{1}{\omega_1 l'}-\frac{1}{\omega_1 l_s}\right)e^{-\frac{t}{\tau'}}\right.$$
$$\left. + \left(\frac{1}{\omega_1 l''}-\frac{1}{\omega_1 l'}\right)e^{-\frac{t}{\tau''}}\right]v_{s0}\sin\left(\omega_1\omega_{10}t+\varphi_0\right) \quad (4.46)$$

A few remarks are in order:

- ω_{10} is the norm in radians per second but ω_1 is the angular frequency in (pu) with respect to ω_{10}.
- The short-circuit current shows a large but fast decaying, nonperiodic component and a sinusoidal component whose amplitude attenuates with two time constants, τ' and τ''.
- The maximum peak value of $i_A(t)$ occurs apparently for $\Psi_0 = \pi/2$.
- By curve fitting, with known $i_A(t)$, τ_a, τ', τ'', l'', l', and l_s, can be identified. The similitude with the SM is clearly visible but the steady-state short-circuit current is, as expected, zero, for the IM.

Example 4.3: Sudden Short Circuit

Calculate and represent graphically (in pu) the $\bar{i}_s(I_d, I_q)$ and $i_A(t)$ for the three-phase sudden short circuit of the IM with the following data: $v_{s0} = 1$ pu, $l'' = 0.20$ (pu), $l' = 0.35$ (pu), $l_s = 4.0$ (pu), $\tau_a = 0.05$ s, $\tau' = 0.1$ s, $\tau'' = 0.05$ s, $\Psi_0 = \pi/6$, $\omega_1 = 1.0$ (pu), and $\omega_{10} = 2\pi 50$ rad/s. Also derive the expression of torque during the short circuit.

Solution:

We go straight to Equation 4.46 to obtain

$$i_A(t) = \frac{1\times e^{-t/0.05}}{1\times 0.2}\cos\frac{\pi}{6} - \left[\left(\frac{1}{1\times 0.35}-\frac{1}{1\times 4.0}\right)e^{-t/0.1}\right.$$
$$\left. + \left(\frac{1}{1\times 0.2}-\frac{1}{1\times 0.35}\right)e^{-t/0.05}\right]\cdot 1\times\sin\left(2\pi 50t+\frac{\pi}{6}\right) \quad (4.47)$$

(Continued)

Example 4.3: (Continued)

The space phasor, $\bar{i}_s$, in synchronous coordinates (Equation 4.45) is

$$\bar{i}_s = i_d + ji_q$$

Its representation as I_d (I_q) is shown in Figure 4.7a with i_A (t) in Figure 4.7b.

To calculate the flux space phasor, Ψ_s (s), we use Equation 4.34

$$\bar{\Psi}_s(s) = l_s\bar{i}_{s0} + \frac{(-v_{s0})e^{j\Psi_0}}{\frac{s}{\omega_{10}}\left(\frac{s}{\omega_{10}} + \frac{1}{\tau_a\omega_{10}} + j\omega_1\right)} \tag{4.48}$$

and thus

$$\bar{\Psi}_s(t) \approx l_s\bar{i}_{s0} - \frac{v_{s0}e^{j\Psi_0}}{\frac{1}{\tau_a\omega_{10}} + j\omega_1}\left(-e^{-\left(\frac{1}{\tau_a\omega_{10}} + j\omega_1\right)t\omega_{10}} + 1\right) \tag{4.49}$$

The flux transients are described by only one (small) time constant, τ_a.

The torque, t_e, during the short-circuit period is

$$t_e(t) = \mathrm{Re}\left(j\bar{\Psi}_s(t)\bar{i}_s^*(t)\right) \tag{4.50}$$

with $\bar{i}_s(t)$ obtained from Equation 4.45. It turns out that the peak torque can reach values of 5–6 (pu). Finally, after the transient process, both the flux and the torque go to zero, as expected.

4.7.1 Transient Current at Zero Speed

A large inertia-load IM may be considered at standstill in the first two to four voltage periods after connection to the power grid. In this case, $\omega_{r0} = \omega_b = 0$, and $\bar{i}_{s0} = 0$ (at $t = 0$). According to Equations 4.34 and 4.41, for $\omega_r = 0$

$$i_s(s) = \frac{\bar{v}_s(s)}{r_s + \frac{s}{\omega_{10}}l(s)}; \quad \Psi_s(s) = l(s)i_s(s); \quad t_e(t) = \mathrm{Re}\left(j\Psi_s i_s^*\right) \tag{4.51}$$

with

$$\bar{v}_s(t) = v_{s0}e^{j(\Psi_0 + \omega_{10}t)} \tag{4.52}$$

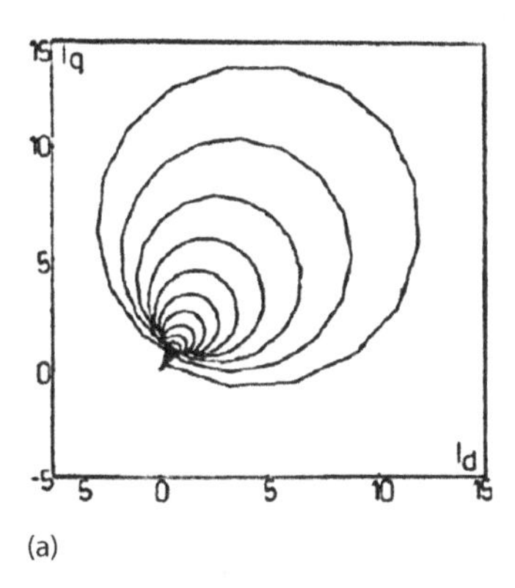

(a)

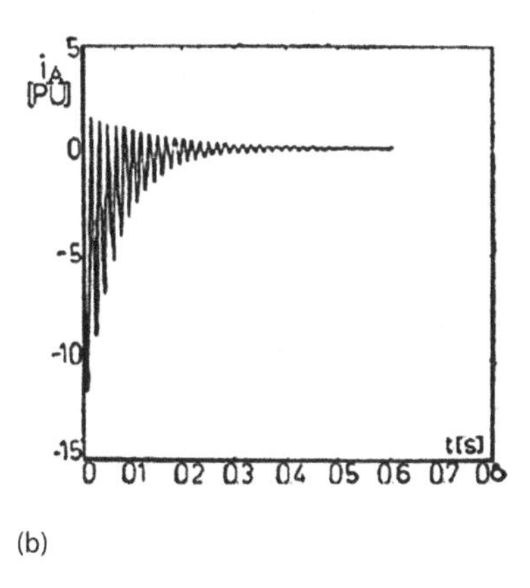

(b)

FIGURE 4.7 Sudden short-circuit current (a) $I_d(I_q)$ and (b) $i_A(t)$ in pu.

During such a process, the current peak value can surpass 10 (pu). For such high values of current, the leakage flux paths saturate, and thus the peak current values increase further. Consequently, in industry, either leakage flux path magnetic saturation is considered (by reducing l_{sl}, l'_{rl1}, l_{rl2}) or experiments at full voltage are required to calculate (measure) correctly the peak starting current and torque, which are so important in safety-critical starting applications (such as in heat-exchanger pump motors in nuclear power plants).

4.8 SMALL-DEVIATION ELECTROMECHANICAL TRANSIENTS

In induction motors, quite frequently, there are small variations in load torque, supply frequency, and voltage amplitude. Also, linearization of the IM model is required for control design.

For linearization, we consider the dq model (Equations 4.9 through 4.11) of a single-cage IM with

$$i_d = i_{d0} + \Delta i_d; \quad i_q = i_{q0} + \Delta i_q \tag{4.53}$$

to obtain

$$|A|s|\Delta X| + B|\Delta X| = |C||\Delta V| + |D|\Delta T_{\text{load}} + |E|\Delta\omega_1 \tag{4.54}$$

with

$$|\Delta X| = |\Delta i_d \Delta i_q \Delta i'_{dr} \Delta i'_{qr} \Delta\omega_r|^T$$

$$|\Delta V| = |\Delta V_d \Delta V_q\, 000|^T$$

$$C = [11000]$$

$$D = [0000-1]^T$$

$$E = \begin{bmatrix} L_s I_{q0} + L_m I_{qr0} - (L_s I_{d0} + L_m I_{dr0}) & L_m I_{q0} + L'_r I_{qr0} \\ -(L_m I_{dr0} + L'_r I_{dr0}) & 0 \end{bmatrix}^T \tag{4.55}$$

$$|A| = \begin{vmatrix} L_s & 0 & L_m & 0 & 0 \\ 0 & L_s & 0 & L_m & 0 \\ L_m & 0 & L_{r'} & 0 & 0 \\ 0 & L_m & 0 & L_{r'} & 0 \\ 0 & 0 & 0 & 0 & -\dfrac{J}{p_1} \end{vmatrix} \tag{4.56}$$

$$|B| = \begin{vmatrix} R_s & -\omega_{10}L_s & 0 & -\omega_{10}L_m & 0 \\ \omega_{10}L_s & R_s & \omega_{10}L_m & 0 & 0 \\ 0 & -\omega_{20}L_m & R'_r & -\omega_{20}L'_r & -(L_m i_{q0} + L_{r'} i'_{qr0}) \\ \omega_{20}L_m & 0 & \omega_{20}L'_r & R'_r & -(L_m i_{d0} + L'_r i'_{dr0}) \\ -\dfrac{3}{2}p_1 L_m i'_{qr0} & \dfrac{3}{2}p_1 L_m i'_{dr0} & \dfrac{3}{2}p_1 L_m i_{d0} & -\dfrac{3}{2}p_1 L_m i_{q0} & 0 \end{vmatrix} \tag{4.57}$$

where

ω_{10} is the initial frequency of the stator (synchronous coordinates),
$\omega_{20} = \omega_{10} - \omega_{r0}$ is the initial rotor slip frequency,
ω_{r0}—initial speed,
$L_s = L_{sl} + L_m$ and $L'_r = L'_{rl} + L_m$

For the initial, steady-state situation; the dq model equation yields

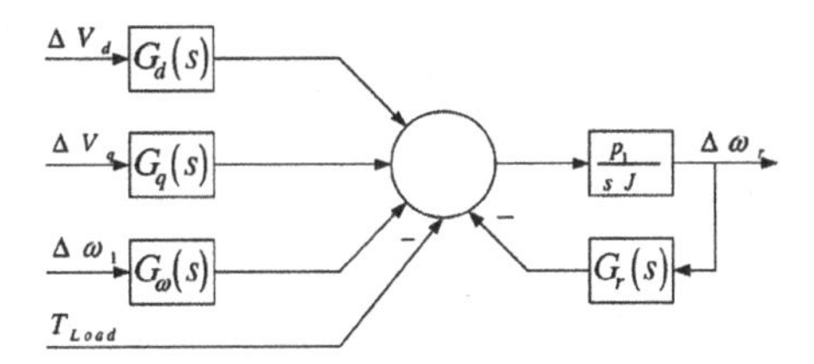

FIGURE 4.8 IM small-deviation speed transfer function.

$$\begin{vmatrix} V_{d0} \\ V_{q0} \\ 0 \\ 0 \end{vmatrix} = \begin{vmatrix} R_s & -\omega_{10}L_s & 0 & -\omega_{10}L_s \\ \omega_{10}L_s & R_s & \omega_{10}L_m & 0 \\ 0 & -\omega_{20}L_m & R_r' & -\omega_{20}L_r' \\ \omega_{20}L_m & 0 & \omega_{20}L_r' & R_r' \end{vmatrix} \begin{vmatrix} i_{d0} \\ i_{q0} \\ i'_{dr0} \\ i'_{qr0} \end{vmatrix} \quad (4.58)$$

Equations 4.55 refers to a fifth-order system with ΔV_d, ΔV_q, $\Delta\omega_1$, and ΔT_{load} as the inputs and Δi_d, Δi_q, $\Delta i'_{dr}$, $\Delta i'_{qr}$, and $\Delta\omega_r$ as the variables. This can be treated with the methods typical for the linear systems.

If we manage to eliminate the current variables from Equations 4.54, we end up with a small perturbation speed transfer function of the form

$$\frac{J}{p_1}s\Delta\omega_r = -G_r(s)\Delta\omega_r + G_d(s)\Delta V_d + G_q(s)\Delta V_q - \Delta T_{\text{load}} + G_\omega(s)\Delta\omega_1 \quad (4.59)$$

A structural diagram, as in Figure 4.8, illustrates Equation 4.59.

To apply the structural diagram to the actual machine, we have to add

$$V_d + jV_q = \frac{2}{3}\left(V_A + V_B e^{j\frac{2\pi}{3}} + V_C e^{-j\frac{2\pi}{3}}\right)e^{-j\theta_b}$$
$$\frac{d\theta_b}{dt} = \omega_1 = \omega_{10} + \Delta\omega_1 \quad (4.60)$$

It should be noticed that even after linearization the fifth (sixth)-order of the system does not permit simple analytical solutions of transients; however, the stability analysis is facilitated by the possibility of using the linear system theories heritage (eigenvalue theory, etc.).

4.9 LARGE-DEVIATION ELECTROMECHANICAL TRANSIENTS/ LAB 4.2

Large voltage sags, starting by direct connection to the grid, large load-torque perturbations, accidental disconnection and reconnection to the power grid, and load dumping for autonomous generator mode represent large-deviation transients; for these cases, the complete dq model should be used.

It is preferable to use flux-linkage variables for the numerical solution of the dq model equations; in stator coordinates, we have

$$\begin{aligned}
\frac{d\Psi_d}{dt} &= V_d - R_s i_d; \quad i_d = C_{11}\Psi_d - C_{12}\Psi'_{dr}; \quad C_{11} = \frac{L'_r}{\sigma L_m^2} \\
\frac{d\Psi_q}{dt} &= V_q - R_s i_q; \quad i_q = C_{11}\Psi_q - C_{12}\Psi'_{qr}; \quad C_{12} = \frac{1}{\sigma L_m} \\
\frac{d\Psi'_{dr}}{dt} &= -R_r i'_{dr} - \omega_r \Psi'_{qr}; \quad i'_{dr} = -C_{12}\Psi_d + C_{22}\Psi'_{dr}; \quad C_{22} = \frac{L_s}{\sigma L_m^2} \\
\frac{d\Psi'_{qr}}{dt} &= -R_r i'_{qr} + \omega_r \Psi'_{dr}; \quad i'_{qr} = -C_{12}\Psi_q + C_{22}\Psi'_{qr}; \quad \sigma = \frac{L_s L_r}{L_m^2} - 1 > 0 \\
\frac{d\omega_r}{dt} &= \frac{p_1}{J}\left[T_e - T_{load}\right]; \quad T_e = \frac{3}{2} p_1 \left(\Psi_d i_q - \Psi_q i_d\right)
\end{aligned} \tag{4.61}$$

As long as the stator phase voltages are given as a functions of time, in stator coordinates

$$\begin{aligned}
V_{A,B,C}^{(t)} &= V_1\sqrt{2}\cos\left(\omega_1 t + \varphi_0 - (i-1)\frac{2\pi}{3}\right) \\
V_d + jV_q &= \frac{2}{3}\left(V_A + V_B e^{j\frac{2\pi}{3}} + V_C e^{-j\frac{2\pi}{3}}\right) \\
&= V_1\sqrt{2}\cos(\omega_1 t + \varphi_0) - jV_1\sqrt{2}\sin(\omega_1 t + \varphi_0)
\end{aligned} \tag{4.62}$$

Now the frequency, ω_1, might also be variable, together with the voltage amplitude, V_1, as long as the voltages are symmetric (balanced). An additional rotor cage means two more equations in Equation 4.62, to account for the skin effect in the rotor cage. Numerical methods, such as the Runge–Kutta–Gill method, are now available as toolboxes in MATLAB®/Simulink® that can be used to solve Equation 4.61.

Example 4.4: Starting Transients

Let us consider an IM with the following data: $R_s = 0.063\ \Omega$, $R'_r = 0.083\ \Omega$, $p_1 = 2$ pole pairs, $L_m = 29$ mH, $L_s = L'_r = 30.4$ mH, $T_{load} = 6$ Nm (low torque load), $\frac{J}{p_1} = 0.06$ kg m^2, $V_1 = 180/\sqrt{2}$ V(RMS), $f_1 = 60$ Hz. Find the speed and torque versus speed during starting by direct connection to the power source.

Solution:

The above model with $\varphi_0 = 0$ is solved numerically for T_e, ω_r, Ψ_d, and Ψ_q, and then for i_d and i_q. Finally, to find the phase current (in stator coordinates)

$$i_A(t) = i_d(t) \tag{4.63}$$

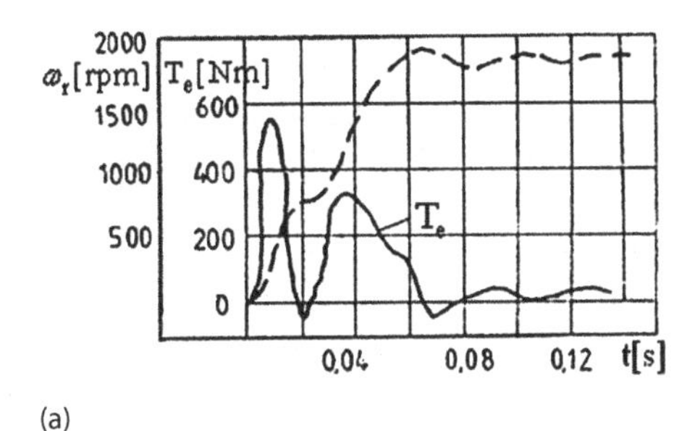

(a)

600
Te[Nm]
400
200
0
500 1000 1500 2000
ωr[rpm]

(b)

FIGURE 4.9 (a) IM speed and torque build up and (b) transient torque vs. speed

The results are shown in Figure 4.9a and b. Observe that all the initial values of the variables are zero in this case.

Due to the reduced inertia and small load, transients in torque and speed are significant. The difference of the transient torque/speed curve (Figure 4.9.b) from the smooth, single maximum point, steady-state, torque/speed curve of IM is an indication of fast mechanical and torque transients. Speed oscillations are also typical for such cases.

4.10 REDUCED-ORDER DQ MODEL IN MULTIMACHINE TRANSIENTS

When a large group of IMs are simulated for transients, the calculation of the fifth order of the dq model may be too time consuming. For synchronous coordinates ($\omega_b = \omega_1$), it is then tempting to neglect the stator transients:

$$\begin{gathered}
\frac{d\Psi_d}{dt} = \frac{d\Psi_q}{dt} = 0 \\
V_d = R_s i_d - \omega_1 \Psi_q; \quad V_q = R_s i_q + \omega_1 \Psi_d \\
\frac{d\Psi'_{dr}}{dt} = -R_r i'_{dr} + (\omega_1 - \omega_r)\Psi'_{qr}; \quad \frac{d\Psi'_{qr}}{dt} = -R_r i'_{qr} - (\omega_1 - \omega_r)\Psi'_{dr} \\
\frac{d\omega_r}{dt} = \frac{p_1}{J}\left[\frac{3}{2} p_1 \left(\Psi_d i_q - \Psi_q i_d\right) - T_{load}\right]
\end{gathered} \tag{4.64}$$

A third-order system has been obtained. However, the fast transients due to flux and current amplitude variations are ignored, even in torque. But, the speed response tends to be usable. Other simplified (modified first order) models, adequate for small or large IMs, have been proposed [1,2], but they have to be used cautiously. A good part of industrial electric loads is represented by induction motors, from a few kilowatts to 20–30 MW/unit. A local (industrial) transformer supplies, in general, a group of such IMs, through pertinent power switches; random load perturbations and turning on and off may occur together with bus transfer. During the time interval between the turn off from one power grid and turn on of the emergency power grid, the group of IMs, with a common reactance feeder and with a parallel capacitive bank (for power factor compensation), exhibits residual voltage, which dies slowly until reconnection takes place. The attenuating rotor currents of IMs produce stator emfs depending on their speed (inertia) and parameters such that some machines may act as motors and some as generators, until their stored magnetic energy runs out.

To treat such complex phenomena the complete fifth-order dq model is the obvious choice. But, neglecting stator transients greatly simplifies the problem. During turn off (Figure 4.10)

$$\sum_{j=1}^{n}\overline{i_{sj}}+\overline{i_c}=0;\quad \frac{d\overline{V_s}}{dt}=\frac{\overline{i_c}}{C} \tag{4.65}$$

where i_c is the capacitor current.

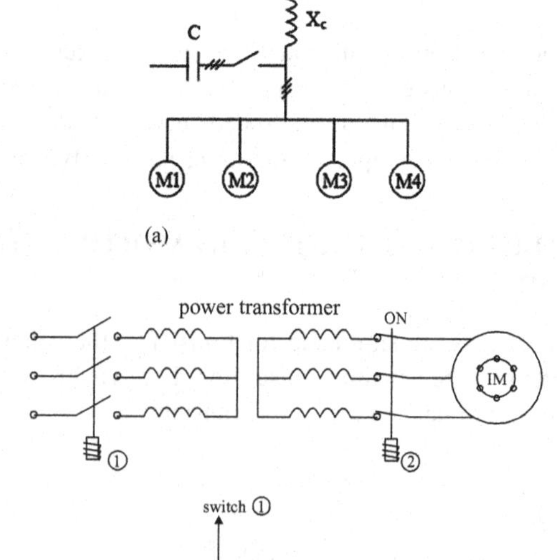

FIGURE 4.10 Grid connection: (a) Group pf IMs with common feeder and (b) transformer feeds an IM.

A unique synchronous reference system is used as the stator frequency for the whole group. The third-order model (Equation 4.64) allows to calculate the residual stator voltages during an IM turn off. It has been shown that the third-order model can predict the residual stator voltage well but experiments show a sudden jump in this voltage right after turn off. A possible explanation of this anomaly might be the influence of magnetic saturation, which maintains the self-excitation in some IMs with larger inertia for some time; after this, de-excitation of those that act as motors might take place. Magnetic saturation should be considered in dq model mandatorily.

4.10.1 Other Severe Transients

There are many other severe electromechanical transients that lead to very large peak currents and torque, related to turn on and turn off and reconnection on-the-fly of IMs. The case of turn off and turn on of the power switch in the primary of a transformer that supplies an IM (Figure 4.10b) seems to produce occasionally very large transients (up to 25 p.u. torque for 6 kW IMs and of only 12 p.u. torque for 400 kW IMs [3]). For a certain turn-off interval (30–40 ms), peak torques of 35 p.u. may occur. This explains premature aging and mechanical defects in IMs after such situations. Only the complete fifth-order dq model can predict transients as those mentioned here; again magnetic saturation of main and leakage flux paths have to be considered.

Elastically couplings of large IMs may produce severe torsional torque transients [4, Chapter 13] while series capacitors, to avoid voltage sags during their direct turn on starting, may produce synchronous resonance [4]. Such transients may again be treated using the dq (space phasor) model of IM.

4.11 m/N_r ACTUAL WINDING MODELING OF IMS WITH CAGE FAULTS

By an m/N_r winding model, we mean an IM with m stator windings and N_r actual (rotor slot count) rotor windings [5]. Each rotor bar and end-ring segment is modeled as such. Stator winding faults such as local short circuits or open coils lead to dedicated self- and mutual inductance expressions as defined by using the winding function method [6,7].

For symmetrical stators, self- or mutual (stator to bar loop circuits) inductances have simpler expressions [4]. There are N_r loops (bars) in the rotor plus one more equation for the end ring current. So the model has $m + N_r + 1 + 1$ equations; the last one is the motion equation. However, it is not easy to define practical expressions for stator-to-rotor loop mutual inductances, while measuring them is not practical (or is it?). Field distribution methods may be used to solve the problem.

For the symmetrical stator, the phase equations are

$$V_{A,B,C} = i_{A,B,C} R_s + \frac{d\Psi_{A,B,C}}{dt} \tag{4.66}$$

The rotor cage structure and unknowns are illustrated in Figure 4.11.

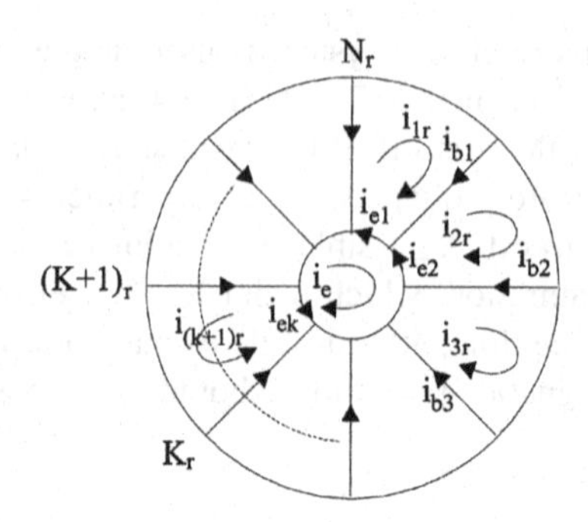

FIGURE 4.11 Rotor cage with rotor loop currents.

For the k_r^{th} rotor loop (in rotor coordinates)

$$0 = 2\left(R_b + \frac{R_e}{N_r}\right)i_{kr} + \frac{d\Psi_{kr}}{dt} - R_b\left(i_{k-1,r} + i_{k+1,r}\right) \tag{4.67}$$

For one end ring segment (R_e, Le-parameters of one end ring segment)

$$0 = R_e i_e + L_e\frac{di_e}{dt} - \sum_{1}^{N_r}\left(\frac{R_e}{N_r}i_{kr} + \frac{L_e}{N_r}\frac{di_{kr}}{dt}\right) \tag{4.68}$$

For a healthy cage, the end ring total current is $i_e = 0$. The bar/ring currents are related by

$$i_{bk} = i_{k,r} - i_{k+1,r}; \quad i_{ek} = i_{k,r} - i_e \tag{4.69}$$

This explains why only $N_r + 1$ independent variables remain.

The stator self- and mutual phase inductances are

$$L_{AA} = L_{BB} = L_{CC} = L_{sl} + L_{0s}; \quad L_{AB} = L_{BC} = L_{CA} = -\frac{L_{0s}}{2}$$

and (Chapter 5), Vol. 1 (companion book)

$$L_{0s} = \frac{4\mu_0\left(w_s k_{ws}\right)^2 \tau l_i}{\pi^2 k_c g\left(1 + k_s\right)p_1} \tag{4.70}$$

where

k_s is the saturation coefficient,
L_{sl} is the leakage stator phase inductance,
τ is the pole pitch,

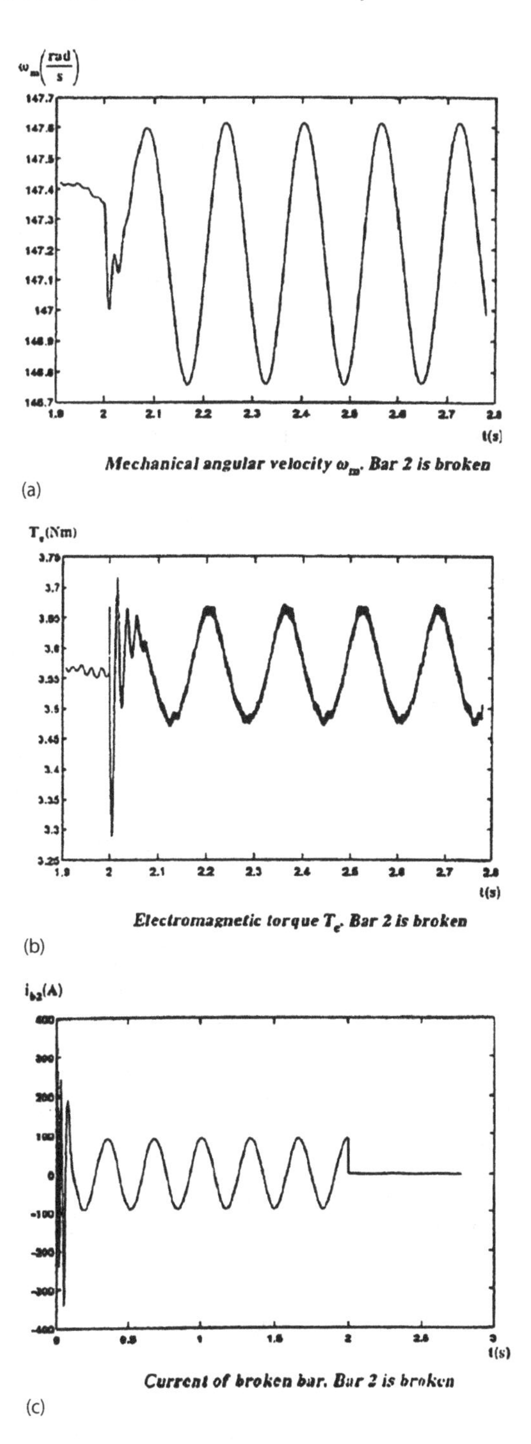

FIGURE 4.12 Rotor bar no. 2 broken at 3 Nm load torque: (a) speed, (b) torque, and (c) broken bar current.

g is the airgap,
p_1 is the pole pair,
k_c Carter coefficient,
w_s is the turns per phase,
k_{ws} is the winding factor, and
l_i is the stack length.

The self-inductance of a rotor loop (based on its area), $L_{kr,kr}$, is

$$L_{k_r,k_r} = \frac{2\mu_0 (N_r - 1) p_1 \tau l_i}{N_r^2 k_c g (1 + k_s)} \tag{4.71}$$

An average mutual inductance between two rotor loops is

$$L_{k,k_r+1} = -\frac{2\mu_0 p_1 \tau L}{N_r^2 k_c g (1 + k_s)} \tag{4.72}$$

The stator-to-rotor loop mutual inductances are

$$\begin{aligned} L_{A k_r}(\theta_{er}) &= L_{sr} \cos\left[\theta_{er} + (k_r - 1)\alpha\right] \\ L_{B k_r}(\theta_{er}) &= L_{sr} \cos\left[\theta_{er} + (k_r - 1)\alpha - \frac{2\pi}{3}\right] \\ L_{C k_r}(\theta_{er}) &= L_{sr} \cos\left[\theta_{er} + (k_r - 1)\alpha + \frac{2\pi}{3}\right] \\ \alpha = p_1 \frac{2\pi}{N_r};\quad \theta_{er} &= p_1 \theta_r;\quad L_{sr} = -\frac{-(w_s k_{ws}) \mu_0 2 p_1 \tau l_i}{4 p_1 k_c g (1 + k_s)} \sin\frac{\alpha}{2} \end{aligned} \tag{4.73}$$

Adding up Equations 4.67 through 4.73 in a matrix form, we obtain

$$\begin{aligned} [V] &= [R_s][i] + \left|L'(\theta_{er})\right| \frac{d}{dt}[i] + \left|\frac{\partial L'(\theta_{er})}{\partial \theta_{er}}\right| [i] \frac{d\theta_{er}}{dt} \\ [V] &= \begin{bmatrix} V_A & V_B & V_C & 0 & 0 & \ldots & 0 \end{bmatrix}^T \\ [I] &= \begin{bmatrix} i_A & i_B & i_C & i_{1r} & i_{2r} & \ldots & i_{N_r} & i_e \end{bmatrix} \end{aligned} \tag{4.74}$$

$[L'(\theta_{er})]$ is an $(m + N_r + 1) \times (m + N_r + 1)$ matrix and so is $[R]$; $[R]$ is much sparser, however. Their components are easy to add up using Equations 4.67 through 4.74. The motion equations can be added with torque, T_e:

$$\begin{aligned} T_e = p_1 L_{sr} \Bigg[&\left(i_A - \frac{1}{2} i_B - \frac{1}{2} i_C\right) \sum_1^{N_r} i_{k_r} \sin\left(\theta_{er} + (k_r - 1)\alpha\right) \\ &+ \frac{\sqrt{3}}{2}(i_B - i_C) \sum_1^{N_r} i_{k_r} \cos\left(\theta_{er} + (k_r - 1)\alpha\right) \Bigg] \end{aligned} \tag{4.75}$$

We end up with an extended phase variable model with many rotor-position-dependent inductances. The case of faulty rotor bars or end ring segments is handled by increasing their respective resistances 10^3–10^4 times. For an IM with the following data, $R_s = 10$, $R_b = R_e = 155\ \mu\Omega$, $L_{sl} = 35$ mH, $L_{0s} = 378$ mH, $w_s = 340$ turns/phase, $k_{ws} = 1$, $N_r = 30$ rotor slots, $p_1 = 2$ pole pairs, $L_e = L_b = 0.1\ \mu$H, $L_{sr} = 0.873$ mH, $\tau =$ 62.8 mm (pole pitch), $l_i = 66$ mm (stack length), $g = 0.37$ mm (airgap), $J = 5.4 \times 10^{-3}$ kg m^2, $f = 50$ Hz, $V_{line} = 380$ V (star connection), $P_n = 736$ W, and $I_{0n} = 21$ A, broken bars situations have been simulated [8].

Numerical results for broken bar no. 2 ($R_{b2} = 200R_b$), with the machine under a load, $T_L = 3.5$ Nm are shown in Figure 4.12a through c for the speed, torque and respective bar current [7]. The rotor bar no. 2 breaks at $t_0 = 2$s.

Slight pulsations in speed and more notable pulsations in torque occur. More broken bars, or end ring segments, produce notable speed and torque pulsations. They also produce $2Sf_1$ frequency pulsations in the stator current. All this information can be processed to diagnose broken bars or end ring segments.

In the above model, the inter-bar electrical resistance to the slot wall has been considered infinite, so there are no inter-bar currents. This is hardly true. Inter-bar rotor currents tend to diminish broken bar effects [9]. As expected, coupled finite-element-circuit 3D methods can also be used to simulate complex transients as those aforementioned [10], but they take more CPU time [11].

4.12 TRANSIENTS FOR CONTROLLED MAGNETIC FLUX AND VARIABLE FREQUENCY

Due to its ruggedness, the cage rotor IM is very suitable for variable speed drives or as electric generator at variable speed. The speed, ω_r, of the IM is

$$\omega_r = \omega_1 (1 - S) = \omega_1 - \omega_2; \quad S = \frac{\text{rotor cage losses}}{\text{electromagnetic power}} \tag{4.76}$$

It is evident that during operation at variable speed, ω_r, the smaller the slip S (in cage rotor IMs), the lower the rotor cage losses. So the smaller the slip frequency (rotor currents frequency), $\omega_2 = S\omega_1$, the better; $\omega_2 > 0$ for motoring and $\omega_2 < 0$ for generating. In addition, quick responses in torque (in motors) and in active power (in generators) are needed. In such situations, magnetic flux constancy is beneficial since it also keeps magnetic saturation under control.

Frequency (speed) control at a constant flux (Ψ_s, Ψ_m, or Ψ_r') thus becomes imperative. Two main speed (power) control strategies have become dominant: rotor flux and stator flux orientation (or vector) control. Investigating the IM transients for vector control follows. The space phasor model in synchronous coordinates is most suitable for the scope of this scheme. Let us reproduce the IM equations with the stator current, $\overline{i}_s$, and the rotor flux, $\overline{\Psi}_r'$, as variables.

4.12.1 Complex Eigenvalues of IM Space Phasor Model

The voltage equations of an IM in the space phasor form (Equations 4.13 and 4.14) may be written with the stator current, $\overline{i}_s$, and the rotor flux, $\overline{\Psi}_r'$, as complex-state variables, in general coordinates (Figure 4.13):

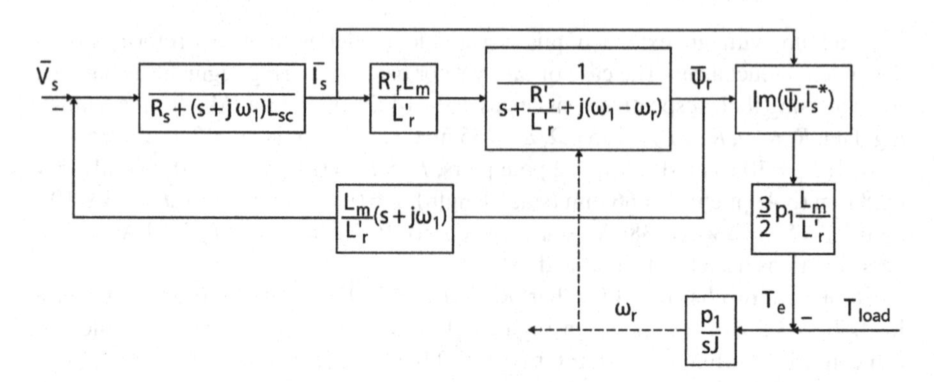

FIGURE 4.13 IM structural diagram with $\overline{i}_s$ and $\overline{\Psi}_r$ as complex variables in synchronous coordinates.

$$\begin{gathered}
\left[R_s+(s+j\omega_b)L_{sc}\right]\overline{i}_{sc}+(s+j\omega_b)\frac{L_m}{L'_r}\overline{\Psi'_r}=\overline{V_s}; \\
S=1-\frac{\omega_r}{\omega_b}; \quad -\frac{R'_r}{L'_r}\overline{i}_s+\left[s+\frac{R'_r}{L'_r}+j(\omega_b-\omega_r)\right]\frac{\overline{\Psi'_r}}{L_m}=\frac{\overline{V'_r}}{L_m}
\end{gathered} \tag{4.77}$$

For given slip S, to investigate Equation 4.77 means to treat electromagnetic (constant speed) transients. As we can see from Equation 4.77, the space phasor formalism implies only a second-order characteristic equation with two complex eigenvalues:

$$\left(R_s+(s+j\omega_b)L_{sc}\right)\left(s+\frac{R'_r}{L'_r}+j(\omega_b-\omega_r)\right)\frac{1}{L_m}+\frac{R'_r}{L'_r}(s+j\omega_b)\frac{L_m}{L'_r}=0 \tag{4.78}$$

The complex eigenvalues, $s_{1,2}$, as solutions of Equation 4.78 depend on slip, S, the frequency, ω_1, and machine parameters. So, direct analytical solutions exist for constant slip, S, and ω_1 transients, when the voltages, $\overline{V}_s$ and $\overline{V}'_r$, vary in amplitude or phase. The stator current, $\overline{i}_s$, and the rotor flux, Ψ'_r, are of the form

$$\overline{i}_s=\underline{A}+\underline{B}e^{s_1 t}+\underline{C}e^{s_2 t} \tag{4.79}$$

with $s_{1,2}$ from

$$\begin{gathered}
s^2L_{sc}+s\left[R_s+R'_r\left(\frac{L_m}{L'_r}\right)^2+L_{sc}\left(\frac{R'_r}{L'_r}+j(\omega_b-\omega_r)+j\omega_b\right)\right] \\
+R_s\left[\frac{R'_r}{L'_r}+j(\omega_b-\omega_r)\right]+j\omega_b L_{sc}\left[\frac{R'_r}{L'_r}+j(\omega_b-\omega_r)\right]+R'_r\frac{L_m}{L'^2_r}j\omega_b=0
\end{gathered} \tag{4.80}$$

A simple numerical computation routine is required to solve Equation 4.80.

4.13 CAGE ROTOR IM CONSTANT ROTOR FLUX TRANSIENTS AND VECTOR CONTROL BASICS

A constant rotor flux means that $s\overline{\Psi}_r = 0$ (in synchronous coordinates), in Equation 4.77:

$$\begin{gathered}\left[R_s + (s + j\omega_1)L_{sc}\right]\bar{i}_s + j\omega_1 \frac{L_m}{L_r'}\overline{\Psi'_r} = \overline{V_s} \\ -\frac{R_r'}{L_r'}\bar{i}_s + \left[\frac{R_r'}{L_r'} + j(\omega_1 - \omega_r)\right]\frac{\overline{\Psi'_r}}{L_m} = 0\end{gathered} \tag{4.81}$$

It is now evident that only one complex eigenvalue remains. The fifth-order system is further reduced:

$$\left[R_s + (s + j\omega_1)L_{sc}\right]\left[\frac{R_r'}{L_r'} + j(\omega_1 - \omega_r)\right]\frac{1}{L_m} + j\frac{R_r'}{L_r'}\omega_1\frac{L_m}{L_r'} = 0 \tag{4.82}$$

Consequently, s is

$$\begin{aligned}\underline{s} = &-\left\{\frac{\omega_1 \frac{L_m{}^2}{L_r'}(\omega_1 - \omega_r)\frac{L_r'}{R_r'}}{L_{sc}\left[1 + (\omega_1 - \omega_r)^2 \frac{L_r'^2}{R_r'^2}\right]} + \frac{R_s}{L_{sc}}\right\} \\ &-j\omega_1\left[\frac{L_m^2}{L_r'L_{sc}}\frac{1}{\left(1 + (\omega_1 - \omega_r)^2\frac{L_r'^2}{R_r'^2}\right)} + 1\right]\end{aligned} \tag{4.83}$$

As expected, at $\omega_r = 0$

$$(s)_{\omega_r = 0} \approx -\frac{R_s + R_r\left(L_m / L_r^{\prime}\right)^2}{L_{sc}} - j\omega_1 \tag{4.84}$$

Even now s depends on speed and machine parameters but $R_e(s) > 0$ at any speed in the motoring $(\omega_1 - \omega_r > 0)$, which means stable behavior, and this explains partly the commercialization of rotor flux orientation control. At small speeds and small frequencies, ω_1, and the generation mode, the real part of s may become positive and instability may occur, and, hence, special measures, through closed-loop control, would be required to mend the trouble. The machine model structural diagram gets simplified considerably, to that in Figure 4.14, using Equation 4.81.

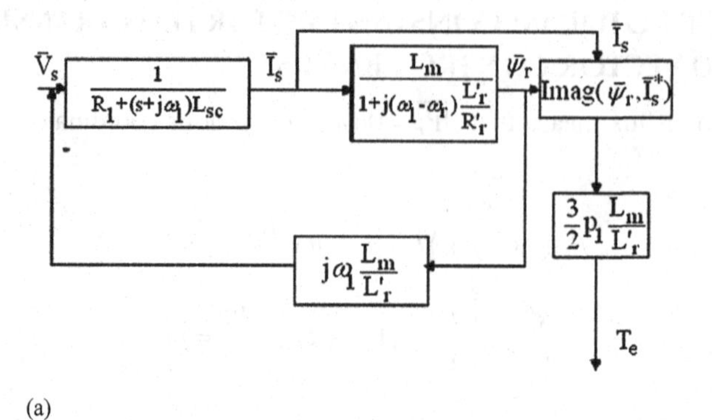

(a)

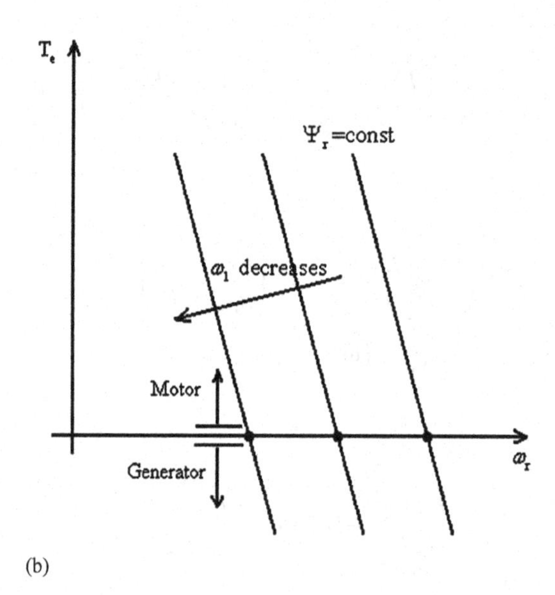

(b)

FIGURE 4.14 IM structural diagram (a) with constant rotor flux in synchronous coordinates ($\omega_b = \omega_1$; steady state is d.c.) and (b) steady-state torque/speed straight line.

The second equation in Equations 4.81 suggests that the stator current is related to the constant (imposed) rotor flux, $\Psi_r^* = \Psi_{dr}^*$:

$$\bar{I}_s = I_M + jI_T;\quad I_M = \frac{\Psi_r^*}{L_m};\quad I_T = I_M S\omega_1\frac{L'_r}{R'_r};\quad S = 1 - \frac{\omega_r}{\omega_1} \tag{4.85}$$

where

I_M is the flux component and

I_T is the torque component.

The rotor current is such that

$$\Psi'_{qr} = 0 = L'_r I'_{qr} + L_m I_T \tag{4.86}$$

Hence

$$\bar{I}'_{\mathrm{r}} = jI'_{qr} = -j\frac{L_{\mathrm{m}}}{L'_{\mathrm{r}}}I_{\mathrm{T}}; \quad L_{\mathrm{sc}} = L_{\mathrm{s}} - \frac{L_{\mathrm{m}}^2}{L_{\mathrm{r}}} \tag{4.87}$$

And therefore

$$\bar{\Psi}_{\mathrm{s}} = L_{\mathrm{s}}\bar{I}_{\mathrm{s}} + L_{\mathrm{m}}\bar{I}'_{\mathrm{r}} = L_{\mathrm{s}}I_{\mathrm{M}} + jL_{\mathrm{sc}}I_{\mathrm{T}} \tag{4.88}$$

These are considerable simplifications in building the space phasor model of the machine, all due to the constant rotor flux constraint. The torque expression becomes

$$T_{\mathrm{e}} = \frac{3}{2}p_1\,\mathrm{Re}\left(j\bar{\Psi}_{\mathrm{s}}\bar{I}_{\mathrm{s}}^*\right) = \frac{3}{2}p_1\left(L_{\mathrm{s}} - L_{\mathrm{sc}}\right)I_{\mathrm{M}}I_{\mathrm{T}} = \frac{3}{2}p_1\frac{\Psi'^2_{\mathrm{r}}}{R'_{\mathrm{r}}}\left(\omega_1 - \omega_{\mathrm{r}}\right) \tag{4.89}$$

Under steady state ($s = 0$ in Equation 4.81)

$$\bar{V}_{\mathrm{s}} = R_{\mathrm{s}}\bar{I}_{\mathrm{s}} + j\omega_1\bar{\Psi}_{\mathrm{s}} \tag{4.90}$$

The above developments lead to the following remarks:

- For constant rotor flux, the order of the space phasor model is reduced to a single complex eigenvalue for constant speed.
- The stator current space phasor, $\bar{I}_{\mathrm{s}}$, can be divided into two orthogonal (d, and q) components: I_{M} (flux component) and I_{T} (torque component). For a constant I_{M}, I_{T} may be varied by varying the frequency ω_1, for given speed. Thus, the torque may be modified in a decoupled way with respect to the flux. This is the essence of vector control; it is similar in behavior to a d.c. brush machine with a separate excitation control.
- For constant rotor flux, the torque expression (Equation 4.89) reveals that the IM operates as a reluctance SM where L_d is L_{s} and L_q is $L_{\mathrm{sc}} \ll L_{\mathrm{s}}$. A large apparent magnetic saliency is obtained because the d axis of the dq model falls along the rotor flux axis and the rotor current space phasor lies along the q axis, to cancel the rotor flux component along the q axis, that is, rotor flux orientation.

To illustrate the usefulness of the above equations, a basic (indirect) vector control scheme is presented in Figure 4.15.

We should recognize Equation 4.85 in Figure 4.15. The rotor flux speed, ω_1^*, is calculated based on the rotor speed, ω_{r}, and the slip speed, $(S\omega_1)^*$. Then the rotor flux angle, $\Theta_{\Psi\mathrm{r}}$, is calculated by integration of speed ω_1^* with time. The initial value of $\Theta_{\Psi\mathrm{r}}$ is irrelevant due to a constant airgap of the IM. Then the Park (or space vector) inverse transformation leads to reference phase currents. AC current controllers are used to PWM control of the PWM inverter and realize the desired (reference) stator currents. It is implicitly admitted that the a.c. current controllers are capable of "executing" almost instantly the reference currents and that the machine parameters are constant and known. As shown by Equation 4.83, there is still a small time constant in the current $\left(\bar{I}_{\mathrm{s}}\right)$ response even at constant rotor flux. For more on rotor flux (vector) control of IMs, see [12, Chapter 9].

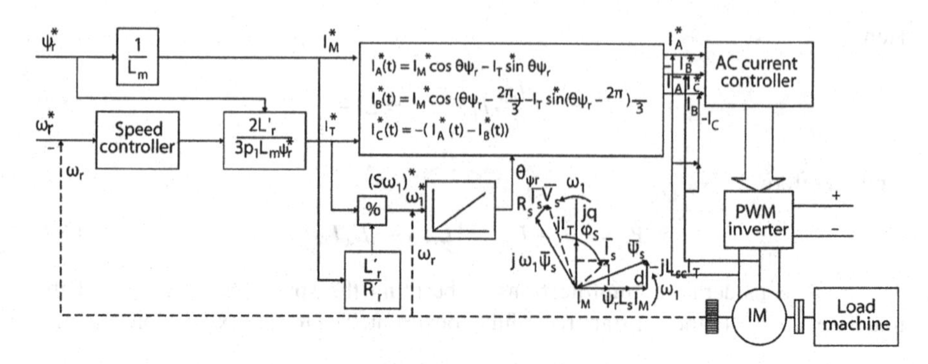

FIGURE 4.15 Basic (indirect) vector control of IM for constant rotor flux.

Example 4.5: Steady State of the Constant Rotor Flux IM

A cage rotor induction motor is controlled at variable frequency and constant rotor flux. Its parameters are $R_s = R_r' = 0.1\,\Omega$, $L_s = L_r' = 0.093$ H, $L_m = 0.09$ H, and $2p_1 = 4$, and it operates at $n = 600$ rpm with $I_M = 20$ A (flux current) and $I_T = \pm 50$ A (torque current), in the space phasor formulation.
Calculate

a. Developed electromagnetic torque,
b. Slip frequency, $S\omega_1$, ω_1,
c. Stator phase voltage, current, and power factor, $\cos\varphi_s$, and
d. For a small flux current, $I_M' = 5$ A, and $I_T = 50$ A at the same voltage as above, calculate $(S\omega_1)'$, ω_1', ω_r', T_e', and $\cos\varphi_s'$.

Solution:

a. According to Equation 4.89 the electromagnetic torque, T_e, is

$$T_e = \frac{3}{2}p_1\left(L_s - L_{sc}\right)I_M I_T = \frac{3}{2}\times 2\times\left(0.093-0.006\right)\times 20\times\left(\pm 50\right) = \pm 261.3\,\text{Nm}$$

with the ⊕ sign for motoring and the ⊖ sign for generating.

b. The slip frequency, $S\omega_1$, from Equation 4.85 is

$$S\omega_1 = \frac{I_T}{I_M\left(L_r'/R_r'\right)} = \frac{50\times 0.1}{20\times 0.093} = 2.688\text{ rad/s}$$

The electrical rotor speed, ω_r, is

$$\omega_r = 2\pi p_1 n = 2\pi\times 2\times 600/60 = 125.6\text{ rad/s}$$

So, for motoring

$$\omega_1 = S\omega_1 + \omega_r = 2.688 + 125.6 = 128.288\text{ rad/s}$$
$$f_1 = 20.42\,\text{Hz}$$
$$\Psi_r = L_m I_M = 0.09\times 20 = 1.8\text{ Wb}$$

c. From Equation 4.90:

$$V_d = R_s I_M - \omega_1 L_{sc} I_T = 0.1\times 20 - 128.288\times 0.006\times 50 = -36.5\ \text{V}$$
$$V_q = R_s I_T + \omega_1 L_s I_M = 0.1\times 50 + 128.288\times 0.093\times 20 = 243.6\ \text{V}$$

So, the phase voltage and current are

$$\left(V_{phase}\right)_{RMS} = \sqrt{\frac{\left(V_d^2+V_q^2\right)}{2}} = \sqrt{\frac{(-36.5)^2+(243.6)^2}{2}} = 174.185\ V$$

$$\left(I_{phase}\right)_{RMS} = \sqrt{\frac{\left(I_M^2+I_T^2\right)}{2}} = \sqrt{\frac{20^2+50^2}{2}} = 38.08\ \text{A}$$

From the space phasor diagram in Figure 4.15

$$\varphi_s = -\tan^{-1}\left(\frac{V_d}{V_q}\right) + \tan^{-1}\left(\frac{I_M}{I_T}\right)$$
$$= -\tan^{-1}\left(\frac{-36.5}{243.6}\right) + \tan^{-1}\left(\frac{20}{50}\right) = 30.32°$$
$$\cos\varphi_s \approx 0.865$$

d. We have to calculate again the slip frequency:

$$\left(S\omega_1\right)' = \frac{I_T'}{I_M'\left(L_r'/R_r'\right)} = \frac{50\times 0.1}{5\times 0.093} = 10.75\ \text{rad/s}$$
$$\Psi_r' = L_m I_M' = 0.09\times 5 = 0.45\ \text{Wb}$$

The voltage equations are again

$$V_d = R_s I_M' - \omega_1' L_{sc} I_T' = 0.1\times 5 - \omega_1'\times 0.006\times 50$$
$$V_q = R_s I_T + \omega_1' L_s I_M = 0.1\times 50 + \omega_1'\times 0.093\times 5$$

To a first approximation, the first terms may be neglected ($R_s \approx 0$):

$$V_d^2+V_q^2 = \left(\omega_1'\right)^2\left[\left(L_{sc}I_T'\right)^2+\left(L_s I_M'\right)^2\right]$$
$$(-36.5)^2+(243.6)^2 = \left(\omega_1'\right)^2\left[(0.006\times 50)^2+(0.093\times 5)^2\right]$$
$$\omega_1' = \frac{246.32}{0.5533} = 445.12\ \text{rad/s}$$

The speed is

$$\omega_r' = \omega_1' - \left(S\omega_1\right)' = 445.12 - 10.75 = 434.37\ \text{rad/s}$$
$$n = \frac{\omega_r'}{2\pi p_1} = \frac{434.37}{2\pi\times 2} = 34.58\ \text{rps} = 2075\ \text{rpm}$$

(Continued)

Example 4.5: (Continued)

A more than three times increase in speed has been obtained for the same voltage by reducing the rotor flux four times.

The new torque value, T'_e, is

$$T'_e = \frac{3}{2} p_1 (L_s - L_{sc}) I'_M I'_T = \frac{3}{2} \times 2 \times (0.093 - 0.006) \times 5 \times 50 = 65.25 \text{ Nm}$$

The actual voltage components are

$$V'_d = 0.5 - 0.3 \times 445.12 = -133 \text{ V}$$
$$V'_q = 5 + 0.465 \times 445.12 = 212 \text{ V}$$

The electromagnetic powers, for the two speeds (600 and 2075 rpm) and the same phase voltage, but different frequencies (20.42 and 70.87 Hz, respectively), and also a 4/1 rotor flux weakening, are

$$P_{elm} = T_e \frac{\omega_r}{p_1} = 261.3 \times \frac{125.6}{2} = 16.409 \text{ W}$$
$$P'_{elm} = T'_e \frac{\omega'_r}{p_1} = 65.25 \times \frac{434.37}{2} = 14.171 \text{ W}$$

That is an almost constant power for a 3.458/1 speed range.

4.13.1 Cage-Rotor IM Constant Stator Flux Transients and Vector Control Basics

For constant stator flux, let us consider stator coordinates ($\omega_b = 0$) and replace $\bar{\Psi}'_r$ by $\bar{\Psi}_s$ in Equation 4.77; constant flux in stator coordinates means constant amplitude of $\bar{\Psi}_s$ at input voltage frequency, ω_1.

$$R_s \bar{I}_s + s\bar{\Psi}_s = \bar{V}_s$$
$$\left[\frac{L_s}{L_m} R'_r + (s - j\omega_r) L_{sc}\right] \bar{I}_s - \bar{\Psi}_s \left[\frac{R_r}{L_m} + \frac{L'_r}{L_m}(s - j\omega_r)\right] = 0 \qquad (4.91)$$

For constant stator flux magnitude (in stator coordinates):

$$\bar{\Psi}_s = \bar{\Psi}_{s0} e^{j\omega_1 t} \qquad (4.92)$$

So, $s\bar{\Psi}_s$ becomes $j\omega_1 \bar{\Psi}_s$ in Equation 4.91:

$$R_s \bar{I}_s + j\omega_1 \bar{\Psi}_{s0} = \bar{V}_s$$
$$\left[\frac{L_s}{L_m} R'_r + (s - j\omega_r) L_{sc}\right] \bar{I}_s - \bar{\Psi}_{s0} \left[\frac{R_r}{L_m} + \frac{L'_r}{L_m} j(\omega_1 - \omega_r)\right] = 0 \qquad (4.93)$$

This is again a single complex eigenvalue system with the characteristic equation:

$$R_s\left[\frac{R_r}{L_m}+\frac{L_r'}{L_m}j(\omega_1-\omega_r)\right]+\left[\frac{L_s}{L_m}R_r'+(s-j\omega_r)L_{sc}\right]j\omega_1=0 \tag{4.94}$$

and

$$\underline{s}=-\left[\frac{L_sR_r'}{L_mL_{sc}}+\frac{R_sL_r'(\omega_1-\omega_r)}{\omega_1L_{sc}L_m}\right]+j\left(\omega_r+\frac{R_sR_r}{\omega_1L_{sc}L_m}\right) \tag{4.95}$$

As for a constant rotor flux, the real part of the eigenvalue ($Re(S)$) for current transients is negative for $0 < S < 1(0 < \omega_r < \omega_1)$. With $R_s \approx R_r'$, it follows that even for the generator mode ($\omega_r > \omega_1$) but with $|S| \ll 1$, the real part of s remains negative. This is the case in most situations with a variable frequency control of IMs. The situation is more delicate with large values of slip, S, typical at very low speeds.

This phenomenon, typical for both the constant rotor and the constant stator flux control, in the generator/motor mode, has been observed with variable speed drives at very low speeds. The closed-loop control should solve this matter.

The structural diagram of an IM, corresponding to Equation 4.93, is illustrated in Figure 4.16.

The IM control at constant stator flux should take advantage of the stator equation (Equations 4.93) in stator coordinates (which is rather simple), to estimate the stator flux amplitude and its position when a closed-loop regulates the stator flux amplitude (Figure 4.17).

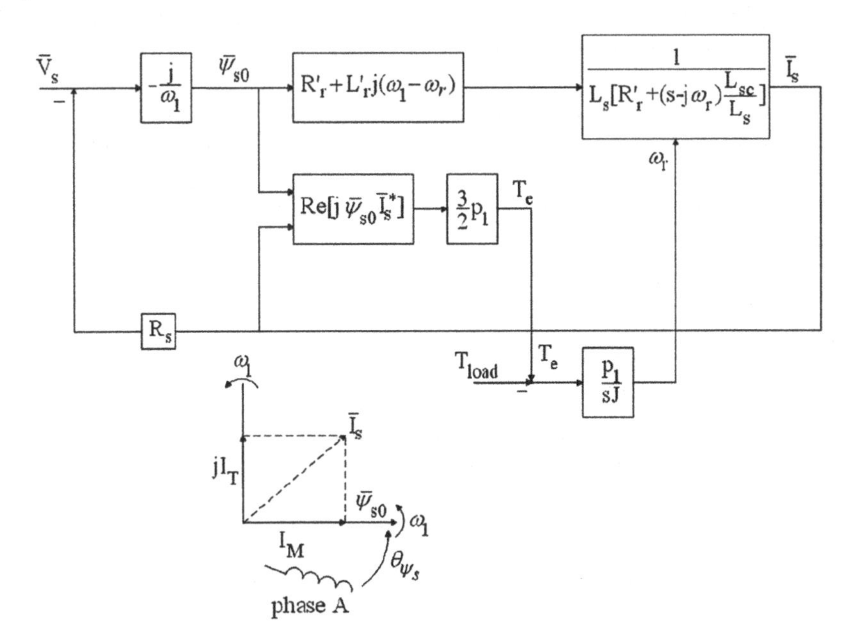

FIGURE 4.16 Structural diagram of IM for constant stator flux amplitude, Ψ_s, in stator coordinates ($\omega_b = 0$).

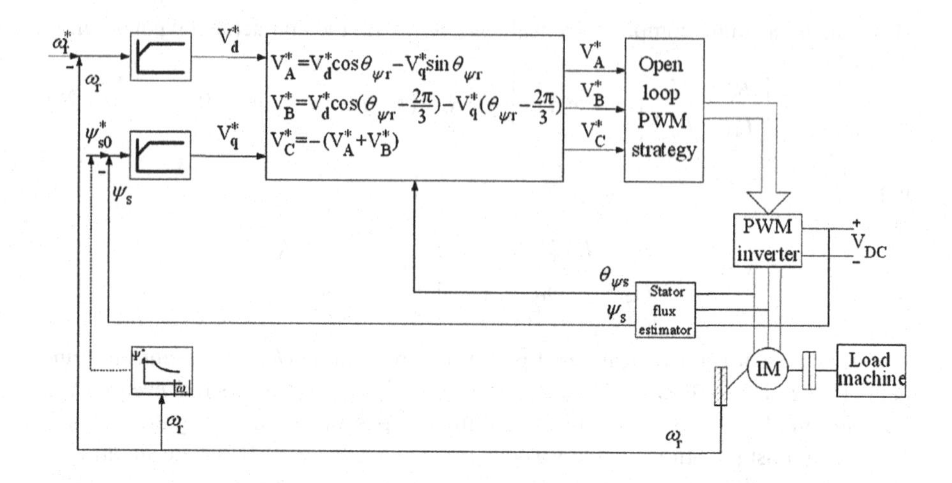

FIGURE 4.17 Basic direct stator flux orientation (vector) control of IM.

The basic (principle) stator flux orientation (vector) control scheme in Figure 4.17 can be characterized by

- A stator flux estimator in stator coordinates (Equation 4.93).
- A flux amplitude closed-loop regulator.
- A voltage vector rotator in stator flux coordinates.
- The a.c. voltages, V_A^*, V_B^*, and V_C^* are "reproduced" by an open-loop PWM (in the PWM inverter) strategy, characteristic to either large power or super high-speed IM drives where the switching frequency (f_{sw}) ratio, f_{sw}/f_1, in the static inverter is low (less than 20–30).
- The apparent absence of any current controller is compensated by the flux closed-loop control and by the limiters at the output of both regulators.

Estimators for the stator (or rotor) flux, eventually also of speed, are problems beyond our scope here but common in modern electric motor or generator controls [12].

4.13.2 Constant Rotor Flux Transients and Vector Control Principles of Doubly Fed IMs

The constant rotor or stator flux transients of doubly fed IMs are characterized by the same eigenvalues as for cage rotor IMs. Equations 4.81, with $s\Psi_r'$ in synchronous coordinates as zero, for a constant rotor flux, are still used:

$$\begin{aligned}
&\left(R_s+\left(s+j\omega_1\right)L_{sc}\right)\bar{I}_s+j\omega_1\frac{L_m}{L'_r}\Psi'_r=\bar{V}_s\\
&-\frac{L_mR'_r}{L'_r}\bar{I}_s+\left[\frac{R'_r}{L'_r}+j\left(\omega_1-\omega_r\right)\right]\Psi'_r=\bar{V}'_r=V'_{dr}+jV'_{qr}
\end{aligned}\qquad(4.96)$$

The torque, T_e, is

$$T_e = \frac{3}{2} p_1 \frac{L_m}{L'_r} \Psi'_r I_{qs} \tag{4.97}$$

In rotor flux coordinates,

$$\Psi'_{dr} = \Psi'_r = L_m \cdot I_{ds}; \quad \Psi'_{qr} = L'_r \cdot I'_{qr} + L_m \cdot I_{qs} = 0$$

This time the rotor equation is used for control as $\overline{V}_s$ is in general imposed both in terms of amplitude and frequency (phase).

$$\begin{aligned} V'_{dr} &= \frac{R'_r}{L'_r} \Psi'_r - \frac{L_m}{L'_r} R'_r I_{ds}; \quad \Psi'_r = \Psi'_{dr} = L_m I_{ds} \left(I'_{dr} = 0 \right) \\ V'_{qr} &= \left(\omega_1 - \omega_r \right) \Psi'_r - \frac{L_m}{L'_r} R'_r I_{qs} \end{aligned} \tag{4.98}$$

The active power in the rotor, P_r^r is

$$P_r^r = \frac{3}{2} p_1 S \omega_1 \Psi_r I_{ds} \tag{4.99}$$

Also the rotor reactive power, Q_r^r is

$$Q_r^r = \frac{3}{2} \left(V'_{dr} I'_{dr} + V'_{qr} I'_{qr} \right) \approx \frac{3}{2} p_1 S \omega_1 \Psi_r I_{qs} \tag{4.100}$$

To regulate the rotor (and stator) active power as a generator (or speed as a motor), we need to control I_{qs}, that is, V'_{qr}. For rotor reactive power control, I_{ds} has to be controlled via V'_{dr}. A basic vector control scheme is shown in Figure 4.18, but with P_s and Q_s (stator powers) instead of P_r and Q_r closed-loop controllers. We call this scheme primitive as, in principle, it is implementable as it is. Prescribing the rotor voltages is not easy as they tend to be small around the standard synchronous speed (zero slip: $\omega_r = \omega_1$). So, the rotor current closed-loop control should be more robust. But this is beyond our scope here. See [13, Chapter 2] for more on the DFIM as a variable speed generator.

4.14 DOUBLY FED IM AS A BRUSHLESS EXCITER FOR SMS

The doubly fed IM may be a.c.-fed in the stator at constant frequency, ω_1, but for variable voltage (by using a thyristor Soft Starter). The rotor is rotated at speed $-\omega_r$, opposite to the stator mmf traveling speed, ω_1. So, the rotor emf has the slip frequency ω_2 (Figure 4.19).

$$\omega_2 = \omega_1 + \omega_r > \omega_1 \tag{4.101}$$

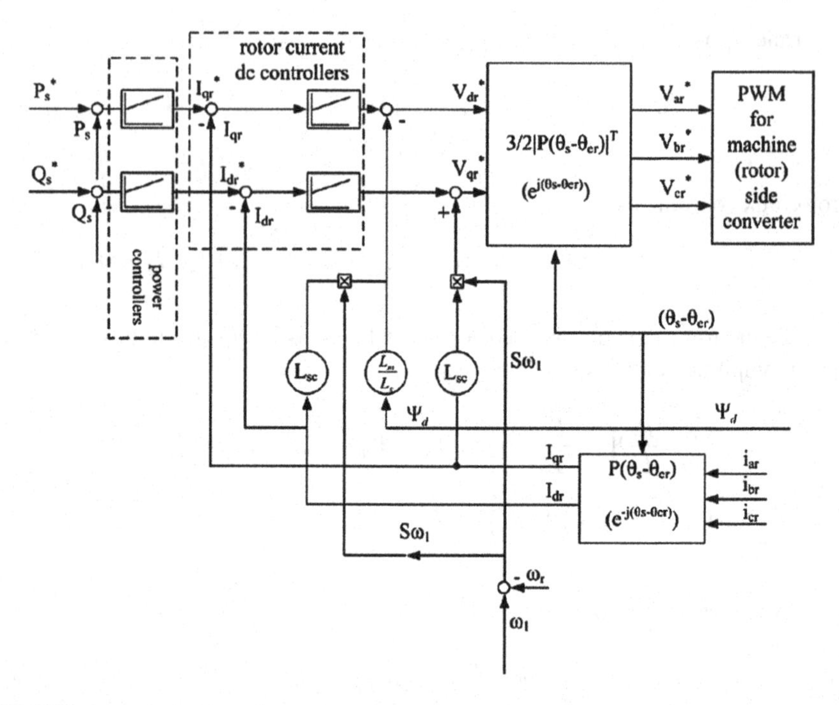

FIGURE 4.18 Primitive direct vector control of doubly fed induction generator in rotor coordinates, for controlled rotor flux.

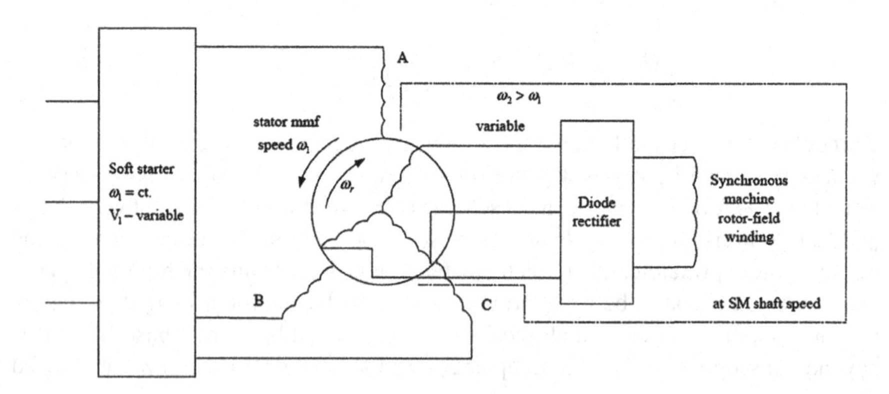

FIGURE 4.19 Doubly fed IM as brushless exciter for SMs.

At zero speed, the rotor emf is produced solely by "transformer" action. When ω_r increases, the rotor emf is more and more produced by motion. If $\omega_r = 4\omega_1$ (this situation occurs when the number of pole pairs of the DFIM, p_1, is four times larger than that of the SM (p_{1SG}), which is excited through the diode rectifier placed also on the rotor), up to 80% of the d.c. excitation power is produced by motion (from the mechanical shaft power) and 20% by the "transformer" action. The space phasor equations in stator coordinates are again

$$\begin{aligned}\overline{V}_s &= R_s\overline{I}'_s + \frac{d\overline{\Psi}'_s}{dt}\\ \overline{V}'_r &= -R'_r\overline{I}'_r + \frac{d\overline{\Psi}'_r}{dt} + j\omega_r\Psi'_r\end{aligned} \tag{4.102}$$

For steady-state and sinusoidal rotor currents (the diode rectifier at a high frequency, ω_2, is characterized often by such situations), $d/dt = j\omega_1$ (stator coordinates).

$$\begin{aligned}\overline{V}_s^s &= R_s\overline{I}_s^s + j\omega_1\overline{\Psi}_s^s;\quad \overline{\Psi}_s^s = L_s\overline{I}_s + L_m\overline{I}_r'^s\\ \overline{V}_r^{sl} &= -R'_r\overline{I}_r^{sl} + j\left(\omega_1+\omega_r\right)\overline{\Psi}_r^{sl};\quad \overline{\Psi}_r^{sl} = L'_r\overline{I}_r'^s + L_m\overline{I}_s\end{aligned} \tag{4.103}$$

$\overline{V}_r'$, as in Equation 4.103, has the frequency ω_1, but in reality (in rotor coordinates) $\overline{V}_r^{rl}$ is

$$\overline{V}_r'^r = \overline{V}_r'^s \cdot e^{j\omega_r t} = \overline{V}_r' \cdot e^{j\left(\omega_1+\omega_r\right)\cdot t} \tag{4.104}$$

and thus has the frequency: $\omega_2 = \omega_1 + \omega_r$. In synchronous coordinates, however, it is all d.c. in steady state. But the steady-state equations (Equation 4.103) keep the same aspect in all coordinates.

We can solve Equation 4.103 to get $\overline{I}_s$ and $\overline{I}_r'$. Let us consider a pure three-phase resistive load of this particular "generator" (the ideal diode rectifier can be assimilated to the unity power factor load):

$$\overline{V}_r' = \overline{I}_r' \cdot R_{load} \tag{4.105}$$

So

$$\overline{I}_s = \frac{jI'_r\left(R'_r + j\omega_2L'_r + R_{load}\right)}{\omega_2 L_m} \tag{4.106}$$

$$\overline{I}_r' = \frac{\overline{V}_s}{\dfrac{j\left(R'_r + j\omega_2L'_r + R_{load}\right)\left(R_s + j\omega_1 L_s\right)}{\omega_2 L_m} + j\omega_1 L_m} \tag{4.107}$$

The stator apparent power is

$$\overline{S}_s = P_s + jQ_s = \frac{3}{2}\left(\overline{V}_s I_s^*\right) \tag{4.108}$$

while the rotor power, P_r, is

$$P_r = \frac{3}{2}\text{Re}\left(\overline{V}_r'\overline{I}_r'^*\right) = \frac{3}{2}R_{load}\left(I'_r\right)^2 \tag{4.109}$$

Example 4.6: Doubly Fed IM (DFIM) as Brushless SM Exciter

Let us consider a DFIM used to excite a large synchronous generator (SG). Its parameters are $R_s = R_r' = 3.815\ \Omega$, $L_{sl} = L_{rl}' = 9.45 \times 10^{-5}$ H, $L_m = 2.02 \times 10^{-3}$ H, stator frequency $f_1 = 60$ Hz, and rotor speed $n = 1800$ rpm (4-pole SG). The number of DFIM pole pairs is $p_1 = 6$. The ratio of the rotor/stator turns is $a_{rs} = 1$ and $V_{snl} =$ 440 V (line voltage, RMS).
Calculate

a. The rotor frequency and the ideal no-load actual rotor voltage, V_{r0}', at $n = 1800$ rpm and $n = 0$ rpm
b. R_{load}, V_r^r, and P_r (all in the rotor) at zero speed. For $I_r' = 1000$ A (phase RMS)
c. The stator voltage V_s, I_s, P_s, Q_s, and P_r for the same R_{load} and current $I_r' = 1000$ A but at $n = 1800$ rpm

Solution:

a. The rotor frequency, ω_2, is, from Equation 4.102,

$$\omega_2 = \omega_1\left(1 + \frac{p_1}{p_{1SG}}\right) = \omega_1\left(1 + \frac{6}{2}\right) = 4\omega_1$$

Consequently, $f_{2n} = 4f_1 = 240$ Hz.
The rotor ideal no-load voltage, V_{r0}', is

$$\left(V_{r0l}'\right)_{n=1800\text{ rpm}} = V_s a_{rs}\frac{\omega_2}{\omega_1} = 1 \times 440 \times \frac{4}{1} = 1760\text{ V (line, RMS)}$$

At zero speed

$$\left(V_{r0l}'\right)_{n=0\text{ rpm}} = V_s a_{rs}\frac{\omega_1}{\omega_1} = 1 \times 440 \times \frac{1}{1} = 440\text{ V (line, RMS)}$$

$$\left(V_{ro}'\right)_{\text{phase,RMS}} = \frac{440}{\sqrt{3}} \approx 254\text{ V}$$

b. For $n = 0$ (zero speed), from Equation 4.107, with $\omega_2 = \omega_1$, $I_r' = 1000\sqrt{2}$ A, and $V_s = \left(440/\sqrt{3}\right)\sqrt{2}$ V, we find the load resistance $R_{load} = 0.226\ \Omega$/phase. The rotor phase voltage is

$$\left(V_{ro}'\right)_{\text{phase, RMS}} = R_{load}\left(I_r'\right)_{\text{phase, RMS}} = 0.226 \times 1000 = 226\text{ V}$$

So the voltage regulation, at zero speed, is

$$\Delta V = \frac{V_s - V_r}{V_s} = \frac{254 - 226}{254} = 0.1102 = 11.02\%$$

The rotor (output) power at zero speed, $(P_r)_{n=0}$, is

$$\left(P_r\right)_{n=0} = 3V_r I_r = 3 \times 226 \times 1000 = 678\text{ kW}$$

c. For n = 1800 rpm, again from Equations 4.106 and 4.107, but with $\omega_2 = 4\omega_1$, we now calculate V_s, as R_{load} and I_r' are given:

$$\left(\bar{V}_s\right)_{phase,\ RMS} = -64.06 - j72.1;\quad V_s = 96.8\left(\text{RMS per phase}\right)$$

The stator current, I_s, (from Equation 4.106) is

$$\bar{I}_s = -1046 + j75.3;\quad I_s = 1049.3\ \text{A} > I_r = 1000\ \text{A}$$

because machine magnetization is done from the stator.

The stator powers (Equation 4.108) are

$$P_s + jQ_s = 184.752\ \text{kW} + j240.751\,\text{kVAR}$$

The delivered rotor power is the same as for zero speed, P_r = 678 kW. So, the difference in powers comes from the shaft (from the mechanical power of the SG).

Discussion:

- At standstill, all output rotor power comes from the stator and the DFIM operates as a transformer with rectifier load. The required stator power and the voltage are maximum.
- As the speed increases, less and less active power is required from the stator as more and more of output (rotor) power is extracted from the shaft mechanical (SG) power.
- The DFIM may serve as a brushless exciter for SMs (SGs) from zero speed and is less sensitive to SG (SM) terminal voltage sags (due to grid faults), because most power is produced mechanically.
- The voltage regulation is small because the "internal reactance" of the DFIM is the short-circuit reactance, $\omega_1 L_{sc}$. Only an a.c. Variac (thyristor Soft Starter) is required to control the SG (SM) field current.

With

$$\bar{\Psi}_r' = \frac{L_m}{L_s}\bar{\Psi}_s + L_{sc}\bar{I}_r' \tag{4.110}$$

the rotor voltage equation (Equation 4.103) becomes

$$\bar{V}_r' = j\omega_2\frac{L_m}{L_s}\bar{\Psi}_s - \left(R_r' + j\omega_2 L_{sc}\right)I_r' = \bar{E}_r' - Z_{DFIM}I_r' \tag{4.111}$$

For $R_s \approx 0$

$$\bar{E}_r' = j\omega_2\frac{L_m}{L_s}\bar{\Psi}_s \approx \bar{V}_s\frac{L_m}{L_s}\frac{\omega_2}{\omega_1} \tag{4.112}$$

So the rotor emf, $\bar{E}_r'$, varies with the stator voltage, V_s, and with the rotor frequency, ω_2 (or speed ω_r). The short-circuit reactance, $\omega_1 L_{sc}$, is evident as Z_{DFIM}.

Voltage regulation is much smaller than that in an inverted-configuration synchronous auxiliary generator on a shaft as an SG d.c. exciter (stator d.c. excitation with a three-phase rotor winding and a diode rectifier), which extracts practically all output power mechanically, and thus may not operate from zero speed. Zero speed operation is required in large variable speed synchronous motor drives.

In view of the above merits, no wonder why the DFIM is used as a brushless exciter for large-power synchronous machines by major global manufacturers in this field.

4.15 PARAMETER ESTIMATION IN STANDSTILL TESTS/LAB 4.3

By parameters we mean

- The magnetization curve, $\Psi_m^*(I_m)$, with the magnetization inductance, $L_m(I_m) = \Psi_m^*(I_m)/I_m$, and the transient magnetization inductance $L_m(I_m) = d\Psi_m^*(I_m)/dI_m$; I_m is the magnetization current.
- The resistances and leakage inductances of the stator, R_s and L_{sl}, and of single (or double, or triple) rotor circuits, R'_{r1}, R'_{r2}, R'_{r3}, L'_{rl1}, L'_{rl2}, and L'_{rl3}, reduced to the stator. These parameters are required in the investigation of steady state and transient performance and for control or monitoring or system design.

While the main flux path magnetic saturation appears in no-load to load operation, the leakage flux path saturation occurs at overcurrents (above 2–3 I_{rated}), unless closed slots are used on the rotor, in order to reduce noise, vibration, and stray-load losses.

Testing of three-phase IMs is highly standardized (see IEC-34 standard series, NEMA 1961–1993 standards for large IMs). Only standstill flux decay and frequency response tests for parameter identification are detailed here.

4.15.1 Standstill Flux Decay for Magnetization Curve Identification: $\Psi_m^*(I_m)$

At standstill, the rotor cage IM is d.c. supplied at an initial value, I_{A0}, with phase A in series with phases B and C in parallel (Figure 4.20).

The arrangement in Figure 4.20 implies a few important constraints:

$$I_B = I_C = -I_A/2; \quad I_A + I_B + I_C = 0$$
$$V_B = V_C; \quad V_A = -2V_B, \text{ because } V_A + V_B + V_C = 0 \tag{4.113}$$
$$V_{ABC} = V_A - V_B = \frac{3}{2}V_A(t)$$

The current and voltage space phasors in stator coordinates are

$$\bar{I}_s(t) = \frac{2}{3}\left[I_A + I_B e^{j2\pi/3} + I_C e^{-j2\pi/3}\right] = I_A(t)$$
$$\bar{V}_s(t) = \frac{2}{3}\left[V_A + V_B e^{j2\pi/3} + V_C e^{-j2\pi/3}\right] \tag{4.114}$$
$$= V_A(t) = \frac{2}{3}V_{ABC}^{(t)}$$

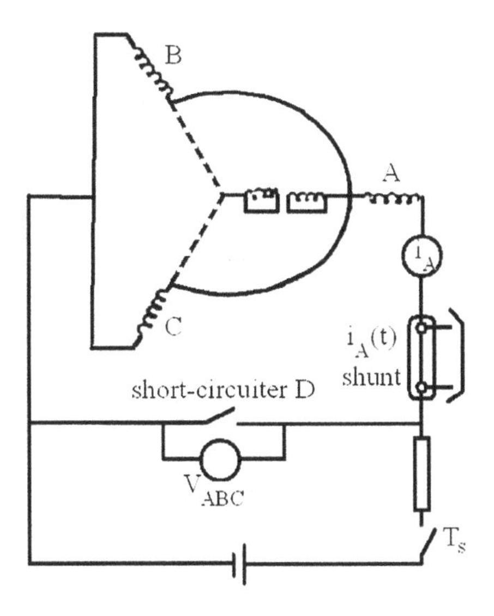

FIGURE 4.20 Standstill flux decay test of IM.

Now, once the d.c. current, I_{A0}, is "installed," the switch, T_s, is turned off and the stator current continues to flow until extinction (decay) through the freewheeling diode D (as for the SM tests). The stator equation after T_s is turned off is

$$-\frac{2}{3}V_{\text{diode}}^{(+)} = \overline{I}_s R_s + L_{sl}\frac{d\overline{I}_s}{dt} + \frac{d\overline{\Psi}_m}{dt} \tag{4.115}$$

The rotor equations for a dual cage rotor configuration are

$$\begin{aligned} 0 &= \overline{I}'_{r1}R'_{r1} + L'_{rl1}\frac{d\overline{I}'_{r1}}{dt} + \frac{d\overline{\Psi}_m}{dt} \\ 0 &= \overline{I}'_{r2}R'_{r2} + L'_{rl2}\frac{d\overline{I}'_{r2}}{dt} + \frac{d\overline{\Psi}_m}{dt} \end{aligned} \tag{4.116}$$

We may use these equations in two ways:

- By integrating the stator equation (only) to find the magnetization curve, $\Psi_m(I_m)$, $I_m = I_{A0}$:

$$\Psi_m(I_m) = L_m(I_m)I_m = R_s\int I_A(t)\,dt + \frac{2}{3}\int V_{\text{diode}}^{(+)}\,dt - L_{sl}I_m \tag{4.117}$$

The leakage inductance, L_{sl}, has to be already known from the design or from the stalled standard frequency (a.c.) test.

By gradually increasing the initial current, I_{A0}, value, the entire magnetization curve can be obtained. To avoid errors caused by temperature in R_s, the latter may be calculated as $R_s = (2V_{ABC0})/(3I_{A0})$, before each current decay test (which should last 1–2 s).

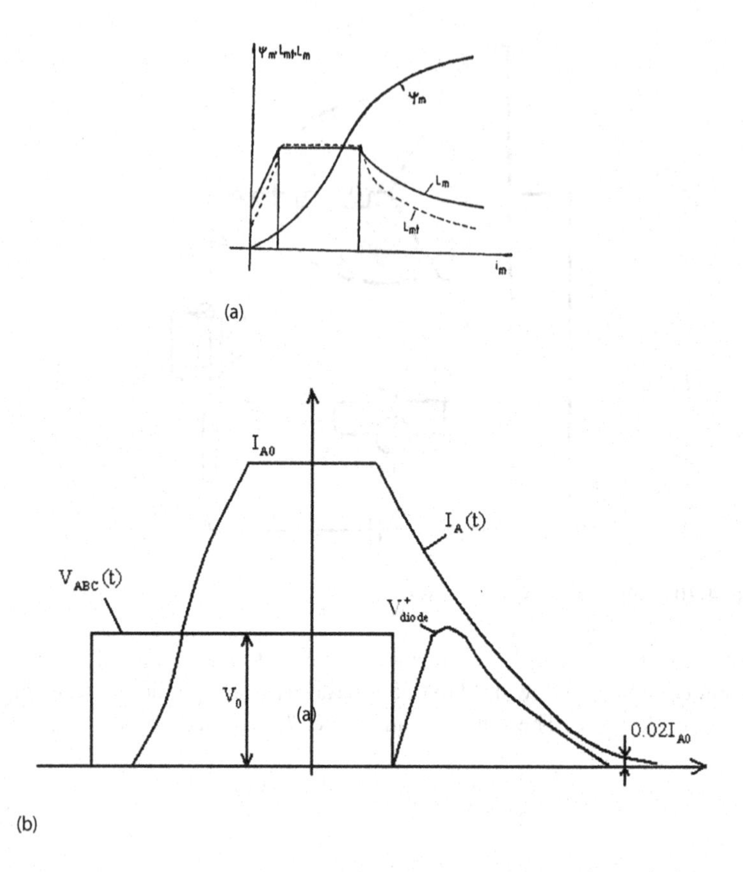

FIGURE 4.21 IM flux decay test outputs: (a) The magnetization curve, $\Psi_m^*(I_m)$, and normal and transient magnetization inductances, L_m and L_{mt}, and (b) current, $I_A(t)$, and V_{diode}^+ during the flux decay test.

Results such as those in Figure 4.21 are obtained.

- By curve fitting, the magnetization curve may be approximated to a differentiable function, and then, from here, $L_{mt}(I_m)$ is calculated as

$$L_{mt}\left(I_m\right) = \frac{d\Psi_m^*}{dI_m} \tag{4.118}$$

The magnetization curve may be obtained from no-load motor tests (as described in Chapter 5, Vol.1 (companion book)) at variable input voltages, but it requires more time and resources to do it. All comparisons between results showed that the standstill flux decay tests produce reliable results for in-speed operation.

As a significant degree of magnetic saturation is allowed by design, to cut machine size, and in the stand-alone induction generator mode (self-excited directly by a parallel+series capacitor or through a PWM inverter), the magnetization curve is a crucial parameter.

4.15.2 Identification of Resistances and Leakage Inductances from Standstill Flux Decay Tests

Returning to Equations 4.115 and 4.116 and with $\overline{I}_s(t) = I_A(t)$, we can process the acquired descending current curve and identify, by curve fitting, $R_{r1}^{'}$, $R_{r2}^{'}$, $L_{rl1}^{'}$, and $L_{rl2}^{'}$ (for a dual cage rotor model) (Figure 4.21b). With d/d$t \to s$ we get

$$\underline{Z}_s \approx \frac{\overline{V}_s(s)}{\overline{I}_s(s)} = -\frac{2}{3}\frac{\overline{V}_{ABC}(s)}{\overline{I}_A(s)} = R_s + s\left(L_{sl} + L_{mt}\right)\frac{(1+s\tau'')(1+s\tau')}{(1+s\tau_0'')(1+s\tau_0')} \tag{4.119}$$

For a step voltage application

$$\overline{V}_s(s) = -\frac{2}{3}V_{ABC}(s) = -\frac{2}{3}\frac{V_0}{s} \tag{4.120}$$

So

$$\overline{I}_s(s) = -\frac{2V_0}{3s\underline{Z}(s)}; \quad L_s = L_{sl} + L_m \tag{4.121}$$

But the time variation of $\overline{I}_s(t)$ can be derived directly from Section 4.7.1 on a sudden short circuit, with $\omega_1 = 0$, $\Psi_0 = 0$, and initial current $I_{s0} = I_{A0} = V_{s0}/R_s$. If, at this time, $\tau_a \approx L_s''/R_s$, then, approximately,

$$\begin{aligned} I_s(t) \approx \frac{V_{s0}}{R_s} - V_{s0}\Bigg[& \frac{\tau'\tau_a}{\tau_a - \tau'}\left(e^{-t/\tau_a} - e^{-t/\tau'}\right)\left(\frac{1}{L'} - \frac{1}{L_s}\right) \\ & + \frac{\tau''\tau_a}{\tau_a - \tau''}\left(e^{-t/\tau_a} - e^{-t/\tau''}\right)\left(\frac{1}{L''} - \frac{1}{L'}\right) \\ & + \frac{\tau_a}{L_s}\left(1 - e^{-t/\tau_a}\right)\Bigg] \end{aligned} \tag{4.122}$$

In this case, the initial value of current is $I_{A0} = (I_s)_{t=0} = V_{s0}/R_s$ while the final value is zero, as expected. Curve-fitting regression methods are applied to find the values of L', L'', and L_s, and τ' and τ''. If the initial current, I_{A0}, is smaller than the rated magnetization current, magnetic saturation is not relevant, and, thus, a nonsaturated value of synchronous inductance, $L_s = L_{sl} + L_m$, is used, as known from design (or from the no-load current value). Due to the special mix of frequencies in step up or down to zero voltage pulses, it is apparent that the frequency content of rotor currents in real running conditions is not well matched in the flux decay tests.

This is how standstill frequency response (SSFR) tests have come into play.

4.15.3 Standstill Frequency Response (SSFR) Tests

SSFR tests for IMs are similar to those for SM, but they are conducted only once, at random rotor position. The experimental arrangement of Figure 4.20 still holds but, this time, a wide variable frequency sinusoidal voltage source is required (from 0.01 to 100 Hz, which will suffice for the most practical cases). This test will be done for each frequency separately, maintaining it for a few periods and measuring the voltage, current amplitudes, and their phase shift angle; $Z(s)$, in Equation 4.119, now becomes $Z(j\omega)$:

$$\underline{Z}(j\omega) = R_s + j\omega(L_{sl} + L_{mt})\frac{(1+j\omega\tau'')(1+j\omega\tau')}{(1+j\omega\tau_0'')(1+j\omega\tau_0')} \tag{4.123}$$

where

$$|\underline{Z}(j\omega)| = \frac{2}{3}\frac{|V_{ABC}|_{RMS}}{|I_{ABC}|_{RMS}}; \arg(\underline{Z}(j\omega)) = \varphi(V_{ABC}, I_{ABC}) \tag{4.124}$$

Typical results from such tests in p.u. are given in Figure 4.22.

For the data in Figure 4.22, using only the curve fitting of the amplitude yields results such as

$$l(s) = 3\frac{(1+s\times0.0125)(1+s\times0.318)}{(1+s\times0.0267)(1+s\times1.073)};[p.u.] \tag{4.125}$$

It is also feasible to use only Arg $(Z(j\omega))$ information and identify the above time constants [14]. The $l(s)$ has two poles and two zeros:$1/\tau_0'$, $1/\tau_0''$, $1/\tau'$, and $1/\tau''$; also $\tau_0'' > \tau''$ and $\tau_0' > \tau'$. Let us denote $\alpha = \tau'/\tau_0' < 1$; this is a lag circuit. The maximum phase lag, φ_c', (Figure 4.22) is

$$\sin\varphi_c' \approx \frac{\alpha - 1}{\alpha + 1} \tag{4.126}$$

The gain change due to the respective zero/pole pair, with α from Figure 4.22, is

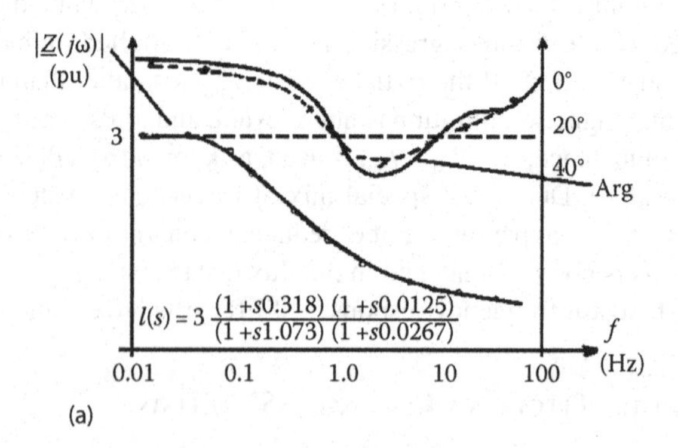

FIGURE 4.22 SSFR of a two-cage rotor IM.

$$\text{gain change} = -20 \log \alpha \tag{4.127}$$

So

$$\tau' = \frac{\tau_0'}{\alpha} \text{ and } \tau_0' = \frac{\sqrt{\alpha}}{2\pi f_c'}. \tag{4.128}$$

After the first zero/pole pair, τ', τ_0' has been calculated, the latter is introduced in Equation 4.125, and then Arg $(Z(j\omega))$ is recalculated versus frequency, and, thus, a new maximum argument is obtained at φ'' (at frequency f_c'') and the computation in Equations 4.127 and 4.128 is done again to find τ'' and τ_0''. This process is continued until no more maxima of Arg $(Z(j\omega))$ occurs.

Additional steps to improve precision in finding f_c' and f_c'', and φ' and φ'' may be taken [14].

It is evident from Figure 4.22 that the two Arg $(Z(j\omega))$ maxima suggest a fair double-cage representation. For more on three-phase IM testing, see [4, Chapter 22].

4.16 SPLIT-PHASE CAPACITOR IM TRANSIENTS/LAB 4.4

The split-phase IM is still used at remarkable performance (efficiency) in driving pumps or compressors for household appliances, in the range of 100 W or more. Many other domestic tools, such as cloth washers, drillers, and sawyers, make use of split-phase capacitor IMs in the range of hundreds of watts to 1 kW. They use a starting self-variable resistor, R_{start}, or a capacitor, C_{start}, in the auxiliary phase. The latter may be kept on during the on-load operation, to increase efficiency, but with a smaller running capacitor, C_{run} ($C_{run} < C_{start}$) (Figurer 4.23a). Yet another application may use a three-phase winding for unidirectional or reversible motion (Figure 4.23b and c).

For some cloth-washing machines, there is a three-phase 12-pole winding used for washing (at a low reversible speed) and a separate 2-pole orthogonal winding for spinning (at a high speed). Switching back and forth for the washing mode and for the drying mode implies important transients.

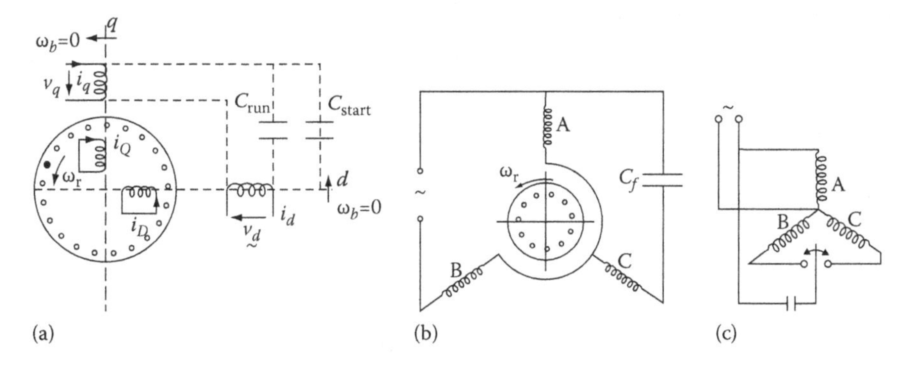

FIGURE 4.23 Split-phase capacitor IMs with (a) auxiliary phase, (b) three-phase, unidirectional, and (c) three-phase, bidirectional.

The three-phase connections (Figure 4.23b and c) may be reduced to orthogonal windings (Figure 4.23a) [4, Chapter 23]. Here, only orthogonal windings are treated.

For better starting performance, 120° space-angle-shifted main and auxiliary windings may be used, but they also may be reduced to two orthogonal ones [4, Chapter 23].

Finally, the general case, when the main and auxiliary windings are orthogonal, but they use different number of slots and copper weights, should be treated in phase variables.

4.16.1 Phase Variable Model

In this case, the rotor is modeled by orthogonal windings, d_r and q_r:

$$\left|I_{m,a,d_r,q_r}\right|\left|R_{m,a,d_r,q_r}\right|-\left|V_{m,a,d_r,q_r}\right|=-\frac{d}{dt}\left|\Psi_{m,a,d_r,q_r}\right| \tag{4.129}$$

$$\begin{aligned}\left|I_{m,a,d_r,q_r}\right|&=\left|I_m,I_a,I_{d_r},I_{q_r}\right|^T\\ \left|I_{m,a,d_r,q_r}\right|&=\mathrm{Diag}\left(R_m,R_a,R_r,R_r\right)\\ \left|V_{m,a,d_r,q_r}\right|&=\left|V_m,V_a,0,0\right|\end{aligned} \tag{4.130}$$

$$\begin{vmatrix}\Psi_m\\ \Psi_a\\ \Psi_{dr}\\ \Psi_{qr}\end{vmatrix}=\begin{vmatrix}L_{ml}+L_m^m & 0 & L_{mr}\cos\theta_{er} & -L_{mr}\sin\theta_{er}\\ 0 & L_{al}+L_m^a & L_{ar}\sin\theta_{er} & L_{ar}\cos\theta_{er}\\ L_{mr}\cos\theta_{er} & L_{ar}\sin\theta_{er} & L'_{rl}+L_m^m & 0\\ -L_{mr}\sin\theta_{er} & L_{ar}\cos\theta_{er} & 0 & L'_{rl}+L_m^m\end{vmatrix}\cdot\begin{vmatrix}I_m\\ I_a\\ I_{dr}\\ I_{qr}\end{vmatrix} \tag{4.131}$$

$$T_e=p_1\left[I\right]\frac{\partial\left|L\left(\theta_{er}\right)\right|}{\partial\theta_{er}}\left[I\right]^T \tag{4.132}$$

$$\frac{J}{p_1}\frac{d\omega_r}{dt}=T_e-T_{load};\quad \frac{d\theta_{er}}{dt}=\omega_r;\quad \omega_r=2\pi pn \tag{4.133}$$

We also should use the constraints

$$\begin{aligned}V_m(t)&=V_m\sqrt{2}\cos\left(\omega_1t+\gamma_0\right)\\ V_m(t)&=V_a(t)+V_C(t)\end{aligned} \tag{4.134}$$

and

$$\frac{dV_C}{dt}=\frac{1}{C}I_a \tag{4.135}$$

When a resistor is added to the auxiliary winding, V_C is replaced by $R_{start}I_a$ and Equation 4.135 is eliminated. The system has a seventh order, with many variable coefficients, and may be solved by numerical methods, albeit within a large CPU time.

This general model gets simplified if the auxiliary winding and the main winding make use of the same copper weight when

$$\begin{aligned} R_a &= R_m a^2; \quad a = w_a k_{wa}/(w_m k_{wm}) \\ L_{al} &= L_{ml} a^2 \\ L_{ar} &= L_{mr} a \\ L_m^a &= L_m^m a^2 \end{aligned} \tag{4.136}$$

In such a case, the dq model in any coordinates may be used because the stator windings become almost symmetric under conditions 4.136, despite different number of turns.

4.16.2 dq Model

Let us use stator coordinates and consider the d axis along the m (main) winding axis and the q axis along the auxiliary winding axis. There is no need to reduce the auxiliary winding to the main winding as long as Equation 4.136 is satisfied.

The dq model equations are straightforward as the stator windings are already orthogonal, and, thus, no Park transformation is needed (for the phase machine connection in Figure 4.23b the latter is instrumental [4, Chapter 23]:

$$\begin{aligned} I_d R_m - V_d &= -\frac{d\Psi_d}{dt}; \quad I_q R_a - V_q = -\frac{d\Psi_q}{dt} \\ I_{dr} R_r &= -\frac{d\Psi_{dr}}{dt} - \omega_r \Psi_{qr}; \quad I_{qr} R_r = -\frac{d\Psi_{qr}}{dt} + \omega_r \Psi_{dr} \\ \Psi_d &= L_{ml} I_d + L_{dm}(I_{dr} + I_d) \\ \Psi_{dr} &= L'_{rl} I_{dr} + L_{dm}(I_{dr} + I_d) \\ \Psi_q &= L_{al} I_q + L_{qm}\left(I_q + \frac{1}{a} I_{qr}\right) \\ \Psi_{qr} &= L'_{rl} I_{qr} + L_{qm}(a I_q + I_{qr}) \end{aligned} \tag{4.137}$$

Again

$$\frac{L_{qm}}{L_{dm}} = a^2 \text{ and } L_{dm} = L_m^m; \quad I_m = I_d; \quad I_a = I_q \tag{4.138}$$

$$V_d = V_m; \quad V_q = V_a = V_m - V_C(t); \quad \frac{dV_C}{dt} = \frac{I_q}{C} \tag{4.139}$$

$$T_e = -p_1(\Psi_{dr} I_{qr} - \Psi_{qr} I_{dr}) = p_1 L_{dm}(a I_q I_{dr} - I_d I_{qr}) \tag{4.140}$$

Adding the motion equations (Equation 4.133), the complete model is obtained. The system's order is now six (θ_{er}, the rotor position, is irrelevant here).

Note: Again, the dq model is valid only if conditions 4.136 are satisfied.

For steady state, the ± space phasor model has already been used in Chapter 5 (Vol.1 (companion book)), and hence it is not repeated here. For more on the subject, see [4, Chapters 24 through 28].

Magnetic saturation can be handled simply in the dq model by the $L_m^m(I_m)$ function, with flux as a variable. Magnetic saturation causes, under steady state, nonsinusoidal stator currents [4, Chapter 25].

4.17 LINEAR INDUCTION MOTOR TRANSIENTS

LIMs are used today in a wide variety of applications, such as urban people movers and propulsion of vehicles on wheels (UTDC in Canada), and with active magnetic suspension by controlled d.c.-fed electromagnets on board (Figure 4.24), (Japan, Korea).

Along each side of the MAGLEV in Figure 4.24b, there are LIMs interleaved with d.c. electromagnets, for suspension control. In this application, both use the same solid iron track (beam) as a back iron core [17].

It should be noticed also that if the normal (vertical) force of the LIM is of attraction type, then it helps the d.c.-controlled electromagnets in producing active magnetic suspension Figure 4.24b. This is the case in well-designed (for good efficiency) LIMs [15, Chapter 3].

The LIM flux control may be performed to produce certain dynamic properties in propulsion, while adding a 20%–30% to suspension via attraction force up to the base speed, U_b, above which flux weakening might be mandatory for propulsion control.

To a first approximation, we may ignore the frequency and saturation effects in the secondary (solid back iron plus aluminum sheet track). We also ignore the longitudinal end effect, though, even for urban people movers (at 20–40 m/s peak speed), the latter is already affecting performance by reducing thrust (propulsion), efficiency, and power factor (Chapter 5). It is possible to use correction coefficients to reduce F_x and L_m and increase R_r (all variable with slip), to account for the longitudinal end effect. The dq model is thus straightforward (as for LSMs in Chapter 3), but now, in synchronous coordinates,

$$\begin{aligned} &\overline{V}_s = R_s\overline{I}_s + jU_s\pi\overline{\Psi}_s/\tau + d\overline{\Psi}_s/dt; \quad U_s \text{ — synchronous speed } (\text{m/s}) \\ &0 = R_r\overline{I}_r' + d\overline{\Psi}_r'/dt + j(U_s - U)\pi\overline{\Psi}_r'/\tau; \quad U \text{ — speed } (\text{m/s}) \end{aligned} \tag{4.141}$$

$$F_x = \frac{3}{2}\frac{\pi}{\tau}\text{Re}\left(j\overline{\Psi}_s\overline{I}_s^*\right)$$

with

$$\overline{\Psi}_s = L_s\overline{I}_s + L_m\overline{I}_r'; \quad \overline{\Psi}_r' = L_r'\overline{I}_r' + L_m\overline{I}_s \tag{4.142}$$

Also, approximately, $L_{rl}' \ll L_{sl}$, so it may be neglected, and thus

$$\overline{\Psi}_r' = \overline{\Psi}_m = L_m\left(\overline{I}_r' + \overline{I}_s\right) = L_m\overline{I}_m \tag{4.143}$$

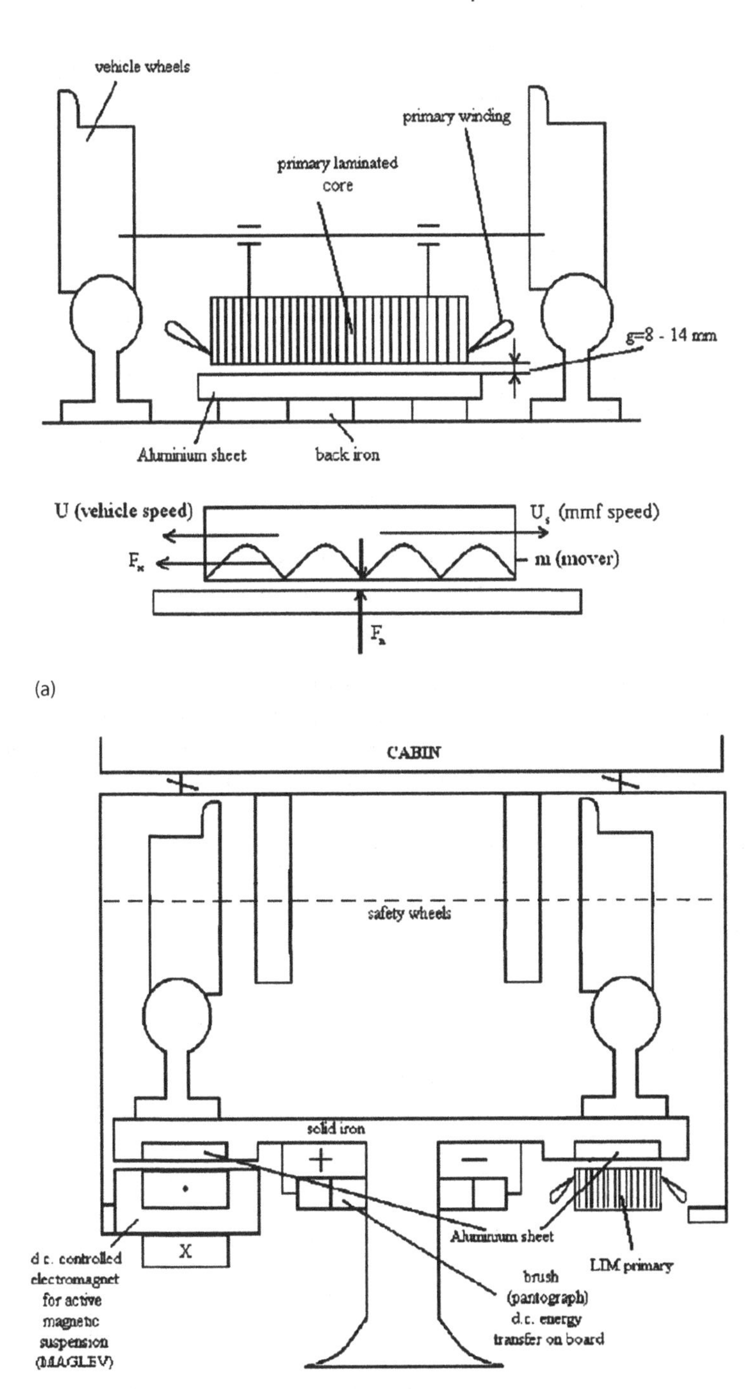

FIGURE 4.24 Single-sided LIM for people movers: (a) Vehicle on wheels and (b) MAGLEV.

So, the secondary flux linkage, $\overline{\Psi}'_r$, is "identical" to the airgap flux, and, thus, its closed-loop control might lead to the normal (suspension) force, F_n, control.

We should be careful when applying Park transformation to primary variables, because the primary is the mover:

$$\overline{I}_s = \frac{2}{3}\left(I_A + I_B e^{j\frac{2\pi}{3}} + I_C e^{-j\frac{2\pi}{3}}\right)e^{-j\Theta_s}; \quad \frac{d\theta_s}{dt} = -\frac{\pi}{\tau}x \tag{4.144}$$

where x is the linear motion variable.

To a first approximation (ignoring the repulsive force component), the normal (suspension) attraction force, F_n is

$$F_n \approx \frac{3}{2} I_m^2 \frac{\partial L_m}{\partial g} \tag{4.145}$$

Vector control (for "rotor" flux orientation) may be used for propulsion, while the normal force, $\hat{F}_n$, may be estimated and then controlled as desired, to reduce noise and vibration, and help active magnetic suspension (Figure 4.25).

Let us recall that not only we ignored the frequency effects in the secondary (upon R'_r) and magnetic saturation in the solid back iron (on R'_r and L_m), but also that L_m (magnetization inductance) varies with the airgap due to track irregularities and that a dynamic error of 20%–25% is allowed in the magnetic suspension airgap (height) control, in order to limit peak kVAs in the PWM converters used to control the d.c. electromagnets, for MAGLEVs.

There are two ways to handle such a situation:

- Better modeling
- More robust propulsion and suspension control

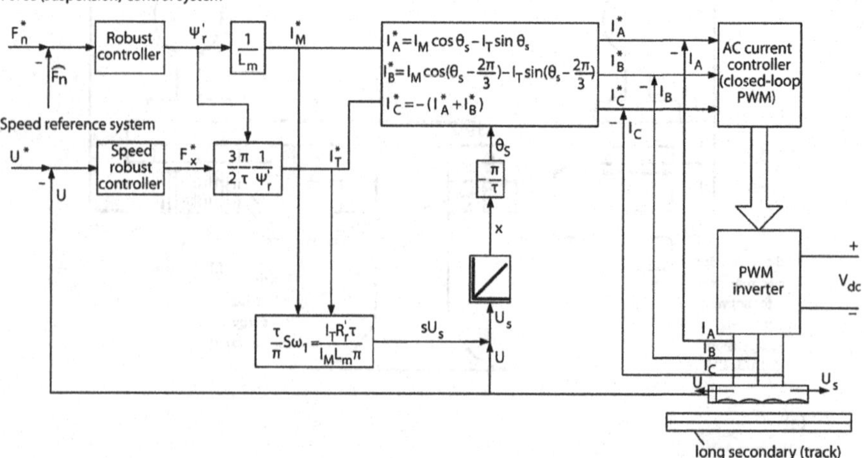

FIGURE 4.25 Vector control of LIM.

Both are worthy of investigation but the second method looks more practical and might deserve "the first shot". However, 3D FE circuit models or a ladder secondary circuit model having a large number of bars (as in this chapter: the mN_r model, Section 4.11) have been tried to simulate transient performance, including end effects, for transients. For control system design, simpler models to account for longitudinal end effect, frequency and saturation effects in secondary [16] and more practical airgap, secondary flux, $\bar{\Psi}'_r$, and speed, U, estimators are all still due.

4.18 LINE-START SELF-SYNCHRONIZING PREMIUM EFFICIENCY IMs

As more than half of all induction electric motors operate at rather constant speed as line-start connected to a.c. power grids, better efficiency is a paramount goal.

Superior classes of efficiency IE – 3, 4, 5 have been proposed recently [20–24].

However, in cage-rotor induction motors higher efficiency and limited starting current are rather conflicting goals. Optimal design (even FEM - only) methodologies have proved that notable 1-2% efficiency increase in efficiency imposes a higher starting current (6.8–7.5 p.u. instead of 6-6.4 p.u. in IE – 1, 2 motors).

This implies larger local power transformers and switching apparatus in the local a.c. power grid.

To avoid such a costly inconvenience three- and one-phase self-synchronizing IMs with a cage rotor provided with PMs "behind" the cage, placed in flux-barriers, have been proposed [Figure 4.26 a, b, c].

There are three stages in the operation of these machines:

- Asynchronous starting
- Self-synchronization
- Synchronous operation under load

During asynchronous operation there are three torque components:

- asynchronous torque produced by the cage: $T_{as}(t)$ which, if the cage is asymmetric (due to the presence of flux-barriers in the rotor) has pulsations, even for three phase symmetric windings in the stator. Magnetic saturation due to large cage currents during asynchronous starting influences not only the asynchronous torque but also the other two-torque components [24]
- PM braking torque T_{aPM} produced by the stator currents losses at rotor frequency $f_r = f_1 - sf_1$, produced by the voltages induced in the stator windings [23–24] via motion by PMs.

When the rotor during asynchronous starting reaches a certain, not too large, slip S_{ss} = (0.04-0.06 in general) the PM and reluctance torques T_{PMs}, T_{Rels}, act much as in the self-synchronization of d.c. excited SMs when the d.c. excitation circuit is commuted from an additional resistance (10 R_F) to the d.c. source.

The self-synchronization is a complex transient that depends on the load-torque level, inertia and the peak PM and reluctance synchronous torques (T_{PMs} and T_{Rels}),

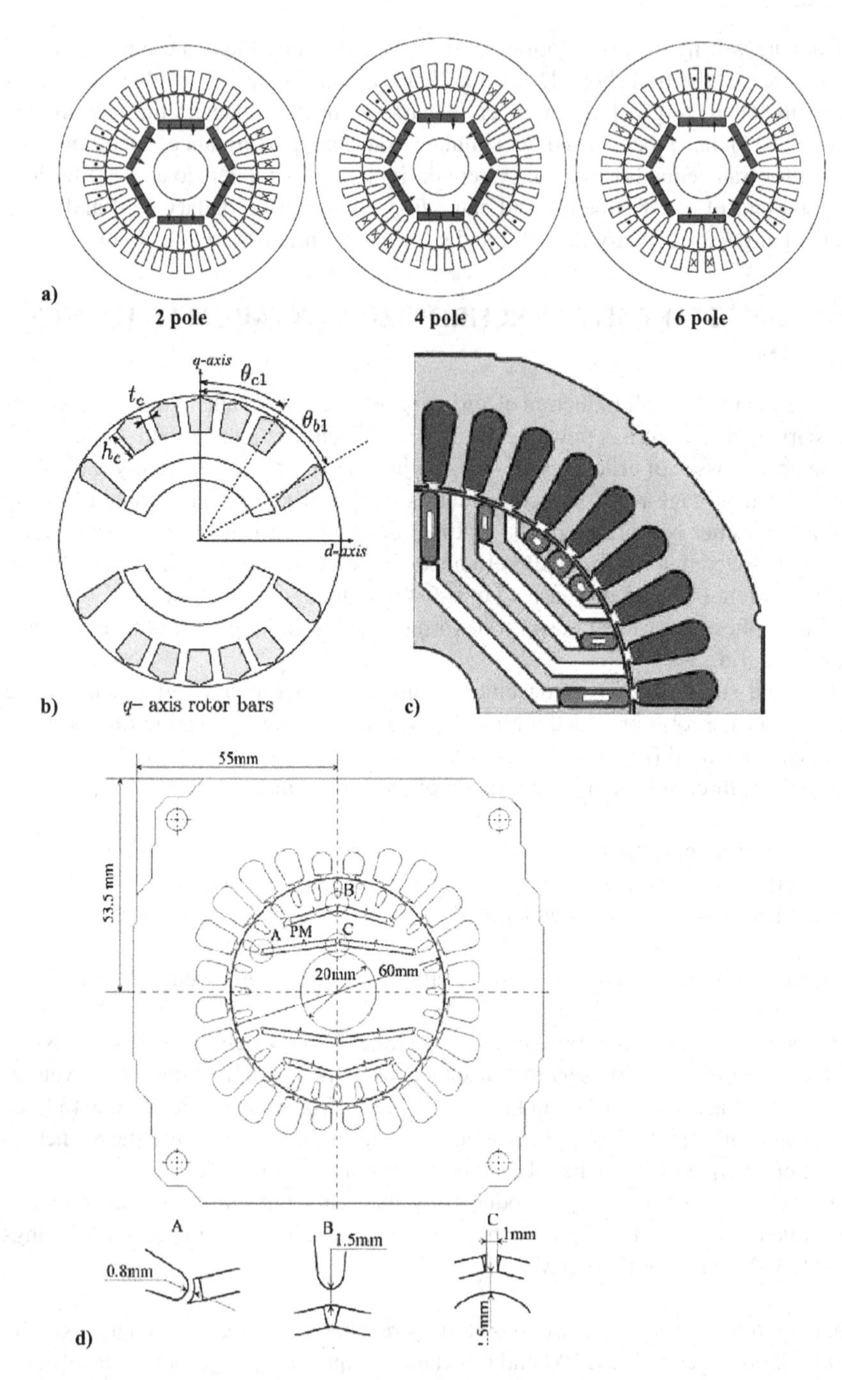

FIGURE 4.26 Line-start self-synchronizing PM (or/and reluctance) cage rotor IMs: a) three phase (PM cage rotor with different number of poles), b) three phase/(PM cage rotor with flux barriers), c) three phase/cage rotor with flux barriers ($L_d/L_q = 3.9$), d) single-phase/cage rotor with PMs in flux barriers.

but also on the slip Sss (which in turn, depends on rotor resistance R_r). A large rotor resistance provides for a large starting asynchronous torque to overcompensate the load torque (if any) and the PM asynchronous braking torque (T_{aPM}), for a safe starting. But Sss will be large and thus self-synchronization may be in jeopardy; plus, the stator-produced space harmonics fields may induce notable additional cage losses. In the absence of PMs the reluctance synchronous torque alone produces self-synchronization and synchronous operation.

Thus the asynchronous starting is improved ($T_{aPM} = 0$), but the self-synchronization and synchronous operation performance depend heavily on the rotor saliency (on L_d/L_q ratio). It is now evident that complex optimal design methodologies—analytical or FEM based—are required to mitigate between such strong constraints of notable conflictual nature. The situation is even more involved for one-phase source split-phase capacitor such small motor during self-synchronization when it has a cage rotor with flux barriers filled (or not) with assisting PMs [23]. The latter is treated in some detail in what follows to bring a feeling of magnitudes that should characterize any engineering investigation.

4.18.1 Line Start One-Phase—Source Split-Phase Capacitor Self-Synchronizing Induction Motor with PMs in the Rotor's Flux Barriers

The topology of interest here is the one already shown in Figure 4.26d. the aim of investigation is to optimally design such a motor "reconciling" the starting, self-synchronization and synchronous operation performance with an eye on active materials cost.

In our configuration the rotor cage is rather symmetric though it is not necessarily so as an asymmetry in the cage may help self-synchronization.

For steady-state asynchronous operation the asynchronous torque (performance) may be calculated using the direct and inverse (+ and −) sequence decomposition that produces the direct and inverse average torques [23] T_{asd}, T_{asq}:

$$T_{asd} = \mathrm{R}_e\left[\left|\underline{Z}_1\right|\left|\underline{I}_d^2\right|\right]\cdot\frac{p}{\omega_1};\quad T_{asi} = \mathrm{R}_e\left|\underline{Z}_2\left|\underline{I}_i\right|^2\right|;\quad T_{eas} = T_{asd} + T_{asi} \tag{4.146}$$

$$\underline{Z}_{1d} \approx \left(\frac{1}{jX_{dm}} + \frac{1}{\frac{R_{rd}}{S} + jX_{rd\sigma}}\right)^{-1};\ \underline{Z}_{1q} \approx \left(\frac{1}{jX_{qm}} + \frac{1}{\frac{R_{rq}}{S} + jX_{rq\sigma}}\right)^{-1}$$
$$\underline{Z}_{2d} \approx \left(\frac{1}{jX_{dm}} + \frac{1}{\frac{R_{rd}}{2-S} + jX_{rd\sigma}}\right)^{-1};\ \underline{Z}_{2q} \approx \left(\frac{1}{jX_{qm}} + \frac{1}{\frac{R_{rq}}{2-S} + jX_{rq\sigma}}\right)^{-1} \tag{4.147}$$

$$\underline{Z}_1 = \frac{\underline{Z}_{1d} + \underline{Z}_{1q}}{2};\quad \underline{Z}_2 = \frac{\underline{Z}_{2d} + \underline{Z}_{2q}}{2} \tag{4.148}$$

The direct and inverse sequence currents are

$$\left|V_{d,i}\right| = \left|\underline{Z}_{d,i}\right| \cdot \left|I_{di}\right|; \quad Z_{d,i} = \begin{bmatrix} \underline{Z}_1 & 0 \\ 0 & \underline{Z}_2 \end{bmatrix} \tag{4.149}$$

To allow for asymmetric stator windings:

$$\frac{R_m}{R_a} = \frac{L_{ml}}{L_{al}} \approx \frac{1}{a^2}; \quad a = \frac{M_a K_{wa}}{M_m K_{wm}} \tag{4.150}$$

The stator resistances and leakage inductances are externalized from the (+ −) model and thus the new main and auxiliary winding voltages $V_m^{'}$, $V_a^{'}$ and currents are modified:

$$\begin{vmatrix} V_m' \\ V_a' \end{vmatrix} = \left|M_{v02p}\right| \cdot \begin{vmatrix} V_\alpha \\ V_\beta \end{vmatrix}; \quad \left(M_{v02p}\right) = \begin{bmatrix} 1 & 0 \\ 1 & a \end{bmatrix}$$
$$\begin{vmatrix} i_\alpha \\ i_\beta \end{vmatrix} = \left|M_{i02p}\right| \cdot \begin{vmatrix} i_m \\ i_a \end{vmatrix}; \quad \left|M_{i02p}\right| = \begin{bmatrix} 1 & 0 \\ 1 & a \end{bmatrix} \tag{4.151}$$

Now we may search directly for main and auxiliary winding currents I_m and I_a from:

$$\begin{vmatrix} V_m' \\ V_a' \end{vmatrix} = \left|\underline{Z}_0\right| \cdot \begin{vmatrix} \underline{I}_m \\ \underline{I}_a \end{vmatrix}; \quad \left|M_{di}\right| = \frac{1}{\sqrt{2}} \begin{bmatrix} 1 & 1 \\ j & -j \end{bmatrix} \tag{4.152}$$

with

$$\underline{Z}_0 = \left|M_{v02p}\right| \left|M_{di}\right| \cdot \left|\underline{Z}_{di}\right| \left|M_{di}\right|^{-1} \left|M_{i02p}\right| \tag{4.153}$$

Adding the stator resistances and leakage inductances and the capacitor impedance in series with the auxiliary winding yields:

$$\begin{vmatrix} \underline{V}_s \\ \underline{V}_s \end{vmatrix} = \left|\underline{Z}_p\right| \cdot \begin{vmatrix} \underline{I}_m \\ \underline{I}_a \end{vmatrix}; \quad \underline{Z}_p = \underline{Z}_0 + \begin{vmatrix} R_m + j\omega_1 L_{ml} & 0 \\ 0 & R_a + j\omega_1 L_{al} - \dfrac{j}{\omega_1 C_a} \end{vmatrix} \tag{4.154}$$

The PM average braking torque T_{aPM} due to i_{dPM}, i_{qPM} stator currents induced in the shorted stator by PM motion induced emfs at frequency $\omega_{1PM} = 2\pi f_1(1-S)$ may be calculated by a dq synchronous model.

For the case study here the 100 W, 3000 rpm motor parameters are given in the Table 4.1. While its main geometrical data are visible in Table 4.2. However, the synchronous model requires the magnetization inductances L_{dm} and L_{qm} which have to be computed analytically or by FEM.

With the fundamental value of PM airgap flux density B_{gPM1} calculated [23] then the PM-emf, E_{PM}, is, as known:

$$E_{PM} = \omega_1 \cdot B_{gPM1} \cdot \frac{2}{\pi} \cdot \tau \cdot l_{stack} \cdot N_m \cdot k_{wm1} \tag{4.155}$$

TABLE 4.1
Motor parameters 100W, 3000 rpm motor

Circuit Parameters

L_{md}^*	0.57 H	R_m	24 Ω
L_{mq}^*	1.39 H	R_a	20.1 Ω
L_{ml}	0.089 H	a	0.844
L_{al}	0.048 H	C_a	3.3 μF
ψ_{PM}	1.14 Wb	C_s	12 Ca

TABLE 4.2
Motor geometry 100W, 3000 rpm motor

Main Constructive Dimensions

D_{is}	61.2 mm	Inner stator diameter
D_{os}	136.7 mm	Outer stator diameter
l_{stack}	48.5 mm	Stator stack length
δ	0.25 mm	Airgap
h_{PM}	1.4 min	PM thickness
B_r	1.1 T	PM remanent flux density (NdFeB)

The machine is thus considered as a PM-motor/generator with two asymmetric windings at short circuit. The stator currents are calculated and the stator losses P_{waPM} are related to PM averages braking torque T_{aPM} as:

$$T_{aPM} = \frac{P_{WaPM}}{\omega_{1PM}} \cdot p_1 \tag{4.156}$$

Results for the case in point for asynchronous operation are given in Figure 4.27.

The direct and inverse asynchronous PM braking and total asynchronous average torque are all visible in Figure 4.27.

Note: Some additional average and pulsating PM braking torque due to magnetic saturation in the rotor during starting has been uncovered in three-phase self-synchronizing IMs [24]. The situation is even more involved in one phase such IMs and is still to be investigated.

For synchronous operation the dq model in rotor coordinates is used with the stator windings symmetric (4.150) for simplicity (Figure 4.28) [23].

Finally, the dq model in rotor coordinates is again used to calculate the starting and synchronous operation transients (Figure 4.29a, b) [23].

Successful starting and self-synchronization and overloading is visible. Subsequently an Hooke-Jeeves based optimal design code was put in place with the multi-term objective cost function:

$$C_t = C_i \cdot \max\left(P_1, \frac{P_n}{P_{max}}\right) + C_e + C_{pen} \tag{4.157}$$

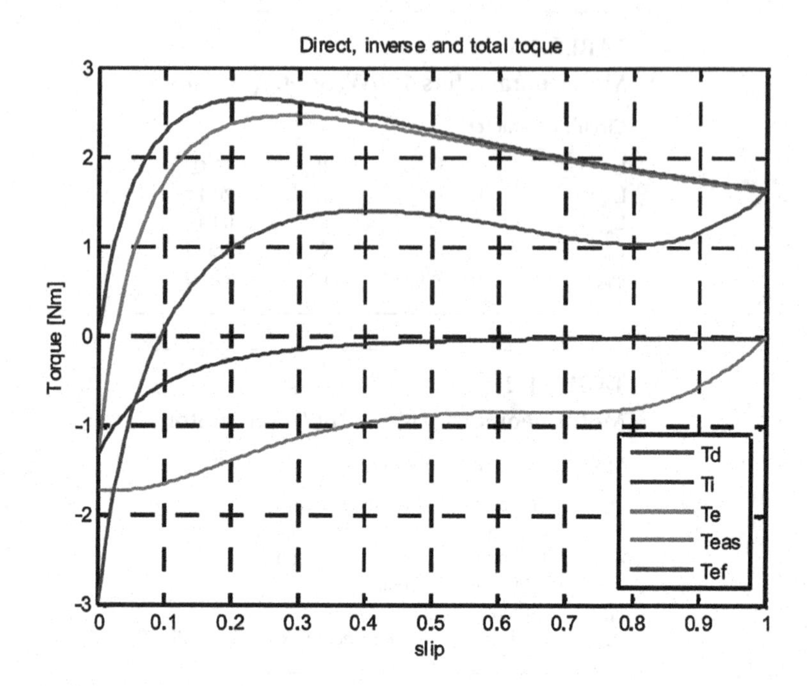

FIGURE 4.27 Asynchronous average torque components.

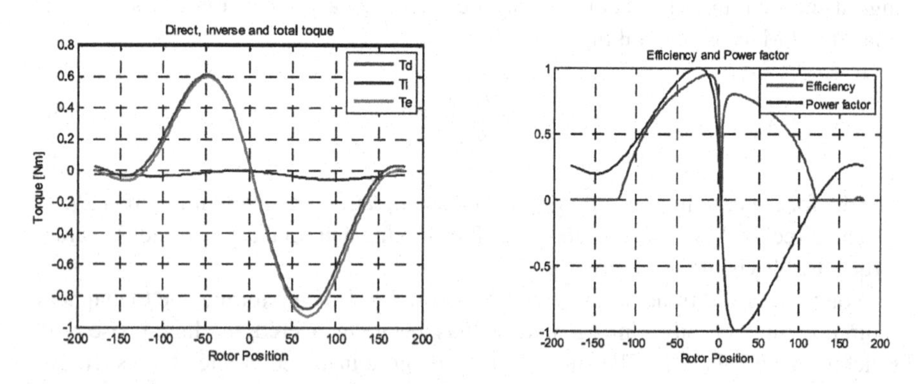

FIGURE 4.28 Synchronous operation: torque, efficiency and power factor versus power angle (rotor position) [23].

where C_e is the energy loss capitalized cost, capacitors (C_s,C_p) costs and motor materials costs are lumped together in C_i (initial cost). C_{pen} is the penalty cost (related to minimum starting torque) and starting abilities of the motor in most critical conditions (of inertia, load torque, etc).

Skipping the details an efficiency of 89% was measured (0.9 calculated) at 0.3 Nm torque (rated) for a 4,22 kg motor (100 W, 3000 rpm) for C_s=40μF and C_f=3.3 μF as starting and running capacitors [23]. The computation time for optimal design on a single contemporary desktop computer, with FEM embedded key validation in the

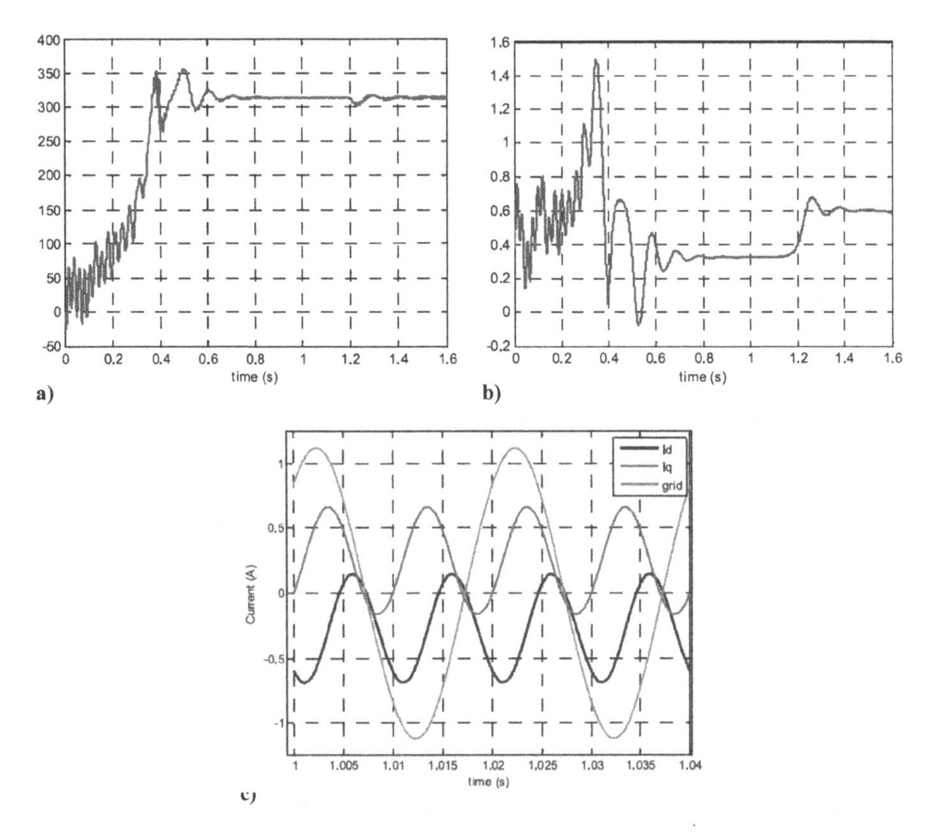

FIGURE 4.29 Starting under rated (0.3 Nm) torque with 0.6 Nm load from t=1.2s, a) speed versus time, b) torque versus time, c) corresponding motor input and output power [23].

optimal design code lasted only 4500 seconds. Such a higher efficiency (+3% above a similar IM) is to be gauged against the additional costs but it may probably be "worth the trouble" by the notable energy bill reduction.

4.19 SUMMARY

- When connected (or reconnected "on the fly") to the power grid, under various grid faults, after load-torque perturbations or voltage sags, or under PWM converter supplies, the induction motor voltages and current amplitude and frequency, and speed vary in time. That is, they undergo transients.
- The phase variable model, with stator/rotor circuit mutual inductances variable with the rotor position, is the most general circuit model applicable to IM transients. Unfortunately, it implies, for numerical solutions, large CPU time.
- For three-phase symmetric stator and rotor windings and uniform airgap (zero rotor eccentricity), the phase variable IM model may be transformed simply into the dq0 model. The parameter equivalence is very simple: the phase resistances and leakage inductances remain unchanged and the magnetization inductance, L_m, is the cyclic one ($L_m = 1.5L_{11m}$).

- The zero sequence stator and rotor current equations relate only to stator (or rotor) resistances and leakage inductances. They do not contribute to torque but produce additional losses. For a star connection, they are zero.
- For 6-phase IMs, two dq0 models are required.
- Magnetic saturation can be simply introduced in the dq model by $L_m (I_m) = \Psi_m/I_m$ and $L_{mt} = d\Psi_m/dI_m$ as functions of the resultant (magnetization) current, I_m [18].
- The rotor slip frequency (skin) effects may be introduced in the dq0 model through additional (fictitious) rotor cage constant parameter circuits in parallel. Core loss windings may be also introduced in the dq model [18].
- Steady state of space phasor model in general coordinates, ω_b, means a frequency of $\omega_1 - \omega_b$ for all variables (ω_1 stator voltage frequency). In synchronous coordinates $\omega_b = \omega_1$, the variables are d.c. in the steady state.
- Under steady state, for cage rotor conditions, the rotor flux linkage, $\overline{\Psi}'_{r0}$, and rotor current, $\overline{I}'_{r0}$, space phasors are orthogonal.
- Even during transients, if the amplitude of the rotor flux remains constant, $\overline{\Psi}'_r$ and $\overline{I}'_r$ remain orthogonal.
- For the motor mode, the stator space phasor, $\overline{I}_s$, is ahead of $\overline{\Psi}_s$ in the direction of motion; the opposite is true for the generator mode.
- For a zero rotor current in the doubly fed IM $\left(V_r' \neq 0\right)$, the ideal no-load speed is obtained at the slip, $S_0 \neq 0$, which is positive or negative; consequently, motor and generator operations are feasible both for $S < 0$ and $S > 0$ when the rotor is fed at a frequency, $\omega_2 = S\omega_1$, through a bidirectional PWM a.c.–a.c. converter.
- In a doubly fed IM (DFIM), the machine magnetization can be done from the rotor source or from the stator source. Note that in the DFIM, the stator frequency and voltage are, in general, constant, while V_r' and its frequency, ω_2, in the rotor are variable such that $\omega_2 = \omega_1 - \omega_r$; it may be supposed that the DFIM operates as a synchronous machine with a.c. rotor excitation; the truth is that there is an additional power component in the stator and rotor that is asynchronous [13].
- Even variable stator frequency ω_1 besides ω_2 has been introduced in DFIG with diode rectified output to reduce cost and provide constant d.c. output voltage for more than 2/1 speed variation range [19].
- For a DFIM in subsynchronous operation ($S > 0$), the power enters on one side and exits through the other for motor and generator operation modes.
- In the DFIM supersynchronous mode ($S < 0$), the power enters or exits from both sides for motor and generator operations. For a limited speed range, $|S_{max}| < 0.25$, the rotor side PWM bidirectional converter is sized at $|S_{max}| P_n$, and thus it costs less. This is applied in most modern wind generator systems.
- Electromagnetic transients mean constant speed transients. For constant parameters, operational (Laplace) parameters may be defined for rotor coordinates in the dq model, as done for synchronous machines.
- The three-phase sudden short circuit may be approached in Laplace formulation. For the two rotor circuit model, the short-circuit current transients exhibit three time constants: one due to the stator and two due to the rotor circuits. It

may be used to identify, by curve fitting, the machine inductances, L'', L', and L_s and the time constants τ'', τ'', τ', τ_0', and τ_a.

- The small-deviation theory is used to linearize the dq model of IM for electromechanical transients (ω_r is also variable); a fifth-order system is obtained and its eigenvalues may be used to check the stability conditions.
- For large-deviation transients, the direct usage of the dq model (with magnetic saturation and the rotor frequency effect included, if necessary) is required. Direct connection to the power grid or large load-torque perturbations are typical of large-deviation transients. For small inertia, the transient torque/speed curves are far away from steady-state curves, exhibiting up to 10/1 current peaks and 5–6 torque peaks, and speed and current oscillations.
- For multimachine transients, a reduced order dq model is required. Ignoring the stator transients ($d\Psi_d/dt = d\Psi_q/dt = 0$) in synchronous coordinates, $\omega_b = \omega_1$, is typical.
- Basic vector control schemes are given to illustrate the principles and relate to modern "electric drives," which is, anyway, a separate subject.
- Doubly fed IMs can also be controlled, as a generator/motor, for a limited speed range, by vector control in the rotor, but regulating the stator active and reactive powers separately. This possibility is also described in this chapter.
- The model and performance of a DFIM as a brushless exciter to SMs, with rotor output voltage at a frequency, $\omega_2 = |\omega_1| + |\omega_r| > |\omega_1|$, with most part of output power extracted from shaft (mechanical) power, is introduced; the output is controlled from the stator by a Soft Starter (Variac) at constant ω_1 and variable voltage (decreasing with speed). This is a widely used industrial solution, as its costs are reasonable, it works from zero speed, has small voltage regulation, and fast SM field current response due to the large stator voltage ceiling reserve at the rated SM speed.
- Standstill tests to identify IM parameters are presented in detail as they are not standard in many places. Flux decay tests are used to determine both the magnetization curve and the machine time constants (by curve fitting).
- Standstill frequency response (SSFR) tests from 0.01 to 100 Hz (in general) are used to determine the operational impedance, $Z(j\omega)$, amplitude, and phase angle. By curve fitting of $|Z(j\omega|$, the machine time constants, visible in the operational inductance, are determined. Alternatively, the maxima of the $Z(j\omega)$ argument (angle) frequencies $f_c', f_c'',\ldots$ and values φ', φ'', … lead to a very simple calculation routine of time constant pairs, $\tau' < \tau_0'$ and $\tau'' < \tau_0''$.
- Disconnection from the power grid and immediate reconnection (before the residual stator voltage has dropped notably) in the primary of the feeding local grid transformer have shown very high transients in torque and current, especially in small IMs. Such events may endanger the integrity of IMs and should be avoided by adequate protection mechanisms.
- To investigate broken rotor bars or end-ring segments, each rotor loop (bar) has to be modeled, with $m + N_r + 1 + 1$ (m—stator phases, N_r—rotor bars, 1—ring current I_e ($I_e = 0$ for healthy rings), 1—for motion equation) equations (variables). The various stator/rotor loop coupling inductances are defined analytically and they depend on the rotor position.

- The non-infinite rotor bar to the core wall resistance leads to inter-bar currents that tend to reduce broken bar effects, and thus make the former's diagnosis more difficult.
- Controlled rotor and stator fluxes are typical in variable speed motor and generator control via PWM inverter supplies.
- If the space phasor model is used (with single rotor circuit model), with stator current, $\bar{I}_s$, and rotor, $\bar{\Psi}_r'$, or stator, $\bar{\Psi}_s$, flux space phasors as variables, at constant speed (slip) electromagnetic transients, only two complex eigenvalues occur. Their real part depends on the speed (slip) and their imaginary part (frequency) depends on the coordinate system's speed, ω_b. It means that all electromagnetic transients can be treated with analytical solutions in space phasors.
- For constant stator or rotor flux, only one complex eigenvalue remains for constant speed (slip) transients. This is a formidable simplification that leads, for constant rotor flux, to a straight line torque/speed curve as in the separately excited d.c. brush machine. The stator current may be decomposed into two components, one for rotor flux and one for torque, which can be controlled separately. This type of vector control is characterized by fast (millisecond range) torque control for wide speed range and good performance.
- Split-phase capacitor IMs are still extensively used in home appliances as only a single-phase a.c. power supply is available. A phase variable model, for two orthogonal (but essentially different in slot occupancy and copper weight) stator windings (the main and the auxiliary (starting) windings) and a cage rotor, is introduced. When the copper weights of the main and auxiliary windings are the same, the dq model may be used in any coordinate to investigate transients. Still, a six-order system is obtained.
- The linear induction motor (three phase) with an aluminum sheet on solid iron track is used in urban (suburban) transportation, on wheels and in MAGLEVs (with active magnetic suspension). LIMs have secondary skin and saturation effects and the longitudinal end effect (Chapter 5, Vol.1, companion book). If they are neglected, the dq model can be used with thrust, F_x, in place of torque, and linear speed, U (in m/s), instead of angular rotor speed, ω_r (in rad/s). The single-sided LIM develops a normal (suspension) force, which is of attraction type at a small slip frequency, $S\omega_1$, in well-designed LIMs. This normal force may provide 20%–25% of total suspension (vehicle weight) force if the thrust is capable of 1 m/s^2 vehicle acceleration. Vector control is thus feasible. The constant rotor flux reference is related to suspension control which is supplemented by additional d.c. electromagnets. As there are quite a few LIMs and d.c. electromagnets on a MAGLEV vehicle, a complex system has to be modeled and controlled for good steady state and dynamic performance.
- Power electronics frequency control has transformed the IM (both with copper cage and with wound rotor (as generator)) from the work horse to the race horse of industry. This trend is here apparently to stay, even after the spectacular entry of PMSMs for variable speed drives.

4.20 PROPOSED PROBLEMS

4.1 Write the phase variable equations of a two orthogonal phase stator cage-rotor IM.
Hints: Check Section 4.1 but notice the $\sin\theta_{er}$ and $\cos\theta_{er}$ dependence of the mutual stator/rotor inductances.

4.2 The magnetization inductance, L_m, of an IM has the dependence on the magnetization current, I_m, as

$$L_m(I_m) = \frac{L_{m0}}{a + bI_m}$$

Calculate the transient inductance function, $L_{mt}(I_m)$.
Hints: Check Equation 4.24 for the L_{mt} expression and use it.

4.3 A cage rotor induction motor has the following parameters: $R_s = 1.0\ \Omega$, $R'_r = 0.7\ \Omega$, $L_m = 0.2$ H, $L_{sl} = L'_{rl} = 6$ mH, and $p_1 = 2$ pole pairs. It operates at 1800 rpm with $S = 0.02$ and rotor flux, $\Psi'_{r0} = 1$ Wb.
Calculate
a. Rotor current, $\overline{I}'_{r0}$, and ω_1 (stator frequency)
b. Magnetization, $\overline{\Psi}_m$, and stator flux, $\overline{\Psi}_{s0}$
c. Stator current, $\overline{I}_{s0}$
d. The stator voltage phasor, $\overline{V}_{s0}$
Hints: Check Example 4.1, but start with Equation 4.28, then Equation 4.29, etc.

4.4 A doubly fed IM has the following parameters: $R_s = R_r = 0.02\ \Omega$, $X_{sl} = X'_{rl} = 0.2\ \Omega$, $X_m = 15\ \Omega$ at $f_{1n} = 60$ Hz, $2p_1 = 4$, and operates at a slip $S = +0.25$ as a motor, with the stator power factor in the stator $\cos\varphi_s = 0.93$ (lagging) at a stator phase current $I_{nph} = 1200$ A and $V_{line} = 4200$ V (line voltage, RMS).
Calculate
a. Stator flux vector, $\overline{\Psi}_{s0}$
b. Rotor flux vector, $\overline{\Psi}'_{r0}$
c. Rotor voltage vector, $\overline{V}'_{r0}$
d. Stator and rotor active and reactive powers, P_s, Q_s, P_r^r, and Q_r^r
Hints: Check Example 4.2 and notice the non-zero stator current phase angle, φ_s, with respect to voltage.

4.5 For the single-cage rotor IM with the data as given in Example 4.3 at zero speed, calculate the flux, $\Psi_s(t)$, and $T_e(t)$ transients and represent them in graphs, for a direct connection to the power grid, to $V_{en} = 220$ V (RMS)-star connection.
Hints: Use directly Equations 4.49 through 4.51.

4.6 The single-cage rotor IM in Problem 4.3 operates at steady state and a slip frequency $\omega_{20} = 2\pi$ rad/s, $\omega_{10} = 2\pi \times 60$ rad/s, and voltage $V_1 = 380$ V (line voltage).
a. Calculate the space phasors, $\overline{I}_{s0}$, $\overline{I}'_{r0}$ (I_{d0}, I_{q0}, I'_{dr0}, I'_{qr0}), and the torque, $T_{e0} = T_{L0}$ (Example 4.1).

b. Calculate the eigenvalues of the linearized dq model system around the steady-state point of (a) based on the equation:

$$|A|s|\Delta X|+|B||\Delta X|=0$$

with $|A|$ and $|B|$ from Equations 4.57 and 4.58.

4.7 Based on theory and the motor data given in Section 4.11, write a MATLAB Simulink program and simulate 1, 2, 3, 4 broken bars, plotting stator current, broken bars currents, and torque versus time.

4.8 Calculate, based on Equation 4.80, the complex eigenvalues of IM for $R_s = R_r = 0.1\ \Omega$, $L_s = L_r' = 32$ mH, $L_m = 30$ mH, for $S = 1, 0.02, -0.02$, at a frequency $\omega_1 = 2\pi \times 60$ rad/s, in stator coordinates ($\omega_b = 0$) and synchronous coordinates ($\omega_b = \omega_1$). Discuss the results.

4.9 Calculate the eigenvalues, s, for the IM in Problem 4.8, for constant rotor flux, at a low frequency, $\omega_1 = 2\pi \times 1.2$ rad/s, and $s = \pm 0.7$ (low speed) and discuss the results.
Hints: Use Equation 4.83 in synchronous coordinates.

4.10 A cage rotor induction motor operates at constant rotor flux $R_r = R_r' = 0.1\ \Omega$, $L_s = L_r' = 0.093$ H, $L_m = 0.09$ H, $2p_1 = 4$, $n = 120$ rpm with $I_m = 20\ A$ and slip frequency $S\omega_1 = \pm 2\pi \times 1$ rad/s.
Calculate
a. Developed torque
b. Frequency, ω_1
c. Stator phase voltage, current, and power factor
Hints: Follow closely Example 4.5.

4.11 Find from Equation 4.95, the eigenvalues, s, for the IM in Problem 4.8 for a constant stator flux at low frequency $\omega_1 = 2\pi \times 1.2$ rad/s and $S = \pm 0.7$ (low speed full torque) and synchronous coordinates. Discuss the results.

4.12 The doubly fed IM in Problem 4.4 operates at a constant rotor flux $\Psi_r' = 10$ Wb and $S = -0.25$, as a motor for $T_e = 42.3$ kNm, $\omega_1 = 2\pi \times 50$ rad/s, and at unity stator power factor.
Calculate
a. Stator current I_{d0} and I_{q0}
b. Rotor current space phasors, $I_r' = jI_{qr}'$; $\bar{\Psi}_r' = \Psi_{dr0}' = L_m I_{ds0}$
c. Rotor voltage components, V_{dr} and V_{qr}
d. Rotor active and reactive power, P_r^r and Q_r^r
e. Stator voltage, $\bar{V}_s$ (phase voltage, RMS), stator powers, P_s and Q_s, and total power, $P_{tot} = P_r^r + P_s$
f. Angle between $\bar{V}_s$ and $\bar{V}_r'$
Hints: See Example 4.2 and Section 4.13 and notice the unity power factor in the stator, when calculating the stator voltage, V_s.

4.13 The SSFR (standstill frequency response) of an induction machine impedance (in p.u.) shows an amplitude of 3 p.u. inductance ($l(s)$) at zero frequency. Its phase has two maxima, one of $\varphi_c' = -30°$ at $f_C' = 7$ Hz and the other of $\varphi_c'' = -15°$ at $f_C' = 60$ Hz. Making use of the phase method, identify the time constants, τ', τ_0' and τ'', τ_0'', and finally write $l(s)$ (in pu)

(Equation 4.123) with $r_s = 0.02$ (in p.u.) and compare with the results in Figure 4.22.

Hints: Check and use Equations 4.125 through 4.128 for the case in point.

4.14 A split-phase induction motor has two orthogonal windings that use the same copper weight (they fulfill Equation 4.136) and has the following parameters: $R_m = 20\ \Omega$, $L_{ml} = 0.2$ H, $L_{dm} = 10L_{ml}$; $p_1 = 1$ pole pair, $a = w_a k_{wa}/w_m k_{wm} = 1.2$, and $J = 10^{-4}$kgm², $R_{rm}=0.7R_m$, $L_{rml}'=L_{ml}$.

Calculate

a. Auxiliary winding parameters, R_a, L_{al}, and L_{qm}
b. Write a MATLAB code to investigate starting and other transients using the dq model for a resistance, R_{start}, for start and a capacitor for running. The switching between the two is instantaneous and takes place at a desired moment in time. Also, provide for the possibility to introduce load-torque variations in time and with speed. Debug and run the program for $R_{start} = 3R_a$ and $C_{run} = 4 \times 10^{-6}$ F and $V_m = 120\sqrt{2}\cos(2\pi 60t)$.

Hints: Use Equations 4.136 through 4.139.

4.15 A three-phase single-sided linear induction motor for an urban MAGLEV has the following parameters: pole pitch $\tau = 0.25$ m (number of poles $2p_1 = 8$). The longitudinal end effect and secondary frequency effects are neglected, rated mechanical airgap is $g = 10$ mm, aluminum sheet thickness, $h_{Al} = 5$ mm, and the back iron permeabilities (both in the primary and in the track) are infinite; it operates at constant rotor flux $(L'_{rl} = 0): L_m = L'_r$. The LIM operates at a speed of U = 30 m/s and at a slip $S = 0.1$, $\Psi'_r = \Psi_m = 1.5$ Wb, $I_T/I_M = 1.6/1$, base thrust, $F_{xn} = 12$ kN; and $R_s = R_r/2.5 = 0.0592\ \Omega$.

Calculate

a. Primary frequency, $f_1(\omega_1)$
b. Secondary resistance, R'_r, stator resistance, R_s, and magnetization inductance, L_m: $L_{sl} = 0.35L_m$
c. Normal (attraction) force considering that L_m varies inversely proportional to $(g + h_{Al})$ and the repulsive normal force is neglected.
d. The stator flux components in rotor flux orientation and finally the stator phase voltage, current, cos φ_s, efficiency (only winding losses are considered)

Hints: Follow Equations 4.140 through 4.144 for F_x and F_n expressions and then Example 4.5 for the rest of the questions. Expected results: f_s = 66.66 Hz, $I_M = 265$ A, $R_r = 0.148\ \Omega$, $L_m = 5.625 \times 10^{-3}$ H, $V_s \approx 920$ V (phase peak value), cos $\varphi_s \approx 0.61$, $\eta \approx 0.85$, $F_n = 39.8$ kN (this is slightly more than three times F_{xn}), $P_{elm} = F_{xn}U = 360$ kW.

REFERENCES

1. S. Ahmed-Zaid and M. Taleb, Structural modeling of small and large induction machines using integral manifolds, *IEEE Trans*. EC-6, 1991, 529–533.
2. M. Taleb, S. Ahmed-Zaid, and W.W. Phia, Induction machine models near voltage collapse, *EMPS*, 25(1), 1997, 15–28.

3. M. Akbaba, A phenomenon that causes most source transients in three phase IMs, *EMPS*, 12(2), 1990, 149–162.
4. I. Boldea and S.A. Nasar, *Induction Machine Design Handbook*, CRC Press, Boca Raton, FL, 2nd edition, 2010.
5. S.A. Nasar, Electromechanical energy conversion in n m-winding double cylindrical structures in presence of space harmonics, *IEEE Trans.*, PAS-87, 1968, 1099–1106.
6. P.C. Krause, *Analysis of Electric Machinery*, McGraw-Hill, New York, 1986 and new (IEEE) edition.
7. H.A. Tolyiat and T.A. Lipo, Transient analysis of cage IMs under stator, rotor bar and end ring faults, *IEEE Trans.*, EC-10(2), 1995, 241–247.
8. S.T. Manolas and J.A. Tegopoulos, *Analysis of squirrel cage with broken bars and end rings, Record of IEEE-IEMDC*, 1997.
9. I. Kerszenbaum and C.F. Landy, The existence of inter-bar current in three phase cage motors with rotor bar and (or) end ring faults, *IEEE Trans.*, PAS-103, 1984, 1854–1861.
10. S.L. Ho and W.H. Fu, Review and future application of FEM in IMs, *EMPS J.*, 26(1), 1998, 111–125.
11. J.F. Bangura and M.A. Demerdash, Performance characterization of torque ripple reduction in IM adjustable speed drives using time-stepping coupled FE state space techniques, *Record of IEEE-IAS*, 1, 1998, 218–236.
12. I. Boldea and S.A. Nasar, *Electric Drives*, 2nd Edition, CRC Press, Boca Raton, FL, 2005, 3rd Edition, 2016.
13. I. Boldea, *Electric Generator Handbook, Vol. 2, Variable Speed Generators*, CRC Press, Boca Raton, FL, 2005, 2nd Edition, 2016.
14. A. Watson, A systematic method to the determination of SM parameters from results of frequency response tests, *IEEE Trans.*, EC-15(4), 2000, 218–223.
15. I. Boldea and S.A. Nasar, *Linear Motion Electromagnetic Devices*, Taylor & Francis Group, New York, 2001.
16. I. Boldea and S.A. Nasar, *Linear Motion Electric Machines*, Chapter 4, John Wiley & Sons, New York, 1976.
17. F. Gieras, *Linear Induction Drives*, Oxford University Press, Oxford, U.K., 1994.
18. I. Boldea and S.A. Nasar, Unified treatment of core losses and saturation in the orthogonal axis model of electrical machines. *Proc. IEE*, 134(6), 1987, 355–363.
19. M. Niraula. L. Maharjan, B. Fahimi, M. Kiani, I. Boldea, "Variable stator frequency control of stand-alone DFIG with diode rectified output", *5th International Symposium on Environment-Friendly Energies and Applications (EFEA)*, 2018, Rome, Italy.
20. D. Mingardi, N. Bianchi, M. Dai Prè, "Geometry of line start synchronous motors suitable for various pole combinations", *IEEE Trans*, IA-53(5), 2017, 4360–4367.
21. D. Mingardi, N. Bianchi "Line-start PM—assisted synchronous motor design, optimization and tests", *IEEE Trans*, IE-64(12), 2017, 9739–9747.
22. H. C. Liu, J. Lee, "Optimum design of an IE-4 line start synchronous reluctance motor considering manufacturing process loss effect", *IEEE Trans*, IE-65, (4), 2018, 3104–3114.
23. L. N. Tutelea, T. Staudt, A. A. Popa, W. Hoffman, I. Boldea, "Line-start 1 phase-source split phase capacitor cage—PM rotor – Relsyn motor: modeling, performance, and optimal design with experiments", *IEEE Trans*, IE-65(2), 2018, 1772–1780.
24. A. Takahashi, S. Kikuchi, K. Miyata, A. Binder, "Asynchronous torque of line starting PMSMs", *IEEE Trans*, EC-30(2), 2015, 498–506.

5 Essentials of Finite Element Analysis (FEA) in Electromagnetics

5.1 VECTORIAL FIELDS

Though apparently a rather abstract concept, "field," a form of matter, plays a key role in explaining electric and magnetic phenomena with deep implications on the design of electric machines [1–17]. Mathematically, there are scalar fields such as those of temperatures, pressures, and current densities, and vectorial fields such as electric fields, magnetic fields, and mechanical stresses fields in solid bodies. For scalar fields, a scalar is assigned to each point in space. For vectorial fields, a vector is linked to every point in space. The relations between scalar and magnetic fields are based on the fundamental laws of electromagnetics, known as Maxwell equations. But first the main properties of operations with vectors are introduced.

5.1.1 Coordinate Systems

In *Cartesian coordinates* (Figure 5.1), vector $\bar{A}$ is defined by its projections along the orthogonal coordinate axes:

$$\bar{A} = A_x \cdot \bar{u}_x + A_y \cdot \bar{u}_y + A_z \cdot \bar{u}_z \tag{5.1}$$

where $\bar{u}_x$, $\bar{u}_y$, and $\bar{u}_z$ are unitary vectors aligned to the orthogonal axes x, y, and z.

For *cylindrical coordinates* (Figure 5.2), the application point of vector, $P(r, \theta, z)$, is defined by the cylinder radius, r, its angle, θ, with axis x, and the height, z. The unitary vectors along the coordinate axes are $\bar{u}_r$, $\bar{u}_\theta$, and $\bar{u}_z$.

The transformation matrix between Cartesian and cylindrical coordinates is given by Equation 5.2:

$$\begin{pmatrix} A_r \\ A_\theta \\ A_z \end{pmatrix} = \begin{pmatrix} \cos(\theta) & -\sin(\theta) & 0 \\ \sin(\theta) & \cos(\theta) & 0 \\ 0 & 0 & 1 \end{pmatrix} \cdot \begin{pmatrix} A_x \\ A_y \\ A_z \end{pmatrix} \tag{5.2}$$

The cylindrical coordinates are useful where the investigated field shows a cylindrical symmetry as in the case of a radial electric field produced by a straight-line

DOI: 10.1201/9781003216018-5

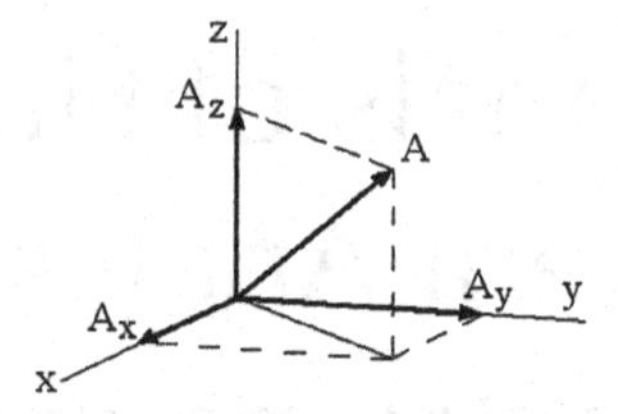

FIGURE 5.1 Cartesian coordinates.

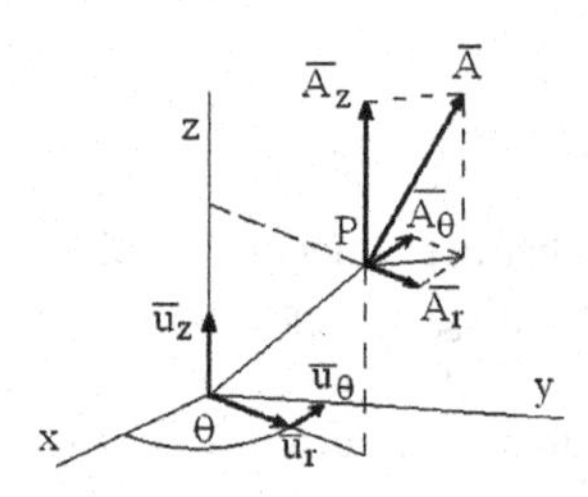

FIGURE 5.2 Cylindrical coordinates.

conductor, which is electrostatically loaded ($\overline{E}_r \neq 0$, $\overline{E}_\theta = 0$, and $\overline{E}_z = 0$), or the magnetic field of the same conductor flowed by current ($\overline{B}_r = 0$, $\overline{B}_\theta \neq 0$, and $\overline{B}_z = 0$). For the points on the z-axis of cylindrical coordinates, the unitary vectors, $\overline{u}_\theta$ and $\overline{u}_z$, are not defined.

Spherical coordinates represent a set of curve-line coordinates used to naturally describe the position on a sphere. The coordinates are given by r, the distance to the origin; θ the azimuthal angle (the angle between the position vector projection in xoy plan and the positive semiaxis, "x"); and ϕ, that is, the zenithal angle (the angle of the position vector with the positive semiaxis, "z"). The unitary vectors of vector $\overline{A}$ (Figure 5.3) are $\overline{u}_r$, $\overline{u}_\theta$, and $\overline{u}_\phi$, and their orientation dependence on the application point coordinates is portrayed in Equation 5.3:

$$\begin{pmatrix} \overline{u}_r \\ \overline{u}_\theta \\ \overline{u}_\phi \end{pmatrix} = \begin{pmatrix} \cos(\theta)\sin(\phi) & \sin(\theta)\sin(\phi) & \cos(\phi) \\ -\sin(\theta) & \cos(\theta) & 0 \\ \cos(\theta)\cos(\phi) & \sin(\theta)\cos(\phi) & -\sin(\phi) \end{pmatrix} \cdot \begin{pmatrix} \overline{u}_x \\ \overline{u}_y \\ \overline{u}_z \end{pmatrix} \tag{5.3}$$

The unitary vectors, $\overline{u}_r$, $\overline{u}_\theta$, and $\overline{u}_\phi$, are not univoquely defined at the point of origin.

The transformation matrix between Cartesian and spherical coordinates is given by Equation 5.4:

$$\begin{pmatrix} A_r \\ A_\theta \\ A_\phi \end{pmatrix} = \begin{pmatrix} \cos(\theta)\sin(\phi) & -\sin(\theta) & \cos(\theta)\cos(\phi) \\ \sin(\theta)\sin(\phi) & \cos(\theta) & \sin(\theta)\cos(\phi) \\ \cos(\phi) & 0 & -\sin(\phi) \end{pmatrix} \cdot \begin{pmatrix} A_x \\ A_y \\ A_z \end{pmatrix} \tag{5.4}$$

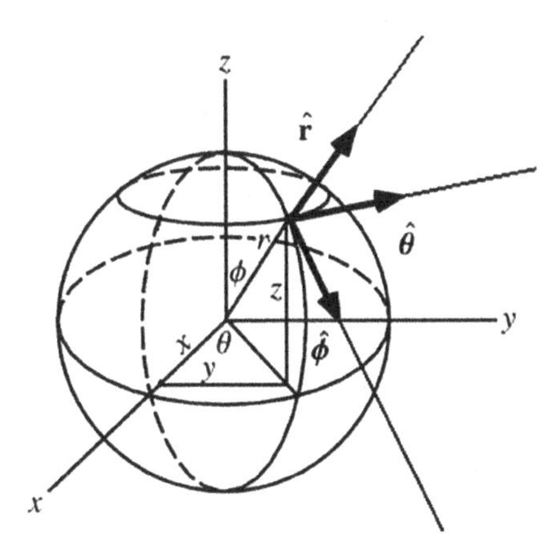

FIGURE 5.3 Spherical coordinates.

The spherical coordinates are very useful for fields having spherical symmetry, such as the electrostatic field produced by a point-shaped charge where only the radial component of the field is nonzero ($\bar{E}_r \neq 0$, $\bar{E}_\theta = 0$, and $\bar{E}_\phi = 0$).

5.1.2 Operations with Vectors

Let us consider two vectors $\bar{A}$ and $\bar{B}$ with amplitudes $|A|$ and $|B|$, and phase shift α. Their scalar product is defined as

$$\bar{A}\cdot\bar{B} = |A|\cdot|B|\cdot\cos(\alpha) \tag{5.5}$$

In Cartesian coordinates the scalar product is

$$\bar{A}\cdot\bar{B} = A_x B_x + A_y B_y + A_z B_z \tag{5.6}$$

The vectorial product is

$$\bar{A}\times\bar{B} = |A|\cdot|B|\cdot\sin(\alpha)\cdot\bar{n}_{AB} \tag{5.7}$$

where $\bar{n}_{AB}$ is the unitary vector that is normal to the plane of vectors $\bar{A}$ and $\bar{B}$, with the direction given by the right-hand rule.

In Cartesian coordinates the vectorial product is

$$\begin{aligned}\bar{A}\times\bar{B} &= \begin{vmatrix} \bar{u}_x & \bar{u}_y & \bar{u}_z \\ A_x & A_y & A_z \\ B_x & B_y & B_z \end{vmatrix} \\ &= (A_y B_z - A_z B_y)\bar{u}_x + (A_z B_x - A_x B_z)\bar{u}_y + (A_x B_y - A_y B_x)\bar{u}_z \end{aligned} \tag{5.8}$$

There are some properties of operations with vectors such as

1. Absolute value:

$$\bar{A}\cdot\bar{A}=|A|^2 \tag{5.9}$$

2. Commutation of scalar product:

$$\bar{A}\cdot\bar{B}=\bar{B}\cdot\bar{A} \tag{5.10}$$

3. Anticommutation of vectorial product:

$$\bar{A}\times\bar{B}=-\bar{B}\times\bar{A} \tag{5.11}$$

4. Distributivity of scalar product for addition:

$$\bar{A}\cdot\left(\bar{B}+\bar{C}\right)=\bar{A}\cdot\bar{B}+\bar{B}\cdot\bar{C} \tag{5.12}$$

5. Distributivity of vector product for addition:

$$\bar{A}\times\left(\bar{B}+\bar{C}\right)=\bar{A}\times\bar{B}+\bar{A}\times\bar{C} \tag{5.13}$$

6. Distributivity of double vector product:

$$\bar{A}\times\left(\bar{B}\times\bar{C}\right)=\left(\bar{A}\cdot\bar{C}\right)\bar{B}-\left(\bar{A}\cdot\bar{B}\right)\bar{C} \tag{5.14}$$

5.1.3 Line and Surface (Flux) Integrals of a Vectorial Field

By the *line integral* of a vectorial field we mean the integral of the scalar product between the respective vector and the unitary vector that is tangent to that line (curve) at every point:

$$L_{12}=\int_c \bar{A}\cdot\overline{\mathrm{d}l} \tag{5.15}$$

In Cartesian coordinates the line integral is

$$L_{12}=\int_c \bar{A}\cdot\overline{\mathrm{d}l}=\int_{P1}^{P2}\left(A_x\mathrm{d}x+A_y\mathrm{d}y+A_z\mathrm{d}z\right)=\int_{x1}^{x2}A_x\mathrm{d}x+\int_{y1}^{y2}A_y\mathrm{d}y+\int_{z1}^{z2}A_z\mathrm{d}z \tag{5.16}$$

The line integral of a vectorial field has a well-defined physical meaning. For a field of forces it is the mechanical work, while for the magnetic field it is the number of ampere turns (magnetomotive force) required to produce the field between the two points. If the result L_{12} does not depend on the shape of the line, but relies only on the initial and final points, then the field is called *conservative*. A field is conservative when and only when its integral on any closed line (curvature) is zero. The flux of a vectorial field through a surface S is given by the surface integral of the scalar product between vector $\bar{A}$ and its unitary vector normal (at 90°) to the surface S:

$$\phi = \int_S \bar{A} \cdot \overline{da} = \int_S \left(\bar{A} \cdot \bar{n} \right) da \tag{5.17}$$

The expression of ϕ in Cartesian coordinates is

$$\phi = \int \bar{A} \cdot \overline{da} = \int_S A_x dydz + \int_S A_y dxdz + \int_S A_z dxdy \tag{5.18}$$

For a fluid, the surface integral of its speed vector represents the volumic flow rate through that surface.

5.1.4 Differential Operations

The *gradient* of a scalar field, φ, is a vectorial field, whose vectors show at every point in space the direction along which the variation of the scalar field is maximum, $\bar{u}_{max}$; its amplitude is equal to the scalar field derivative along that direction:

$$\text{grad}\left(\vartheta\right) = \nabla\vartheta = \max\left(\frac{\partial\vartheta}{\partial l}\right)\bar{u}_{\text{max}} \tag{5.19}$$

In Cartesian coordinates the gradient is

$$\text{grad}\left(\vartheta\right) = \nabla\varphi = \frac{\partial\vartheta}{\partial x}\bar{u}_x + \frac{\partial\vartheta}{\partial y}\bar{u}_y + \frac{\partial\vartheta}{\partial z}\bar{u}_z \tag{5.20}$$

The *surface gradient* is defined for surfaces along which the scalar field is discontinuous with the normal direction, $\bar{n}_\Sigma$, to the surface and the amplitude equal to the difference between scalar field values on the two faces of the surface:

$$\text{grad}_\Sigma\left(\vartheta\right) = \left(\varphi_2 - \varphi_1\right)\bar{n}_\Sigma \tag{5.21}$$

The gradient of a *scalar field* is always a conservative field. The "*rotor*" (vortex) of a vectorial field is defined as the limit of the ratio between a closed surface integral of a *vectorial product* between the respective vector and the normal to the surface unit vector, to the volume closed by the surface when the latter tends to zero:

$$\text{rot}\left(\bar{A}\right) = \nabla \times \bar{A} = \lim_{V\to 0}\frac{1}{V}\oint_\Sigma \bar{n} \times \bar{A}\, dS \tag{5.22}$$

The projection of the rotor of a vectorial field on a given direction, $\bar{n}$, is equal to the limit of the field integral along a closed line (curve) of the surface closed by the respective line when the latter tends to zero; the closed line (curve) is situated in a plane defined by its normal, $\bar{n}$:

$$\bar{n}\cdot\text{rot}\left(\bar{A}\right)=\lim_{S\to 0}\frac{1}{S}\oint_{C}\bar{A}\cdot\overline{dl} \tag{5.23}$$

In Cartesian coordinates the rotor expression is

$$\begin{aligned}\left(\bar{A}\right)=\nabla\times\bar{A}&=\begin{vmatrix}\bar{u}_x & \bar{u}_y & \bar{u}_z\\ \dfrac{\partial}{\partial x} & \dfrac{\partial}{\partial y} & \dfrac{\partial}{\partial z}\\ A_x & A_y & A_z\end{vmatrix}\\ &=\left(\frac{\partial A_z}{\partial y}-\frac{\partial A_y}{\partial z}\right)\bar{u}_x+\left(\frac{\partial A_x}{\partial z}-\frac{\partial A_z}{\partial x}\right)\bar{u}_y+\left(\frac{\partial A_y}{\partial x}-\frac{\partial A_x}{\partial y}\right)\bar{u}_z\end{aligned} \tag{5.24}$$

If the field is not continuous through a surface, then its surface rotor is defined as the vectorial product between the normal to the surface, $\bar{n}_\Sigma$, and the difference between vector values on the two sides of the discontinuity surface:

$$\text{rot}_\Sigma\left(\bar{A}\right)=\bar{n}_\Sigma\times\left(\bar{A}_2-\bar{A}_1\right) \tag{5.25}$$

The "*divergence*" of a field is a scalar (associated to every point) that shows the tendency of the field to spring from $\left(div\left(\bar{A}\right)>0\right)$ or to $\left(div\left(\bar{A}\right)<0\right)$ converge to a respective point. The divergence of a field is the limit of the ratio between the field flux through a closed surface and the volume closed by that surface when the latter tends to zero:

$$\text{div}\left(\bar{A}\right)=\nabla\cdot\bar{A}=\lim_{V\to 0}\frac{1}{V}\oint_{\Sigma}\bar{A}\cdot\overline{dS} \tag{5.26}$$

In Cartesian coordinates the divergence is

$$\text{div}\left(\bar{A}\right)=\nabla\cdot\bar{A}=\frac{\partial A_x}{\partial x}+\frac{\partial Ay}{\partial y}+\frac{\partial A_z}{\partial z} \tag{5.27}$$

If the field is discontinuous on the surface, a surface divergence is defined as

$$\text{div}_\Sigma\left(\bar{A}\right)=\left(\bar{A}_2-\bar{A}_1\right)\cdot\bar{n} \tag{5.28}$$

The fields whose divergence is zero at every point are called *solenoidal fields*; the flux density field is a solenoidal field.

The *Laplacian* is a second-order derivative operator, denoted by Δ (or ∇^2). The fields whose Laplacian is zero at every point are called *harmonic fields*. The *Laplacian of a scalar field* is also a scalar field equal to the divergence of its gradient:

$$\nabla^2\varphi = \nabla\cdot(\nabla\varphi) = \text{div}\left(\text{grad}(\varphi)\right) \tag{5.29}$$

In Cartesian coordinates the Laplacian of a scalar field is

$$\nabla^2\varphi = \frac{\partial^2\varphi}{\partial x^2} + \frac{\partial^2\varphi}{\partial y^2} + \frac{\partial^2\varphi}{\partial z^2} \tag{5.30}$$

The *Laplacian of a vectorial field* is

$$\nabla^2\bar{A} = \nabla\left(\nabla\cdot\bar{A}\right) - \nabla\times\left(\nabla\times\bar{A}\right) = \text{grad}\left(\text{div}\left(\bar{A}\right)\right) - \text{rot}\left(\text{rot}\left(\bar{A}\right)\right) \tag{5.31}$$

The Laplacian of a vectorial field in Cartesian coordinates is written as

$$\begin{aligned}\nabla^2\bar{A} &= \nabla^2 A_x\bar{u}_x + \nabla^2 A_y\bar{u}_y + \nabla^2 A_z\bar{u}_z \\ &= \left(\frac{\partial^2 A_x}{\partial x^2} + \frac{\partial^2 A_x}{\partial y^2} + \frac{\partial^2 A_x}{\partial z^2}\right)\bar{u}_x + \left(\frac{\partial^2 A_y}{\partial x^2} + \frac{\partial^2 A_y}{\partial y^2} + \frac{\partial^2 A_y}{\partial z^2}\right)\bar{u}_y \\ &+ \left(\frac{\partial^2 A_z}{\partial x^2} + \frac{\partial^2 A_z}{\partial y^2} + \frac{\partial^2 A_z}{\partial z^2}\right)\bar{u}_z\end{aligned} \tag{5.32}$$

5.1.5 Integral Identities

The *first theorem of the gradient* says that the line integral of a field of gradients between points P_1 and P_2 equals the difference between the scalar field values φ_1 and φ_2 at the two points:

$$\varphi_{12} = \int_{P1}^{P2} \text{grad}(\varphi)\cdot\overline{\text{d}l} = \varphi_2 - \varphi_1 \tag{5.33}$$

The *second theorem of the gradient* says that the volume integral of a field of gradients is equal to the scalar field integral at the origin over the surface that closes the respective volume:

$$\int_V \text{grad}(\varphi)\,\text{d}V = \oint_\Sigma \varphi\,\overline{\text{d}S} \tag{5.34}$$

This theorem of the gradient is applied particularly in electrostatics.

The *rotor* (Kelvin–Stokes) *theorem* states that the closed-line integral of a vectorial field is equal to the flux produced by the rotor of the respective vectorial field through a surface S bordered by the respective closed line:

$$\int_c \bar{A} \cdot \overline{\mathrm{d}l} = \int_S \mathrm{rot}\left(\bar{A}\right) \cdot \overline{\mathrm{d}S} \tag{5.35}$$

The *divergence* (Ostrogradsky–Gauss) *theorem* says that the flux of a vectorial field through a closed surface is equal to the integral of its divergence on the volume closed by the respective closed surface:

$$\int_\Sigma \bar{A} \cdot \overline{\mathrm{d}S} = \int_V \mathrm{div}\left(\bar{A}\right) \mathrm{d}V \tag{5.36}$$

This theorem is particularly important in formulating the laws of electric and magnetic fluxes.

The *first theorem of Green* is the equivalent of the integrals by parts for scalar and vectorial fields, respectively; the *second Green's theorem* is a direct application of the first theorem for symmetric expressions.

The first theorem of Green for scalar fields is

$$\int_\tau U\mathrm{div}\left(k\ \mathrm{grad}\left(V\right)\right) + k\ grad\left(U\right) \cdot grad\left(V\right) \mathrm{d}\tau = \oint_S kU \frac{\partial V}{\partial n} \mathrm{d}\bar{S} \tag{5.37}$$

The second theorem of Green is

$$\int_\tau Udiv\left(k\ grad\left(V\right)\right) - Vdiv\left(k\ grad\left(U\right)\right) \mathrm{d}\tau = \oint_S k\left(U\frac{\partial V}{\partial n} - V\frac{\partial U}{\partial n}\right) \mathrm{d}\bar{S} \tag{5.38}$$

For vectorial fields, the first theorem of Green is

$$\int_\tau k\ \mathrm{rot}\left(\bar{A}\right) \cdot \mathrm{rot}\left(\bar{B}\right) - \bar{A} \cdot \mathrm{rot}\left(k\ \mathrm{rot}\left(\bar{B}\right)\right) \mathrm{d}\tau = \oint_S k\bar{A} \times rot\left(\bar{B}\right) \cdot \mathrm{d}\bar{S} \tag{5.39}$$

while the second one is

$$\int_\tau \bar{B} \cdot \mathrm{rot}\left(k\ \mathrm{rot}\left(\bar{A}\right)\right) - \bar{A} \cdot \mathrm{rot}\left(k\ \mathrm{rot}\left(\bar{B}\right)\right) \mathrm{d}\tau = \oint_S k\left[\bar{A} \times \mathrm{rot}\left(\bar{B}\right) - \bar{B} \times \mathrm{rot}\left(\bar{A}\right)\right] \cdot \mathrm{d}\bar{S} \tag{5.40}$$

The Green's theorems are used in the finite element method (FEM) to determine the expressions of the coefficients of the algebraic equation system that substitutes the partial derivative field equations with boundary conditions.

5.1.6 Differential Identities

The differential operators for the product of a constant k and a scalar field U, and, respectively, a vectorial field, $\bar{A}$, produce results identical with the product of the constant k and the respective field operator:

$$grad(kU) = k\ grad(U) \tag{5.41}$$

$$\text{rot}(k\bar{A}) = k\ \text{rot}|\bar{A}| \tag{5.42}$$

$$div(k\bar{A}) = kdiv(\bar{A}) \tag{5.43}$$

$$\nabla^2(k\bar{A}) = k\nabla^2(\bar{A}) \tag{5.44}$$

The differential operators for the summation of two scalar fields, U and V, and, respectively, two vectorial fields, $\bar{A}$ and $\bar{B}$, are equal to the summation operators applied separately to the two terms:

$$grad(U+V) = grad(U) + grad(V) \tag{5.45}$$

$$\text{rot}(\bar{A}+\bar{B}) = \text{rot}(\bar{A}) + \text{rot}(\bar{B}) \tag{5.46}$$

$$div(\bar{A}+\bar{B}) = div(\bar{A}) + div(\bar{B}) \tag{5.47}$$

$$\nabla^2(U+V) = \nabla^2(U) + \nabla^2(V) \tag{5.48}$$

For the product of two fields, these rules do not generally apply. For the gradient, however,

$$grad(UV) = Ugrad(V) + Vgrad(U) \tag{5.49}$$

$$\text{rot}(U\bar{A}) = U\text{rot}(\bar{A}) + grad(U) \times \bar{A} \tag{5.50}$$

$$div(U\bar{A}) = U\ div(\bar{A}) + grad(U) \cdot \bar{A} \tag{5.51}$$

$$div(\bar{A} \times \bar{B}) = -\bar{A} \cdot \text{rot}(\bar{B}) + \text{rot}(\bar{A}) \cdot \bar{B} \tag{5.52}$$

$$\nabla^2(UV) = U\nabla^2 V + V\nabla^2 U + 2\ grad(U)\ grad(V) \tag{5.53}$$

The gradient of composed fields is

$$grad(U(V)) = \frac{\partial U}{\partial V} grad(V) \tag{5.54}$$

The rotor of the fields of gradients is always zero:

$$\text{rot}\left(grad\left(U\right)\right) = 0 \tag{5.55}$$

The divergence of a field of rotors is also always zero:

$$div\left(\text{rot}\left(\bar{A}\right)\right) = 0 \tag{5.56}$$

The last two identities bear a special importance on electromagnetic fields, by allowing to define the scalar and vectorial magnetic potential concepts.

5.2 ELECTROMAGNETIC FIELDS

5.2.1 Electrostatic Fields

Electrostatics investigates the electric field produced by electric charges in the absence of magnetic field variations.

The electric field, $\bar{E}$, is nonrotor type and is called electrostatic:

$$\text{rot}\left(\bar{E}\right) = 0 \tag{5.57}$$

Consequently, the electrostatic field is a field of gradients, derived from a scalar field called the electrostatic potential, V:

$$\bar{E} = -grad\left(V\right) \tag{5.58}$$

The displacement vector, $\bar{D} = \varepsilon\bar{E}$, through a closed surface, Σ, is equal to the electric charge within the closed surface (the Gauss law):

$$\oint_{\Sigma} \varepsilon\bar{E}\text{d}S = \int_{V} \rho\text{d}v \tag{5.59}$$

The local form of the Gauss law is

$$div\left(\varepsilon\bar{E}\right) = \rho \tag{5.60}$$

By substituting Equation 5.58 in Equation 5.60, the electric potential (V) equation is obtained:

$$div\left(\varepsilon\ grad\left(V\right)\right) = -\rho \tag{5.61}$$

Making use of differential identities in Section 5.1.6 yields

$$\varepsilon\nabla^2 V + \nabla\varepsilon\cdot\nabla V = -\rho \tag{5.62}$$

In most cases, the electric permittivity (ε) of a medium is constant, and thus ∇ε in Equation 5.62 is zero, and, thus, the Poisson equation is obtained:

$$\nabla^2 V = -\frac{\rho}{\varepsilon_0} \tag{5.63}$$

In charge-less domains (ρ = 0), the Laplace equation holds:

$$\nabla^2 V = 0 \tag{5.64}$$

5.2.2 Fields of Current Densities

The motion of electric charges is described by the current density vector, $\bar{J}$:

$$\bar{J} = qn\bar{V}_{\mathrm{d}} = \rho\overline{V}_{\mathrm{d}} \tag{5.65}$$

where

q is the particle charge,
n is the particle count per unit volume, and
$\bar{V}_{\mathrm{d}}$ is the average (drift) speed.

The electric current through a surface is given by the surface integral of the current density over the surface:

$$I = \int_S \bar{J} \cdot \overline{\mathrm{d}S} \tag{5.66}$$

The total electric current through a closed surface is equal to the time variation of electric charge within the closed surface.

The local form of the law of electric charge conservation is expressed by the continuity equation:

$$div\left(\bar{J}\right) = -\frac{\partial \rho}{\partial t} \tag{5.67}$$

It is possible to define the total current density as the summation of conduction and displacement components:

$$\bar{J}_{\mathrm{tot}} = \bar{J} + \frac{\partial \bar{D}}{\partial t} \tag{5.68}$$

The continuity equation of the total current density thus becomes

$$div\left(\bar{J}_{\mathrm{tot}}\right) = 0 \tag{5.69}$$

so the total current density is a solenoidal field.

The relationship between the current density, $\bar{J}$, and the electric field, $\bar{E}$, in conductors is

$$\bar{J} = \sigma\bar{E} \quad (5.70)$$

where σ is the electric conductivity.

5.2.3 Magnetic Fields

The magnetic field is a component of the electromagnetic field that exerts forces on electric charges in motion. The magnetic field is characterized by two vectorial variables (concepts) that are the magnetic field, $\bar{H}$, and the magnetic flux density, $\bar{B}$:

$$\bar{B} = \mu\bar{H} \quad (5.71)$$

where μ is the permeability of the medium in henry per meter (H/m).

The flux density field is solenoidal:

$$div\left(\bar{B}\right) = 0 \quad (5.72)$$

So the magnetic field may be derived from a vector potential, $\bar{A}$:

$$\bar{B} = \mathrm{rot}\left(\bar{A}\right) \quad (5.73)$$

The magnetic field is produced by electric charges in motion, and the relationship between the forces of the magnetic field and the respective field is given by *Ampere's law*. Ampere's law states that the closed-line integral of the magnetic field is equal to the electric current through the surface that subtends the closed line.

$$\oint_{\Gamma} \bar{H} \cdot \overline{dl} = i \quad (5.74)$$

The local form of Ampere's law is written as

$$\mathrm{rot}\left(\bar{H}\right) = \bar{J}_{\mathrm{tot}} \quad (5.75)$$

The relationship between the magnetic potential, A, and the magnetic field sources is

$$\mathrm{rot}\left(\frac{1}{\mu}\mathrm{rot}\left(\bar{A}\right)\right) = \bar{J}_{\mathrm{tot}} \quad (5.76)$$

5.2.4 Electromagnetic Fields: Maxwell Equations

The electromagnetic field is a physical field produced by electric charges in motion, which influences the behavior of electric charges. It may be seen as a combination of the electric and the magnetic fields. Maxwell has put together in a set of equations the electric charges, the field production, and the interaction:

$$\begin{aligned} div\left(\bar{E}\right) &= \frac{\rho}{\varepsilon_0} \\ div\left(\bar{B}\right) &= 0 \\ \text{rot}\left(\bar{E}\right) &= -\frac{\partial \bar{B}}{\partial t} \\ \text{rot}\left(\bar{B}\right) &= \mu_0 \bar{J} + \mu_0 \varepsilon_0 \frac{\partial \bar{E}}{\partial t} \end{aligned} \tag{5.77}$$

where

ε_0 is the vacuum permittivity and

μ_0 is the vacuum permeability.

These equations are valid also for other mediums with constants ϵ and μ. For the domains free of electric charges ($\rho = 0$) and without electric current ($J = 0$), the equations of electromagnetic waves are obtained:

$$\nabla^2 \bar{E} = \frac{1}{c^2} \cdot \frac{\partial^2 \bar{E}}{\partial t^2} \tag{5.78}$$

$$\nabla^2 \bar{B} = \frac{1}{c^2} \cdot \frac{\partial^2 \bar{B}}{\partial t^2} \tag{5.79}$$

$$c = \frac{1}{\sqrt{\varepsilon_0 \mu_0}} \tag{5.80}$$

where c is the speed of light in vacuum.

The waves equations (Equation 5.78) are useful in designing antennas, and wave guides in telecommunications, and in investigating the electromagnetic compatibility of electric devices.

5.3 VISUALIZATION OF FIELDS

A few methods for representing and visualizing fields are introduced here in order to facilitate the understanding of the intricate physical phenomena related to fields. For scalar fields, the direct methods of the equipotential surface (equipotential field lines in 2D space) are used, in general. As an example, Figure 5.4 represents the electric

FIGURE 5.4 Electric potential of two electrostatically charged conductors.

(scalar) potential produced by two straight conductors of infinite lengths charged electrostatically with opposite polarities.

A similar method uses colored maps to represent 2D scalar fields or sections in 3D space. To every point in plane a distinct color is assigned. To visualize vectorial fields, the vectors of the field are shown by a few points. The orientation of the arrows coincides with the orientation of the vectors of the field and their length is proportional to their amplitude (see, e.g., Figure 5.5).

A widespread method for the visualization of vectorial fields consists of showing the field lines (paths). The concept of field line (path) was introduced by Faraday. The field line (path) is an imaginary line that is tangent at every point to the field vector at that point.

Figure 5.6 illustrates the magnetic field paths produced by two (opposite polarity) parallel electric currents of conductors of infinite lengths. The geometric description of field lines means describing that field. For magnetic (solenoidal) fields, the field lines are dense in the zones with an intense (strong) field; valuable information on field amplitude distribution is obtained in this way. For nonsolenoidal fields, there are points where field lines appear or disappear. For 2D-domain magnetic fields, the representation of lines is performed by drawing the magnetic potential contour lines (the components perpendicular to the plane of symmetry). When the scalar potential concept is used to solve the field problem, in order to represent the magnetic field lines, a scalar variable equivalent to the normal component of the magnetic potential is defined. For most commercial FEM software, it is difficult to do such a thing. We may, however, represent the field by arrows associated with field vectors, together with the equipotential lines of the applied field potential. Figure 5.7a and b illustrates comparatively the field vectors and the equipotential lines for an electrostatic and a magnetostatic field, respectively.

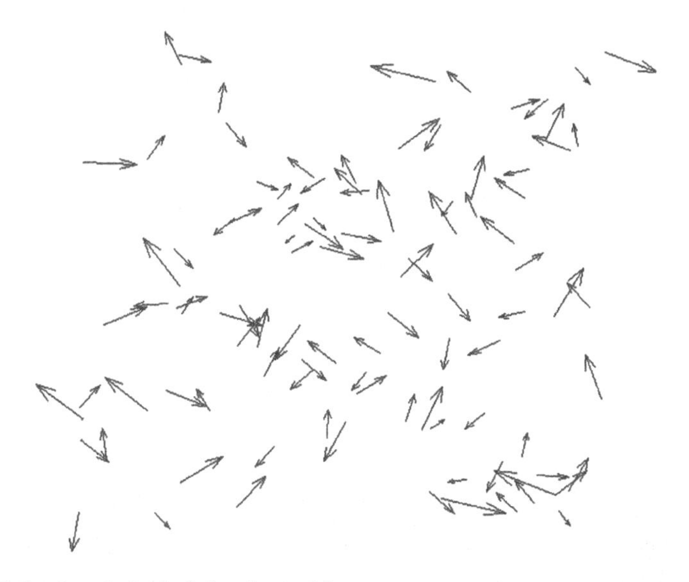

FIGURE 5.5 Speeds field of chaotic particles.

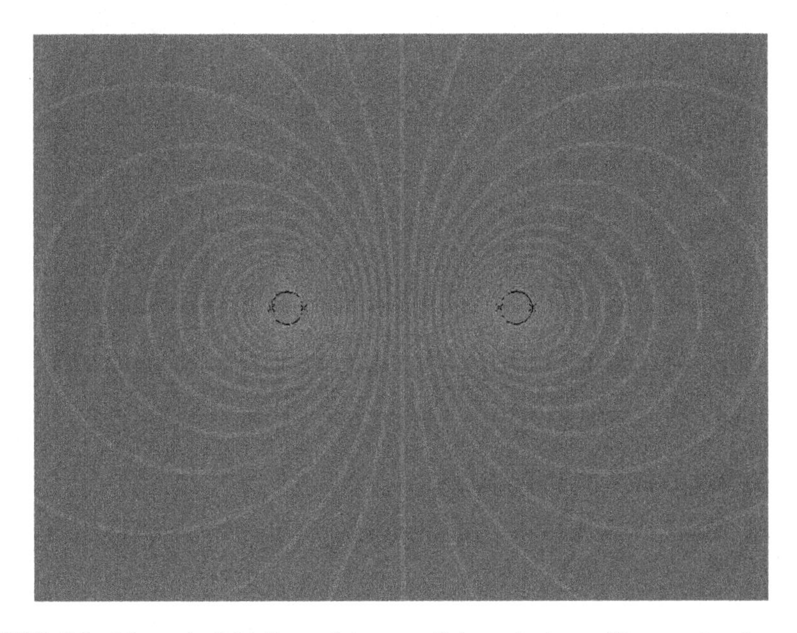

FIGURE 5.6 Magnetic field lines of two parallel conductors with currents of opposite polarity.

The vectorial fields may be represented by equipotential lines or the color map, for the vectors' amplitudes and projections along a given direction. The distribution of the magnetic flux density amplitude produced by two currents (Figure 5.8) implies the need to add a legend for the color/amplitude correspondence.

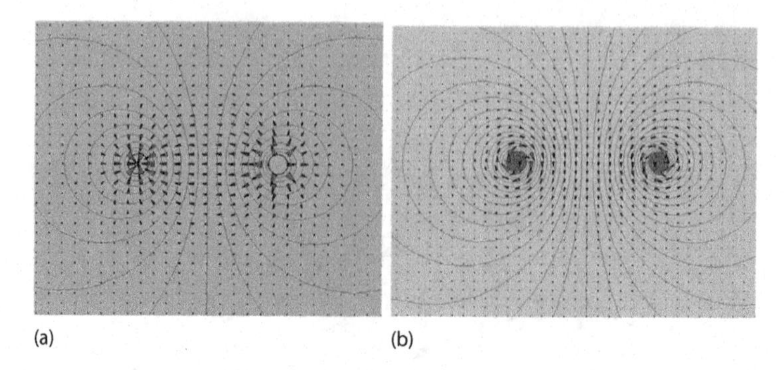

FIGURE 5.7 Comparative representation of (a) the electrostatic field (field of gradients) and (b) the magnetostatic field (field of rotors).

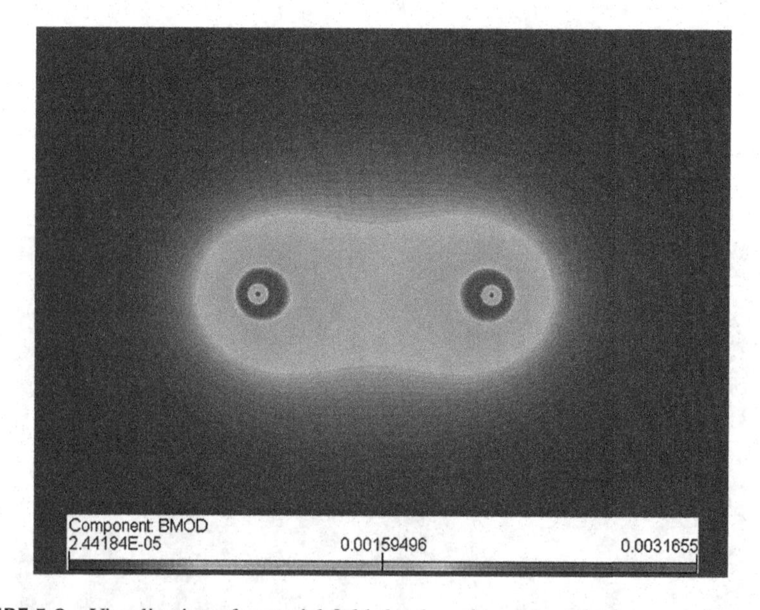

FIGURE 5.8 Visualization of vectorial fields by the color map method.

5.4 BOUNDARY CONDITIONS

Differential and partial derivative equations accept a unique solution only if correct initial boundary conditions are given.

5.4.1 Dirichlet's Boundary Conditions

Dirichlet's boundary conditions imply that the magnetic potential of the investigated field is specified along the boundary.

$$\Phi = \Phi_f \tag{5.81}$$

where Φ is the generic vector potential.

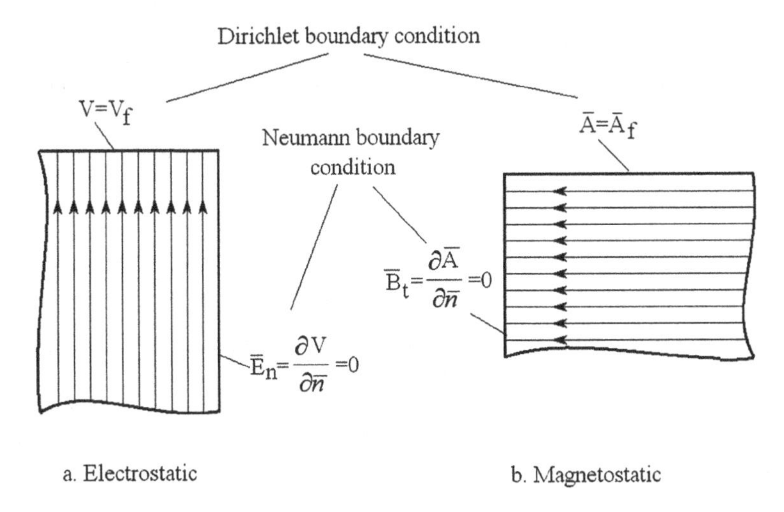

FIGURE 5.9 Dirichlet's and Neumann's boundary conditions.

If $\Phi_f = 0$ on the boundary, homogenous Dirichlet's conditions are obtained. For a unique (single) field solution, Φ_f has to be specified at least at one point on the boundary. For such a magnetic field, the field lines are tangent to the surface, while for an electrostatic field, they are normal to that surface (boundary) (Figure 5.9).

5.4.2 Neumann's Boundary Conditions

Neumann's boundary conditions specify the field potential derivative along the normal to the surface of the boundary. If this derivative is equal to zero, Neumann's conditions are called homogenous:

$$\frac{\partial \Phi}{\partial n} = 0 \tag{5.82}$$

For such a magnetic field, the field lines are normal to the surface, while for an electrostatic field, they are tangent to the surface (boundary) (Figure 5.9).

5.4.3 Mixed Robin's Boundary Conditions

$$\frac{\partial \Phi}{\partial n} + k\Phi = \Phi_g \tag{5.83}$$

If the composite magnetic potential, ϕ_g, is zero we again have homogenous conditions. Robin's boundary conditions allow for defining boundary impedances through which the effect of fields outside the boundary is considered.

5.4.4 Periodic Boundary Conditions

Periodic boundary conditions allow to reduce the computation effort (zone) by the use of symmetry.

The field potential on the two boundaries Γ_1 and Γ_2 is the same for homologous points as in Figure 5.10b (even symmetry), or has equal and opposite values as in Figure 5.10c (odd symmetry):

$$\Phi_{\Gamma_1} = \pm\Phi_{\Gamma_2} \tag{5.84}$$

The symmetry conditions may be imposed only for identical boundaries (those that may be overlapped by translation or rotation).

The shape of the boundaries may be quite intricate (Figure 5.11), which allows for the utilization of symmetric boundary conditions even if a part of the geometry moves.

5.4.5 Open Boundaries

Open boundaries are used in the absence of natural boundaries, such as in the case of core-less solenoidal coils, when the magnetic potential decreases to zero at infinity. Three main solutions for such problems are recommended in the literature:

5.4.5.1 Problem Truncation

The method of problem truncation implies the selection of an arbitrary boundary sufficiently away from the zone of interest. The distance to the arbitrary boundary has to be at least five times longer than the radius of the zone of interest, to secure good precision results [6]. The main disadvantage of this method consists in the large number of discretization elements outside the zone of interest, which leads to a larger computation effort (time).

5.4.5.2 Asymptotical Boundary Conditions

Outside the external boundaries there is no field source, and thus the field potential converges to zero when the radius goes to infinity. For 2D magnetic field problems, admitting the potential A on the circular boundary to be dependent on an angle θ, an analytical solution exists:

$$A(r,\theta) = \sum_{k=1}^{\infty} \frac{a_k}{r^k} \cos(k\theta + \alpha_k) \tag{5.85}$$

where a_k and α_k are coefficients calculated such that the analytical solution (valid outside the boundary), on the boundary, coincides (fits) with the finite element solution inside the boundary.

As high harmonics attenuate rapidly, only one term, n, may be retained:

$$A(r,\theta) \approx \frac{a_n}{r^n} \cos(n\theta + \alpha_n) \tag{5.86}$$

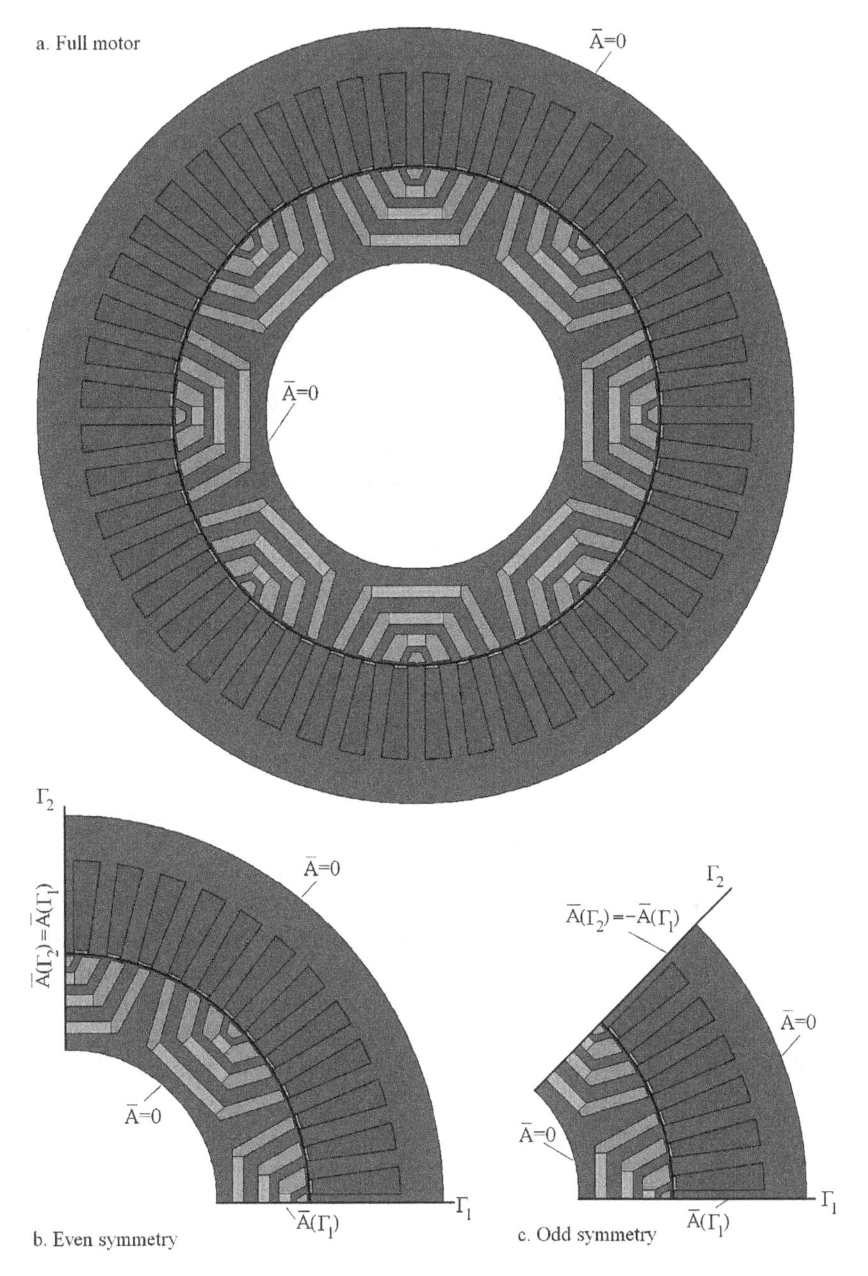

FIGURE 5.10 Symmetric boundary conditions for (a) full motor, (b) even symmetry, and (c) odd symmetry.

The derivative of Equation 5.86 along the radius r (normal to the boundary) yields

$$\frac{\partial A}{\partial r} + \frac{n}{r} A = 0 \tag{5.87}$$

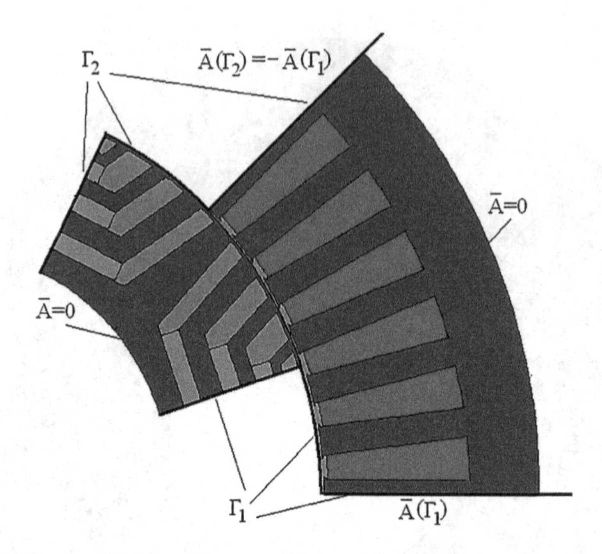

FIGURE 5.11 Symmetrical boundary conditions and rotor displacement.

It should be noticed that Equation 5.87 is equivalent to Robin's condition (Equation 5.83) where

$$k = \frac{n}{r}; \quad \Phi_g = 0 \tag{5.88}$$

By imposing mixed boundary conditions the precision of field solution increases, without additional computation effort [6].

5.4.5.3 Kelvin Transform

For remote field regions, the field is rather homogenous, because in general there are no field sources there and the medium is air or vacuum. In such cases, the magnetic potential observes Laplace's equation. The Kelvin transform (Equation 5.89) maps the points, r, outside the circle of radius r_0 into points inside the circle, R (Figure 5.12a), while the Laplace equation (Equation 5.90) takes a similar form with respect to r or R:

$$R = \frac{r_0^2}{r} \tag{5.89}$$

$$\frac{1}{r}\frac{\partial}{\partial r}\left(r\frac{\partial A}{\partial r}\right) + \frac{1}{r^2}\frac{\partial^2 A}{\partial \theta^2} = 0 \tag{5.90}$$

By the r to R variable change, the boundary at infinity is mapped to the center of the circle, r_0, and thus

$$A_R\left(R = 0\right) = 0 \tag{5.91}$$

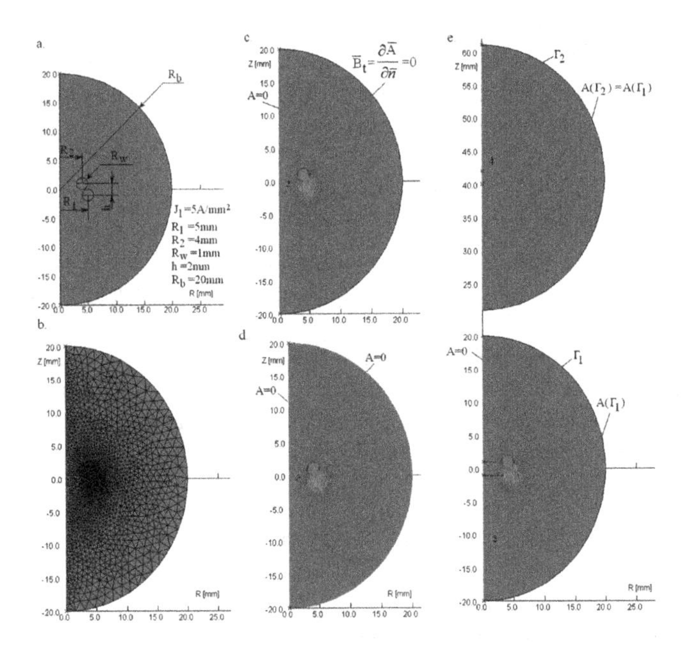

FIGURE 5.12 Boundary conditions when the field extends to infinity. (a) Dual conductor geometry, (b) discretization mesh, (c) magnetic field lines for Neumann's boundary conditions, (d) Dirichlet's boundary conditions, and (e) symmetrical boundary conditions.

Symmetrical conditions are imposed on the boundaries of the investigated and created domains:

$$\frac{\partial A}{\partial r} = -\frac{\partial A}{\partial R} \tag{5.92}$$

Figure 5.12 illustrates the single-turn coil current produced magnetic field. Cylindrical symmetry of the problem leads to the use of cylindrical coordinates in solving the magnetic field problem. As the FEM can investigate finite domains, the investigated infinite domain is mapped into a sphere of radius R_b (Figure 5.2a), which is four times larger than the coil radius. Neumann's boundary conditions (Figure 5.12c) apply for infinite permeability outside the sphere. Dirichlet's boundary conditions (Figure 5.12d) hold for zero permeability (superconducting material) outside the sphere R_b. By using the Kelvin transform, the exterior of sphere R_b is reduced to a sphere with the same radius as the investigated domain, but without field sources. By imposing symmetrical conditions (Figure 5.2e) on the two spheres, the correct solution of the field problem is obtained. The field lines are quite different near the boundary for the three situations. However, as will be shown in Section 5.6, the self and mutual inductances of coils 1 and 2 are about the same, mainly because the domain radius ratio $R_b/r_0 = 4$.

5.5 FINITE ELEMENT METHOD

To simplify the mathematics, the system of differential equations, which describes a phenomenon, is written in the operational form:

$$L\Phi(P,t) = f(P,t) \tag{5.93}$$

where L is a generic operator that symbolizes the system of equations with partial derivatives, ϕ is a space, and function P represents the coordinates of a point and t is time.

The FEM solves approximately partial derivative and integral equations [3–5]. With this method, the field domain is divided into a finite number of simple elements (triangles, plane rectangles, tetrahedrons, prisms, and parallelepipeds) and the solution of the field problem is approximated by simple functions:

$$\Phi^*(P,t) = \sum_{j}^{N} \Phi_j v_j(P,t) \tag{5.94}$$

where v_j are interpolations or expansions of basic functions and ϕ_j are coefficients still to be determined. Physically, ϕ_j represents the magnetic potential in the nodes of the discretization mesh.

The equivalent system of equations should be stable, in the sense that the errors from the input data and those from intermediate calculations should not accumulate and produce gross (useless) results.

5.5.1 Residuum (Galerkin's) Method

Galerkin's method operates directly with differential equations. The exact solution ϕ, which satisfies Equation 5.93, is approximated by a simple function ϕ^*, which satisfies the relationship:

$$L\Phi^* - f = r \tag{5.95}$$

The smaller the r the better the approximation. A globally good approximation over the entire domain is targeted. The residuum, r, is integrated over the entire domain, after multiplication by a weighting function, w_i, and this integral is forced to zero:

$$R_i = \int_D w_i \left(L\Phi^* - f\right) \mathrm{d}\tau = 0 \tag{5.96}$$

The most used residuum method is Galerkin's method where the weighting functions are chosen equal to the expansion functions, v_j (Equation 5.94):

$$R_i = \int_D \left(v_i \left(\sum_{j=1}^{N} \Phi_j v_j \right) - v_j f \right) \mathrm{d}\tau \tag{5.97}$$

By forcing all terms in Equation 5.97 to zero, a system of matrix equations is obtained:

$$S \cdot \Phi = T \tag{5.98}$$

where

S is the system (or stiffness) matrix, because the FEM was first used in mechanical engineering

ϕ is the column matrix of the vector potential in the mesh nodes

T is a column matrix, whose terms depend on function f

$$t_i = \int_D v_i f \, \mathrm{d}\tau \tag{5.99}$$

The elements of the stiffness matrix, S, depend on the interpolation functions, v_i and v_j:

$$S_{ij} = \frac{1}{2} \int_D \left(v_i L v_j + v_j L v_i \right) \mathrm{d}\tau \tag{5.100}$$

The matrix S is a sparse matrix. Additionally, if Equation 5.97 holds, the matrix S is symmetric:

$$\begin{gathered} \langle L\Phi, \varphi \rangle = \langle \Phi, L\varphi \rangle \\ S_{ij} = S_{ji} = \int_D v_i L v_j \mathrm{d}\tau \end{gathered} \tag{5.101}$$

In such a case, the algebraic system of equations by Galerkin and variational methods are identical.

5.5.2 Variational (Rayleigh–Ritz) Method

Starting again from the system of differential equations, a functional $F(\phi)$ is defined such that its minimum with respect to ϕ is the solution of the system [1];

$$F(\Phi) = \frac{1}{2} L \langle \Phi, \Phi \rangle - \frac{1}{2} \langle \Phi, f \rangle - \frac{1}{2} \langle f, \Phi \rangle \tag{5.102}$$

To define the functional, the inner product of two functions is expressed as the integral of one function multiplied by the complex conjugate of the second function:

$$\langle \Phi, \varphi \rangle = \int_D \Phi \bar{\varphi} \mathrm{d}\tau \tag{5.103}$$

The function ϕ is substituted with ϕ^* in Equation 5.94. The ϕ_j coefficients are calculated by zeroing the derivative of F, with respect to the former:

$$\frac{\delta F}{\delta \Phi_j} = 0; i = 1,2,3,\ldots,N \tag{5.104}$$

For the variational method, condition (5.101) has to hold. In such conditions, the function that minimizes F is a system (Equation 5.93) solution [1]. The stiffness matrix is always symmetric and the system of algebraic equations is identical to that of Galerkin's method. When the boundary conditions are nonhomogenous, Equation 5.101 is not satisfied. In such a case, the functional definition has to consider the boundary nonhomogenities:

$$F(\Phi) = \frac{1}{2}\langle L\Phi,\Phi\rangle - \frac{1}{2}\langle L\Phi,\Psi\rangle + \frac{1}{2}\langle \Phi,L\Psi\rangle - \frac{1}{2}\langle \Phi,f\rangle - \frac{1}{2}\langle f,\Phi\rangle \tag{5.105}$$

where Ψ is any particular function that satisfies the nonhomogeneous boundary conditions.

5.5.3 Stages in Finite Element Method Application

5.5.3.1 Domain Discretization

The investigated field domain is divided into N elements. If the domain is 1D, the respective curve is divided into linear segments. For 2D problems, the domain is a surface and every subdomain is a polygon, in general, a triangle or a rectangle. Finally, for 3D problems, the volumic domain is divided into tetrahedrons and prisms with triangular and parallelepipedic bases.

The precision of this method strongly depends on the number of elements, N, and on how the mesh was built.

5.5.3.2 Choosing Interpolation Functions

The interpolation functions approximate the unknown function in every subdomain (finite element). The general form of the interpolation is given by Equation 5.94, where P represents the coordinates of a point in a subdomain. The first- and second-order polynomials are, in general, used as interpolation functions.

5.5.3.3 Formulation of Algebraic System Equations

The residuum or variational methods yield the algebraic system equations of Equation 5.98, where the unknown vector refers to the scalar or the vector potential in the nodes of the mesh.

5.5.3.4 Solving Algebraic Equations

When the material in the field domain is linear, the coefficients of the algebraic equations are constant, and the Gauss–Seidel method is used often to solve the algebraic equations. If the material (iron core) is nonlinear, the coefficients in the stiffness matrix are not constant anymore, and the Newton and Raphson method [2] is often used to solve the algebraic equations.

5.6 2D FEM

Many field problems in electrical engineering can be solved two dimensionally by observing the parallel plane symmetry in rotary electrical machines, in electrostatic fields around in power transmission lines, or the axial symmetry in solenoidal coils and other tubular, linear electric actuators. For electric machines with plane symmetry, the magnetic fluxes are calculated for unit depth (stack length) and then multiplied by the stack length to find the actual magnetic flux in the machine. It is also necessary to calculate separately the coils' end-connection inductances by analytical or equivalent 2D FEM methods. The axial variation of the magnetic field is neglected and so is the effect of the various frame parts. With the exception of small-stack-length electric machines, the 2D FEM gives satisfactory results in the presence of the plane symmetry. For 2D ($x,y;r,\theta$) magnetic fields, the magnetic potential, $\bar{A}$, has only one nonzero component (along the z-axis) thereby reducing the computation effort by one order of magnitude.

In what follows, the derivation of the algebraic equations for the magnetostatic 2D fields is pursued.

Equation 5.96, for $\bar{A} \to A_z$, can be written as

$$\frac{1}{\mu}\frac{\partial^2 A_z}{\partial x^2}+\frac{1}{\mu}\frac{\partial^2 A_z}{\partial y^2}=-J_z \tag{5.106}$$

The first-order interpolation function for the steady state is

$$A(x,y)=a+bx+cy \tag{5.107}$$

where a, b, and c are constants whose values may be determined by imposing the value of A_z in the nodes of the discretization mesh:

$$\begin{pmatrix} 1 & x_1 & y_1 \\ 1 & x_2 & y_2 \\ 1 & x_3 & y_3 \end{pmatrix}\cdot\begin{pmatrix} a \\ b \\ c \end{pmatrix}=\begin{pmatrix} A_1 \\ A_2 \\ A_3 \end{pmatrix} \tag{5.108}$$

The rotor of A_z inside a finite element does not depend on the chosen point.

$$\mathrm{rot}\left(\bar{A}\right)=\mathrm{rot}\left(0,0,A_z\right)=\left(c,-b,0\right) \tag{5.109}$$

The expression of a functional whose steady-state solution is the field solution from the m mesh elements becomes

$$\begin{aligned} F_m &= \int_{Q_m}\left(\frac{1}{2}\bar{B}\cdot\bar{H}-\bar{J}\cdot\bar{A}\right)\mathrm{d}S=\frac{1}{2\mu}\int_{Q_m}\left(rot\left(\bar{A}\right)\cdot rot\left(\bar{A}\right)\right)\mathrm{d}S-J_z\int_{Q_m}A_z\mathrm{d}S \\ &= \frac{1}{2\mu}Q_m\left(b^2+c^2\right)-J_zQ_m\left(a+bx_m+cy_m\right) \end{aligned} \tag{5.110}$$

where

x_m and y_m are centroid coordinates of triangle m
Q_m is the area of subdomain m

Coefficients a, b, and c depend on the coordinates of the nodes of the domain m and on the magnetic potential in these nodes; consequently, in matrix form, Equation 5.105 becomes

$$F_m = \frac{1}{2} A_{123}^t S_m A_{123}^t - A_{123}^t T_m \tag{5.111}$$

where

A_{123} is the column vector of A_z in the nodes of element m,
S_m is the stiffness matrix, and
T_m is the column vector electric currents imposed as the field source.

The elements of these matrices, s_i and t_i, are

$$s_{ij} = \frac{q_i q_j + r_i r_j}{4\mu Q_m} \tag{5.112}$$

$$t_i = -\frac{Q_m}{3} J_z \tag{5.113}$$

where q_1, q_2, and q_3 and r_1, r_2, and r_3 [1] are obtained by solving Equation 5.108:

$$\begin{pmatrix} q_1 \\ q_2 \\ q_3 \end{pmatrix} = \begin{pmatrix} 0 & 1 & -1 \\ -1 & 0 & 1 \\ 1 & -1 & 0 \end{pmatrix} \begin{pmatrix} y_1 \\ y_2 \\ y_3 \end{pmatrix} \tag{5.114}$$

$$\begin{pmatrix} r_1 \\ r_2 \\ r_3 \end{pmatrix} = \begin{pmatrix} 0 & -1 & 1 \\ 1 & 0 & -1 \\ -1 & 1 & 0 \end{pmatrix} \begin{pmatrix} x_1 \\ x_2 \\ x_3 \end{pmatrix} \tag{5.115}$$

The functional over the entire field domain is

$$F = \sum_{m=1}^{M} F_m = \frac{1}{2} A^{\mathrm{T}} S A - AT \tag{5.116}$$

with the condition

$$\frac{\partial F}{\partial A_i} = 0 \quad \text{for } i = 1, 2, 3, \ldots, N \tag{5.117}$$

The algebraic equations system is thus obtained, and is ready to be solved:

$$SA - T = 0 \tag{5.118}$$

5.7 ANALYSIS WITH FEM

After the field solution is obtained, it has to be verified. This is done first by visualizing the field lines. This way all analytical symmetries are verified. The magnetic field lines never meet or intersect each other. If it happens, it may be that the graphic resolution is too small or the field solution is wrong. The magnetic flux lines are always closed. Sometimes they close through the boundary and then part of them is not visible. The total electric current through the area closed by a flux line should be nonzero with the exception that the flux lines cross regions with permanent magnetization. For a parallel plane symmetric geometry, the magnetic flux through any surface that intersects two flux lines is the same irrespective of the surface shape of points of intersection, and is equal to the difference between the magnetic potential attributed to the two field lines.

Visualization of a magnetic flux density map provides valuable information, in the sense that the designer may reduce or enlarge some portions of the magnetic circuit. Also, the consolidation elements, ventilation channels, will be placed in regions with very low flux density. For rotary and linear electric machines, the distribution of the radial component of airgap flux density along the rotor periphery is very important in predicting torque pulsations, vibration, and noise. The total magnetic flux in a coil with a magnetic core is made of main and leakage components. Using FEM the total flux in the coil may be calculated by applying Equation (5.35). For an infinitely thin-side coil with plane symmetry, the total flux is

$$\phi_{\mathrm{b}} = (A_2 - A_1) l_z \tag{5.119}$$

where

A_1 and A_2 are the magnetic potentials in the points in the plane intersected by the coil sides and

l_z is the coil length.

If the magnetic vector potential is not constant along the transverse cross-section (because of leakage flux of coils placed in electric machine slots), an average flux per coil, ϕ_{b}, is calculated:

$$\phi_{\mathrm{b}} = \left(\frac{1}{S_2} \int_{S_2} A\mathrm{d}S - \frac{1}{S_1} \int_{S_1} A\mathrm{d}S \right) l_z \tag{5.120}$$

By using Equation 5.120, the distribution factor of coils in slots can be considered, coil by coil.

The current density in the coil wires is considered constant along the cross-section and so no skin or proximity effect is considered.

The self-inductance of a coil may be calculated either from the magnetic field energy, E_m, or from the coil flux with current, I_b:

$$L_b = \frac{2E_m}{I_b^2} \quad (5.121)$$

$$L_b = \frac{\phi_b}{I_b} \quad (5.122)$$

The two formulae have to produce the same result.

With the flux density, $\bar{B}$, and the magnetic field, $\bar{H}$, expressed as functions of the magnetic potential, $\bar{A}$, and applying Green's theorem (Equation 5.39), the magnetic energy, E_m, formula becomes

$$2E_m = \int_V \bar{B} \cdot \bar{H} dv = \int_V \frac{1}{\mu} \text{rot}(\bar{A}) \cdot \text{rot}(\bar{A}) dv = \int_V \bar{A} \cdot \text{rot}\left(\frac{1}{\mu} \text{rot}(\bar{A})\right) dv$$
$$+ \oint_S \frac{1}{\mu} \bar{A} \times \text{rot}(A) \cdot \overline{ds} \quad (5.123)$$

Substituting Equations 5.73 and 5.76 in Equation 5.123, E_m becomes

$$2E_m = \int_V \bar{A} \cdot \bar{J} dv + \oint_S \bar{A} \times \bar{H} \cdot \overline{ds} \quad (5.124)$$

If all boundary conditions are homogenous, then the second term in Equation 5.124 is zero and

$$2E_m = \int_V \bar{A} \cdot \bar{J} \, dv \quad (5.125)$$

Using E_m from Equation 5.125 and ϕ_b from Equation 5.120 in Equations 5.121 and 5.122, for a geometry with plane parallel symmetry and no skin effect, the two formulae of L_b become equivalent.

For coils made of massive conductors (with skin effect), the coil flux, ϕ_b, becomes

$$\phi_b = \frac{1}{I_b} \left(\int_{S_2} AJ dS - \int_{S_1} AJ dS \right) l_z \quad (5.126)$$

For the mutual L_{12} inductance between coils 1 and 2, the definition in relation to mutual flux, ϕ_{12}, and current solely in the first coil, I_1, is used:

$$L_{12} = \frac{\phi_{12}}{I_1} \quad (5.127)$$

Let us now illustrate the computation of magnetic energy, E_{m}, and flux, ϕ_i, of the coil in Figure 5.12.

The computation of E_{m} from the volume integral is reduced to a plane surface integral because of symmetry around the rotation axis:

$$E_{\mathrm{m}}(B,H) = \int_S \pi r B H \mathrm{d}s \tag{5.128}$$

$$E_{\mathrm{m}}(A,J) = \int_S \pi r A J \mathrm{d}s \tag{5.129}$$

The computation of magnetic flux linkage in turns 1 and 2 is performed by integrating the magnetic potential over their surfaces:

$$\phi_i = \frac{1}{S_i} \int_{S_i} 2\pi r A \mathrm{d}s \tag{5.130}$$

The results of these calculations are shown in Table 5.1.

Observations: Due to discretization errors, the values in Table 5.1 are not identical; the same is true for the self-inductance of turn 1 calculated from E and ϕ, but the errors are acceptable due to the large distances from the boundary. The energy and inductances are larger for Neumann's conditions (infinite permeability beyond the border) and smallest for Dirichlet's conditions. Reducing the domain (boundary) radius, R_{b}, leads to larger errors.

5.7.1 Electromagnetic Forces

An important objective of FEM analysis is the computation of electromagnetic forces. There are three main methods to calculate them, as follows.

5.7.1.1 Integration of Lorenz Force

By this method the force exerted on conductors with electric currents is calculated:

$$\bar{F} = \int_V \bar{J} \times \bar{B} \mathrm{d}v \tag{5.131}$$

For problems with a plane or axial symmetry, the volume integral turns into a surface integral with the force vector situated in the symmetry plane:

$$\bar{F} = l_z \int_S \bar{J} \times \bar{B} \mathrm{d}s \tag{5.132}$$

TABLE 5.1
Energy and Inductances

Boundary Type	$E_m(BH)$ [μJ]	$E_m(AJ)$ [μJ]	Φ_1 [nWb]	Φ_2 [nWb]	$L_{11}(E_m)$ [nH]	$L_{11}(\Phi)$ [nH]	L_{12} [nH]
Neumann	1.49	1.5086	192.07	78.7989	12.329	12.278	5.037
Dirichlet	1.4613	1.4796	188.395	76.4771	12.093	12.043	4.889
Kelvin transform	1.4804	1.4963	190.517	77.8844	12.229	12.179	4.979

$$\bar{F} = \int_S 2\pi r \bar{J} \times \bar{B} \mathrm{d}s \tag{5.133}$$

For cylindrical symmetry, the resultant force has only one nonzero component along the axis of symmetry. Some commercial FEM software (such as "Vector Field") calculates two force components even for a cylindrical symmetry. A knowledge of radial force is required to verify the mechanical stress of the electrical conductors of the coil. The Lorenz force procedure is precise, but can be applied only for electric conductors in air. In most electrical machines, however, the coils are placed in slots where the actual force is exerted on the slot walls and not on the electric conductors, and thus the Lorenz force procedure is inoperable.

5.7.1.2 Maxwell Tensor Method

The resultant field force upon the objects found within a closed surface is calculated by integrating the Maxwell stress tensor (Figure 5.13) along the closed surface:

$$\bar{F} = \oint_\Sigma \mathrm{d}\bar{F} = \mu_0 \oint_\Sigma \left(\left(\bar{H} \cdot \bar{n} \right) \bar{H} - \frac{1}{2} H^2 \bar{n} \right) \mathrm{d}s \tag{5.134}$$

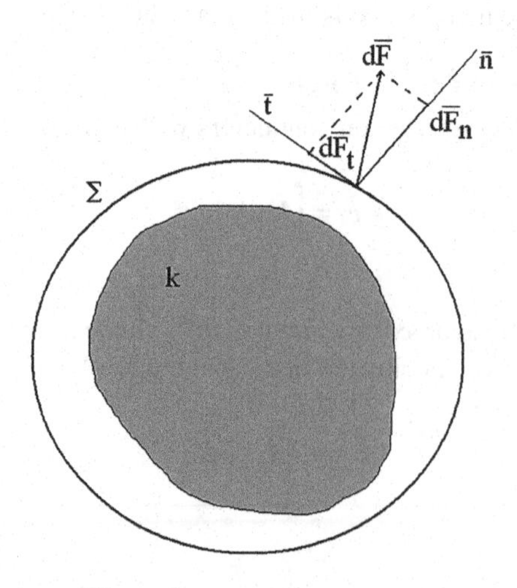

FIGURE 5.13 Integration of Maxwell tensor.

The normal and tangential forces to the surface, dF_n and dF_t, per unit area are

$$dF_t = \mu_0 H_t H_n ds \tag{5.135}$$

$$dF_n = \frac{1}{2}\mu_0\left(H_n^2 - H_t^2\right)ds \tag{5.136}$$

To reduce computation errors, refined discretization of the domain with 60° (equilateral) triangles with their nodes in a linear medium, is required.

5.7.1.3 Virtual Work Method

The field force computation is based on the total energy conservation principle

$$dw_{mec} + dw_m = dw_{el} \tag{5.137}$$

where

dw_{mec} is the mechanical work,
dw_m is the magnetic energy increment, and
dw_{el} is the received electric energy from outside (a source).

A coenergy of the system, w'_m, is defined as the difference between the received electric energy, w_{el}, and the energy stored in the magnetic field, w_m:

$$w'_m = w_{el} - w_m \tag{5.138}$$

The projection of this force along the direction of motion is calculated as the magnetic coenergy derivative with respect to the motion variable:

$$\bar{F} \cdot \bar{1}_{\delta l} = \frac{\delta w'_m}{\delta l} \tag{5.139}$$

This method implies the computation of magnetic field and magnetic coenergy in two adjacent positions. As the solution to the field problem is also numerical, the numerical derivative will amplify the computation errors (Figure 5.14).

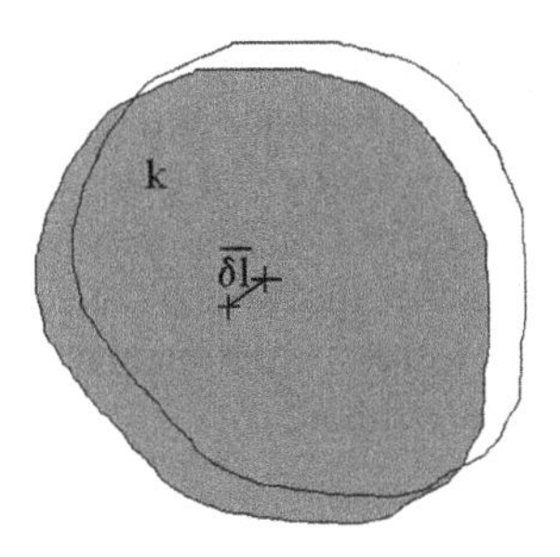

FIGURE 5.14 Virtual displacements.

5.7.2 Loss Computation

Losses in a conductive medium with electric currents are Joule losses, P_{co}:

$$P_{co} = \int_V \rho J^2 dv \quad (5.140)$$

For a magnetostatic regime, the current density is constant and the Joule losses imply only current and resistance calculations. For alternative or transient current operation modes in a massive conductive medium, the current density distribution depends on the solution of the field problem, and thus the volume integral in Equation 5.140 is required.

5.7.2.1 Iron Losses

The iron loss computation is based on the volume integral of the iron loss volumic density (W/m^3) over the magnetic core volume; alternatively, the Steinmetz approximation formulae are used. Finally, the complex permeability concept in the field solution is utilized.

$$P_{iron} = \int_V p_{Bs,fs} \left(\frac{B_m}{B_s}\right)^{\alpha_B} \left(\frac{f}{f_s}\right)^{\alpha_f} \gamma_{Fe} dv \quad (5.141)$$

where

$p_{Bs,fs}$ are the specific (volumic) losses at a given flux density B_s and frequency f_s, respectively,
B_m is the maximum a.c. flux density in volume element dV,
f is the working frequency,
$\alpha_B = (1.6 - 2.2)$ is the exponent for iron loss dependence on B_m,
$\alpha_f = (1.5 - 1.7)$ is the exponent for iron loss dependence on frequency, and
γ_{Fe} is the mass density.

For iron core coils or single-phase electric transformers, the maximum flux density in every point is related to the maximum magnetization current. For polyphase rotary machines (even single-phase), the maximum flux density cannot be determined by solving the field problem at a single moment in time, but for more than one moment within a period.

Even in this case, the calculated loss value is not exact because there are higher time harmonics in magnetic flux due to space and time flux density harmonics. If the flux density pulsates, is traveling (circular), or elliptical in type, in various points in the magnetic core, the computation of iron loss becomes more complex. Finally, the mechanical machining of magnetic silicon–iron sheets (laminations) in electrical machines adds additional core losses; experimental work is necessary to validate almost any iron loss computation method.

REFERENCES

1. N. Bianchi, *Electrical Machine Analysis Using Finite Elements*, CRC Press, Taylor & Francis Group, Boca Raton, FL, 2005.
2. S.J. Salon, *Finite Element Analysis of Electrical Machines*, Kluwer Academic Publishers, Norwell, MA, 2000.
3. D.A. Lowther and P.P. Silvester, *Computer-Aided Design in Magnetics*, Springer-Verlag, Berlin, Germany, 1986.
4. C.W. Steele, *Numerical Computation of Electric and Magnetic Fields*, Van Nostrand Reinhold Company, New York, 1987.
5. P.P. Silvester and R.L. Ferrari, *Finite Elements for Electrical Engineers*, Cambridge University Press, London, 1983.
6. D. Meeker, *Finite Element Method in Magnetics*, User's Manual, Version 4.0, January 8, 2006.
7. G. Arfken, *Circular Cylindrical Coordinates*, 3rd edn., Academic Press, Orlando, FL, pp. 95–101, 1985.
8. W.H. Beyer, *CRC Standard Mathematical Tables*, 28th edn., CRC Press, Boca Raton, FL, 1987.
9. G.A. Korn and T.M. Korn, *Mathematical Handbook for Scientists and Engineers*, McGraw-Hill, New York, 1968.
10. C.W. Misner, K.S. Thorne, and J.A. Wheeler, *Gravitation*, W.H. Freeman, San Francisco, CA, 1973.
11. P. Moon and D.E. Spencer, *Circular-Cylinder Coordinates*. Table 1.02 in field theory handbook, including coordinate systems, differential equations, and their solutions, 2nd edn., Springer-Verlag, New York, 1988.
12. P.M. Morse and H. Feshbach, *Methods of Theoretical Physics, Part I*, McGraw-Hill, New York, p. 657, 1953.
13. J.J. Walton, Tensor calculations on computer: Appendix. *Comm. ACM*, 10, 1967, 183–186.
14. J.S. Beeteson, *Visualising Magnetic Fields*, Academic Press, San Diego, CA, 1990.
15. H.E. Knoepfel, *Magnetic Fields—Comprehensive Theoretical Treatise for Practical Use*, Wiley-Interscience, John Wiley & Sons, New York, 2000.
16. F.E. Low, *Classical Field Theory—Electromagnetism and Gravitation*, Wiley-Interscience, John Wiley & Sons, New York 1997.
17. O.C. Zienkiewicz, R.L. Taylor, and J.Z. Zhu, *The Finite Element Method: Its Basis and Fundamentals*, 6th edn., Elsevier Butterworth-Heinemann, Oxford, U.K., 2005.

6 FEA of Electric Machines Electromagnetics

6.1 SINGLE-PHASE LINEAR PM MOTORS

The FEM analysis of electric machines consists of three major stages:

1. Preprocessor stage—This stage provides a description of the field problem associated with the studied machine and its operation regimes. It contains several steps such as choosing the problem symmetry, making the machine geometry embedded drawing, choosing the frontier conditions, choosing the magnetic field source, generating mesh, choosing the type of the solved problem (magnetostatic, a.c. problem, transients problem, rotating machine problem, linear machine problem, etc.), choosing the settings for the solver (maximum number of iterations, mesh refinement, solution global error, etc.), and creating an input file for the solver.
2. Solving the field problem.
3. Postprocessor stage—In this stage, the field distribution can be shown and the circuit equivalent parameters can be computed.

A single-phase linear PM motor in tubular construction is presented in Figure 6.1. It was proposed [1] as a thermal engine valve actuator.

It contains

- A dual part somaloy passive mover
- A tubular PM made of 6–12 sectors to reduce the eddy current losses in the PMs due to the a.c. mmf of twin coil currents, all in the stator.
- A stainless steel (shaft) coupled to the thermal engine valve

The electromechanical device also contains two mechanical springs that are not shown in Figure 6.1. The two mechanical springs act in opposite directions and their forces are in equilibrium when the mover is in the middle position. The PM magnetic force at zero current acts the mover so as to increase its displacement (Figure 6.2).

This force has an unstable equilibrium point when the mover is in the middle position. The magnetic field produced by coil currents breaks the force equilibrium and orients the mover in a certain direction, according to the current polarity. The magnetic force at zero current is larger than the spring force when the mover is in the extreme position and holds the mover in that position, which produces a good

DOI: 10.1201/9781003216018-6

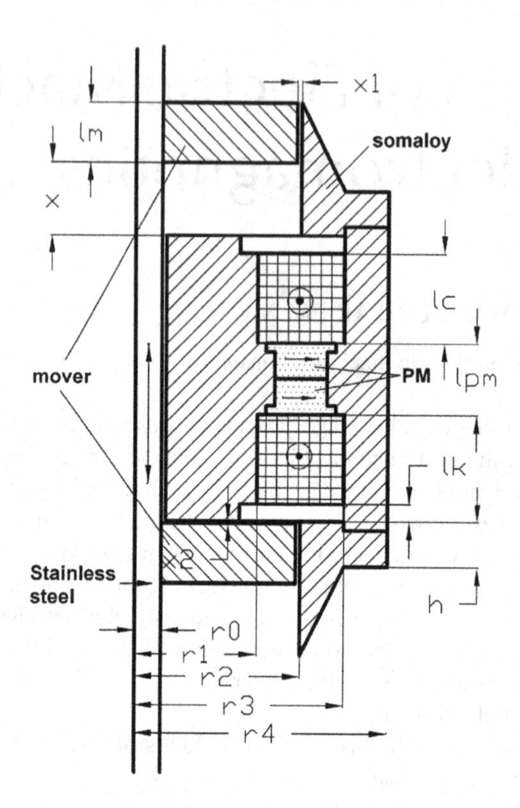

FIGURE 6.1 Electro-valve linear PM actuator.

reduction in copper losses. Reducing the coil currents decreases the electromagnetic force and the springs drive the mover through the equilibrium point.

The main geometric dimensions of the linear actuator of our case study, Figure 6.1, are given in Table 6.1.

The FEM analysis was performed using the Vector Field computer software. The FEM analysis methodology is similar to Flux2D from Cedrat, FEMM, or other commercial software for magnetostatic FEMs.

6.1.1 Preprocessor Stage

The FEM analysis was performed in magnetostatic mode for several mover positions. The axis symmetry with modified vector potential, "r A," was chosen due to the tubular construction of the analyzed device. To introduce the device geometry, it is adequate to use the GUI interface [2,3], or to use "comi scripts" in the Vector Field or "lua scripts" in the FEMM. For a simple problem, when is not necessary to solve the problem repetitively for other geometric dimensions, it is preferable to use the GUI interface. When it is necessary to solve a similar problem repetitively, in order to find better performance or to compute circuit parameters vs. mover position, then it is feasible to write one or more scripts in order to build an automatic parameterized

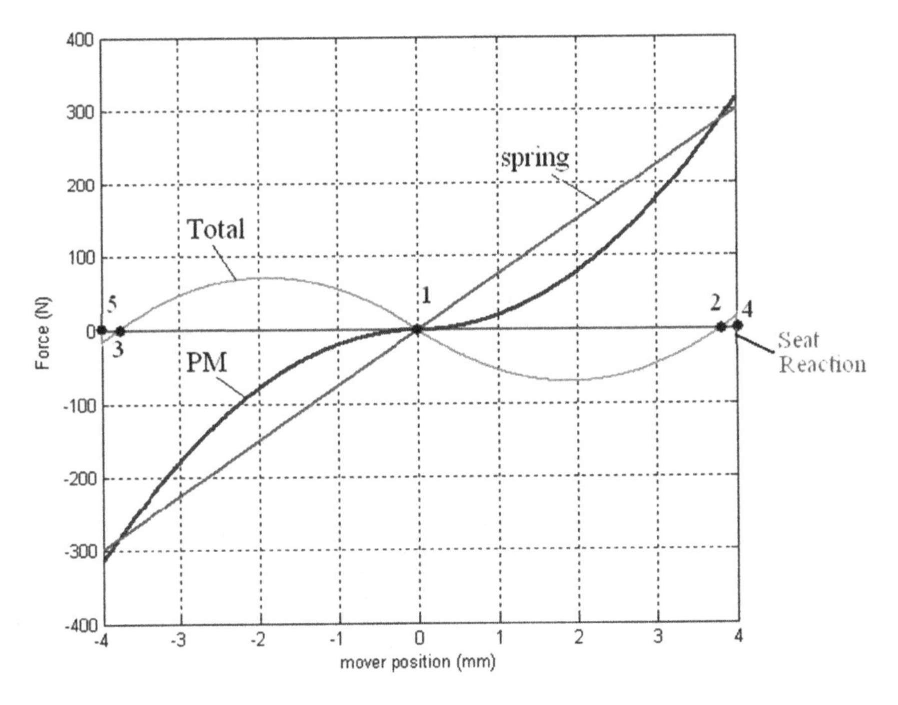

FIGURE 6.2 Mechanical spring and permanent magnet thrust.

TABLE 6.1
Electro-Valve Main Dimensions

No.	Dimensions	Value	Units	Observation
1	r0	3	mm	Mover rod radius
2	r1	14	mm	Outer radius of inner stator core
3	r2	19	mm	Inner radius of stator pole
4	r3	24	mm	Inner radius of stator outer core
5	r4	29	mm	Outer radius of stator core
6	xm	4	mm	Mover displacement magnitude
7	x1	0.5	mm	Radial airgap
8	x2	0.2	mm	Minimum values of axial airgap
9	lm	6.7	mm	Mover pole length
10	lc	10	mm	Coil height
11	lk	2	mm	Distance between coil and stator pole
12	h1	5	mm	Stator core overlength
13	lpm	8	mm	PM height

drawing. In this case it is necessary to compute the electro-valve thrust vs. the mover position, the coil linkage flux vs. the position, and the coil inductance vs. mover position. Consequently, it is necessary to solve a similar problem several times. The computer code is more flexible if it is divided into three scripts: one script contains only

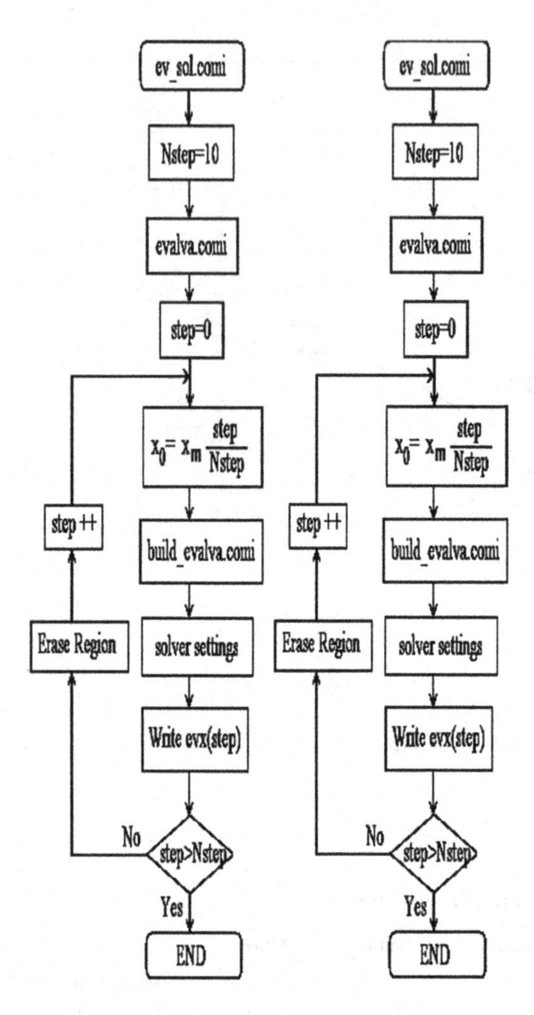

FIGURE 6.3 Block diagram of ev_sol.comi script.

the device dimensions—the input script; the second script contains the geometry drawing description; and the third script is used to set the solution parameters. For our example, this script is called "ev_sol.comi" and its block diagram is shown in Figure 6.3.

In this script, the number of mover positions (Nstep) is set. The "evalva.comi" script is called upon to set the main geometric dimensions as they are presented in Table 6.1. The mover position is set according to the current step and then the "build_evalva.comi" script is called upon to build the electro-valve drawing according to the current mover position. Solver specifications such as tolerance and the maximum number of iterations are set. The "scale factor" is also chosen. For every mover position it is possible to solve the problem for several currents in the coils. In this case, the following scale factor was chosen: 0, 0.25, 0.5, 0.75, 1.25, 1.5, 1.75, 2, −0.25, −0.5, −0.75, −1, −1.25, −1.5, −1.75, −2. A unitary factor scale is chosen implicitly.

The solution will be computed for rated currents and also for rated currents multiplied with a given scale factor. Choosing several scale factors is useful to check the influence of saturation on thrust, linkage flux, and inductance. A large number of scale factors increase the computation time. A zero value scale factor implies a solution where the field is produced only by the permanent magnet. Different behaviors are expected for positive and negative currents due to the prepolarization of the PM magnetic circuit.

Every mover position is saved as an independent magnetostatic problem. If the step does not reach the maximum number of steps, the regions are erased, the step is increased, and a new problem is generated. The script is finished when the maximum number of steps is reached. The last problem regions are not erased, so they remain on the display. In this way, the user can check the region drawing and the mesh. The geometric dimensions unit is set in millimeters and the current density unit in Ampere per square millimeter in the input script, "evalva.comi." The magnetic nonlinear material is also set in this file by associating the material number with a text file containing the magnetization curve. The permanent magnet features are given through its demagnetization curve (Figure 6.4a).

Before starting the geometry drawing it is important to observe if there is geometric symmetry and also if a complex geometry could be decomposed in simple and identical shape components. In this example, the mover rod is made of stainless steel. Considering its permeability to be equal to the permeability of vacuum it could be assimilated with vacuum in magneto-static mode, and, consequently, it could be omitted from the drawing. There is no symmetry in the field due to the mover position and the PM and current magnetic field over position, so the entire field problem will be analyzed. The stator has symmetry around the x-axis and thus it is enough to draw only the stator parts placed over the x-axis. The stator component placed under the x-axis will be constructed using copy command.

The command "Draw" is used in Vector Fields in order to construct the geometry. It is a complex command that is used to assign the material features, the number of subdivisions for each line, the frontier condition, and other such requirements [2].

The electro-valve could be decomposed in the following parts:

- Mover magnetic disks
- Stator inner cylinder
- Permanent magnet
- Coils
- Outer stator part
- Background

The mover disk in RZ plane projection is a simple rectangle and, consequently, a quadrilateral shape with regular subdivision (H shape) could be chosen [2, pp. 2–48]. The upper disk corners are computed:

$$\begin{aligned} X_{12} &= r_2 - x_1 \\ X_{34} &= r_0 \end{aligned} \tag{6.1}$$

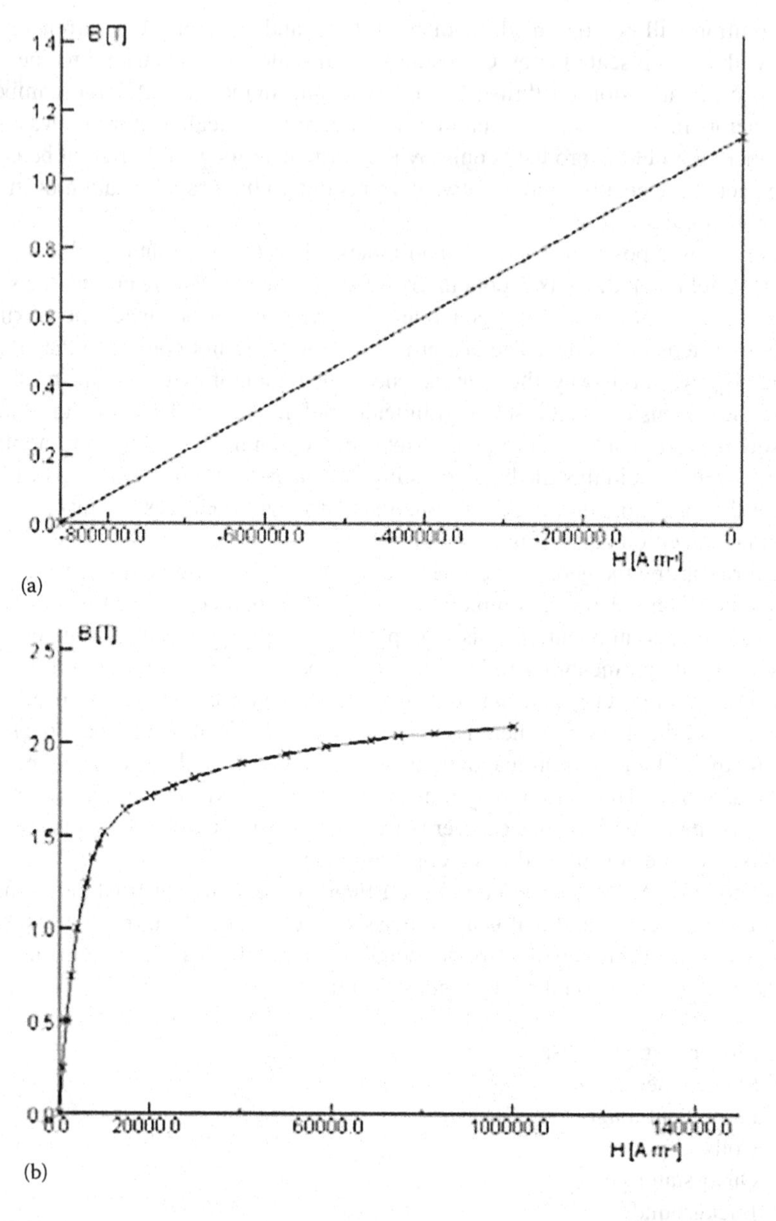

FIGURE 6.4 Magnetization features. (a) PM demagnetization curve and (b) Somaloy 550—magnetization curve.

$$
\begin{aligned}
Y_{14} &= x_0 + \frac{l_{\mathrm{PM}}}{2} + l_c + l_k + x_m \\
Y_{23} &= x_0 + \frac{l_{\mathrm{PM}}}{2} + l_c + l_k + x_m + l_m
\end{aligned}
\tag{6.2}
$$

where x_0 is the mover position. The down-mover disk corner coordinate is computed in a similar way.

$$Y_{14} = x_0 - \left(\frac{l_{\text{PM}}}{2} + l_c + l_k + x_m \right)$$
$$Y_{23} = x_0 - \left(\frac{l_{\text{PM}}}{2} + l_c + l_k + x_m + l_m \right) \tag{6.3}$$

The construction of different parts of the electro-valve by decomposing them in simple geometry parts is shown in Figure 6.5.

An example of a computer code suitable to draw the mover disk is as follows:

```
/Mover magnetic piece
DRAW SHAP=H, MATE=5, PHAS=0, DENS=0,
X12=#r2-#x1, X34=#r0,
Y14=#x0+#lpm/2+#lc+#lk+#xm, Y23=#x0+#lpm/2+#lc+#lk+#xm+#lm,
N1=#nmz, N2=#nmr, F1=NO, F2=NO, F3=NO, F4=NO
```

The “Draw” command is followed by choosing the shape type, the material number, and, finally, the current density.

The borderline between PM regions and coil regions should meet the inner stator core in a mesh node. In order for this to be ensured, it is good to add one point to the borderline on the inner stator when the external medium is discontinuous. In this way, we avoid the preprocessor interrogation when a PM or a coil region is built. An unexpected interrogation (add new points?) from the preprocessor when a script is running is a source of error. Finally, the inner stator core (half of this) is described by six points, despite its rectangular shape. The “Poly” shape is chosen in order to describe the inner and outer stator cores. The first corner, P1, of the inner stator core is set by its Cartesian coordinates (X_1 and Y_1). The second point, P_2, could be given

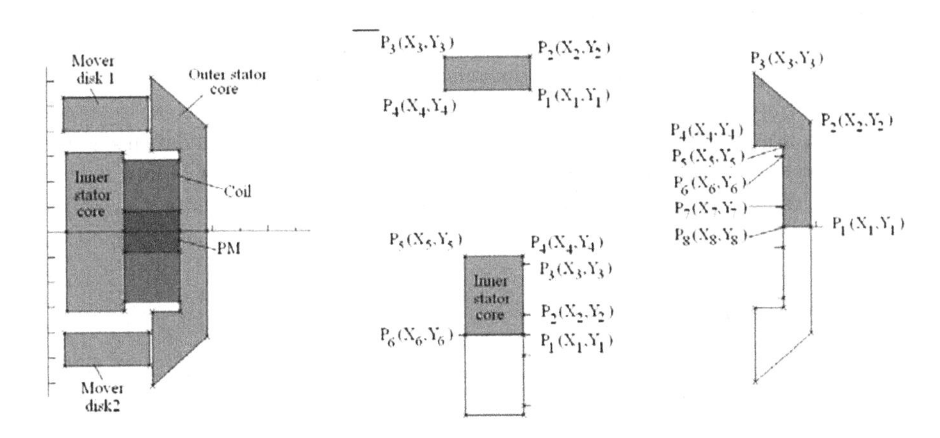

FIGURE 6.5 Building the electro-valve geometry.

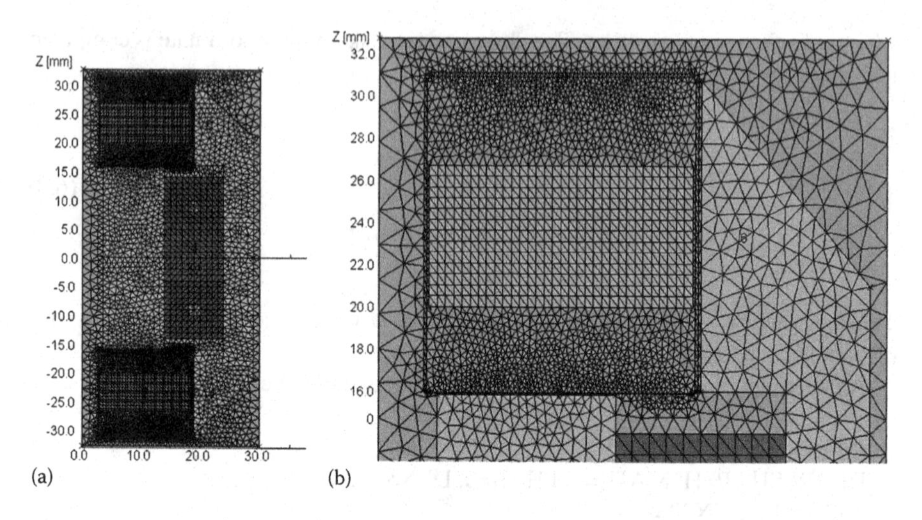

FIGURE 6.6 Electro-valve mesh. (a) Full problem and (b) details around mover.

by a translation along *Y*-axis with half of the permanent magnet length. The preprocessor allows for setting the polygon corner using Cartesian or polar coordinates, as well as translations along the coordinate axis.

A background object that fills all free space between material objects is added. When the interaction forces are computed using the Maxwell tensor, it is better to design the integration path through finite elements that have all nodes in a linear medium without currents, in order to reduce the integration error. This condition could be achieved if the mesh has at least three layers in the airgap. In order to enforce three mesh layers in the airgap we can define an extra region along the airgap with air features. In our case, such a region was defined around the mover disk, placed in such a manner that it does not reach this region when it is moved from one extreme position to the other. In the generated mesh, Figure 6.6, there are three layers in the airgap. Also, the regular mesh of "H" region with mover disks, coils, and permanent magnets can be observed in Figure 6.6.

The preprocessor stage is over after the problem description is written for every mover position. The solver can be run interactively or by using a batch process. When there are several problems to be solved (in our case, one problem for every mover position), it is preferable to add all problems to a batch query and then start the batch process; however, this takes some time. In our electro-valve, for 11 mover positions and 18 currents for each position, 2 h and 40 min on a Pentium 4 at 2.4 GHz with 512 MB of RAM memory were needed.

6.1.2 Postprocessor Stage

The main reason of the postprocessor stage is to extract the performance parameters from the finite element solutions. The results file will be loaded into the postprocessor. The magnetic field lines can be shown immediately. This feature is a common option included in the menu list for most commercial FEM software. In Vector Fields,

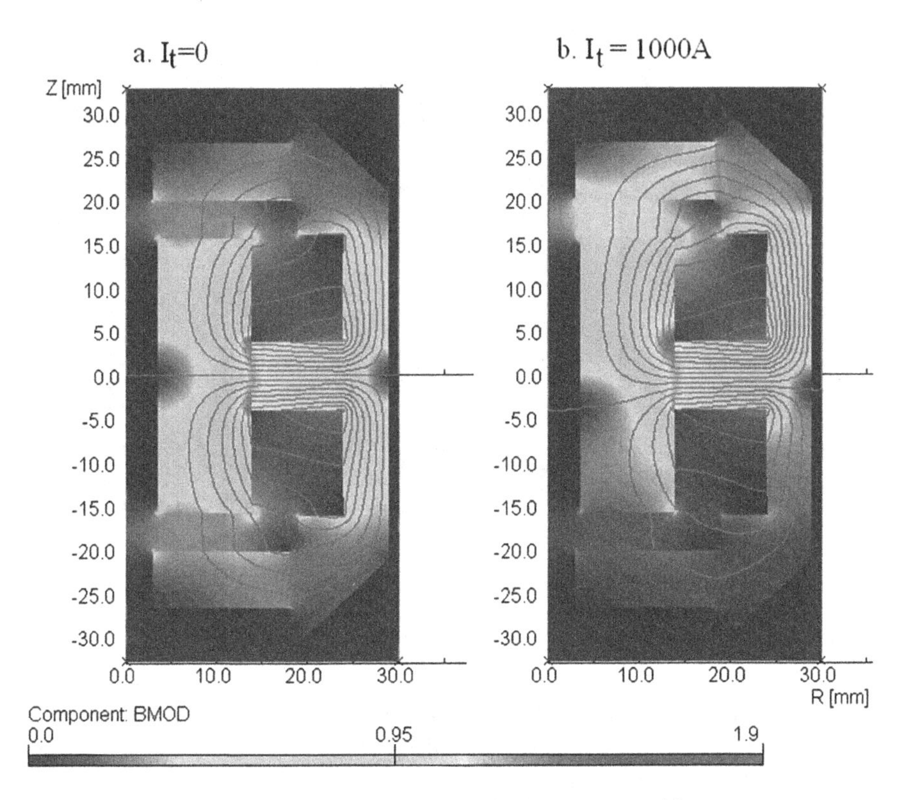

FIGURE 6.7 Flux density and field lines for equilibrium mover position.

we have to select a "Component" that is equal to "POT" and then in the submenu "Contour Plot" we have to choose "Execute." It is possible to set the number of lines, the plotting style, the label type, and whether the drawing needs to be refreshed or not [3]. The flux density module could also be represented directly as a color map. It is recommended to choose a filled zone style with a large number of lines in order to represent the flux density module, and a contour line with a small number of lines in order to represent the field lines. Using the "refresh" switch, the field line could be superimposed on the flux density map as shown in Figures 6.7 and 6.8.

When the mover is in the central position and the current is zero, the field distribution (lines and flux density module) should be symmetric in relation to the X-axis as shown in Figure 6.7a. If the magnetic field from FEM is not symmetric when the geometry and magnetic field source are symmetrical, then the solution is wrong, and, before computing any performance parameter, it is necessary to find the mistake. We have to check the border conditions, the region descriptions, the current and PM values, and the phase.

The flux density module map shows (Figure 6.7), in our case, local and global saturation. When the mover is in the middle position there is a small region with local saturation. The total current is defined as the product between the number of turns and current in each turn, or as an equivalent current density integral over the conductor area. In our example, the equivalent current density is 5 A/mm^2, which means

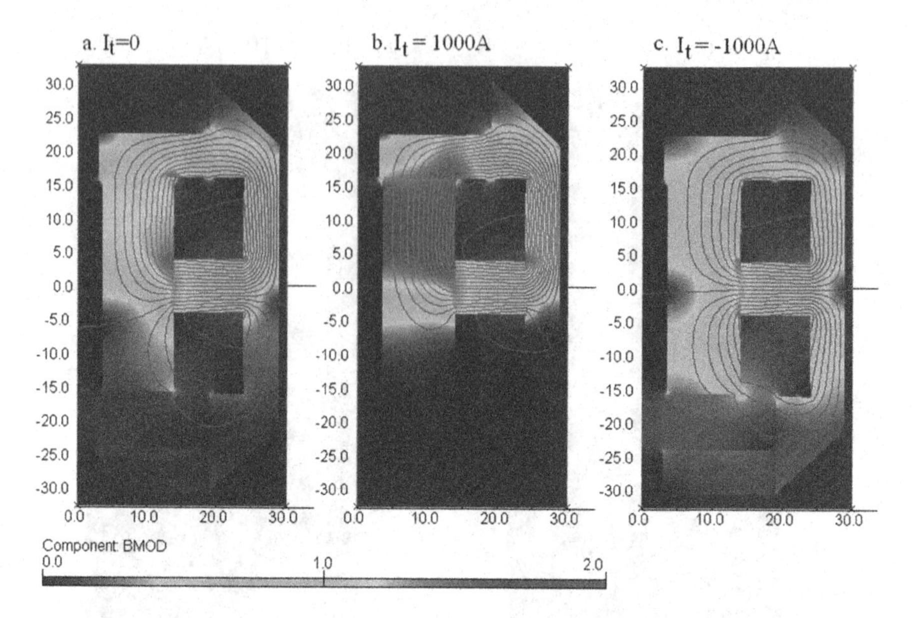

FIGURE 6.8 Flux density and field line for mover position $x_0 = -4$ mm.

about 10 A/mm² on the net conductor area if the filling factor is considered to be around 0.5. This is an acceptable value for the peak current density. It could be doubled, for short-lived transients or when considering a very efficient cooling system. The current field adds to the permanent magnet field in some regions and substracts in other regions. The field distributions become unsymmetrical but without notable global saturation when the mover is in the central position. When the mover is in its maximum displacement, Figure 6.8, the permanent magnet field is asymmetrically distributed. A positive current creates a notable saturation in the upper part of the magnetic circuit.

The magnetic core saturation also depends on the material features. A magnetic flux density of 2 T indicates a heavy saturation for Somaloy as it is used for electro-valve core but it could indicate a moderate saturation for Hiperco laminations. The saturation level could be better appreciated if the magnetic relative permeability is presented as shown in Figure 6.9.

The magnetic permeability decreases in the region where the current field is added to the permanent magnet field; for rated current ($I_t = 1000$), heavy saturation occurs and the permeability is reduced at half of its maximum value (Figure 6.9b). Increasing further the coils current at 2000 A will reduce magnetic permeability by 10 times from its maximum value, Figure 6.9c. The magnetic permeability map is presented only for the soft magnetic core (Somaloy). This is an important option that could be used in order to avoid an unusual representation scale.

The most important performance computed by the FEM is the coil linkage flux and the electromagnetic force. The linkage flux is computed by applying relation (5.129) on the coil surface. The integral is computed by choosing an adequate menu in the Vector Fields software. The user has to divide the integral values given by the

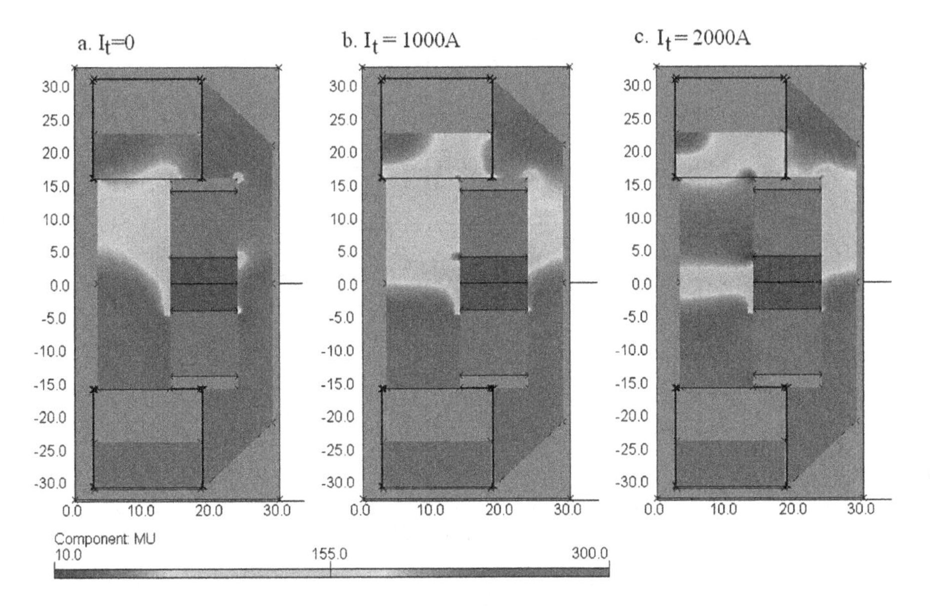

FIGURE 6.9 Relative magnetic permeability in the magnetic core.

software with the coil surface, and it is also necessary to take into account the length unity. For example, if the length unity is in millimeters, then the flux from Equation 5.129 is in mWb. The linkage flux is computed for several mover positions and several currents for each mover position. In this way, it is possible to compute inductance vs. current to prove the presence of magnetic saturation. The data extraction could be mechanized by using a "comi" script in Vector Field or "lua" script in the FEMM software [4,5]. The desired data is stored in a table and then graphic representation and other data processing can be done in dedicated software, developed, for example, in MATLAB®.

The total linkage flux vs. mover position and coil current are shown in Figure 6.10.

The linkage flux is produced by the permanent magnet and the coil current.

$$\psi(x_0,i_c) = \psi_{PM}(x_0,i_c) + L(x_0,i_c) \cdot i_c \tag{6.4}$$

The superposition principle is not perfectly valid in the nonlinear magnetic circuit and thus only an approximate value of linkage inductance can be computed.

$$L(x_0,i_c) = \frac{\psi(x_0,i_c) - \psi_{PM}(x_0,i_c)}{i_c} \cong \frac{\psi(x_0,i_c) - \psi_{PM}(x_0,0)}{i_c} \tag{6.5}$$

This value is useful to compute the linkage flux from the current and permanent magnet components, but, for emf voltage computation, it is better to use the differential inductance approximation:

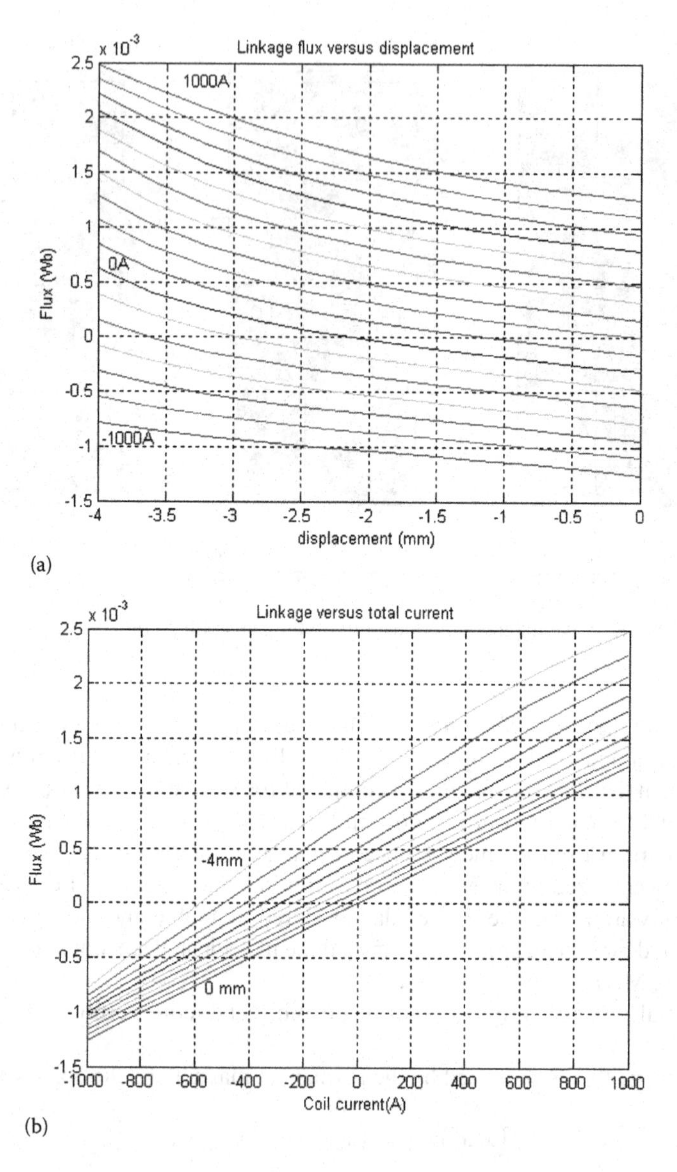

FIGURE 6.10 Linkage total flux for two coils in series and 1 turn/coil: (a) linkage flux vs. displacement and (b) linkage flux vs. total current.

$$L_d\left(x_0,i_c\right)=\frac{\partial\psi\left(x_0,i_c\right)}{\partial i_c}\cong\frac{\psi\left(x_0,i_2\right)-\psi\left(x_0,i_1\right)}{i_2-i_1} \tag{6.6}$$

The static and differential linkage inductance approximations are shown in Figure 6.11.

The real differential inductance should be equal or smaller than static inductance, but, in Figure 6.11, a small violation of this rule is noticed in some points due to

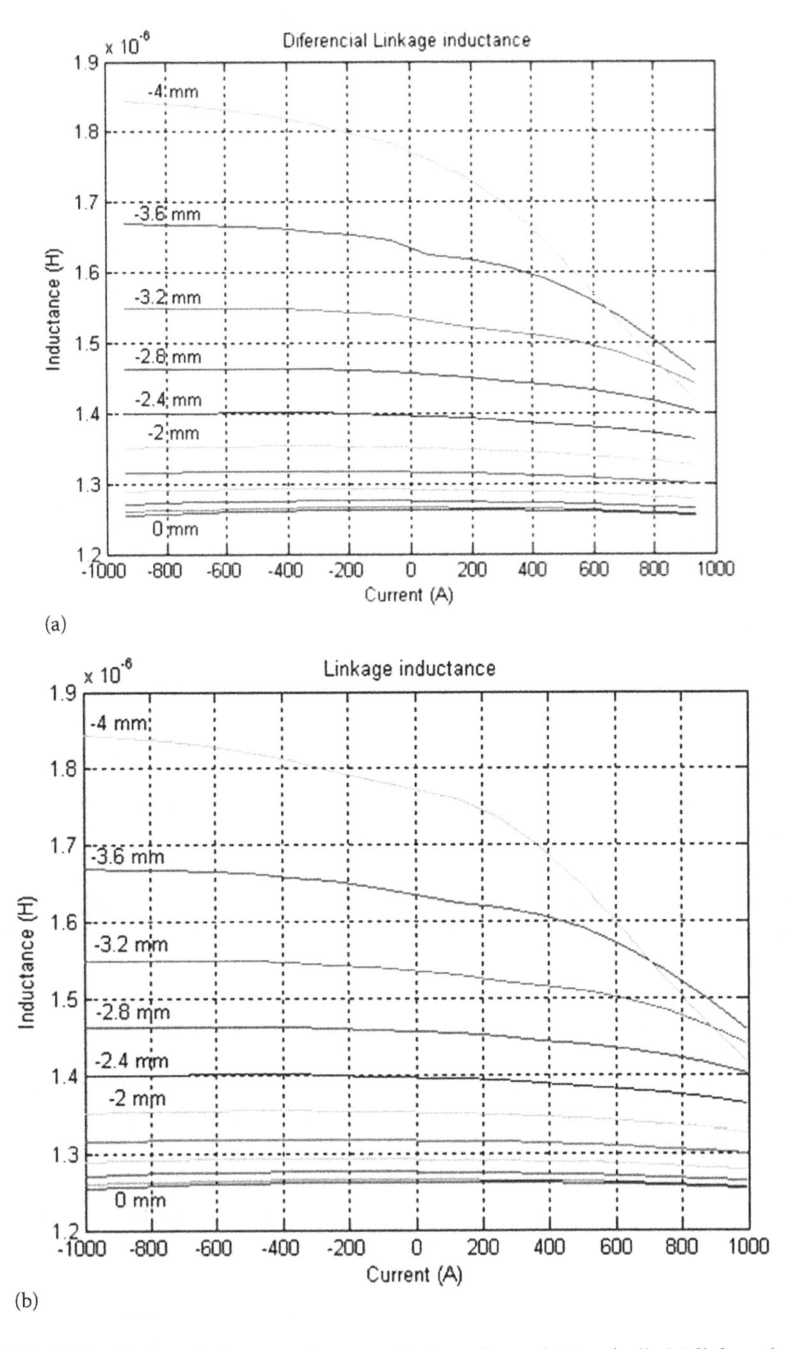

FIGURE 6.11 Linkage inductance for two coils in series and 1 turn/coil: (a) linkage inductance and (b) differential linkage inductance.

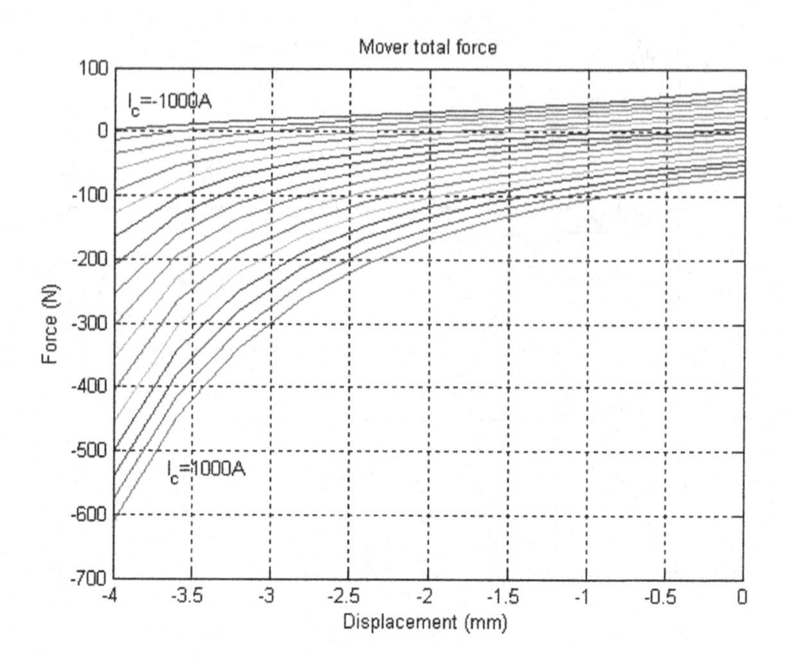

FIGURE 6.12 Mover thrust vs. position.

approximations. However, the static and dynamic inductance values are close to each other for our case study.

The thrust on the mover is computed by the integration of the Maxwell tensor. A family of thrust curves vs. mover position and several currents are shown in Figure 6.12.

The electromagnetic force could be computed for every part of the device using this method. The contribution of the upper mover disk to the total mover thrust is shown in Figure 6.13.

It could be remarked that the thrust on the upper disk is always negative. The thrust is always positive on the lower mover disk. The electromagnetic thrust along the free direction is small, so the mechanical spring is the main contributor in releasing the mover and in accelerating it in the opposite direction. When the mover passes through the middle position, it can have large electromagnetic acceleration, but, if any mechanical shock is to be avoided, the motion should be stopped when it is on that part of its trajectory. Unfortunately, the braking thrust is poor for that part of the trajectory and the spring has to do again the main braking job. In conclusion, the described device has rather poor motor features but it is excellent to use as an electromechanical latch.

The axial force on the coils vs. current is shown in Figure 6.14 for different mover displacements. This force is used for the mechanical design of the coils.

A global characterization of the electro-valve is given by its electromagnetic energy, Figure 6.15, computed with Equation 5.128 for total field-stored energy (permanent magnet and current source), and with relation (5.127) for only field energy stored from current sources in the presence of PM.

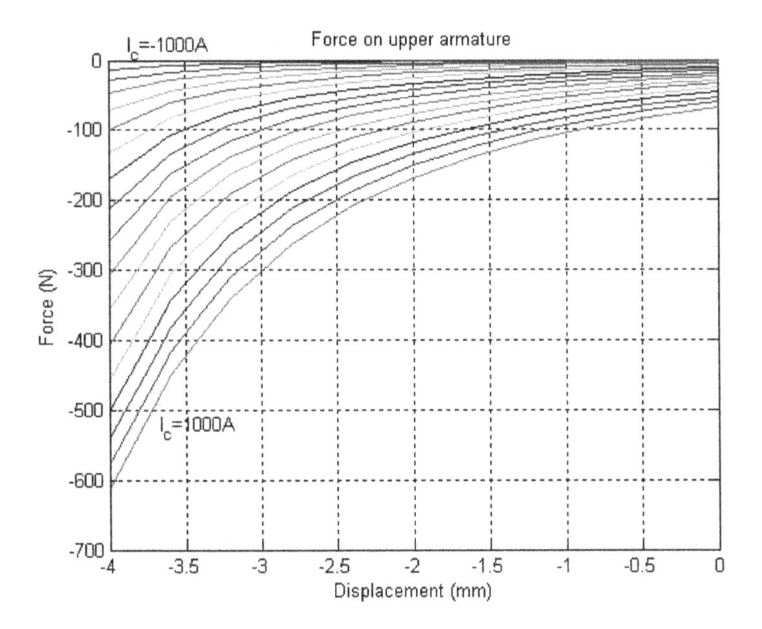

FIGURE 6.13 Contribution of the upper mover disk to the mover thrust.

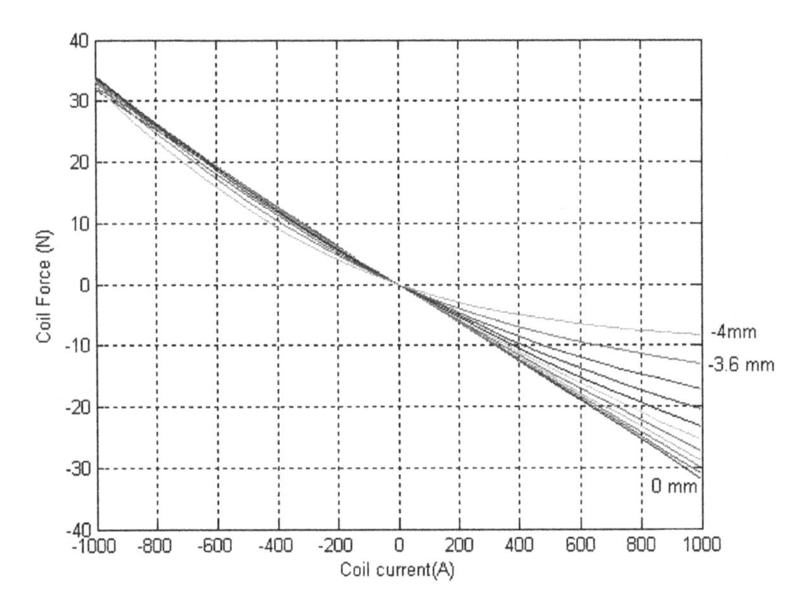

FIGURE 6.14 Resultant axial forces on the coils.

6.1.3 Summary

The magnetostatic field solution by the FEM for a tubular PM linear motor as presented here emphasizes the fundamentals of the FEM application to electric machines, allowing also for thrust and inductance calculations to prepare for high-precision

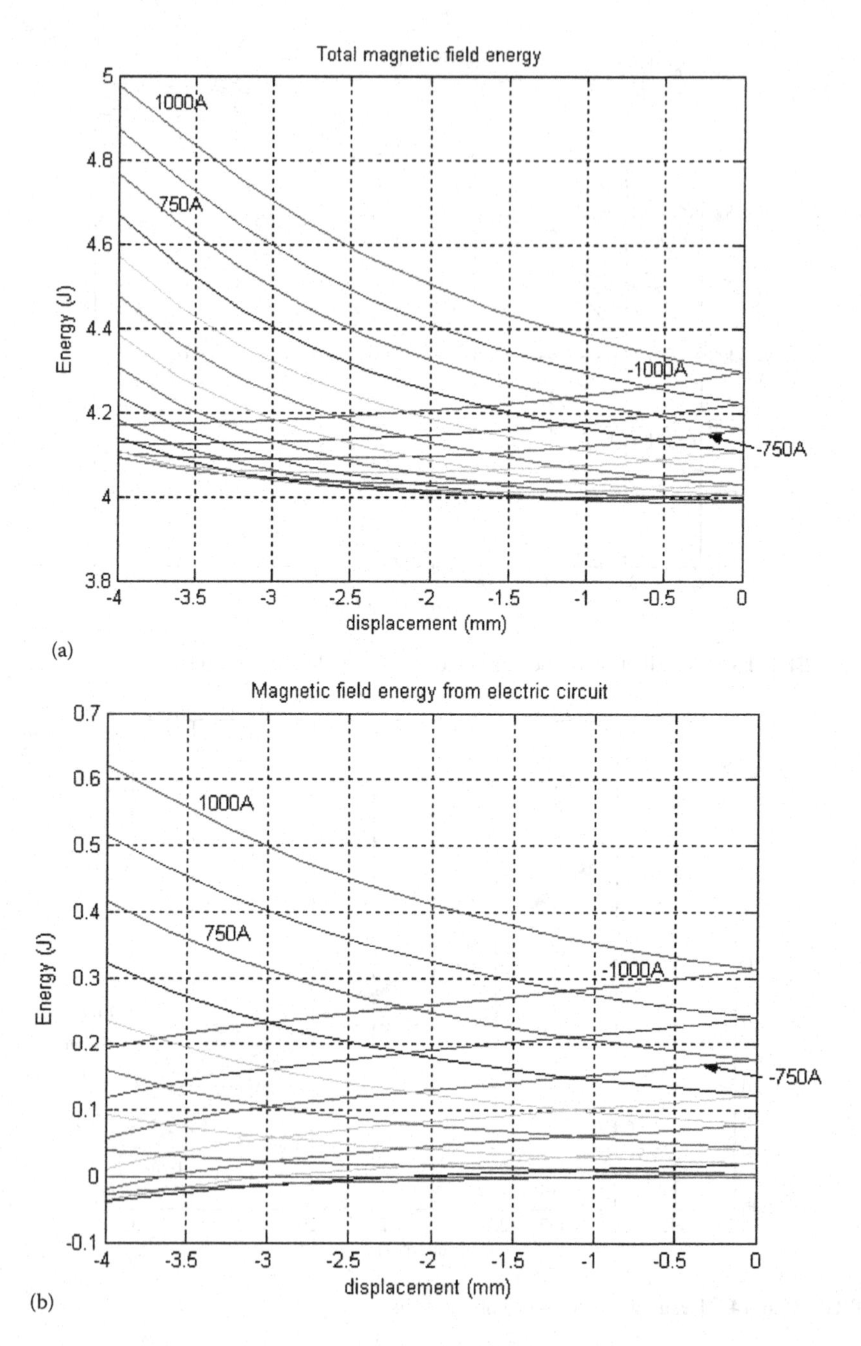

FIGURE 6.15 Magnetic field energy vs. mover position and stator current. (a) Total magnetic field energy and (b) magnetic field energy from the electric circuit.

circuit models needed for the investigation of dynamics and control of electric machines.

6.2 ROTARY PMSMs (6/4)

The rotary permanent magnet synchronous motor (PMSM) with surface permanent magnet 4-pole rotor and six coils in the stator, Figure 6.16, is used to illustrate the finite element methodology applied on rotary machines without current in the rotor.

This motor is used in order to improve the motor efficiency at reasonable manufacturing costs in home appliances and automotive industries. The configuration with three stator coils per two rotor poles is probably less costly but finite element analysis has shown large uncompensated radial forces, especially when the motor is loaded.

The finite element analysis is illustrated on a 200 W, 1500 rpm motor and 300 V d.c. link voltage.

The stator lamination, Figure 6.17 main geometrical dimensions are

- D_{si}—inner stator diameter
- D_{so}—outer stator diameter
- w_{sp}—stator pole width
- h_{sc}—stator yoke width
- a_{sp}—relative stator pole span angle
- h_{s3} and h_{s4}—the heights of the stator slot closure
- R_1—stator pole shoe curvature
- h_{s1}—stator teeth height

The rotor geometrical dimensions, shown in Figure 6.18 are

- Dro—rotor outer diameter
- Dri—rotor inner diameter

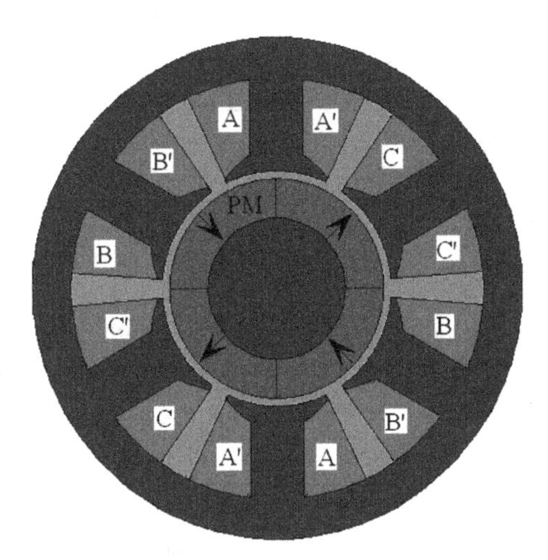

FIGURE 6.16 The 6-slot/4-pole PMS with surface PM rotor.

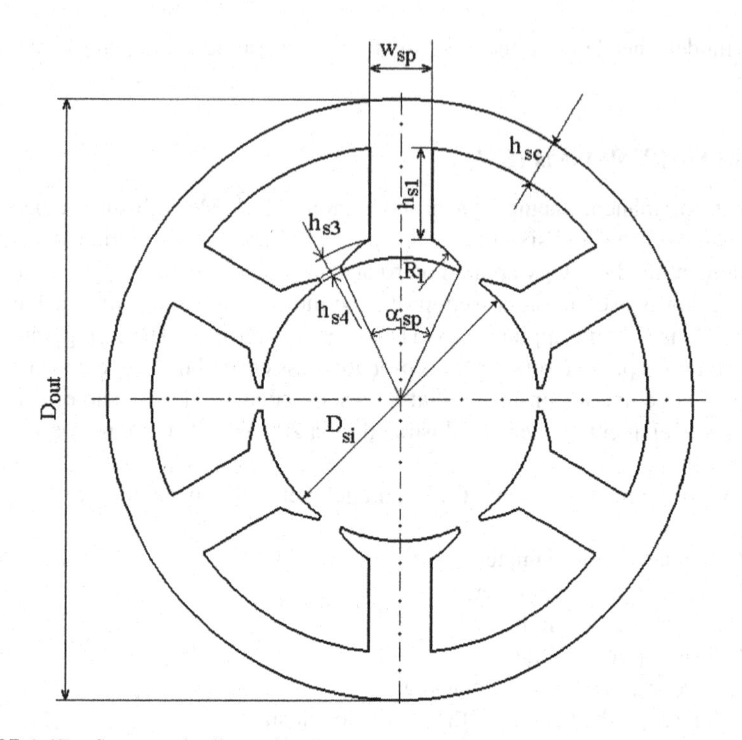

FIGURE 6.17 Stator main dimensions.

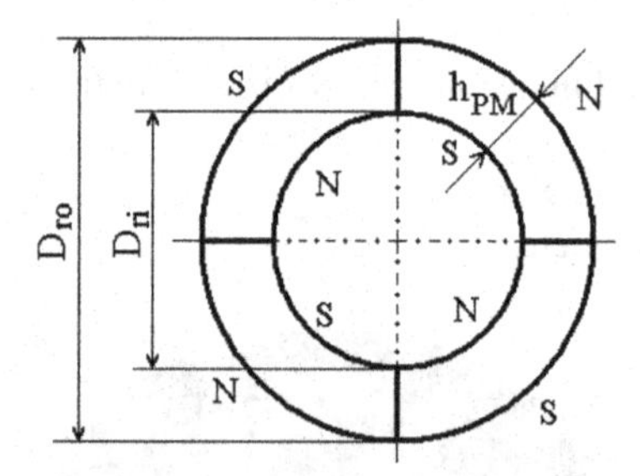

FIGURE 6.18 Rotor main dimensions.

- hPM—PM height

The stator core stack length, "lstack" and airgap length, *g*, are also important geometric dimensions. The main geometric dimensions used in the following simulations are given in Table 6.2.

6.2.1 BLDC Motor: Preprocessor Stage

When a PMSM is controlled with trapezoidal (rather than sinusoidal) currents and two active phases (of three), it is called a brushless d.c. (BLDC) motor.

TABLE 6.2
The Main PMSM Geometric Dimensions

Data Name	Initial Data	Units
D_{out}	68	mm
D_{si}	30	mm
h_{s4}	0.6	mm
h_{s3}	1.4	mm
h_{sc}	4.5	mm
w_{sp}	7	mm
l_{stack}	50	mm
α_{sp}	51	°
h_{PM}	3	mm
g	0.5	mm

The BLDC motor has a plane parallel symmetry if the coil end connections are neglected, and, consequently, the BLDC motor magnetic field is solved as an "*xy* symmetric problem." Suitable length (millimeter) and current density (Ampere per square millimeters) units are chosen. The main geometric dimensions are declared and initialized as user-defined variables in order to mechanize the machine drawing and problem description. The machine drawing is divided into elementary parts and then the "copy" and "rotate" technique is used in order to construct the entire cross section. The ideal 6/4 pole machine has a symmetry over 180°. If we are not interested to study the rotor/stator eccentricity, then it is enough to solve the magnetic field problem only for one-half of the cross-section machine as shown in Figure 6.19. The machine rotor could be automatically rotated using the rotating machine module from Vector Fields.

The complex shape of the stator core is drawn in a simple manner as shown in Figure 6.20. Nine points are necessary to draw the stator core element and only four points to draw the coil side. The points' coordinates are computed from the input geometry dimensions. The user could choose Cartesian or polar coordinates for every point. For the first point, P_1, it is simpler to use polar coordinates:

$$P_1\left(R_P, P_p\right) = P_1\left(\frac{D_{si}}{2}, 90^o\right) \tag{6.7}$$

The coordinates of the second type point are also given in polar coordinates:

$$P_2\left(R_P, P_p\right) = P_1\left(\frac{D_{si}}{2} + \Delta\delta,\ 90° + \alpha_{sp}\right) \tag{6.8}$$

where $\boldsymbol{\Delta\delta}$ is the airgap variation in order to consider geometry with nonuniform airgap. In our case, the airgap is, however, uniform and thus $\boldsymbol{\Delta\delta}$=0. The line curvature specification between points P_1 and P_2 is equal to the inverse of the stator inner radius if the length of the airgap is uniform. The line curvature that links points P_1 and P_2, is set at the same time with the P_2 coordinate (in the "Vector Fields" software). It is not

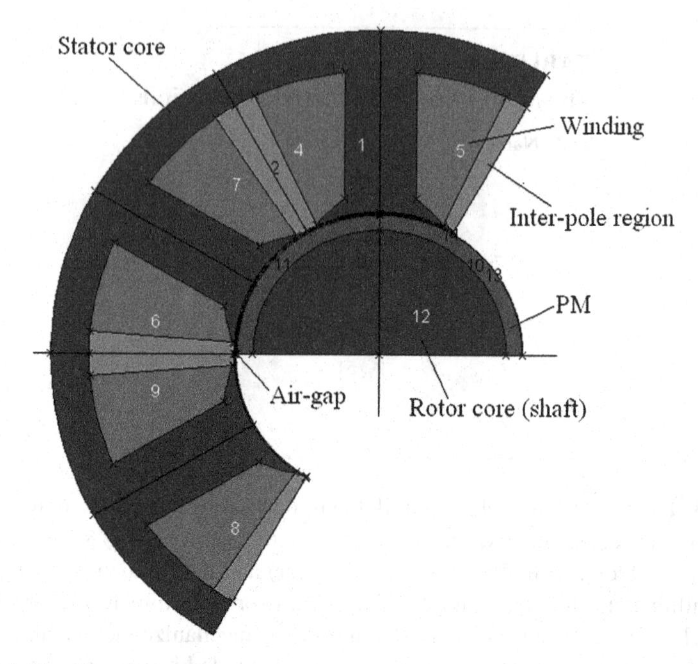

FIGURE 6.19 The BLDC motor problem description using symmetry.

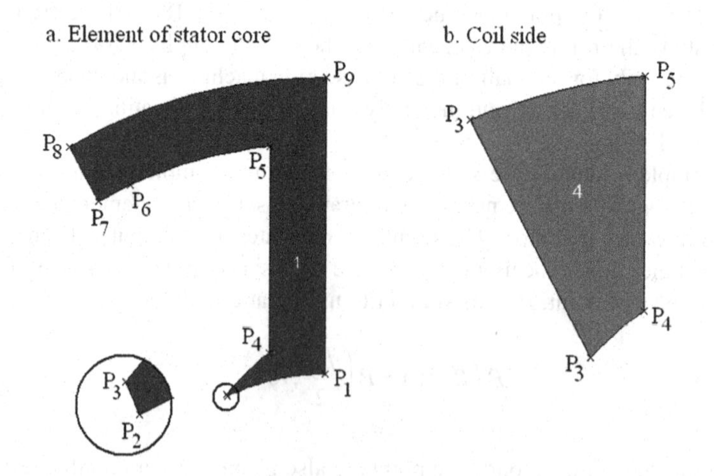

FIGURE 6.20 Stator core and coil elementary shape.

necessary to compute the coordinates for all points that describe the geometry elements. In our example, it is not necessary to compute the P_3 coordinates because it could be introduced as a radius variation with "sh4" length. The straight movement is obtained for zero curvature specification. The point P_4 is given through Cartesian coordinates:

$$P_4\left(X_P,Y_P\right)=P_4\left(-\frac{w_{sp}}{2},y_4\right) \tag{6.9}$$

$$y_4 = \sqrt{\left(\frac{D_{si}}{2} + \text{sh4} + \text{sh3}\right)^2 - \left(\frac{w_{sp}}{2}\right)^2} \tag{6.10}$$

The point, P_5, is introduced as an abscise variation with "sh1" length. The points P_6 and P_7 are introduced in polar coordinates:

$$P_6\left(R_P, P_p\right) = P_6\left(\frac{D_{s0}}{2} - h_{sc},\ 90° + \alpha_{sp}\right) \tag{6.11}$$

$$P_7\left(R_P, P_p\right) = P_7\left(\frac{D_{s0}}{2} - h_{sc},\ 90° + \frac{180°}{N_{sp}}\right) \tag{6.12}$$

Point P_8 is introduced as a radius variation with "hsc," while point P_9 is introduced as a polar angle variation with 180°/Nsc angle in negative direction (clockwise). After the elementary geometry of stator core presented in Figure 6.20 is built, it could be replicated, mirrored, and rotated in order to obtain the stator core drawing as shown in Figure 6.19.

The inter-pole region and an airgap layer close to the stator are built in the same way as the stator element. The inter-pole region and airgap layer close to the stator are obtained by the "replicate" command at the same time with the stator. The first coil side is built in the same way as the stator core elements, but the other coil sides are obtained by the "copy" command instead of the "replicate" command because it is necessary to set different values of current density for each coil side and this is possible only if they represent different regions [2,3].

The rotor is built in the same way as the stator. At first, the permanent magnet region is built. The second permanent region is a copy of the first permanent magnet region in order to avoid opposite polarization settings. The permanent magnet polarization direction can be set later when all geometry is built. The rotor core under first pole is drawn directly, while the whole desired rotor core is made using "replication" of the first part. An airgap layer close to the rotor is built.

The computation of force through the Maxwell tensor is much more accurate if the integration path passes only through linear elements that have all nodes in the same medium. In order to achieve better accuracy for force computation, a third airgap layer bordering the stator airgap layer is defined.

The rotating machine module of "Vector Fields" software allows transients regime with rotation of a part of a machine. A special region, "rotating machine airgap" should be introduced in the description of the problem. The region parameters are the average radius and the symmetry. The circle with the given radius must not pass through any region, so it must be outside of any point on the rotor, and inside of any point on the stator. As a convention, the inner part of this special region rotates. The outer rotor machine could be also studied, despite inner regions rotation, because the electromechanical phenomena are governed by relative position and relative speed. The "symmetry" parameter is an integer value and its absolute value is specifying the number of rotational symmetries of the model. In our case, the symmetry is 2 because half of the machine is simulated and the sign is "plus" because the periodicity conditions are positive (a pair of poles is studied). If the whole machine is modeled, the symmetry value must be 1. It is essential that the value of this parameter matches the symmetry of the model correctly. The outside edge of the rotor and the inside edge of the stator are recommended to have a constant radius and the subdivisions on rotor and stator sides of the gap elements have a similar size. Using the additional airgap layers near the stator and rotor, which in the model are part of the stator and rotor, we can adjust the rotor outer and stator inner radius and make them constant, even when the real airgap is variable, and the number of inner stator subdivisions could also be set equal to the "outer rotor." Finally, the airgap contains four layers as shown in Figure 6.21.

The radial magnetization of permanent magnet means that the magnetic polarization is a function of position and it could be set only after mesh setting, when the coordinate X and Y variables are available. The PM polarization is introduced as an extra condition as shown in the following example:

```
EXTRA
REG 10 C=PHASE, F=atan2d(Y; X)
REG 11 C=PHASE, F=180+atan2d(Y; X)
Quit
```

The current density for each conductor region is also set after regions meshing. At this stage, the current density could be computed from total current per coil and coil cross area, which is a region-intrinsic parameter after the region is meshed. A circuit label number is also set for each conductor at this stage. This label number will be used when the current through the coil is computed from an external circuit or when it is given from an external driven function. By using an external driven function, it is possible to change the phase currents according to the rotor position and simulate

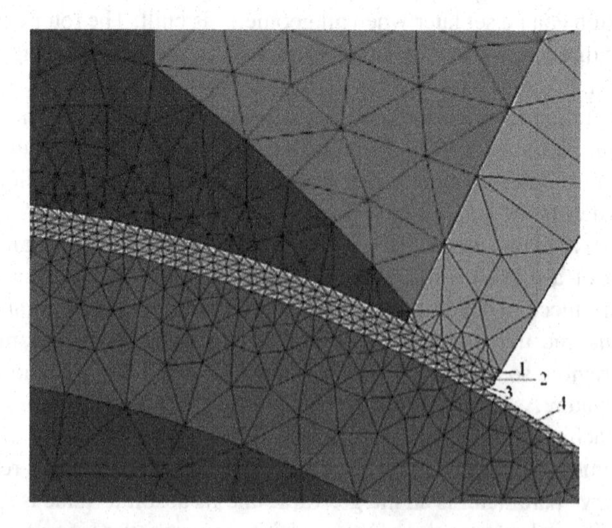

FIGURE 6.21 Airgap mash layers: 1—layer close to the stator, 2—layer containing the integration path of Maxwell tensor, 3—rotating airgap layer, and 4—layer close to the rotor.

the synchronous running of the machine. The current density and circuit label setting, are shown in the following example where the region number is from Figure 6.19.

```
MODI REG1=4 DENS=-#It/area, N=1
MODI REG1=5 DENS=#It/area, N=1
MODI REG1=6 DENS=-#It/area, N=2
MODI REG1=7 DENS=#It/area, N=2
MODI REG1=8 DENS=-#It/area, N=3
MODI REG1=9 DENS=#It/area, N=3
```

The rotor speed could be set as a constant or as a variable speed, when it is given as a table in a file. Also, it is possible to compute the rotor speed directly by the finite element software using the mechanical equation:

$$\frac{1}{J}\cdot\frac{d\Omega}{\mathrm{d}t} = T_{\mathrm{em}} - sign(\Omega)\cdot T_{\mathrm{f}} - T_{\mathrm{l}} - k_{\mathrm{t}}\Omega$$

The inertia moment, J, the friction torque, T_f, the load torque, T_l, and the speed-varying torque coefficient, k_t, are set by the user while the electromechanical torque, T_{em}, is computed from the FEM software. The machine length should be set by the user at the same time with previous parameters. A more complex mechanical equation (with load torque function of rotor position or as an explicit function of time) is possible using a command file. A "logfile" is created when the solver runs for the rotating machine and the desired solution parameters could be stored for every rotor position. An adaptive or fixed-time step could be used. The entire field solution is stored for a given time. There is a simple way to transform the desired rotor position in a time moment when constant speed is used. The stator currents could be maintained at a constant when torque and flux vs. internal angle will be extracted in the postprocessor, or when their values are correlated with rotor position using the "drive functions" when the torque pulsation and linkage flux vs. time is studied at constant internal angle. The "drive function" could be an elementary function such as sine, cosine, step, or exponential function but they could also be read from a table. In our example, the phase currents have a trapezoidal wave shape as in Figure 6.22 and the current waveforms are given as a table vs. time.

The time dependence of the current density is computed by multiplying the current density from each region with the drive function at the current time. In our example, the drive function is given in per unit and its maximum value is unity. In this way, the user definition for the drive function is compatible with embedded drive function (d.c., sine, cosine, ramp, exponential).

6.2.2 BLDC Motor Analysis: Postprocessor Stage

Machine analysis is performed in postprocessors, where field solutions from FEMs are used, to compute the machine circuit parameters, torque capability, and magnetic saturation level. At first, it is necessary to check the FEM solution by observing the magnetic field lines. In the first simulation of the BLDC motor, the machine rotor was rotating at a constant speed (1500 rpm), while the current was maintained

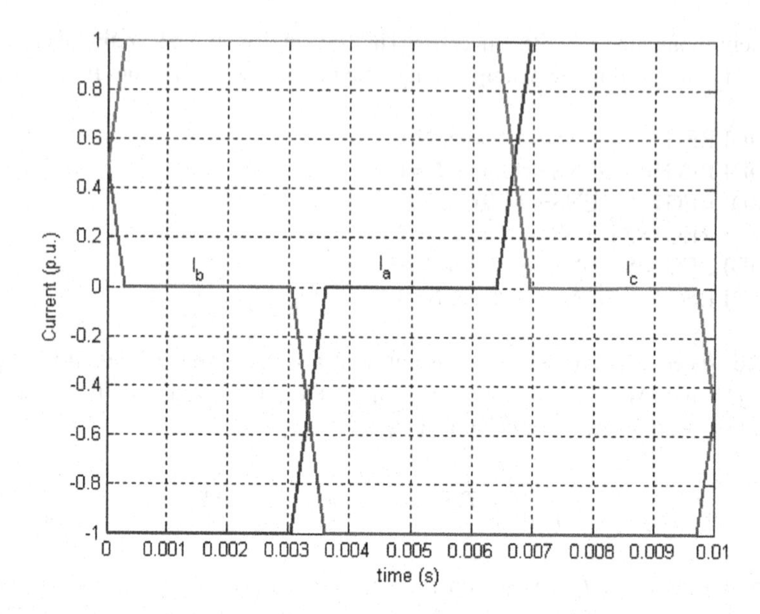

FIGURE 6.22 Drive function for trapezoidal current wave shape.

constant: negative in the first coil, positive in the second coil, and zero in the third coil as shown in the Figure 6.23d.

The magnetic field solutions were stored for 37 rotor positions (time moments) while a quarter mechanical rotor revolution was done. The magnetic field lines and flux density map for three rotor positions (first, median, and the last) are shown in Figure 6.23. Using the problem periodicity, only half the machine was simulated. In this way, the computation effort was dramatically reduced. The solution storage memory was reduced by two times and the computation time was reduced by more than two times. The magnetic field lines are closed through expected paths in all three cases and we can conclude that the periodicity conditions are appropriate.

The magnetic field of the coil is added to the PM field, increasing or decreasing the total magnetic field. For some rotor positions, as in Figure 6.23c, the magnetic fields adding is predominant and resulting field is increased for a large region of the stator core. The current density presented in Figure 6.23d is an average value over all conductor regions. The real current density is computed taking into account the slot-filling factor. Considering a usual filling factor $k_{\text{fill}} = 0.4$, the real current density is computed as

$$J_{\text{cu}} = \frac{J_{\text{reg}}}{k_{\text{fill}}} = \frac{2.7}{0.4} = 6.75 \text{ A/mm}^2 \tag{6.13}$$

This is an acceptable value of current density at peak torque (about two times larger than rated torque).

The electromechanical torque is returned from the solver in the "log-file" (for "Vector Fields") but it is also computed at the postprocessor stage by integrating the

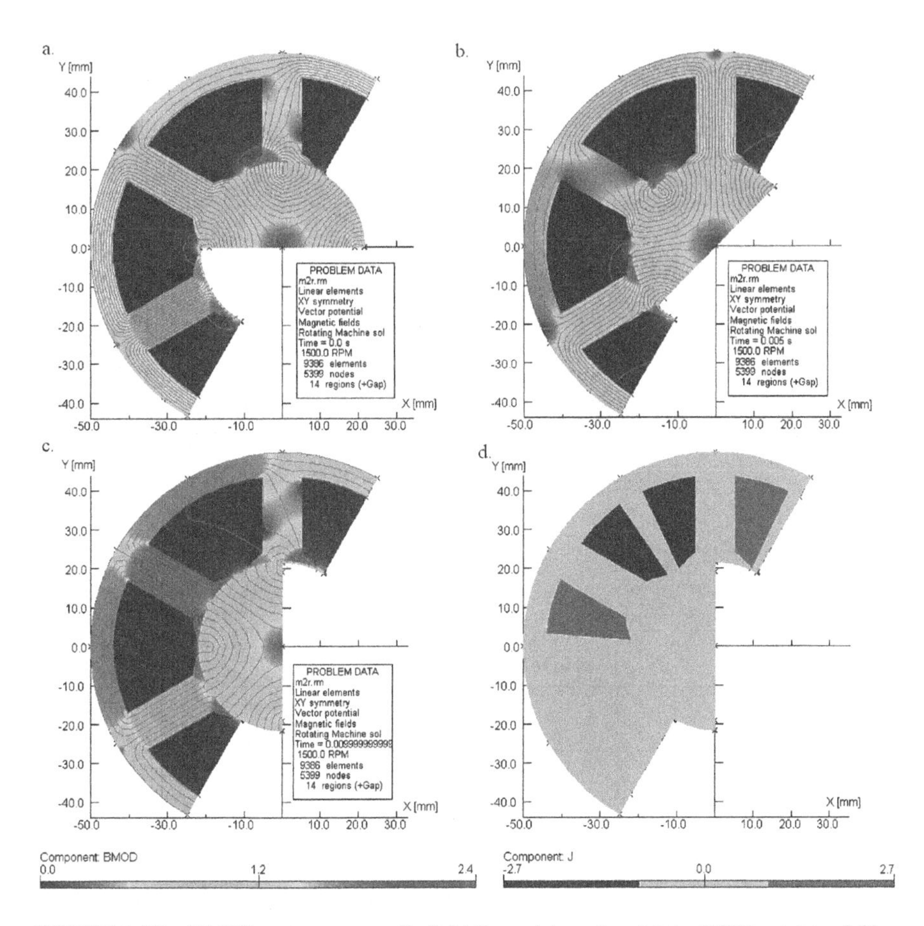

FIGURE 6.23 BLDC motor—magnetic field lines: (a) $t = 0$ s, (b) $t = 0.005$ s, (c) $t = 0.01$ s, (d) current distribution.

Maxwell tensor. The symmetry parameter is automatically considered for the torque from "log-file," but, when the torque is computed from the Maxwell tensor integration, only the interaction through the integration path is considered, and, consequently, in order to obtain total machine torque, it is necessary to multiply the integration results with the symmetry parameter. In our example, the integration path is a circle arc with R_{torq} radius and an angle between 60° and 240°, where the R_{torq} radius is

$$R_{\text{torq}} = \frac{D_{\text{si}}}{2} - 0.375 \cdot g \tag{6.14}$$

The machine core length is not considered in 2D FEM, and the computed torque is in fact a specific torque for a machine length of 1 m as shown in Figure 6.24. The torque could be presented vs. either the rotor position or time. However, these are equivalent at a constant speed.

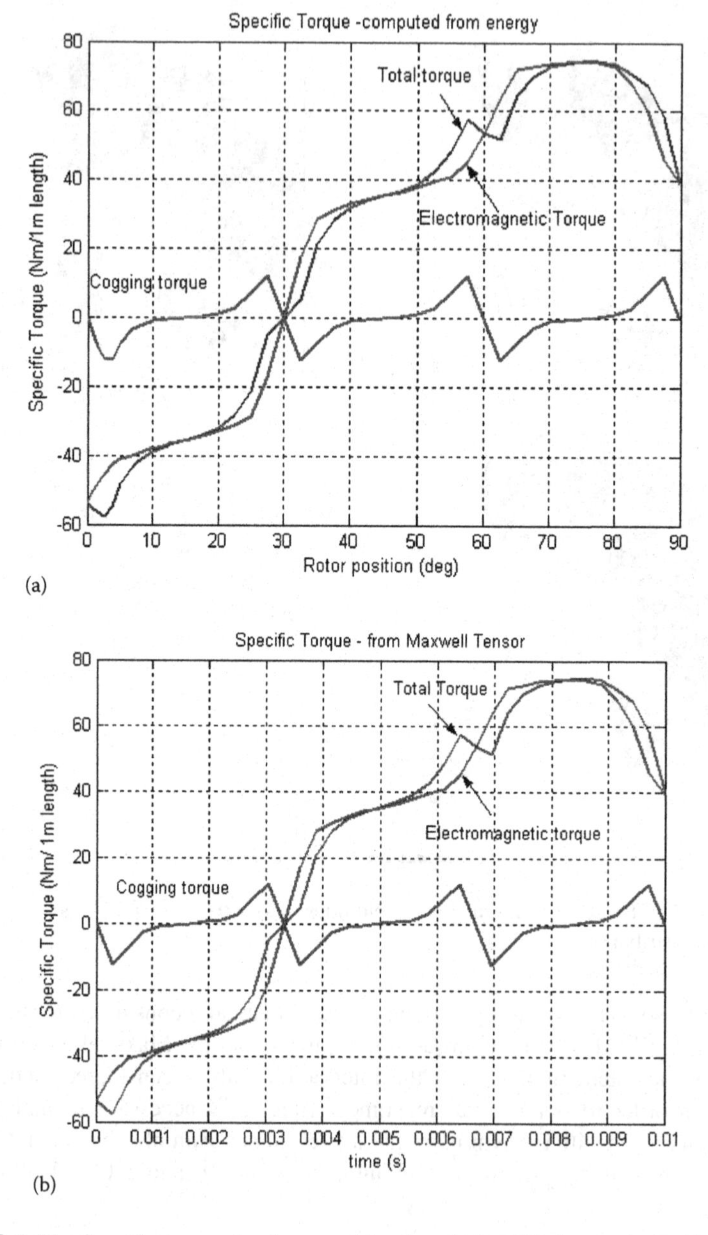

FIGURE 6.24 Specific torque for 1 m machine length for 529 A turns/coil. (a) Specific torque—computed from energy and (b) specific torque—from Maxwell tensor.

It could be noted that there are no notable differences between the torque computed from magnetic energy variation and that computed from Maxwell tensor integration (Figure 6.24).

The machine torque is computed by multiplying the specific torque from the FEM with the machine length. Figure 6.25 shows the machine torque vs. the rotor position.

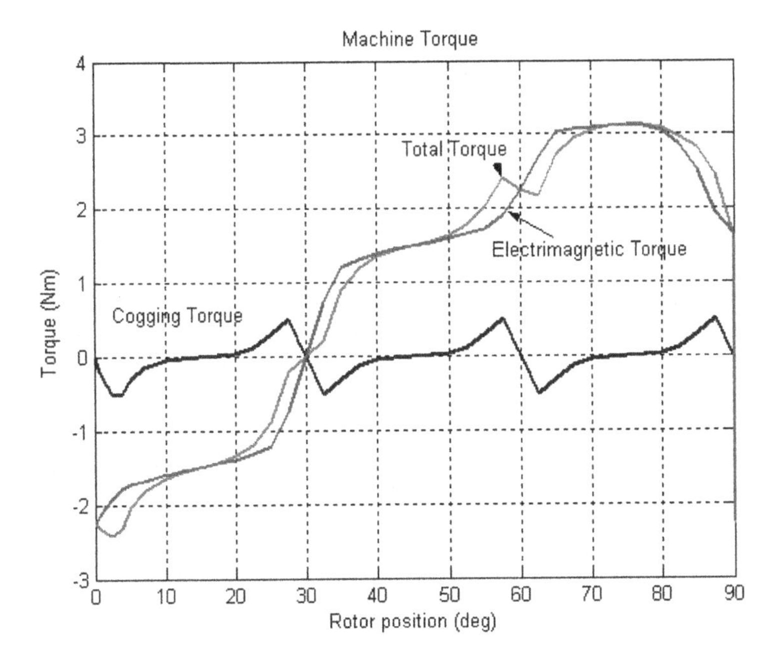

FIGURE 6.25 Machine torque vs. rotor position for 529 A turns/coil.

The electromagnetic torque has maximum values between 60° and 90° of rotor mechanical position; consequently, these angles will be used to switch the phase currents in such a way as to repeat the field configuration from 60° to 90° for all rotor positions (the drive function from Figure 6.22 was chosen in this way).

The phase flux linkage is computed as the average of the vector potential difference between the coil sides (Equation 5.119). The computed flux linkage is for a coil with a single turn and a core length of 1 m. The coil flux linkage is obtained by multiplying the FEM value with the core length and with the number of turns per coil. The flux linkage in the phase with zero current (phase c) is equal to the permanent magnet flux linkage for all rotor positions, Figure 6.26, despite the magnetic phase coupling and core saturation. This feature could be used in control when the voltage of the idle phase is measured.

Observing the flux density distribution (absolute value) from Figure 6.23 it is difficult to choose a value for iron loss computation. The space average of the square flux density on the stator core could be a proper value for iron loss computation:

$$B_{\text{sav}} = \sqrt{\frac{1}{S}\int_{S} B^2 \mathrm{d}s} \tag{6.15}$$

The space average values of the flux density are variable in time, Figure 6.27. For some rotor positions, the total space average flux density is larger than the permanent magnet average flux density while for other rotor positions it is smaller.

The space average flux density depends monotonically on the current for a single a.c. current excitation device and its maximum is reached when the current reaches

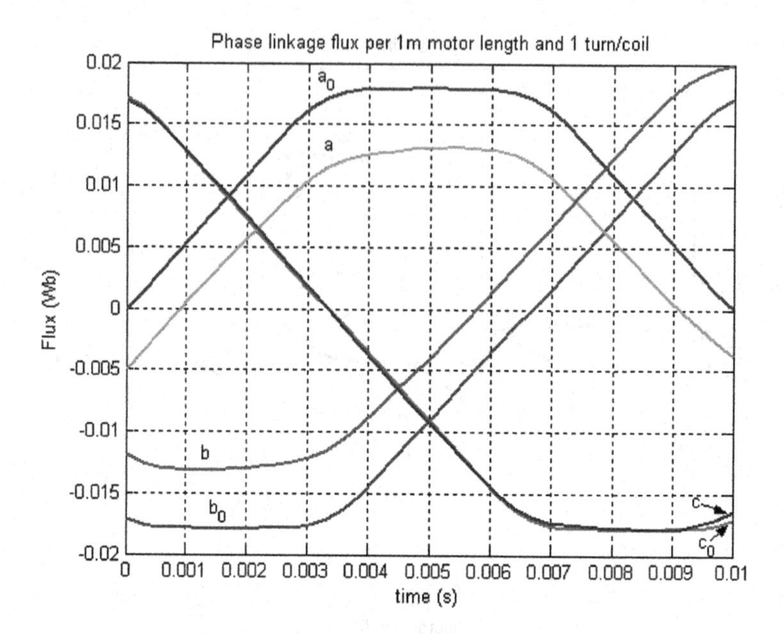

FIGURE 6.26 Phase flux linkages, without (a_0, b_0, c_0) and with i_a= -i_b currents (a, b, c) vs. time

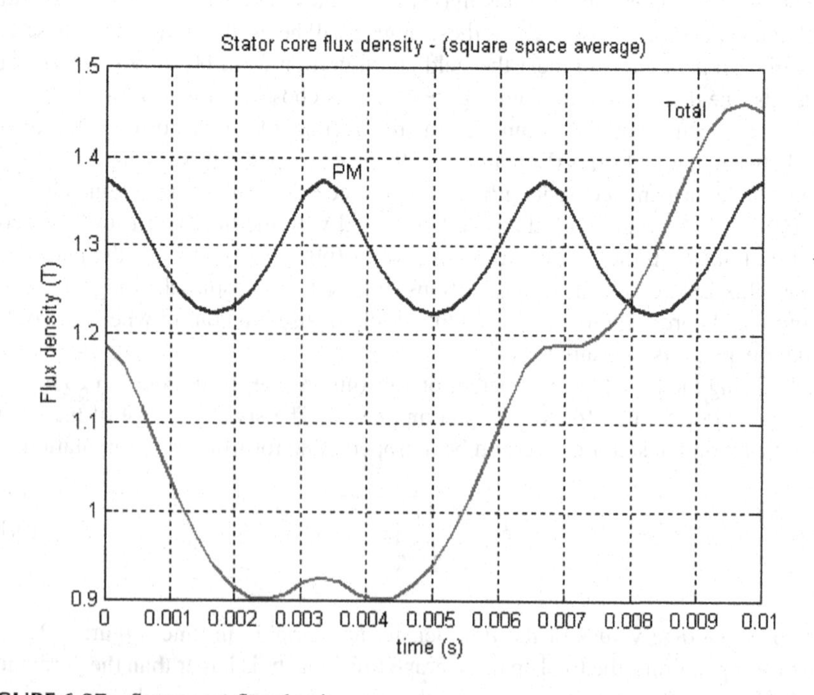

FIGURE 6.27 Stator core flux density.

its maximum. This value should be a proper value for iron loss computation but this is so only for single-phase transformers. The same value is obtained if the rms value of the flux density is multiplied with the crest factor, which is $\sqrt{2}$ for sinusoidal variations:

$$B_{pk} = \sqrt{\frac{2}{T}\int_{T} B_{sav}^{2}\left(t\right)\mathrm{d}t s} \tag{6.16}$$

Assuming that there is a region with pure rotational magnetic field with the flux density magnitude equal to B_{sav}, the iron losses for the rotating field double the iron losses for the pulsation field of the same magnitude. This implies that the equivalent flux density from Equation 6.16 could also be used to compute iron losses for pure rotating fields. In conclusion, we assume that the peak flux density from Equation 6.16 could be used to compute the iron losses for complex field variations because it works for extreme situations, i.e., pulsation fields and rotational fields.

The stator currents commutation according to the drive function is as shown in Figure 6.22. The 6/4 poles BLDC machine was chosen because it is assumed that radial forces are zero when the rotor eccentricity is zero, whereas the 3/2 poles machine has important uncompensated radial forces. The entire machine was studied in order to compute radial forces. The following results are presented for zero eccentricity but it is possible to consider the eccentricity effect on the radial forces, torque pulsations, and flux linkage. All regions that are inside the "rotating airgap" rotate around the origin of the coordinate axis. We can produce an eccentricity by moving the rotor regions or the stator regions along the radius. The airgap layers will be resized according to the new geometry configuration. A full mechanical revolution is necessary to study the eccentricity behavior.

In the following example, the radial forces are computed at zero eccentricity; thus, again, it is enough to study only a quarter of the mechanical revolution but for the entire machine geometry.

The map of the flux density (absolute values) and magnetic field lines are presented in Figure 6.28 at zero current and in Figure 6.29 at rated currents. Two significant moments were chosen to present the flux density and magnetic field lines: the current commutation moment that happens at rotor initial position (Figure 6.29a) and the moment of maximum torque that occurs after 30° electrical degrees, which means a mechanical rotation of 15°.

The current density distribution shown in Figure 6.29c proves that at zero rotor position, the current in phase b is equal to the current in phase c as it was set by the drive function (Figure 6.22) and then the current in phase b becomes zero. The flux density map and magnetic field lines give general information about field configurations for different rotor positions with and without coil currents but it is difficult to appreciate, only from the pictures, the exact field values in the points of interest. We can see that the maximum flux density value in the teeth corner is close to 2.3 T when only PMs produce magnetic fields and it increases to 2.35 T when the coils are supplied with rated current, but it is difficult to appreciate the flux density in the airgap and along the magnetic core path. It is possible to pick the flux density in several

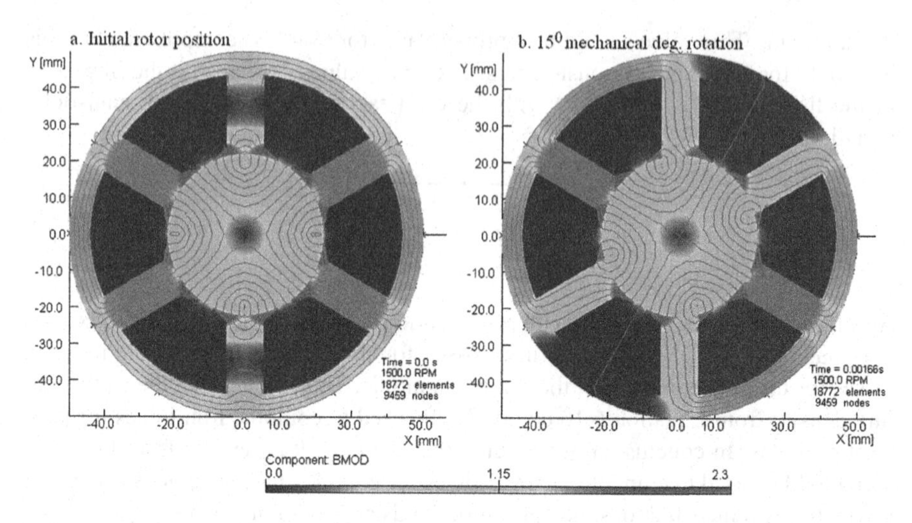

FIGURE 6.28 The flux density map and magnetic field lines produced only by PMs.

points but more organized information is available using the flux density graphs along the path of interest. All classical machines design is based on the airgap flux density and its distribution.

The radial component of the flux density along the airgap is presented in Figure 6.30 for initial rotor position without current, curve "p0"; with current curve "ip0"; and for 15° mechanical rotation without current, curve "p 15" and with current curve "ip 15." There are similar ways to get the flux density graph along a path, in different commercial software. It is only necessary to choose the represented component and to define the path. Sometimes, it is useful to save the desired graph values in a table and then use this offline for more complex analyses, such as harmonics extraction and solution comparisons. The fundamental of the flux density distribution moves with the rotor while the stator slot opening effects remain fixed to the stator coordinate as shown in Figure 6.30.

The flux density along the stator teeth is also an important value in the classical design. Using the field distribution map we can choose one or more paths to represent the field density. In our example, the phase b teeth seem to be heavily saturated and a path along the b teeth axis, as shown in Figure 6.31 (Path 1), was chosen to represent the flux density.

The radial component of the flux density along Path 1 is shown in Figure 6.32.

The tangential flux density component could be also presented but it will be small due to the magnetic field symmetry along the b phase axis for the chosen rotor position. The absolute value of the flux density is practically equal to the radial value in this situation. The equivalent radial flux density in the stator teeth is, in Figure 6.32, an average between that at the inner stator yoke radius and that at the inner stator radius. The curve "p0" shows the flux density along Path 1, for initial rotor position, produced only by the permanent magnets and curve ip0 shows the flux density produced by the permanent magnets and currents. A slight increase in the flux density can be seen when the currents are present. There is no notable difference between

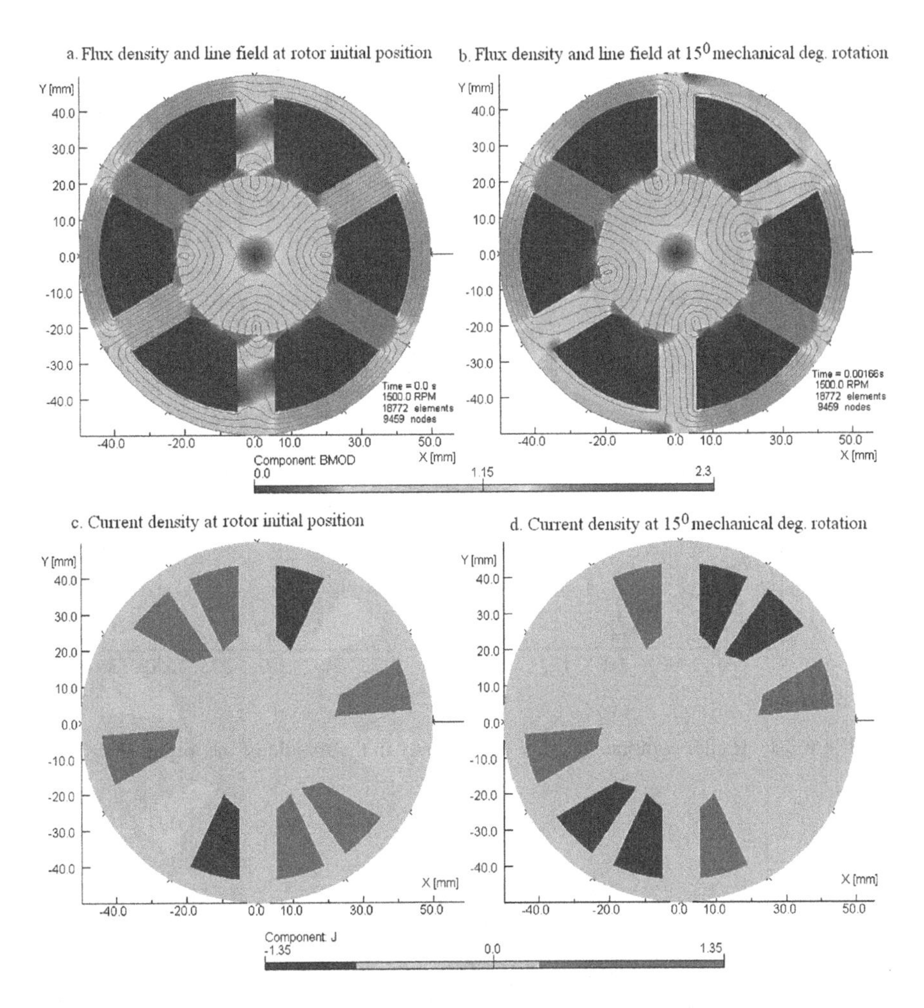

FIGURE 6.29 The flux density map and magnetic field lines at rated current.

radial flux density with and without current, when the rotor mechanical position is 15°. These curves are similar to the ip0 curve, and, consequently, they are not presented in Figure 6.32.

The tangential flux density along the stator yoke is also an important parameter in the machine classical design where an mmf drop is computed for each machine part. The radial flux density component in the yoke does not contribute directly to the mmf drop on the yoke but it could increase the absolute flux density value and thus increase the saturation level. The tangential flux density and the absolute flux density value along the yoke path (Path 2 in Figure 6.31) are presented in accordance to the position angle in Figure 6.33.

Path 2 is a circle placed in the middle of the yoke. The flux density is shown at initial rotor position without current (curve p0) and with current (curve ip0) and for

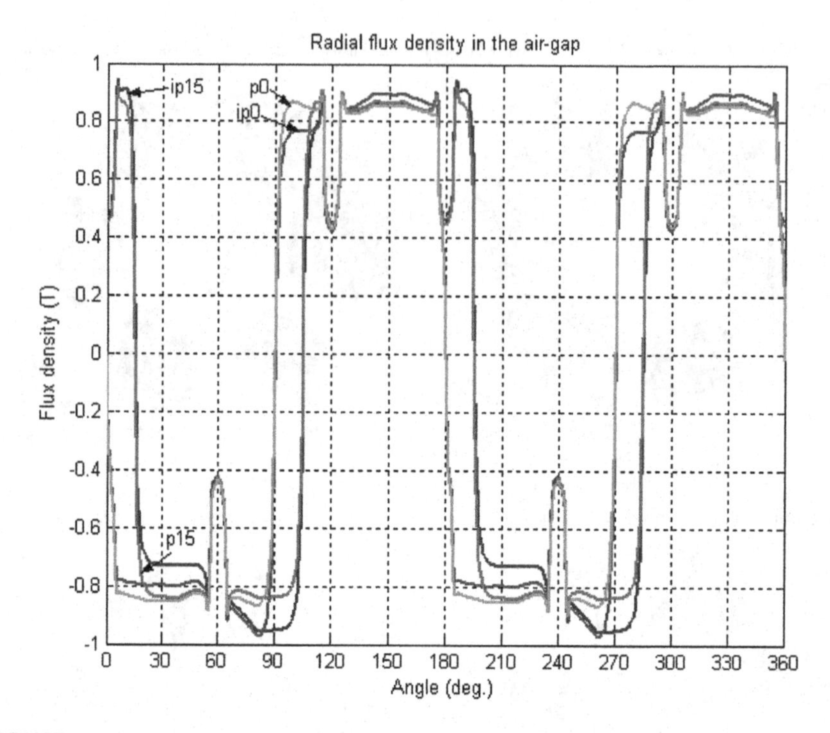

FIGURE 6.30 Radial component of the flux density in the middle of the airgap (R = 21.75 mm).

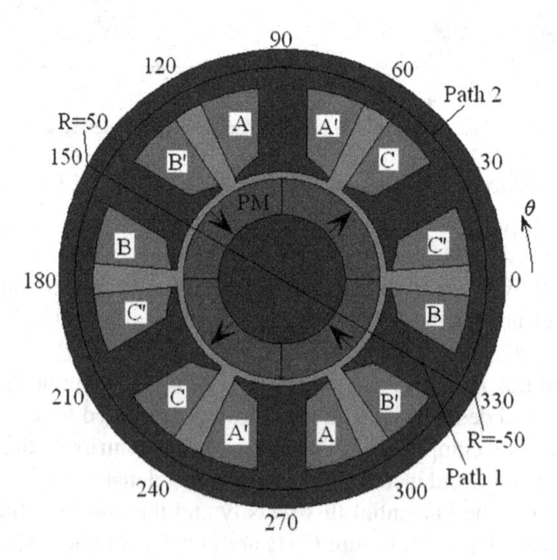

FIGURE 6.31 Path definition for flux density representation.

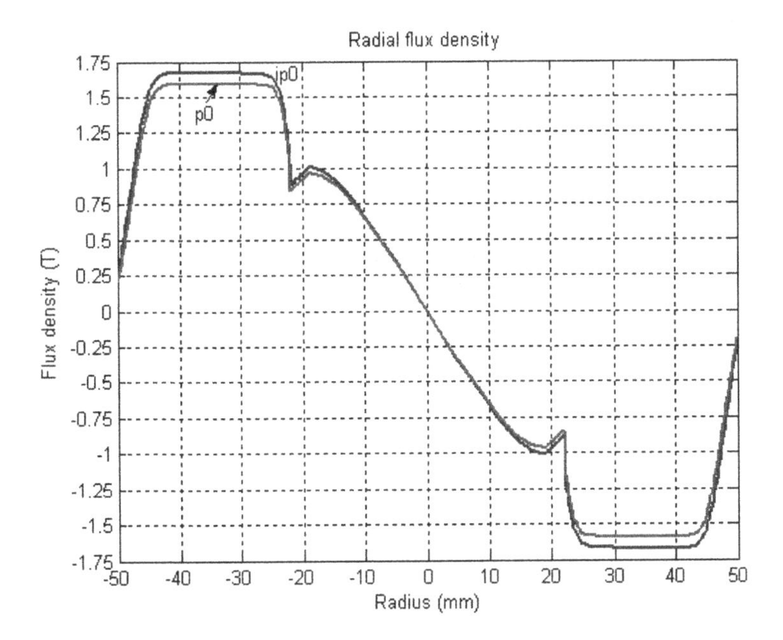

FIGURE 6.32 Radial flux density along b axis (Path 1 in Figure 6.31) for initial rotor position.

15° rotor position, without current (curve p15) and in the presence of current (curve ip15). The large influence of the coil currents' presence can be seen in some yoke regions. The radial component of the flux density produces visible effects on the flux density magnitude, which is different from the tangential component in some zones. The peak value, which is considered in classical design, is only slightly changed by the radial component of the yoke flux density. The flux density along the yoke path moves along with the PM rotor but the six-teeth stator structure has an important effect on the flux density distribution in the stator yoke.

The specific torque per unit core length computed from energy variation is in good agreement with the specific torque computed by the Maxwell tensor integration, Figure 6.34.

A large torque pulsation with 12 pulses per complete revolution can be seen. The torque produced from PM interaction with stator current (electromagnetic torque) at 265 A turns/coil is around half the peak torque in Figure 6.25 for the doubled mmf. This means a negligible reduction of the peak torque due to the magnetic cross saturation.

The specific radial force components on the rotor are negligible under no load and also in the load regime as shown in Figure 6.35 where curves F_{x0} and F_{y0} are the no-load-specific forces while F_x and F_y are the radial force components when the machine is loaded.

The radial forces are presented vs. time and 0.01 s corresponds to a 90° rotor movement. The specific radial forces presented in Figure 6.35 represent numerical computational errors because the magnetic force on the half rotor is around 5000

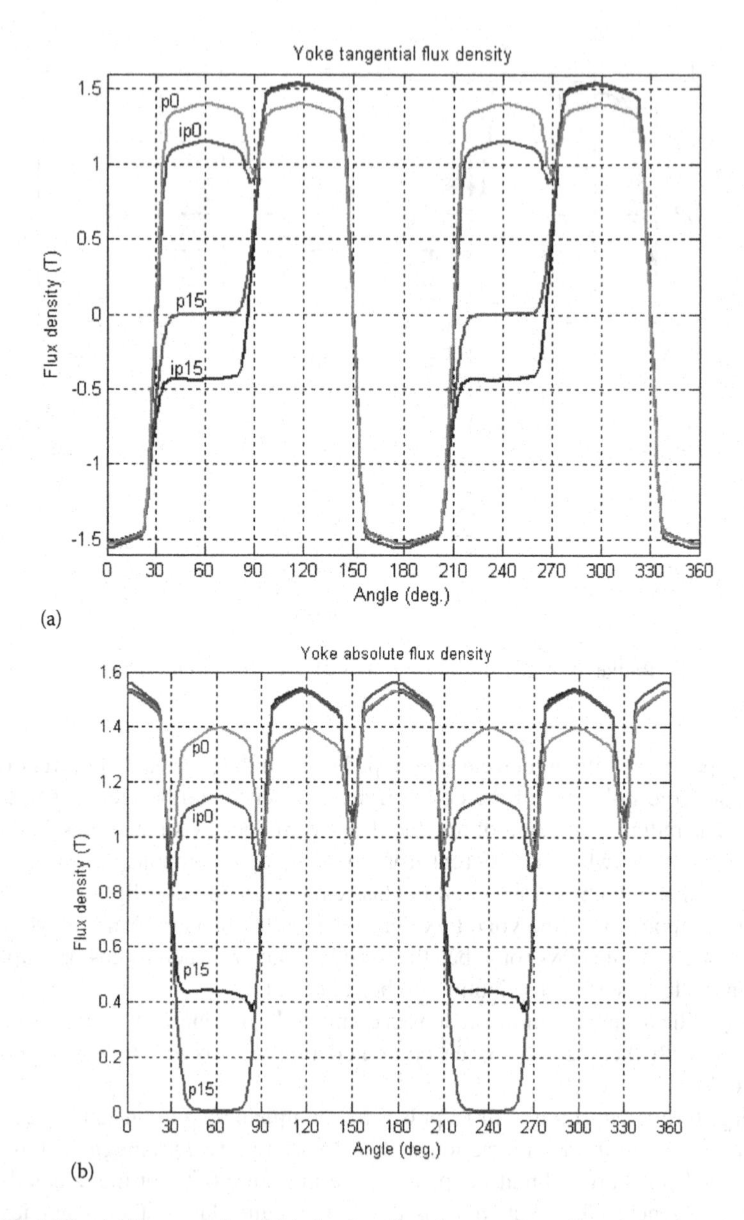

FIGURE 6.33 (a) Tangential and (b) absolute flux density along the yoke path.

N/m. The peak value of uncompensated radial forces from Figure 6.35, 0.06 N/m, shows that the accuracy of radial forces computation is good. For comparison, Figure 6.36 shows the specific radial force of a 3/2 teeth/pole BLDC machine with a smaller inner diameter (15 mm, while in our example the stator inner diameter is 44 mm).

The specific radial forces are around 400 N/m, which is 10,000 larger than in the 6/4 configuration. This large radial force produces motor vibration, noise, and premature bearing wears.

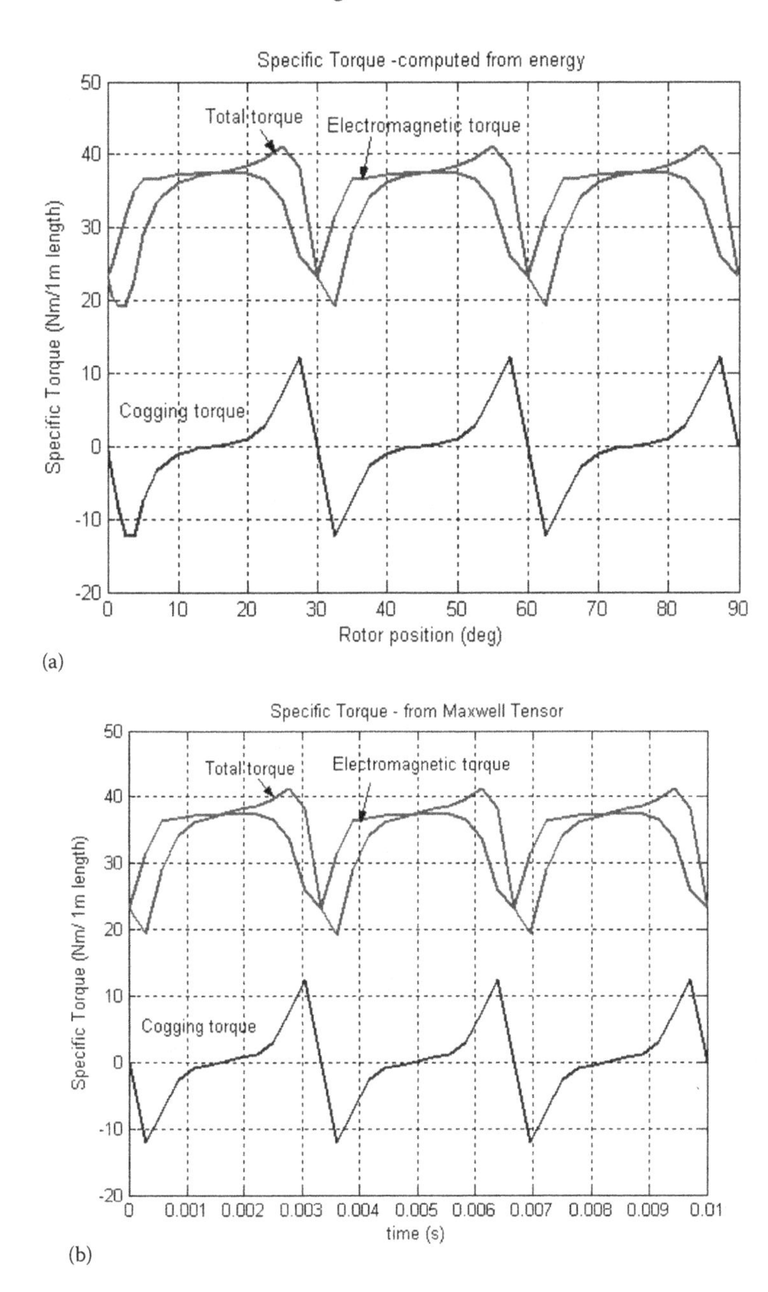

FIGURE 6.34 Specific torque vs. rotor position in synchronous operation at 265 A turns/coil.

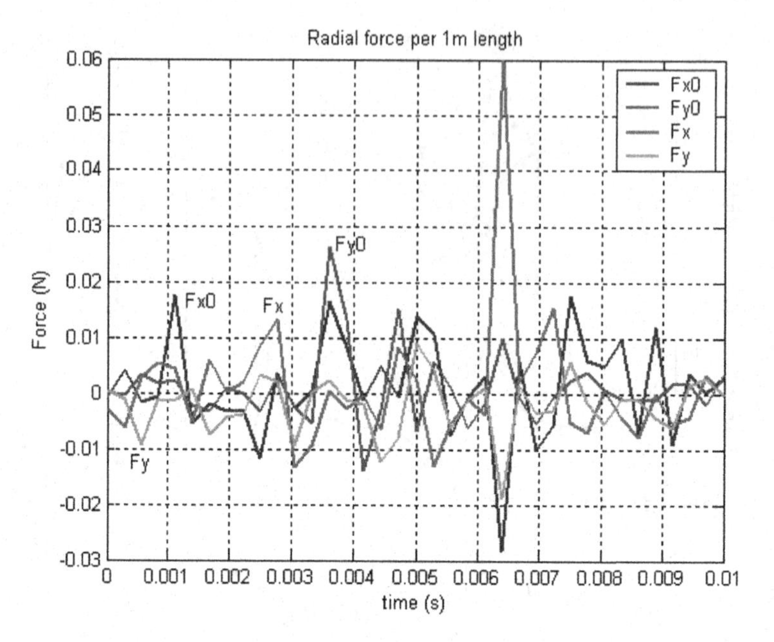

FIGURE 6.35 Radial forces of the 6/4 BLDC motor.

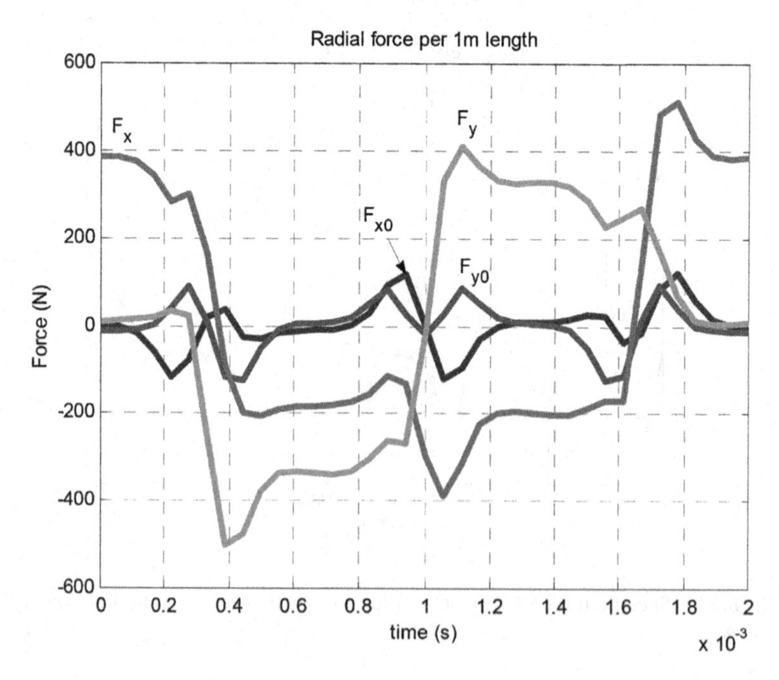

FIGURE 6.36 Radial forces of a BLDC machine with 3/2 stator teeth/rotor pole with Dsi = 15 mm.

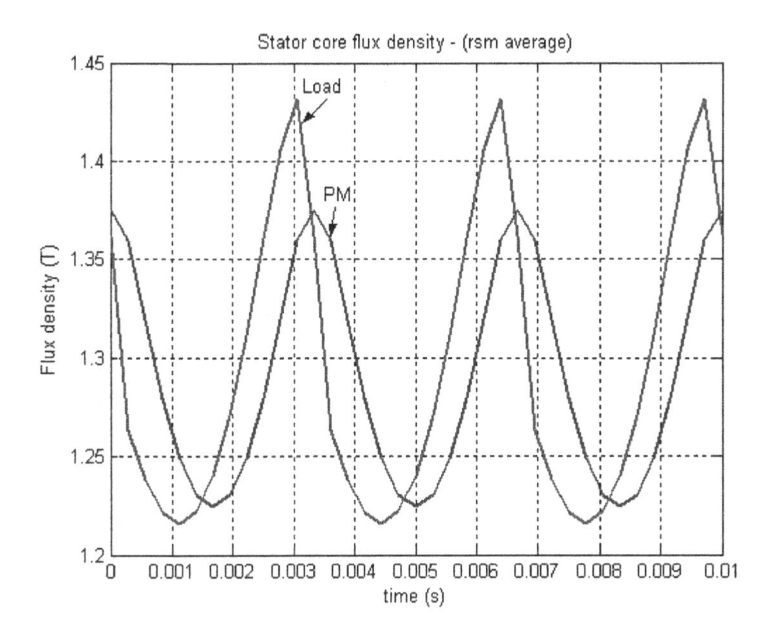

FIGURE 6.37 Core average flux density in synchronous operation.

The core average flux density vs. time is shown in Figure 6.37 for no-load and load regime. The equivalent peak flux density values for core losses computation, calculated with Equation 6.16, are B_{pk0} = 1.8286 T for no load and B_{pk} = 1.8374 T for rated load. The core losses increasing from no load to rated load is small (about 1% of rated power if the core losses are considered in direct ratio with the flux density magnitude). The high order flux density harmonics increase the core losses much more.

The maximum flux density in the teeth from Figure 6.32 is 1.67 T, and in the yoke, from Figure 6.33, is 1.53 T. The peak flux density from FEM is with 14.73% larger-than-average value used in classical design if the square averaging between teeth and yoke flux density is considered for a quick comparison. This increase in the flux density is the effect of nonhomogeneous flux density distributions and local saturation and it increases the iron losses by 31.63%, if we consider only the fundamentals. The average flux variation is presented for half the fundamental period; this means that the flux pulsation from Figure 6.37 is dominated by the sixth harmonic whose magnitude is 0.103 T.

The frequency is six times larger than the fundamental frequency and the iron losses will be 11.39% higher than the iron losses produced by a fundamental field. These losses represent 15% when compared with classical computational methods. Finally, the iron losses are 46% larger when compared with classical computational methods even when the flux density from the classical method is based on FEM. The classical design error is usually corrected by a large manufacturing factor, which increases the computed iron losses by 1.5–2.5 times, to be in agreement with test results.

The circuit parameters and mechanical features are computed using the finite element results. The motor electromagnetic torque is computed by multiplying the

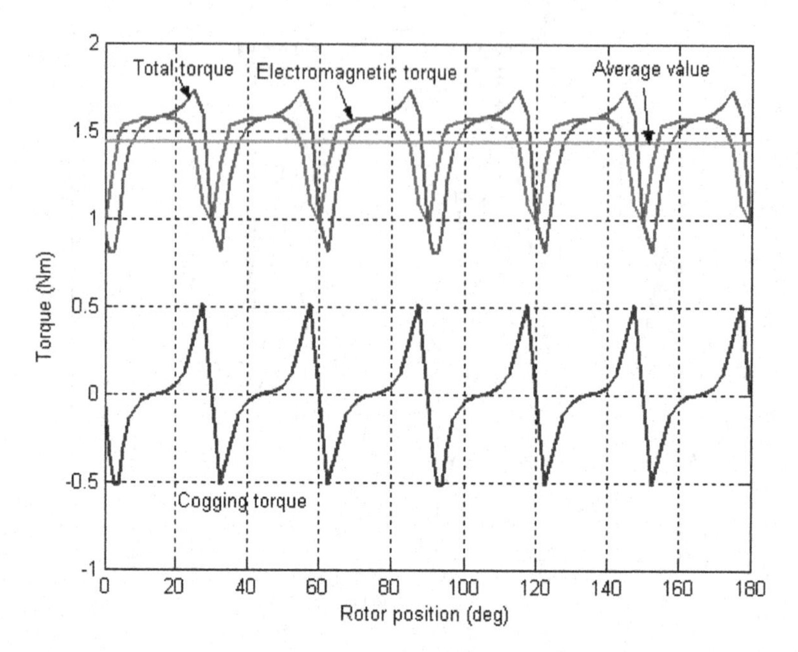

FIGURE 6.38 Motor torque pulsations at 265 A turns/coil.

specific torque from Figure 6.34 with the core axial length. The torque for a complete electrical period could be computed using the periodicity feature. The six torque pulses per electrical period are obtained as shown in Figure 6.38.

Several methods were proposed to reduce the torque pulsations, but only PM step skewing yields acceptable results in terms of total torque and torque pulsation reduction [8]. The average value of the electromagnetic torque at a given current magnitude is computed and then it is used for adjusting the final current magnitude in order to produce the rated torque.

In our case, a linear variation of torque with the current magnitude was proven up to the peak torque, which is around two times larger than the rated torque. In this case, a linear relation is used to adjust the rated current magnitude:

$$I_{tn} = I_{t1} \frac{T_n}{T_{n1}} \tag{6.17}$$

where

I_{tn} is the rated value of total coil current mmf,
I_{t1} is the total coil current mmf used in the model,
T_n is the rated torque, and
T_{n1} is the average torque from simulations.

If the motor torque is not in direct ratio to the stator coil current magnitude due to cross saturation or due to inductance variation, then several finite element runs will be performed at different currents and the rated current will be adjusted by a linear or quadratic interpolation.

The average length of one coil turn is computed keeping in mind the core length and coil end connection length. At every moment, excepting the commutation moments, two phases are in conduction and the current is at its magnitude. The copper losses are computed as

$$P_{\text{co}} = 2\rho_{\text{tc}} \frac{l_{\text{c}}}{k_{\text{fill}} A_{\text{c}}} I_{\text{tn}}^2 \tag{6.18}$$

where

ρ_{tc} is the copper resistivity at working coil T_c temperature,
l_c is the coil turn average length,
A_c is the coil region surface—could be computed from FEM by surface integral, and
k_{fill} is the coil region filling factor.

After the computation of copper losses, the one turn equivalent winding resistance is computed as

$$R_{1\text{turn}} = \frac{P_{co}}{2I_{tn}^2} \tag{6.19}$$

The iron losses could be computed based on the peak flux density using the Steinmetz method or a more complex [10] method. In this example, a simple method based on a single losses coefficient is used:

$$P_{\text{Fe}} = p_{50\text{Hz}1\text{T}} \left(B_{pk}^2 \left(\frac{f_{1n}}{50} \right)^2 + B_{1h}^2 \left(\frac{6f_{1n}}{50} \right)^2 \right) \cdot m_{\text{core}} \tag{6.20}$$

Then the electric efficiency is computed.

The synchronous inductance for the equivalent one turn winding is computed as the ratio between the variation of one coil winding flux and the total current.

The induced voltage in one turn equivalent winding is computed directly by multiplying the core length with the linkage flux derivative, which is available in the rotating machine module of "Vector Fields". The motor control keeps the phase currents in phase with the PM–induced voltage. The phase currents are considered to be constant except for commutation moments. This means that the flux-linkage time derivative does not depend on the currents when two motor phases work together. The numbers of turns in series per phase windings is computed as

$$N_1 = \text{round}\left(\frac{V_{\text{dc}}}{2\left(V_{pk1\,\text{turn}} + R_{1\text{turn}} I_{tn}\right)} \right) \tag{6.21}$$

The real phase current and linkage flux are simple to compute since the number of turns per phase winding are known:

$$I_n = \frac{I_{tn}}{N_1}$$
$$I_a = \frac{I_{ta}}{N_1} \tag{6.22}$$
$$\psi_a = N_1 \cdot l_{stack} \cdot \psi_{as1turn};\ l_{stack}\ \textit{in meters}$$

where $\psi_{as1turn}$ is the specific flux linkage per phase (a single turn and 1 m core length). Similarly, the current and flux linkage are computed for all phases. The phase current waveform vs. rotor position is shown in Figure 6.39 and the flux variation is shown in Figure 6.40.

The curves F_{ia}, F_{ib}, and F_{ic} are the phase flux linkage at rated load, and F_{ia0}, F_{ib0}, and F_{ic0} are the no-load phase flux linkages, respectively (produced only by PMs).

The real values of the resistance and inductance are finally computed as

$$R_s = N^2 \cdot R_{1turn}$$
$$L_s = N^2 \cdot L_{1turn} \tag{6.23}$$

The computed inductance does not include the end coil–connections inductance. This value should be computed analytically or from a separate finite element method.

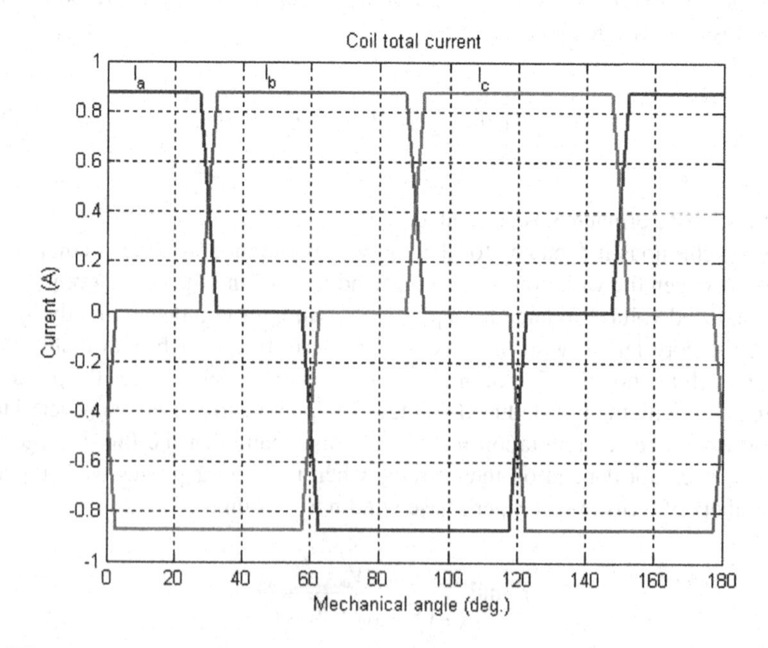

FIGURE 6.39 Phase current waveforms (304 turns/coil).

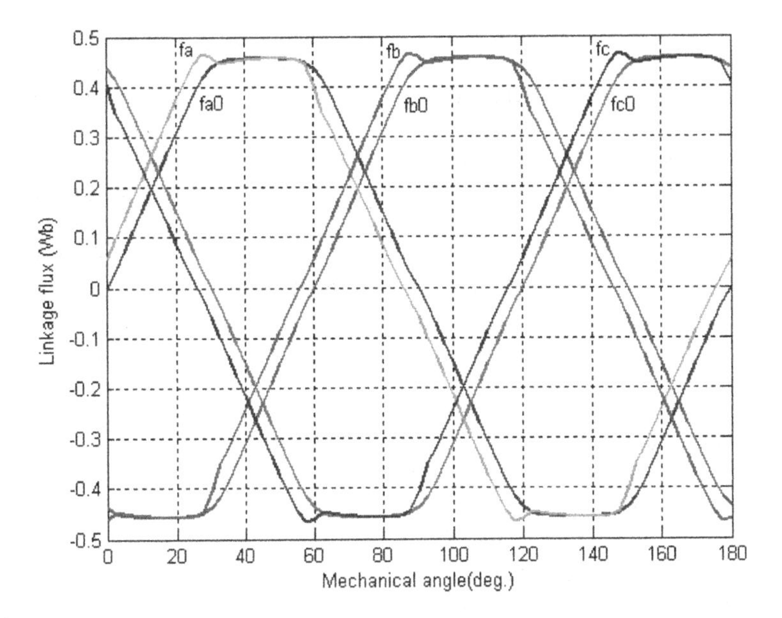

FIGURE 6.40 Phase flux linkage.

The coil conductor's cross section is

$$q_{cu} = \frac{k_{fill} \cdot A_c}{a_1 \cdot N_1} \tag{6.24}$$

where a_1 is the number of parallel paths and A_c is slot area/coil.

6.2.3 Summary

This paragraph shows how the magnetostatic FEM solution for magnetic field distribution allows for a rather comprehensive analysis and design of a brushless d.c. motor (PMSM with trapezoidal currents); core loss calculations are also included.

Moreover, FEM investigations with coupled field circuit solution are already available for a rather complete dynamics analysis of controlled PMSMs (with sinusoidal or trapezoidal current control). But this is beyond our scope here.

FEM analysis of interior PM rotor PMSM with distributed stator windings is presented in [10].

6.3 THE THREE-PHASE INDUCTION MACHINES

The first step in modeling an induction machine using the finite element method is to find a periodic symmetry in order to reduce the computation effort. The maximum value of periodicity symmetry is equal to the largest common divisor of the pole number, stator slot number, and rotor slot number. The entire induction machine is divided in a number of sectors equal to the periodicity symmetry and only one part of this will be modeled using the finite element method. If the studied part contains an odd number of poles, then the symmetry is considered negative (odd) because the magnetic vectors' potential in the paired points along the cutting border has the same value but an opposite sign. If the studied part contains an even number of poles, then the periodicity symmetry is positive (even) and the magnetic vector potential in the paired points along the cutting border has the same value. The required geometric dimensions could be divided in main induction machine dimensions and slots geometric dimensions. The main induction machine dimensions are internal stator diameter—Dsi, outer stator diameter—Dso, outer rotor diameter—Dro, airgap length—g, and core length l_c. The core length does not appear in the two-dimensional finite element model but it is used to compute the main induction machine parameters and performance from the finite element analysis. The number of poles, pp, the number of stator slots, Nss, and the number of rotor slots, Nrs, are also main parameters for building induction machine geometry. The stator slot and rotor slot dimensions depend on the slot type and here we will present the slots geometric dimensions for a few particular cases.

A cross section of induction machines indicating the geometric main dimensions is shown in Figure 6.41, and the main dimension values are given in Table 6.3.

In this example, the largest common divisor of poles, stators slots, and rotor slots is two, and, consequently, only half of the motor is modeled using the finite element method. The model contains a single pole and the symmetry is negative (odd).

A large number of stator and rotor slot geometries were presented in Chapter 5, vol.1 (companion book).

In our study, we choose the stator slot type presented in Figure 6.42 and the rotor slot type presented in Figure 6.43. Their geometry dimensions are given in Tables 6.4 and 6.5, respectively.

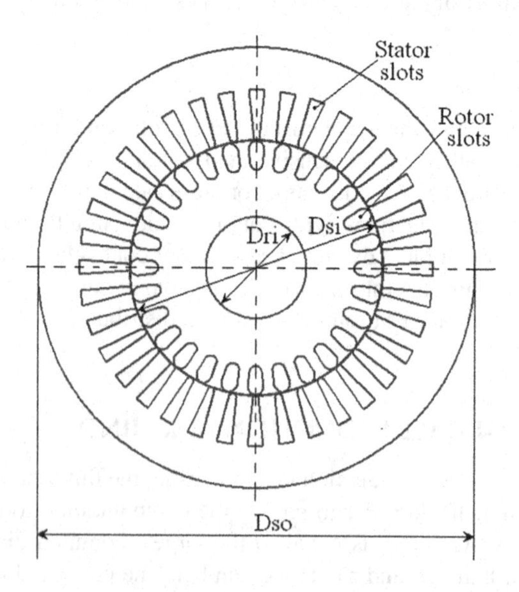

FIGURE 6.41 Cross section of induction machine geometry.

TABLE 6.3
Example of Induction Motors Main Dimensions (22 kW)

No.	Parameter	Value	Measure Unit	Observation
1	Dsi	155	mm	Stator inner diameter
2	Dso	250	mm	Stator outer diameter
3	Dri	50	mm	Rotor inner diameter
4	lc	210	mm	Core length
5	g	0.6	mm	Airgap thickness
6	Poles	2		Number of poles
7	Nss	36		Number of stator slots
8	Nrs	28		Number of rotor slots

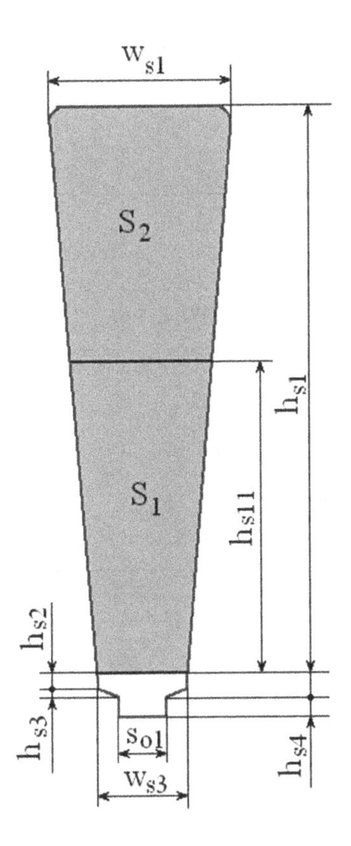

FIGURE 6.42 Stator slot geometry.

The stator slot has a trapezoidal form for small machines, while the body of the stator teeth is rectangular. If the constant width of the stator teeth is given, then the width of the slots is computed using a geometry constraint. When the stator winding has two layers, it is necessary to divide the main height of the slot in two parts S_1, and S_2, which have the same cross area.

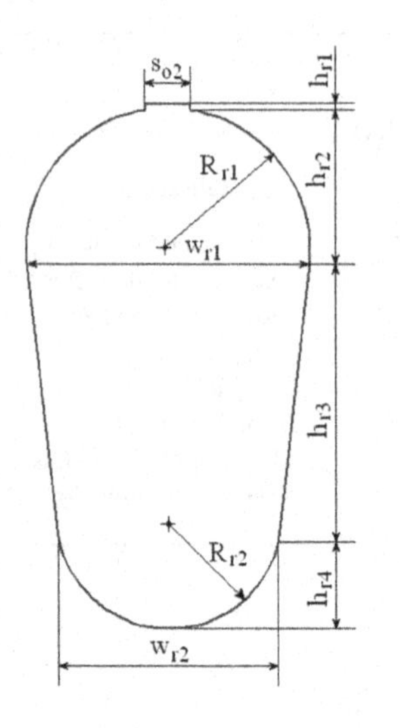

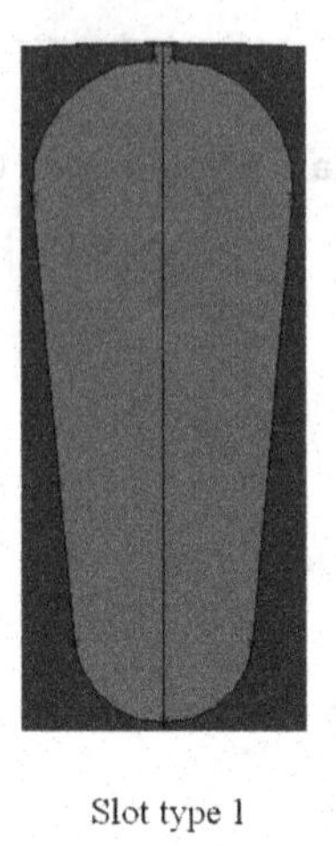
Slot type 1

Slot type 2

FIGURE 6.43 Rotor slot geometry.

TABLE 6.4
Stator Slot Dimensions

No.	Parameters	Value	Units	Observation
1	hs1	20.35	mm	Coils height
2	hs11	11.2	mm	First layer height
3	hs2	0.65	mm	
4	hs3	1.2	mm	
5	hs4	0.8	mm	
6	so1	3.9	mm	Slot opening
7	sbt	6.5	mm	Stator teeth width
8	kfill	0.4		Slot filing factor

TABLE 6.5
Rotor Slot Dimensions

No.	Parameters	Slot Type 1		Slot Type 2	Units
1	hr1	0.5	I	0.5	mm
2	hr2	4.31	C	2	mm
3	hr3	16	I	23	mm
4	hr4	2.54	C	0	mm
5	so2	0.6	I	0.6	mm
6	wr1	8	I	5.2	mm
7	wr2	5.5	I	6.7	mm
8	1/Rr1	0.2492	C	0	1/mm
9	1/Rr2	0.3625	C	0	1/mm

For the rotor slots, Figure 6.43, all geometric dimensions could be independent for a general case. If a rotor slot has the upper and bottom curvature tangent to the rotor teeth lateral wall, then the curvature radii, Rr1 and Rr2, and heights, hr2 and hr4 (curvature parts height), are computed from slot widths wr1 and wr2, and main part height, hr3:

$$\alpha_{rs} = \operatorname{atan}\left(\frac{w_{r1} - w_{r2}}{2h_{r3}}\right) \tag{6.25}$$

$$R_{r1} = \frac{w_{r1}}{2\cos(\alpha_{rs})} \tag{6.26}$$

$$R_{r2} = \frac{w_{r2}}{2\cos(\alpha_{rs})} \tag{6.27}$$

$$h_{r2} = \sqrt{R_{r1}^2 + \frac{s_{o2}^2}{4}} + R_{r1}\sin(\alpha_{rs}) \tag{6.28}$$

$$h_{r4} = R_{r2}\left(1 - \sin(\alpha_{rs})\right) \tag{6.29}$$

If the rotor teeth have plane parallel walls (constant width), then the slots wall angle is equal to half the angle between the two adjacent teeth. The slot width and slot height are not independent in this case and Equation 6.25 could be used to compute one of the dimensions when the other two are given.

The "I" comments in Table 6.5 indicate input data, while the "C" comments indicate computed data in order to satisfy geometry restrictions.

In the rotor slot construction, the curvature of the slots wall (inverse of the radius Rr1 and Rr2) is used. Setting the curvature to zero and hr2 = 0, hr4 = 0, the trapezoidal slot shape is obtained as a particular case. The induction machine geometry description is similar to the PMSM geometry description and it will not be presented in detail. In the rotor with general slot geometry, it is simple to introduce the drawing and the teeth drawing considering that the slots axes, are parallel to the *x*-axis of the Cartesian coordinate system. After this, the slots and teeth drawings are rotated in a clockwise direction with angle α_{rs} in order to reach the desired position. When only a part of the induction motor is studied, it is better to cut the studied part along a tooth axis and not along a slot axis. The induction machine study using the finite element method is more complex than that of PMSM because the rotor currents are produced by induced voltages and they are not directly controlled by the user. Fortunately, the behavior of the induction machine is determined by a circuit model with good accuracy and there is a simple method to extract the circuit parameters from the finite element method. The machine is modeled in FEM for two simple cases: ideal no load, and, separately, for the rotor bar skin effect.

6.3.1 Induction Machines: Ideal No Load

The rotor current could be considered to be equal to zero for the ideal no-load regime of the induction machine. The stator currents are assumed to have sinusoidal time variations. Even in this ideal condition (sinusoidal stator currents and synchronous rotor speed), the rotor bar currents are not equal to zero due to the space harmonics. The rotor current produced by space harmonics could be neglected since the main objective of no-load induction machine simulation by finite element is to compute the magnetization inductance and stator leakage inductance, and to observe the magnetic saturation level in different regions.

The phase current distribution in the stator slots considering a two-layer winding is shown in Figure 6.44 for a coil pitch that is shorter, with one slot pitch, than a pole pitch.

The negative sign in front of the phase name shows the negative current for that coil side. Also, it can be seen from Figure 6.44 that the stator and rotor drawing composition from elementary parts are as described for PMSM.

The magnetic field lines and flux density map are shown in Figure 6.45 for the stator current equal to the rated value of the magnetization current.

The radial component of the flux density in the airgap vs. position is shown in Figure 6.46. The influence of stator and rotor slot opening on the flux density waveform, and also the saturation effect, are visible in Figure 6.46.

The airgap flux density spectrum is shown in Figure 6.47 for a magnetization current equal to 10.67 A (peak value). The harmonics spectrum of airgap flux density depends on the magnetic saturation level. The fundamental and most important harmonic magnitudes vs. phase current are shown in Figure 6.48.

The magnetic saturation occurs around 9 A, and, after this value, the fundamental flux density magnitude increases slowly while the third harmonic, which has small values for nonsaturation current, increases rapidly. When the magnetic saturation is

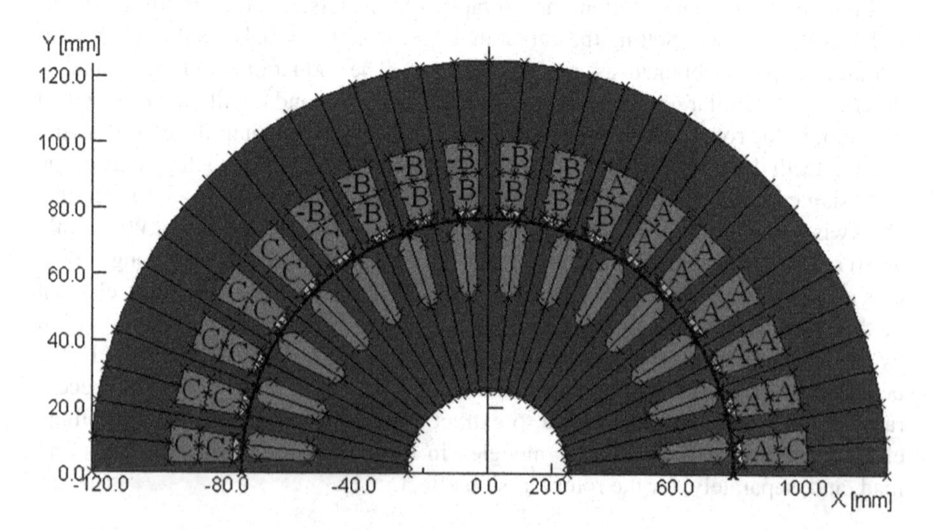

FIGURE 6.44 Induction machine phases distribution in slots.

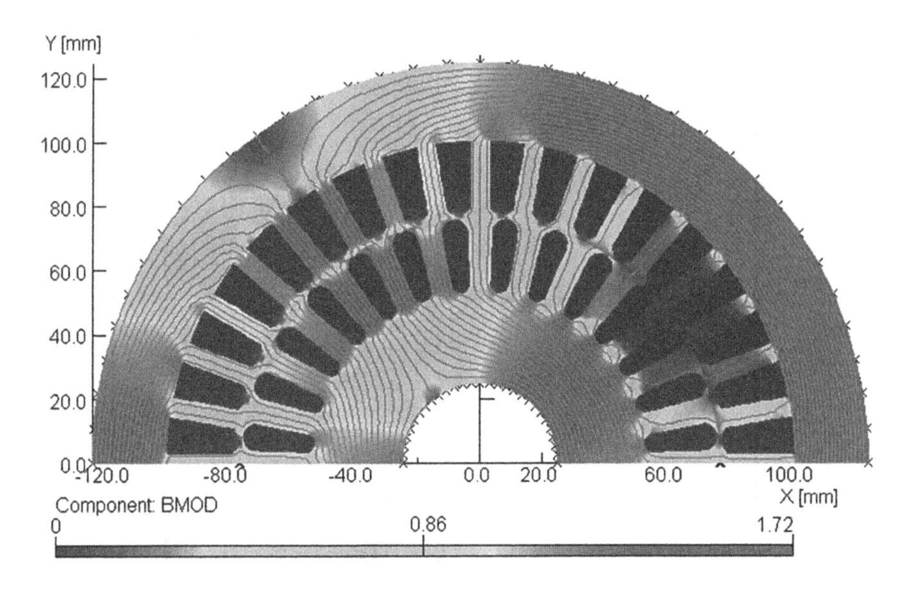

FIGURE 6.45 Flux density map and magnetic field lines at Im = 10.67 A, 5 turns/coil.

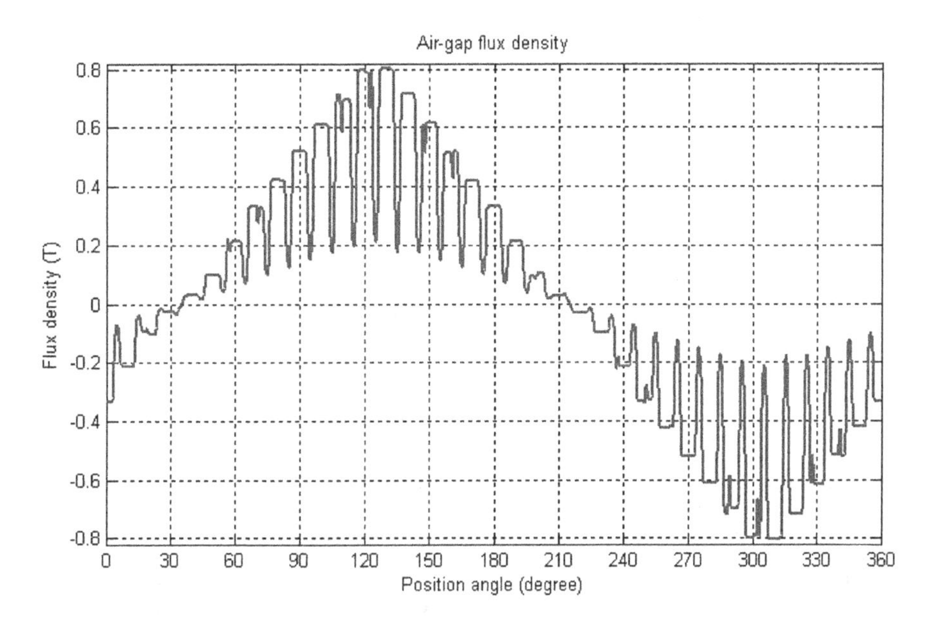

FIGURE 6.46 Airgap flux density distribution at Im = 10.67 A.

very deep, the third harmonic growing trends are reduced; however, it is more than 25% of the fundamental magnitude. The first stator slot harmonic (35) is one of most important harmonics and it increases with the magnetization current in a similar way to the fundamental component. The rotor slot harmonics are also present in the airgap flux density but in this example they have small magnitudes because the rotor slot

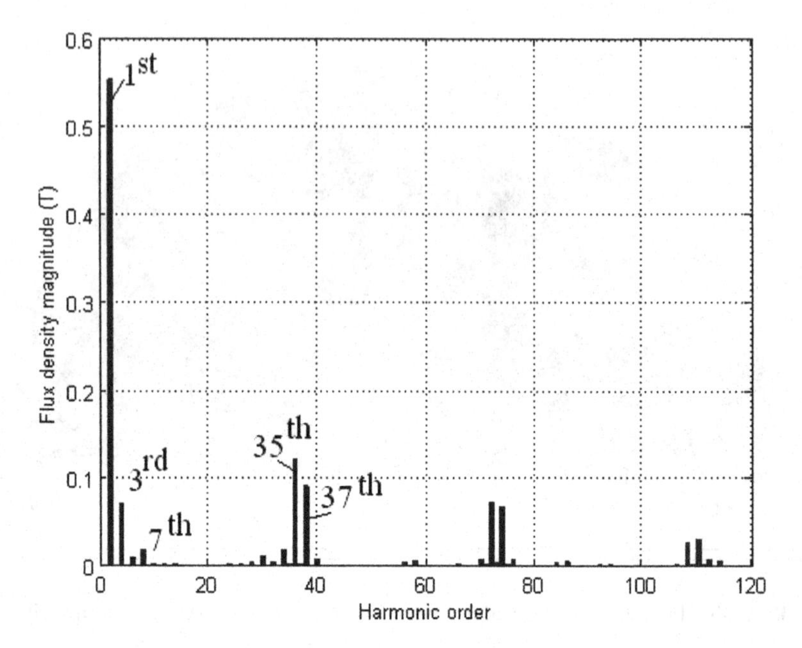

FIGURE 6.47 Airgap flux density spectrum at Im = 10.67 A.

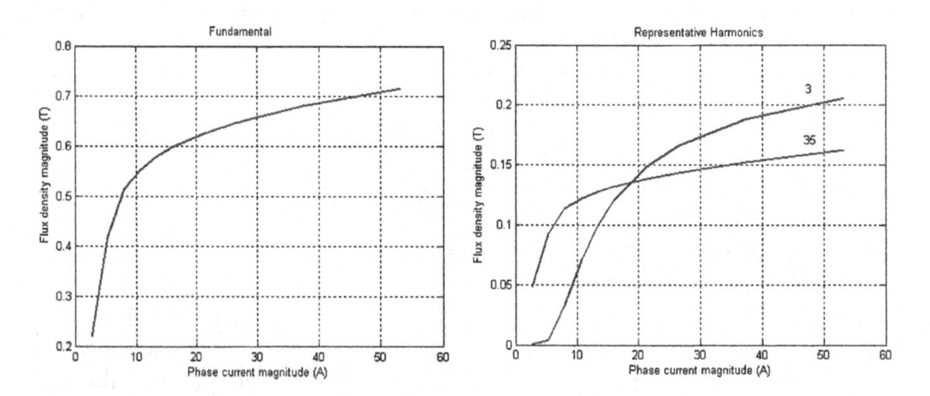

FIGURE 6.48 Fundamental and representative harmonics of airgap flux density magnitude vs. magnetization current.

openings are small. The rotor teeth corners saturation is expected for a large current in the rotor (starting current) and then the equivalent rotor slot opening increases. The rotor slot harmonics increase for large currents but they still remain smaller than stator slot harmonics.

The fundamental linkage flux, Figure 6.49, which represents the magnetization flux or main flux, is computed with the analytical Equation 6.30 where the first harmonic fundamental of airgap flux density comes from finite elements.

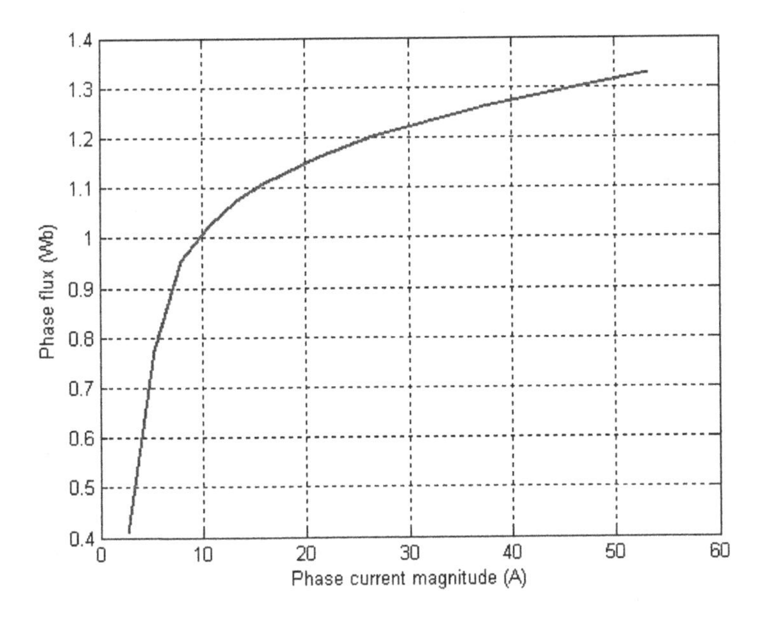

FIGURE 6.49 Main phase flux vs. magnetization current (5 turns/coil).

The main phase (magnetization) inductance, Figure 6.50, is computed dividing the first harmonic flux magnitude to the magnetization current magnitude (Equation 6.31):

$$\Psi_{m1} = \frac{2}{\pi} k_{w1} N_1 \tau l_c B_{ag1} \tag{6.30}$$

$$L_m = \frac{\Psi_{m1}}{I_m} \tag{6.31}$$

where

k_{w1} is the winding factor for the first harmonic,
N_1 is the number of turns in series per phase,
τ is the pole pitch,
l_c is the core length, and
B_{ag1} is the magnitude of the first harmonic of airgap flux density.
The airgap flux density was computed with symmetric currents in all phases.

The total phase flux could be computed by the integration of magnetic vector potential, Equation 5.119, for each phase. In our case, the "phase a" current is equal to current magnitude and the linkage total flux also reaches its magnitude. We can observe that this value is smaller than the magnitude of the main flux. The total linkage flux contains the main flux plus airgap and leakage flux. When the third harmonic

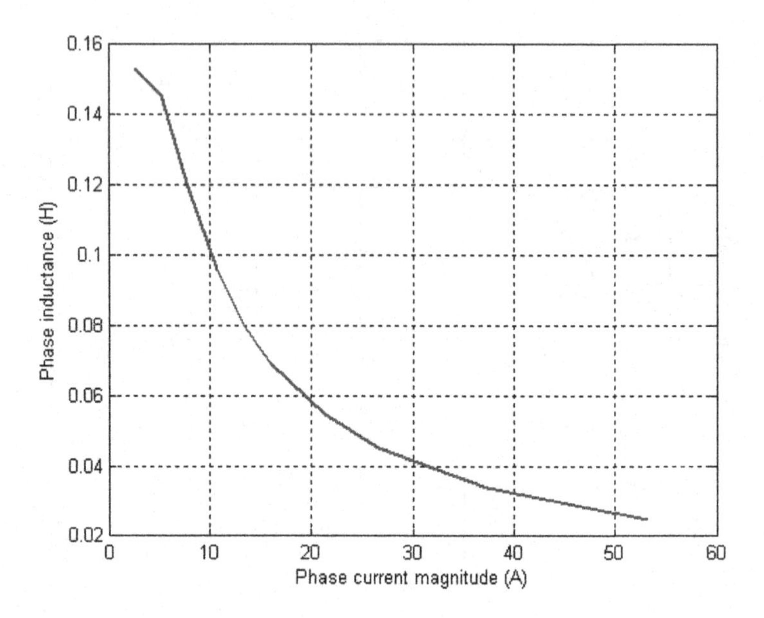

FIGURE 6.50 Magnetization inductance vs. current (5 turns/coil).

alone is larger than 25% of the fundamental harmonic for the main magnetic flux, it is possible that the maximum value of the total flux is smaller than the magnitude of the first harmonic of the magnetization flux. For such large saturation levels, it is not possible to compute the leakage flux as difference between the maximum value of the total flux and the magnitude of the main flux (first harmonic). The leakage flux will be computed as a sum of the leakage flux in different parts of the machine. From the two-dimensional finite element model, it is possible to compute the airgap leakage flux and slots leakage flux. The airgap leakage flux is related with space high order harmonics (Equation 6.33) while the slots leakage flux is related with the magnetic field energy stored in the stator slots. The end connection inductance is computed using analytical models [11].

$$L_{\sigma 1} = L_{\sigma h} + L_{\sigma s1} + L_{\sigma el} \tag{6.32}$$

$$\Psi_{agh} = \frac{2}{\pi} N_1 \tau l_c \sqrt{\sum_{\upsilon=3}^{\upsilon_{max}} \left(\frac{k_{w\upsilon}}{\upsilon} B_{ag\upsilon} \right)^2} \tag{6.33}$$

$$L_{\sigma h} = \frac{\Psi_{agh}}{I_m} = \sqrt{\sum_{\upsilon=3}^{\upsilon_{max}} \left(\frac{k_{w\upsilon} B_{ag\upsilon}}{\upsilon k_{w1} B_{ag1}} \right)^2} \bullet L_m \tag{6.34}$$

where

$k_{w\upsilon}$ is the winding factor for the υth harmonic,
$B_{ag\upsilon}$ is the magnitude of the υth airgap flux density harmonic, and
υ_{max} is the largest harmonic that is considered.

This number υ_{max} should be smaller than half the number of flux density samples per period.

It is difficult to segregate the magnetic energy produced by a phase current in each slot when short winding steps are used. This is simple to compute in FEMs from the magnetic energy stored in all the slots, in the coil parts, and in the free (wedge) parts:

$$E_{ms} = \frac{1}{2} L_{\sigma s1} \left(i_a^2 + i_b^2 + i_c^2 \right) = \frac{3}{4} L_{\sigma s1} I_m^2 \tag{6.35}$$

The slot's magnetic energy is computed only for the simulated part of the machine, E_{mss}. In order to compute the stored energy in all slots, we have to multiply the FEM value with the symmetry number, n_s, which is equal to two in our example because half the motor is modeled. The slot leakage inductance is

$$L_{\sigma s1} = \frac{4}{3} n_s \frac{E_{mss}}{I_m^2} \tag{6.36}$$

A part of the main magnetic field enters into the opening of slots and increases the slot energy, producing large errors on leakage slots inductance. This error could be reduced if the leakage inductance is computed for large currents (rated currents).

The total linkage flux per phase is computed as the magnetization flux plus the leakage flux:

$$\Psi_1 = \Psi_{m1}\left(I_m\right) + L_{\sigma 1} I_m \tag{6.37}$$

The stator resistance is also computed as

$$R_s = \rho_{tc} \frac{N_1 l_{turn}}{q_{co}} \tag{6.38}$$

The required voltage vs. current is

$$V_1 \approx \sqrt{\left(\omega_1 \Psi_1\right)^2 + \left(R_s I_m\right)^2}$$

The no-load ideal current at rated voltage is computed by interpolation and in our example it is $I_{m0} = 9.82$ A.

6.3.2 Rotor Bar Skin Effect

The skin effect is studied on a single rotor bar. In fact, due to the symmetry, the problem could be reduced to half a bar as shown in Figure 6.51.

The "steady-state harmonic (a.c.)" module is used in "Vector Fields" in order to study the eddy current in the rotor bar. The discrete element dimensions should be small enough in regions where the magnetic field or current density has large variations. The automatic mesh refinement is chosen, and, in our example, the number of elements was increased three times (from 938 elements to 2658 elements, and from

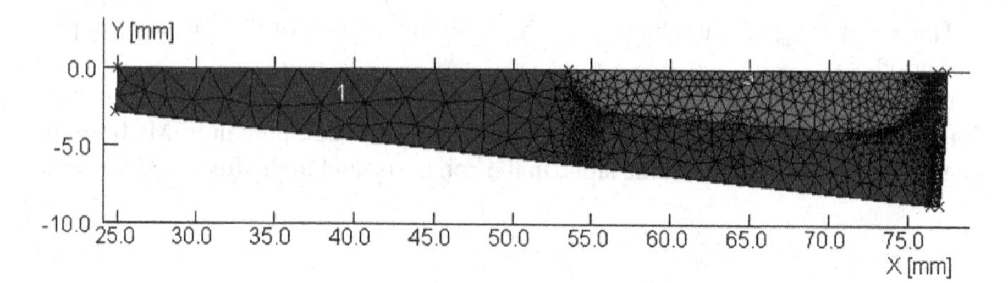

FIGURE 6.51 The rotor bar geometry for skin effect study.

522 nodes to 1424 nodes). This is a reasonable (quite small) number of nodes for a FEM problem but if the entire machine is simulated instead of only half of the rotor bar and half of the rotor teeth, then the number of nodes would increase seriously (about 50 times). The a.c. magnetic problem could be solved for several frequencies and we can then know the rotor parameters (resistance and leakage inductance) at different values of rotor slip.

The size of elements is very small in regions where the current density or the magnetic flux density show large variations, and larger in other regions. This means an efficient distribution of nodes and an efficient use of the computation resources.

The magnetic field and current distribution is shown in Figure 6.52 at 50 and 10 Hz, respectively. The current distribution map and field lines could be presented at

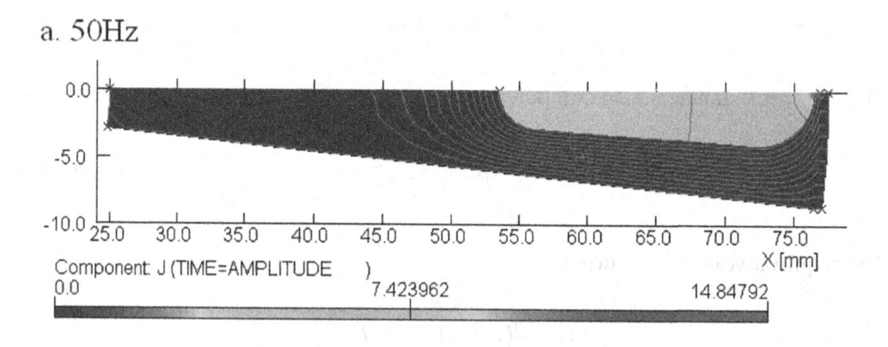

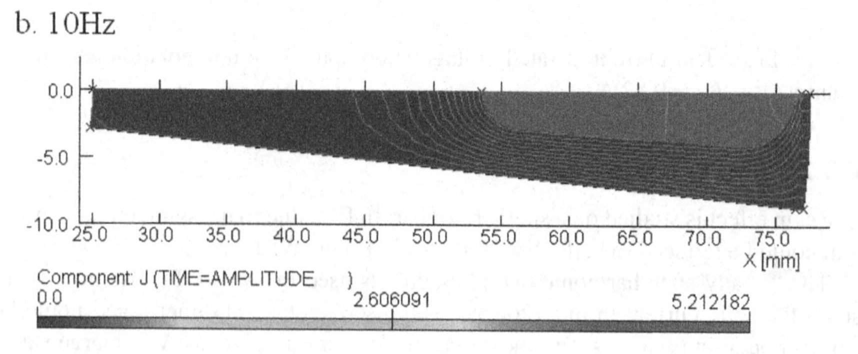

FIGURE 6.52 Current density distribution comparison: (a) at 50 Hz, (b) at 10 Hz.

different times (phase), Figure 6.52, but they could also be presented by the magnitude values as in Figure 6.52 or by the time average values for each point. The magnitude mode representation gives the maximum value of the current density or of the magnetic potential for every point. Usually, the current density magnitude or the vector magnetic potential magnitude does not reach all points at the same time and thus the current distribution and the magnetic potential lines may not satisfy the Maxwell equation in the magnitude or average representation mode.

From Figure 6.53 we can see that current density is not in phase all through. A current density distribution along the rotor slot axis is presented in Figure 6.54 at different times (total current phase). Now it is clear that current distribution in the slots depends on the time moment. The current density phase angle along the rotor slot axis is presented in Figure 6.55 for different current frequencies. The current density phase angle is related to the total current phase through the slot.

A large difference of current density phase angle between the top and the root of the rotor slot is observed at 50 Hz current frequency. Small differences in current density phase due to skin effect are observed even at small frequencies (1 Hz, respectively, 0.5 Hz) where the skin effect on current density magnitude is practically negligible. A comparison of current density amplitude distributions along the rotor slot axis is presented in Figure 6.56.

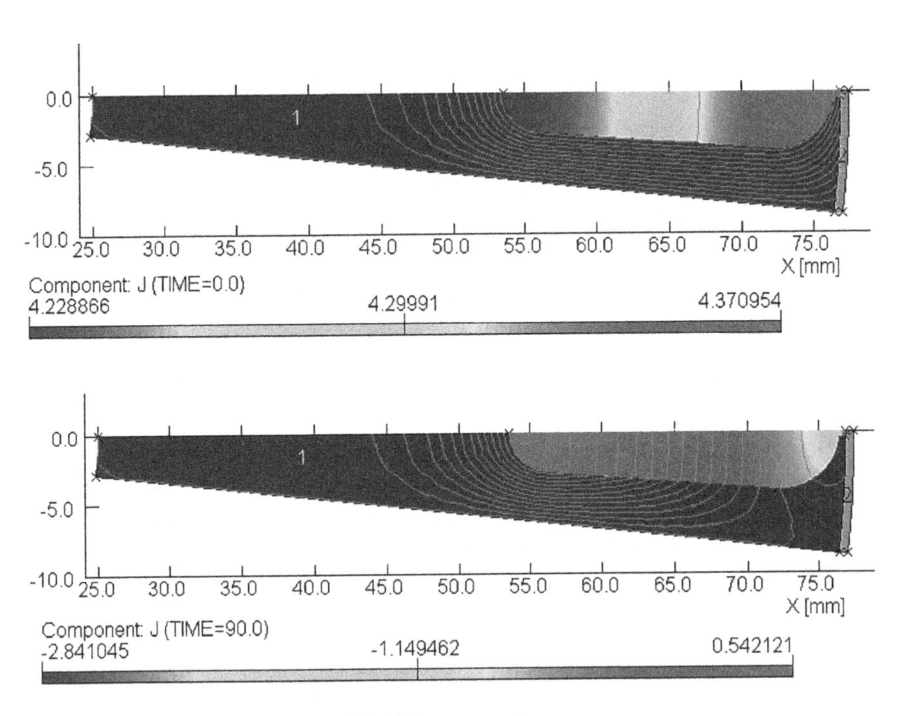

FIGURE 6.53 Current density and field line at two times.

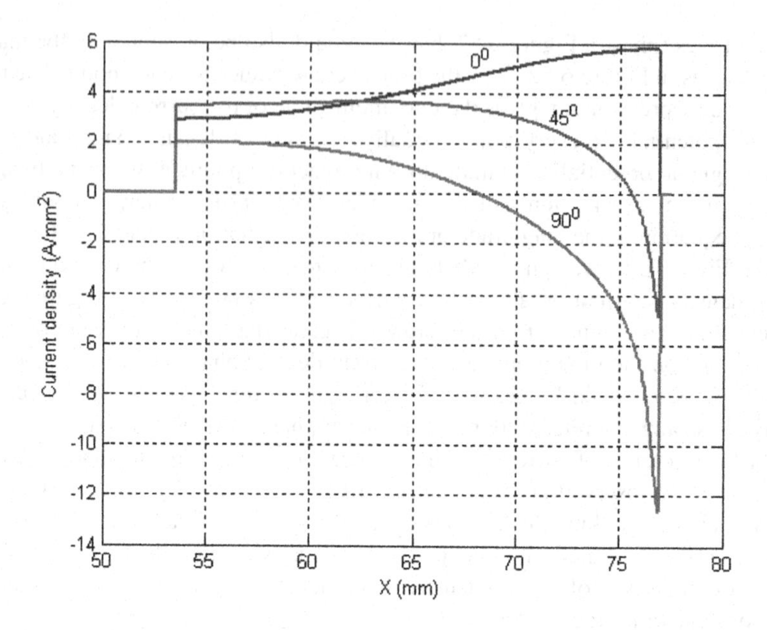

FIGURE 6.54 Current density distribution at different times.

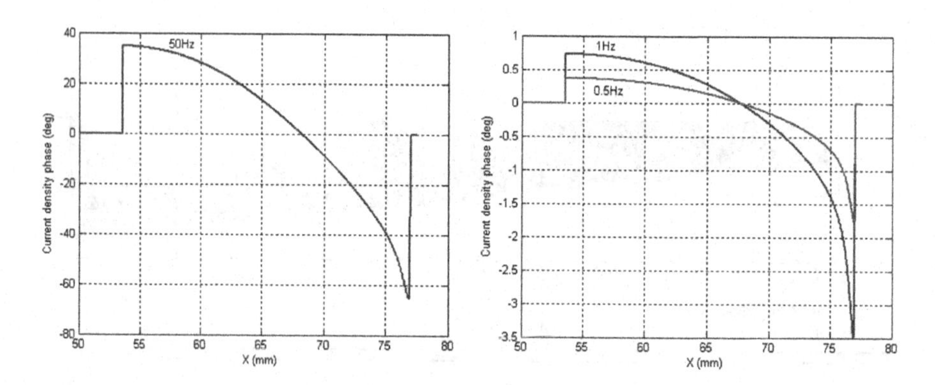

FIGURE 6.55 Current density phase angle along the slot axis.

At low frequency, the current density is uniformly distributed on the slot surface and it has small values, while for industrial frequency, the nonuniform current density distribution produces a large current density peak at the slot opening.

The large current density increases the Joule–Lentz losses that are computed using specific losses integration on the slot volume.

$$p_{co}(t) = \int_V \rho\left(J(t)\right)^2 dV \qquad (6.39)$$

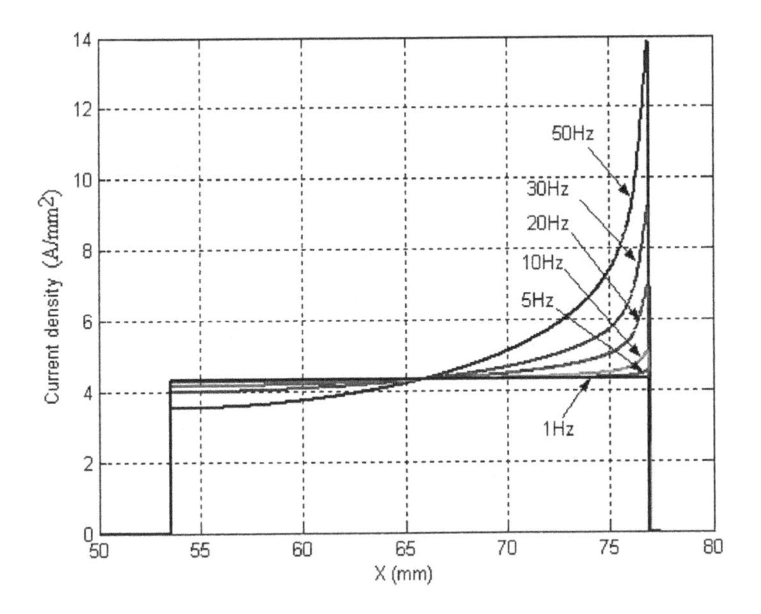

FIGURE 6.56 Current density magnitude along slot axes.

$$p_{co} = \frac{1}{T}\int_{t_1}^{t_1+T} p_{co}(t)\,dt = \frac{1}{2}\rho l_c \int_S J_m^2 dS \tag{6.40}$$

Some commercial FEM software compute the coil losses directly per unity length or for certain core lengths. An equivalent rotor-winding resistance considers the skin effect in the circuit parameter. The rotor bar resistance is computed from average bar losses:

$$R_b = \frac{p_{coh}}{2I_{bh}^2} l_c \tag{6.41}$$

where

p_{coh} is the average losses in half slot and
I_{bh} is the total current on half bar.

The equivalent resistance vs. frequency is shown in Figure 6.57 for two slot types. The slot shapes are presented in Figure 6.43 and their dimensions are provided in Table 6.5.

Nonuniform current density distribution also decreases the slot leakage inductance, compared to d.c. leakage inductance, because a small current in the lower part

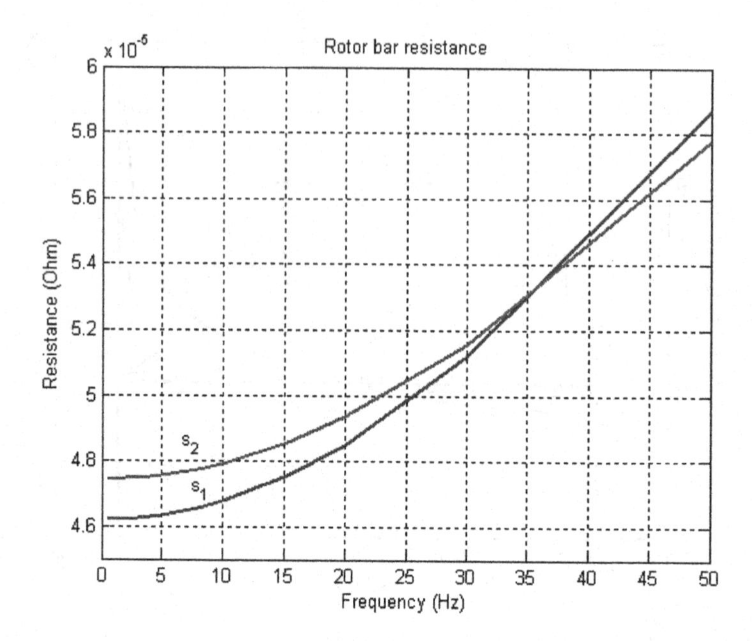

FIGURE 6.57 Rotor bar resistance vs. frequency: s_1—slot type 1; s_2—slot type 2 (Figure 6.43).

of the slot produces a small field in that slot area. The slot leakage inductance is computed from the slot magnetic stored energy as for steady state.

The stator slot inductance vs. frequency is shown in Figure 6.58 for the two same slot types. Small oscillations of the rotor slot inductance with frequency at low frequency reflect the accuracy limitation of the numerical solution. The slots leakage inductance increases slightly when the frequency is increased for the second slot type (Figure 6.58 curve s_2). This is also produced by the accuracy limitation of the numerical solution. However, these errors are small and they have no practical importance.

The skin effect factor, which is the ratio between the actual value of the resistance or inductance and their d.c. values, reflects the skin effect on the circuit parameters better.

The resistance skin effect factor, Figure 6.59, is always larger or equal to unity while the inductance skin effect factor, Figure 6.60, is always smaller or equal to unity. The single reason of this rule violation, in Figure 6.60 is the computation error. However, these errors are small, less than 0.1%, and, again, they could be neglected.

The finite element analysis showed that the skin effect at 50 Hz is larger for slot type s_1, which means larger torque starts at direct connection to the power grid. The saturation of the rotor teeth root is smaller for slots s_1 and also mechanical resistance of rotor slot root is better because they have a larger base and a smaller stress concentrator due to the large adaptation radius. In conclusion, slot s_1 is better than slot s_2.

The total rotor resistance and leakage inductance also contain the short-circuit ring resistance and its inductance. The short-circuit ring parameters as well as the winding stator end-connections inductance cannot be computed directly using

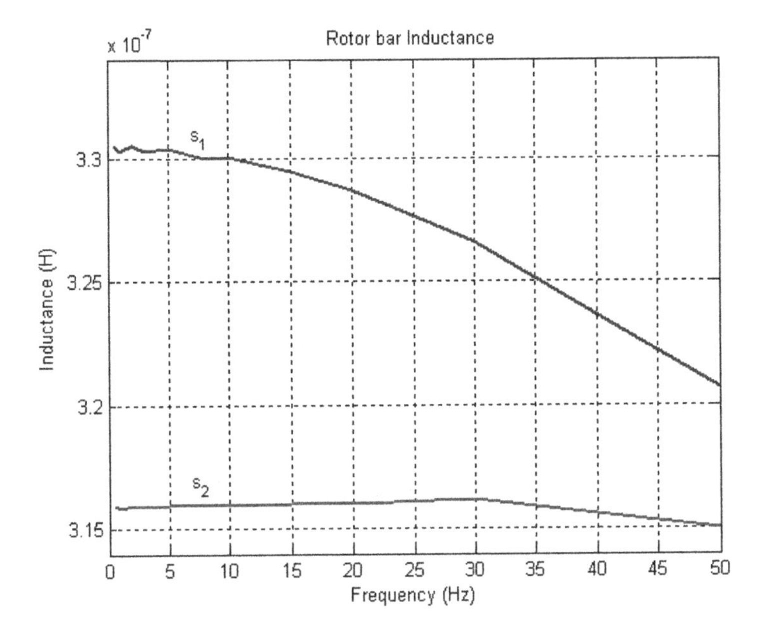

FIGURE 6.58 Rotor bar inductance vs. frequency: s_1—slot type 1; s_2—slot type 2.

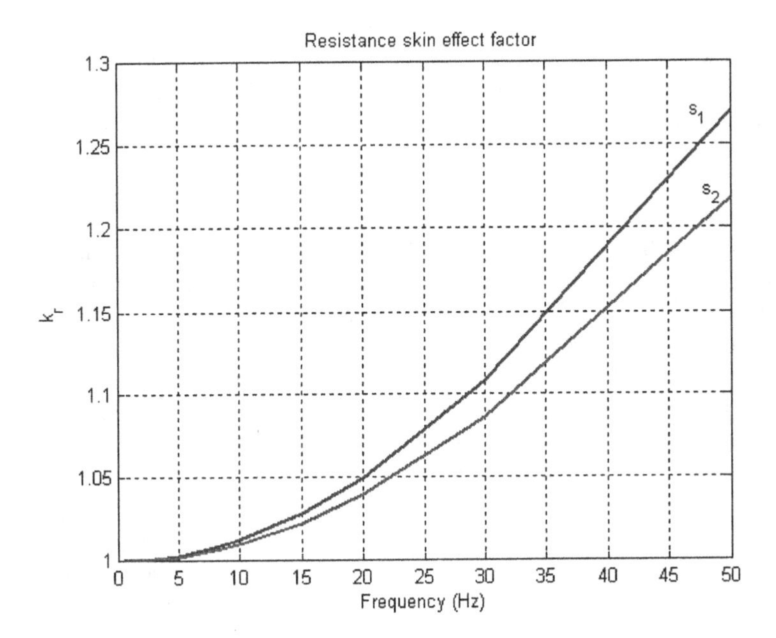

FIGURE 6.59 Resistance skin effect factor: s_1—slot type 1; s_2—slot type 2.

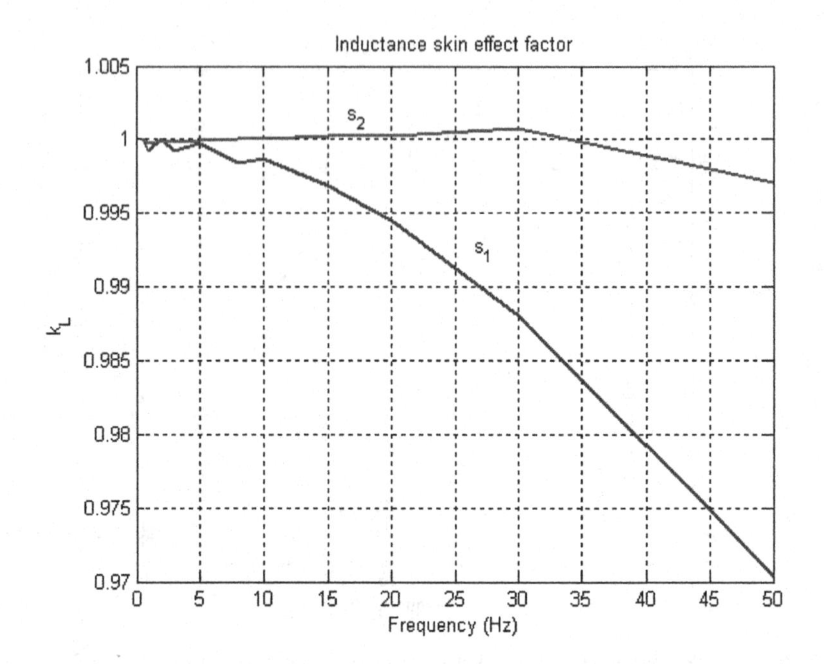

FIGURE 6.60 Inductance skin effect factor: s_1—slot type 1; s_2—slot type 2.

two-dimensional finite element methods. Analytical methods [11] could be used to compute the end ring parameters. The skin effect in the short-circuited end ring is small and the equivalent skin effect of the rotor winding is smaller than that computed for the rotor bar.

6.3.3 Summary

This paragraph shows how to use the FEM in order to compute the circuit parameters of induction machines. The magnetostatic model is used to determine the magnetization inductance and the stator slot leakage inductance. The a.c. model is used to determine the rotor slot leakage inductance and the skin effect factor of similar a.c. models could be used to find the skin effect in the stator winding. The circuit parameters are not constant: the magnetization inductance depends on the magnetization current while the bar resistance and rotor leakage inductance depend on the rotor (slip) frequency. An exhaustive finite element model of induction machines should consider stator windings and rotor bars connected with the external circuit. The torque vs. position in mechanical steady-state regime could also be computed. Torque pulsations and radial forces could also be computed in that case, but the computation effort is much larger. Numerous applications request only circuit parameters, or key performance indexes which could also be simply computed from magnetostatic and steady-state a.c. simulations [12,13].

REFERENCES

1. I. Boldea, S. Agarlita, L. Tutelea, and F. Marignetti, *Novel Linear PM Valve Actuator: FE Design and Dynamic Model*, Record of LDIA, 2007.
2. Opera 2D-Reference Manual, VF-07-02-A2, Vector Field Limited, Oxford, U.K.
3. Opera 2D-User Manual, Vector Field Limited, Oxford, U.K.
4. R. Ierusalimschy, L.H. de Figueiredo, and W. Celes, *Lua 5.1 Reference Manual*, Lua.org, ISBN 85-903798-3-3, 2006.
5. R. Ierusalimschy, *Programming in Lua*, 2nd edn., Lua.org, ISBN 85-903798-2-5, 2006.
6. I. Boldea and L. Tutelea, *Optimal Design of Residential Brushless D.C. Permanent Magnet Motors with FEM Validation*, Record of ACEMP, Bodrum, Turkey, 2007.
7. N. Bianchi, S. Bolognani, and F. Luise, Analysis and design of a brushless motor for high speed operation, *Energ. Convers. IEEE Trans.*, 20(3), 2005, 629–637.
8. V. Grădinaru, L. Tutelea, and I. Boldea, *2.5 kW, 15 krpm, 6/4 PMSM: Optimal Design and Torque Pulsation Reduction via FEM*, Record of OPTIM 2008.
9. C.B. Rasmusen, Modeling and simulation of surface mounted PM motors, PhD thesis, Institute of Energy Technology, Aalborg University, Denmark, 1997.
10. V. Zivotic, W.L. Soong, and N. Ertugrul, *Iron loss reduction in an interior pm automotive alternator*, *IEEE IAS Annual Meeting*, pp. 1736–1743, 2005.
11. I. Boldea and S.A. Nasar, *Induction Machine Design Handbook*, CRC Press, New York, 2nd edn., 2010.
12. L. Alberti, N. Bianchi, S. Bolognani, *Variable speed IM performance computed using FEM*, *IEEE Trans.* Vol I A-47, no 2, 2011, pp. 789–797.
13. D. M. Ionel, M. Popescu, *Ultrafast FEA of Brushless PM Machines Based on Space-Time Transformations*, IBID, pp 744–753.

7 Thermal FEA of Electric Machines

The thermal analysis of the electrical machine is very important because overpaying the maximum allowed temperature of the winding according to the insulation class will reduce the life time of the electric machine. Based on the Arhenius equation which regards the chemical reaction speed versus temperature, the life time will be reduced exponentially having a half-time constant for every 10-degree temperature overpaying. The insulation class refers to 25,000 hours for insulation life time. This means less than 4 years if the machine is running 24 hours per day, 365 days per year at full load and the winding reaches the maximum allowed temperature. According to the same rule, reducing the maximum temperature will increase the life time. For example, reducing the maximum allowed temperature with 20 degrees will increase 4 times the life time, at 100,000 hours. The life time for each insulation class versus temperature is presented in [1].

The machine temperature variation modifies the machine behavior. For example, for the induction motor the mechanical characteristic is changed: the starting torque is decreasing from about 64.7Nm at 20°C to 54.2Nm at 150°C for a 3kW, 4 pole motor [2]. The critical slip is also decreasing notably when the winding temperature is increasing in the same range [2]. The airgap flux density of PM machine is decreasing notably when machine temperature varies in a large range (50°C to 150°C). Consequently, the rated torque is decreasing (for example from about to 6.1Nm to 5 Nm which means 18%), while, in the field weakening speed region, larger torque could be obtained [2].

It is well known that the copper losses depend on the winding temperature but also the iron core losses depend on the temperature; considering this dependence will improve the thermal model [3].

7.1 THERMAL MODELS

An accurate thermal analysis is very complex and beside implies also free or forced fluid flows under a thermal gradient. Also, more simplified models were proposed in the literature: single-body thermal model, two-body thermal model, equivalent thermal circuit 2DFEM, 3DFEM, thermal FEM coupled with fluid flow.

7.1.1 The Single-body Thermal Model

The single-body thermal model is the simplest model and it rather estimates the frame temperature than the winding temperature. The frame maximum temperature should

DOI: 10.1201/9781003216018-7

be smaller than winding maximum temperature according to insulation class. The entire machine is considered as a single homogeneous body with the same area as frame surface. The frame over-temperature is computed with a single simple formula.

$$\Delta\Theta_F = \frac{P_{loss}}{k_h A_c} \tag{7.1}$$

where

P_{loss}—the machine total loss (W),
K_h—convection heat transfer coefficient (W/m²/C),
A_c—frame cooling surface (m²), and
$\Delta\Theta_F$—frame over-temperature (⁰C).

The frame over-temperature $\Delta\Theta_F$ is the difference between frame temperature Θ_F and ambient temperature in natural convection or cooling fluid temperature in forced convection Θ_A.

$$\Delta\theta_F = \theta_F - \theta_A \tag{7.2}$$

The heat convection coefficient is between 5-10 W/m²/C for air natural convection, 10-300 W/m²/C for air forced convection, 50-20000 W/m²/C for liquid forced convection [4]. The forced convection coefficient depends on the fluid speed. For air the convection coefficient is increasing linearly from about 14 to 16 W/m²/C when the speed increase from 0.5 to 6.5 m/s and then from 43 to 98 W/m²/C when air speed is increasing from 8 to 20 m/s [4 –Figure 4.4].

A simple thermal equivalent scheme, Figure 7.1, could be attached to this problem.

The steady-state solution is:

$$\Delta\theta_F = P_{loss} R_{\theta FA} \tag{7.3}$$

Where $R_{\Theta FA}$ is the equivalent thermal resistance between frame and ambient.

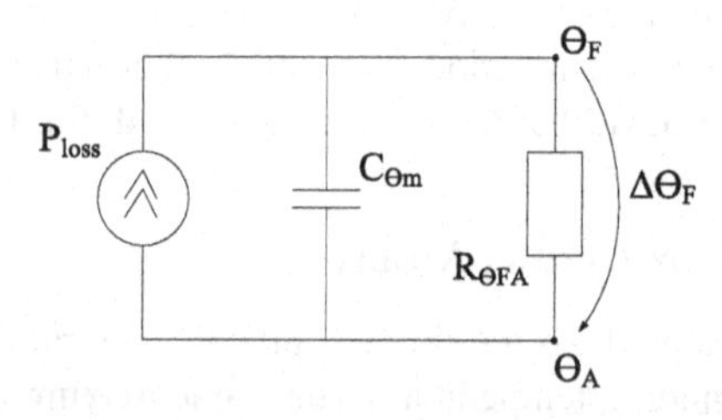

FIGURE 7.1 Single-body thermal model—equivalent scheme.

$$R_{\theta FA} = \frac{1}{k_h\, A_c} \tag{7.4}$$

The transient over-temperature variation is:

$$\Delta\theta_F(t) = P_{loss} R_{\theta FA}\left(1 - e^{-\frac{t}{\tau_\Theta}}\right) \tag{7.5}$$

$$\tau_\theta = R_{\theta FA}\, C_\theta \tag{7.6}$$

Where C_Θ is the thermal capacity of the body.

7.1.2 The Two-body Thermal Model

The two-body model is an improvement of the single-body model in order to approximate the winding temperature. There is assumed to exist two bodies: winding and iron core with their losses thermally coupled. The equivalent thermal scheme for the two-body thermal model is presented in Figure 7.2.

The transients equations are as follows:

$$C_{\theta cu}\frac{d\Delta\theta_{cu}}{dt} = -\frac{1}{R_{\theta cuF}}\Delta\theta_{cu} + \frac{1}{R_{\theta cuF}}\Delta\theta_F + P_{cu} \tag{7.7}$$

$$C_{\theta Fe}\frac{d\Delta\theta_F}{dt} = \frac{1}{R_{\theta cuF}}\Delta\theta_{cu} - \left(\frac{1}{R_{\theta cuF}} + \frac{1}{R_{\theta FA}}\right)\Delta\theta_F + P_{Fe} \tag{7.8}$$

The steady-state over temperatures are as follows:

$$\Delta\theta_{cu} = P_{cu}\, R_{\theta cuF} + \left(P_{cu} + P_{Fe}\right) R_{\theta F} \tag{7.9}$$

$$\Delta\theta_F = \left(P_{cu} + P_{Fe}\right) R_{\theta F} \tag{7.10}$$

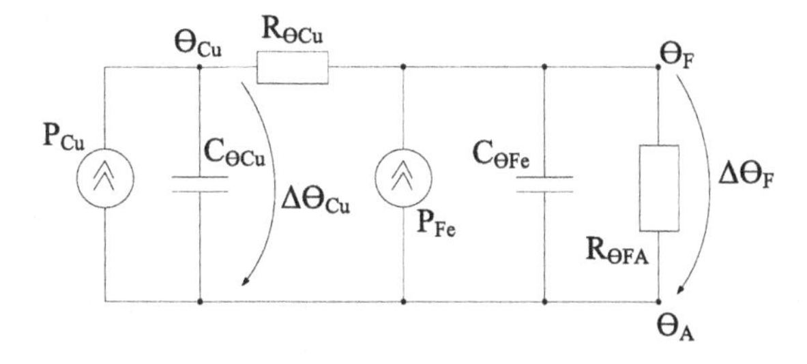

FIGURE 7.2 Two-body thermal model—equivalent scheme.

Numerical integration of differential equation system could be used to obtain transients solutions; for example, they could be analyzed by a simple Matlab Simulink model.

In electrical machines, the losses are dependent on the temperature. The copper losses temperature dependency is through the resistance and through the current, if constant torque is required when the current is increased to compensate the PM flux decrease with temperature. The maximum temperature is computed at rated current that is a constant value. Then only the dependency on the winding resistance remains. Rewriting the winding steady-state temperature considering the copper losses dependence on the winding temperature yields:

$$\Delta\theta_{cu} = P_{cu}(\theta_0)\left(1+\alpha(\theta_{cu}-\theta_0)\right)\left(R_{\theta cuF}+R_{\theta F}\right)+P_{Fe}R_{\theta F}, \tag{7.11}$$

where α is the electric resistivity temperature coefficient.

If the initial copper power losses are computed at the ambient temperature Θ_A then equation (7.11) becomes

$$\Delta\theta_{cu} = P_{cu}(\theta_A)\left(1+\alpha'\Delta\theta_{cu}\right)\left(R_{\theta cuF}+R_{\theta F}\right)+P_{Fe}R_{\theta F} \tag{7.12}$$

where

$$\alpha' = \frac{\alpha}{1+\alpha(\theta_A-\theta_0)}. \tag{7.13}$$

Finally, the over-temperature is

$$\Delta\theta_{cu} = \frac{P_{cu}(\theta_A)\left(R_{\theta cuF}+R_{\theta F}\right)+P_{Fe}R_{\theta F}}{1-\alpha' P_{cu}(\theta_A)\left(R_{\theta cuF}+R_{\theta F}\right)} \tag{7.14}$$

With the notation of winding over-temperature produced only by copper losses, $\Delta\Theta_{cu,pcu}$, the final over temperature will stabilize close to the initial one if

$$\alpha'\Delta\theta_{cu,pcu} \ll 1 \tag{7.15}$$

where

$$\Delta\theta_{cu,pcu} = P_{cu}(\theta_A)\left(R_{\theta cuF}+R_{\theta F}\right) \tag{7.16}$$

The motor over-temperature will never stabilize if the product α' $\Delta\Theta_{cu,pcu}$ is equal or larger than one due to a kind of positive feedback with unitary or supra-unitary gain.

The formula (7.14) is useful also in finite element steady-state analyses to avoid running repeatedly the FEM method to follow over-temperature weakly convergent series.

7.1.3 Equivalent Thermal Circuit (Thermal Network)

This model extends the two-body thermal model by dividing the machine in several components as frame, stator yoke, stator teeth, end coil, slot coil, rotor components with their losses coupled by thermal resistance [4]. Adding the thermal capacity of each component in the nodes allows to compute transient temperatures evolution.

7.2 THERMAL ANALYSIS OF ELECTRIC MACHINES BY FINITE ELEMENT

The thermal analysis of electrical machine by finite element is presented step by step on a case study, by FEMM 4.2 [5], free software. A simple thermal problem could be solved by using only the FEMM GUI (graphic user interface) as it is presented in FEMM Heat Flow Tutorial [6] but, for real problems the scripting files should be used, for a better productivity. Interfaces to Octave & Matlab [7], Scilab, Mathematica and Python are available. In this chapter, the MATLAB scripting files are used.

Stating the finite element analysis assumes that we have all geometric dimensions, power losses distributions, and material thermal features. Only the convection heat exchange between electrical machine and environment is considered in the following example. So, the environment temperature and surface transmission coefficient should be known.

The finite element analysis has three distinct steps:

Describing the problem,

Solving the field partial derivative equations,

Extract and interpret the values of features interest.

In this chapter, only the first and third step will be presented because solving the field equation is embedded in the FEM software.

7.2.1 Equivalent Coil Conductivity—Simplified Thermal Model of the Slot

In small electrical machines, the round conductors are distributed in the slots with a filing factor around 0.4. Winding impregnation is used in order to improve the heat transfer from copper conductors to the slot wall. In the simplified thermal model of the slot a single-coil body with equivalent conductivity is considered in the core slot instead of considering, each conductor and its impregnation space. The equivalent coil thermal conductivity without of considering, conductor insulation [8] is

$$\lambda_{coil} = \lambda_{imp} \frac{(1+k_f)\lambda_{cu} + (1-k_f)\lambda_{imp}}{(1-k_f)\lambda_{cu} + (1+k_f)\lambda_{imp}}$$

where

λ_{coil}—the equivalent coil conductivity on transversal direction,
λ_{cu}—copper conductivity,
λ_{imp}—thermal conductivity of impregnation material, and
k_f—slot filling factor.

7.2.2 Boundary Conditions

The computation effort could be reduced by considering the thermal field symmetry. In the following example a single machine pole is considered, and thus the symmetric boundary condition should be applied on the cutting surface. On the outside part, the convection heat exchange boundary condition is set. The outer surface could be the external lamination smooth surface (also a smooth frame), Figure 7.3a, but in practice most frequently a frame with cooling blades, Figure 7.3b, is used to increase the

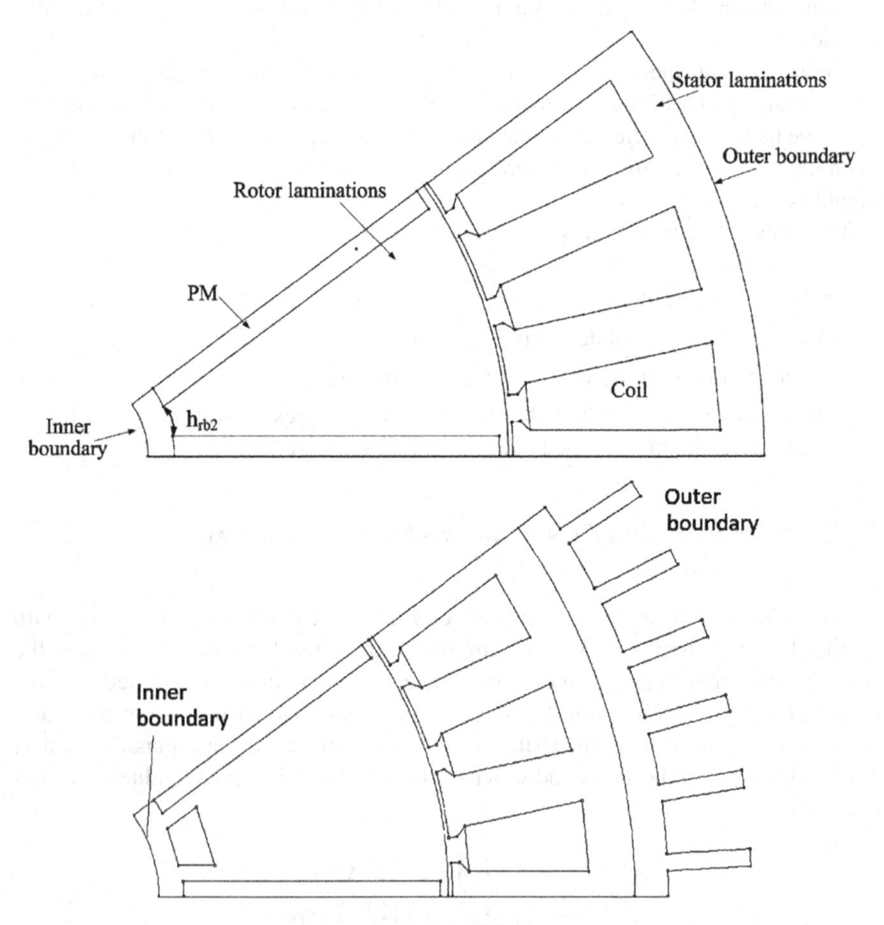

FIGURE 7.3 Motor structure and boundary condition: (a) lamination core smooth surface, (b) frame with cooling blades.

cooOling surface. In general, the frame axial length is larger than stack lamination but this cannot be considered directly in the 2D simulation. Eventually, in the simulation, the blades height could be larger than real values to get the equivalent cooling surface. The inner surface border condition should consider the heat transferred to the shaft and then in the axial direction to the ambient trough the shaft. For a long machine, the axial transferred heat through the shaft could be negligible. In real machines, there is a heat exchange also on two frontal areas (shield area), which could not be considered directly in a 2D FEMM simulations. In the present study case, the inner surface is small, and complex heat transfer trough the shaft is replaced with a convection transfer.

7.2.3 The Input Data

The input data consist in all dimensions necessary to build the FEM model, the materials features, and the power losses. These data are available after the electromagnetic design. In our case study, the 5kW spoke PM motor at 2400 rpm designed in chapter 9 (section 9) is considered. So, all dimensions are taken from that case study. The data are introduced in the model as input files that contain also running specifications besides the problem describing data. The rated power and rated speed are not necessary for the thermal model, but they could be introduced as comments for user's information. An example of an input file is presented below:

```
%General information
poles=10; %—numbers of poles
nphase=3;
q1=1;

%Main geometric information
Dro=101; %mm Rotor outer diameter
lstack=75.5;%mm axial length
Dso=145; %mm Stator outer diameter
Dri=32; %mm Inner rotor diameter
hag=0.5; %mm—Air-gap height
hpm=3.45; %mm—height of the permanent magnet

%Stator core information
wtooth=5.9; %mm tooth width
hsy=5.3; %mm stator yoke height

%Stator slot information
hs4=1; % mm—Stator tooth pole tip height
hs3=1; % mm—height for wedge places
sMs=3; % mm -stator month opening

%Rotor information
hry=3; %mm—rotor yoke
hrb=1; %mm—rotor bridge over PM
```

```
hrb2=1.8; %mm—base rotor bridge
hrb2max=2.5; %mm—maximum value for rotor bridge
hrbarrier=4.5; %mm—height of the rotor barrier
dhpm=0.2; %mm both side distance between PM and lamination on PM high
dlpm=1; %mm both side distance between PM and lamination on PM length

%Coil data
nce=1; %number of elementary conductors in parallel
dco=1.29; %mm diameter of elementary conductor
sb=29; % turns per coil

%Convection heat exchange data
Tamb=40; % deg.C—temperature of cooling fluid
alpha_t=14.2; %W/m^2*deg. thermal transmission coefficient
kv=5; % ventilation factor
%Frame dimension
hframe=4; %mm—frame height -0 means no frame
Nblades=60; %blades number
hblade=10;     %mm blades height
gblade=2;      %mm blades thickness

%Losses
Pcu=220; %W copper losses
Pfe=50.62; %W iron losses
slotVol=7.098e-6; %m^3
coreVol=400.86; %m^3

%Used material
stcore_mat='M19'; % stator laminations features
pm_material='VAC677'; % permanent magnet features
%pm_material='FERRITE_415'; % permanent magnet features
shaft_mat='Steel_1006';
wire_type='Cu'; % allowed: Cu, Al
imp_mat='epoxy'; % impregnation available "epoxy', 'epoxy24quartz
frame_mat='Al2014'; %available 'castIron', 'Al2014'

%simulation settings
simrange=1; %n>0 number of simulated poles; n=-1—single slot
analysesType='c0'; % s -steady state, t transients,
            % c cooling, c0 with stopped fan
%Transient regime settings
ftime=7200; %s -final time
tstep=120; %s -step time

%Output files prefix
fem_file='mth2p5'
```

7.3 STEADY-STATE SIMULATION RESULTS

In steady-state simulation, the temperature over entire simulated surface is available in Kelvin degrees.

The temperature distribution is heavily depending on the slot impregnation material and on the cooling condition. A few cases are presented. For a smooth surface the ventilation factor, k_v, was increased at 5 and that means about 13m/s air flow (according to [4] , page 176. Figure 4.4), in order to keep the conductor's temperature in 'F' insulation class (Figure 7.4).

Adding an aluminum frame with 60 blades with 10 mm height allows to reduce the ventilation factor to 1.5; that means an air flow speed below 7m/s. The wire maximum temperature could be decreased further with about 7 degree if a better impregnation material is used, for example, an epoxy with quartz that increases the thermal conductivity at 0.6 W/(m K) from an usual epoxy with 0.21 W/(m k). This way the difference between peak temperature and average temperature (that could be the measured stator resistance) is decreased.

The comparative temperature distribution is shown in Figure 7.5.

In previous simulations, a perfect impregnation was considered. Sometimes the impregnation is not perfect and thus the coil temperature increases. In the following example, Figure 7.6 and machine M4 from Table 7.1, the worst case is presented: no coil impregnation. The peak temperature was increased at 247.3 C degree.

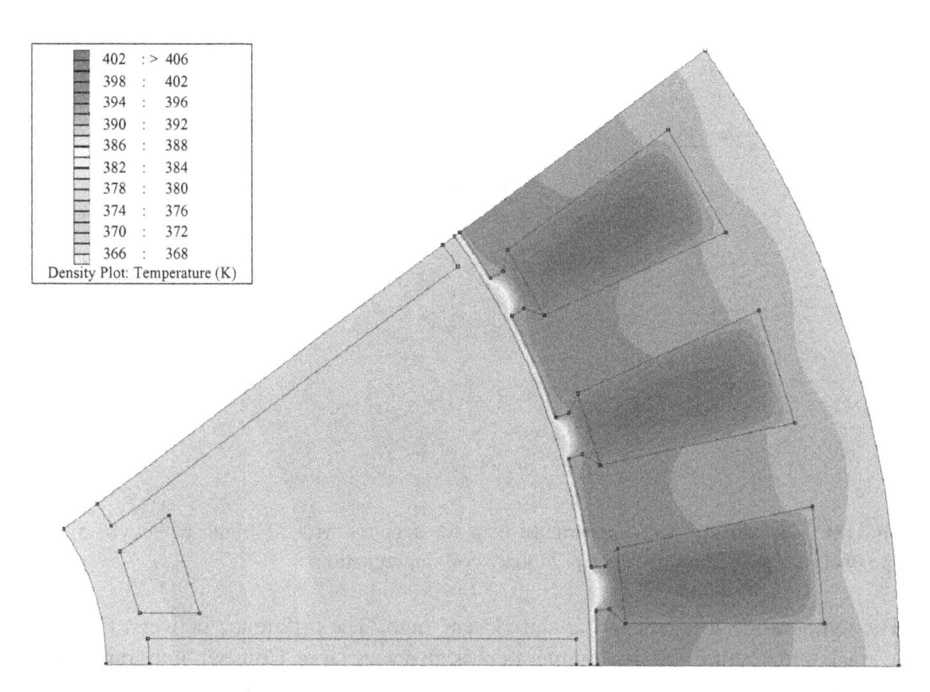

FIGURE 7.4 Steady-state temperature map for a motor without frame with kv=5, epoxy coil impregnation.

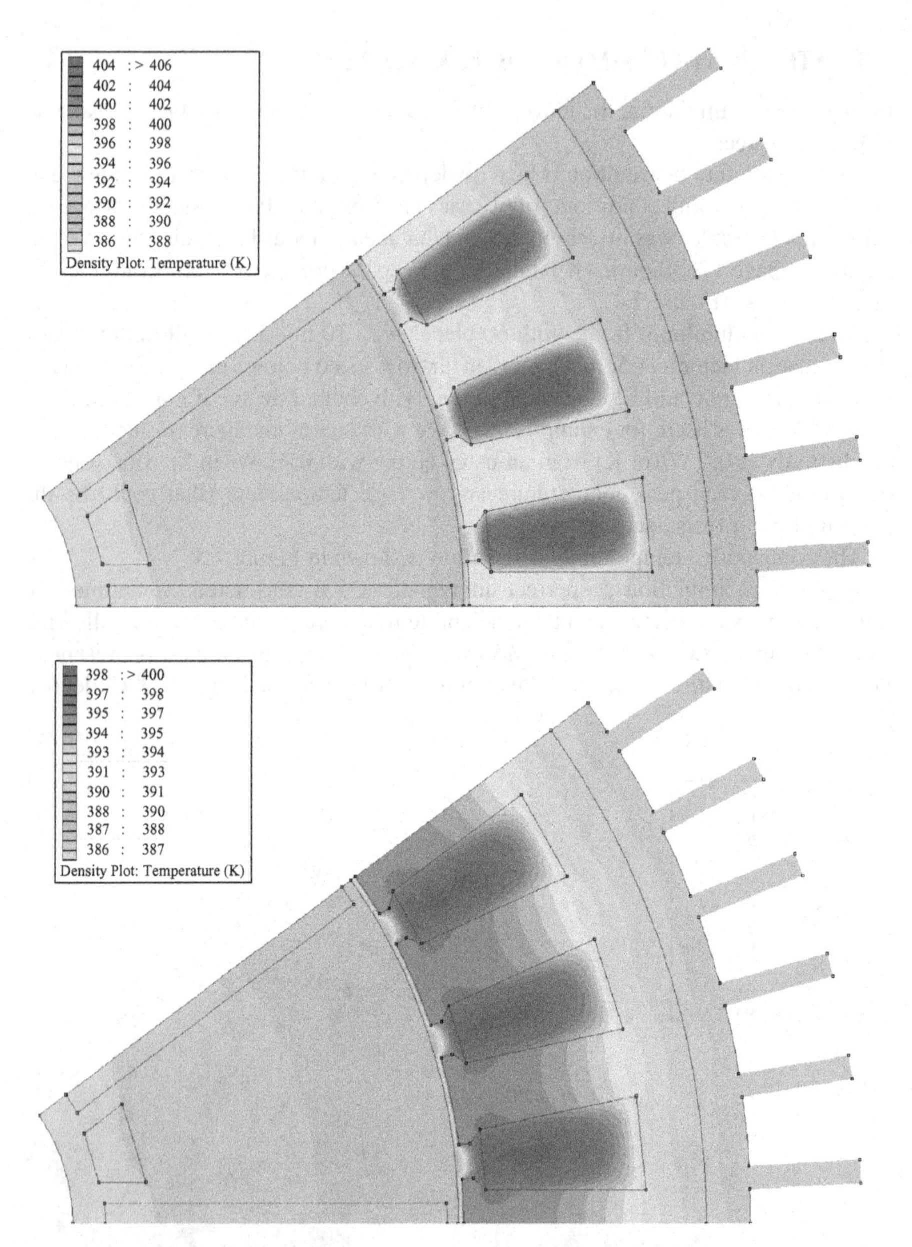

FIGURE 7.5 Steady-state temperature map for a motor with Al frame with kv=1.5, (a) epoxy coil impregnation, (b) epoxy+24quarz coil impregnation.

The temperature gradient in the slot is very high. The difference between the coil average temperature and iron temperature is also very large. Coils without impregnation should never be used in practice, but the simulation shows that a larger difference between coil average temperature and frame temperature (only this could be

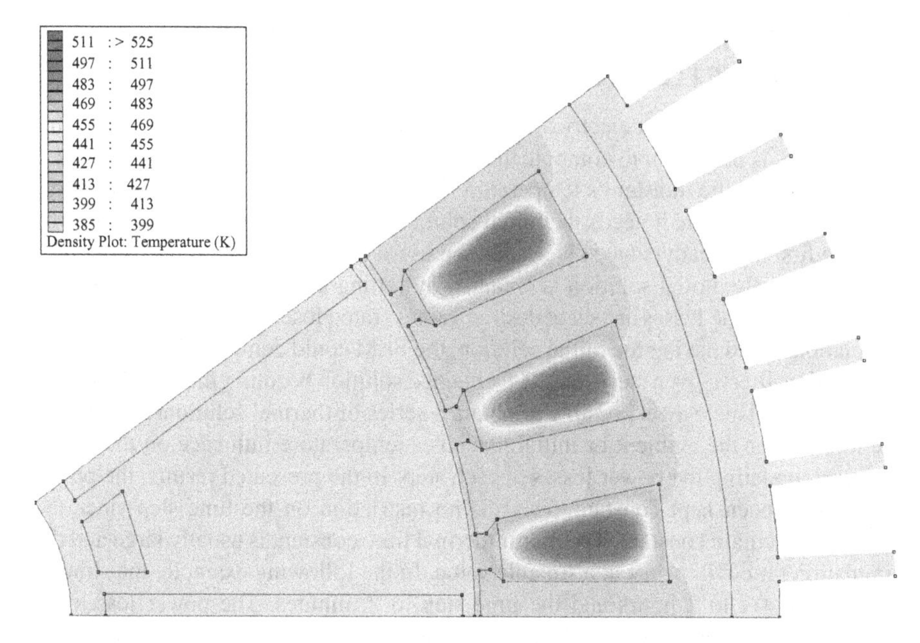

FIGURE 7.6 Steady-state temperature map for a motor with Al frame with kv=1.5, without impregnation (do not use!).

TABLE 7.1
Steady-state Temperature at Rated Current for the Ambient Temperature 40°C

	M1 No frame epoxy	M2 Al frame epoxy	M3 Al Frame epoxy +quartz	M4 Al frame Air Never use!	M5 Cast iron epoxy	Units
Tcu1—copper peak	133.2	133.4	126.2	247.3	134.4	°C
Tcu–copper average	127.7	127.9	123.9	186.8	128.9	°C
Tfe1—iron core peak	122.9	123.6	123.3	123.7	124.6	°C
Tfe—iron core average	120.7	120.7	120.6	120.8	121.8	°C
Tframe1—frame peak	–	117.8	117.8	117.7	118.8	°C
Tframe—frame average	–	117.5	117.5	117.5	118	°C
Simulation time	2.35	8	4.2	4.2	2.8	Sec.

measured or estimated in practice without temperature sensors in the machine) could indicate problems with slot impregnation.

The temperatures of motor with cast iron frame, M5 in Table 7.1, are about only 1 degree larger than the same configuration but with aluminum frame M2.

Tcu1 is measured in the middle of the coil, Tfe1 near to the airgap (at h3 + h4 = 2 mm from the airgap) and Tframe 1 in the frame near to the stator core.

7.4 THERMAL TRANSIENT ANALYSIS OF ELECTRIC MACHINES BY FINITE ELEMENT

In many applications, the electrical machines should run in overload for short time. Therefore, it is important to compute the temperature in transients. The FEMM 4.2 is able to compute the transient's temperature by solving the field differential equation. For a unique solution, it needs the initial solution. The initial solution is obtained by running first the steady-state solver and then use it as initial solution. Considering a cold starting, the initial solution is obtained by setting the copper and iron losses at zero. After that the losses are set at desired values (rated loses values in the following simulations), and having the initial solution, the FEM could compute the temperature distribution after a given step time. The obtained solution becomes the initial solution for next step. This way is possible to obtain a series of thermal solutions. It is possible to consider the ambient or initial machines temperature influence on the power losses by updating the power losses at each step. In the presented results, the power losses have been kept constant. There is no restriction on the time step since the parameters remain constant. The motor thermal time constant is usually large and the simulation time will be set at a suitable value. In the following example, the simulation time is set to 2 hours and the time step to 2 minutes. The power losses are assumed to be constant during to the time step. The temperatures variation during thermal transients for a high ventilation factor are shown in Figure 7.7 for no frame motor, Figure 7.7a, respectively, for an aluminum framed motor, Figure 7.7.b. It could be noticed that maximum winding temperature is below to 80°C. The thermal time constant was decreased (thermal frame ambient equivalent resistance is decreased) and the temperature reached steady-state values. The same aluminum-framed machine transients with reduced ventilation factor is shown in Figure 7.8.a. The temperature is larger and also the thermal time constant is larger. A little larger time constant could be observed for the cast iron frame machine. A comparison of the temperature after 2 hours with the steady-state value, Table 7.2 shows how far is the temperature to steady state.

The transient temperature of a machine without coil impregnation is shown in Figure 7.9. The coil Tcu1 reach 150 C (155 C is the maximum allowed for the F insulation class) in less than 5 minutes.

The transient's simulation allows to compute temperature evolution for any regime. For example, considering as initial solution the steady-state solution and then setting the power losses at zero the thermal shutdown transient regime is simulated. For no frame motor, the temperature goes close to the ambient temperature after 2 hours if the ventilation keeps running while if the ventilation is stopped the thermal time constant is increasing, Figure 7.10. The difference between temperature decreases with ventilation and without ventilation is smaller for the framed motor, Figure 7.11. In the steady-state regime, there is a difference of several degrees between coil, iron, and frame temperatures. The differences are reduced quickly in a few minutes and then all temperature curves evolve close to each other.

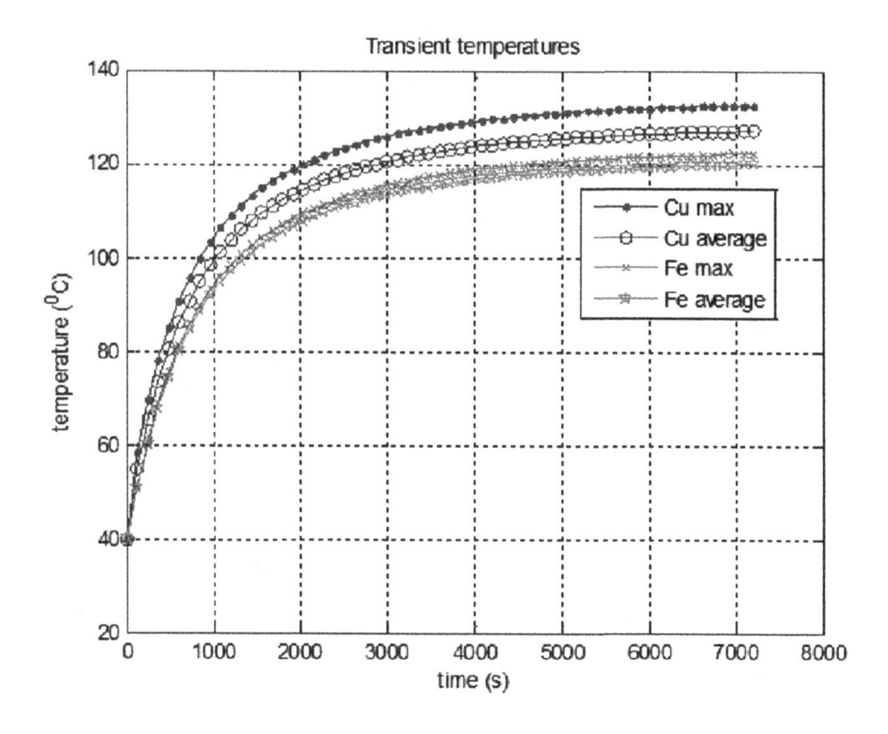

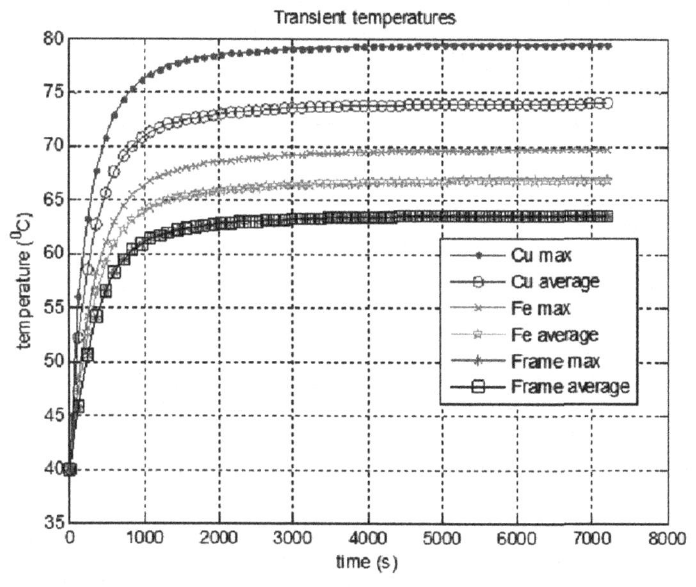

FIGURE 7.7 Temperature transients for high ventilation k_v=5 (a) without frame, and with Aluminum frame, (b).

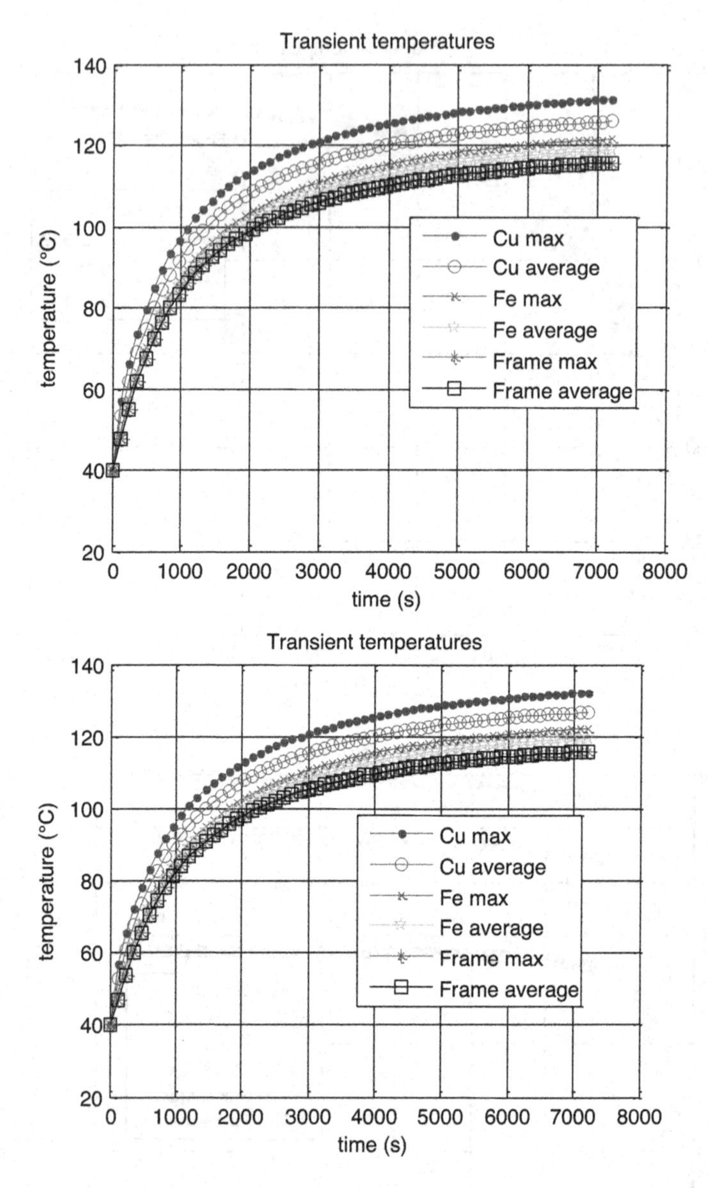

FIGURE 7.8 Temperature transients for moderate ventilation factor $k_v = 1.5$: (a) Aluminum frame, (b) Cast Iron frame.

The temperatures reached after 2 hours, Table 7.3, show again that no-frame motor thermal time constant is smaller but it increases more when the ventilation is shut down.

The running regime with different power losses and stopped time could be combined freely to obtain desired regime.

TABLE 7.2
Temperatures at 2 hours After Start

	M1–no Frame		M2 Aluminum Frame		M5 Iron Cast Frame		Units
	2 Hours	Steady State	2 Hours	Steady State	2 Hours	Steady State	
Tcu1—copper peak	132.6	133.2	131.9	133.4	132.1	134.4	°C
Tcu—copper average	127.1	127.7	125.9	127.9	126.7	128.9	°C
Tfe1—iron core peak	122.3	122.9	121.5	123.6	122.3	124.6	°C
Tfe—iron core average	120.1	120.7	118.8	120.7	119.6	121.8	°C
Tframe1—frame peak		–	115.9	117.8	116.6	118.8	°C
Tframe—frame average		–	115.7	117.5	115.9	118	°C
	117.45	2.35	139.94	8	139.72	2.8	Sec.

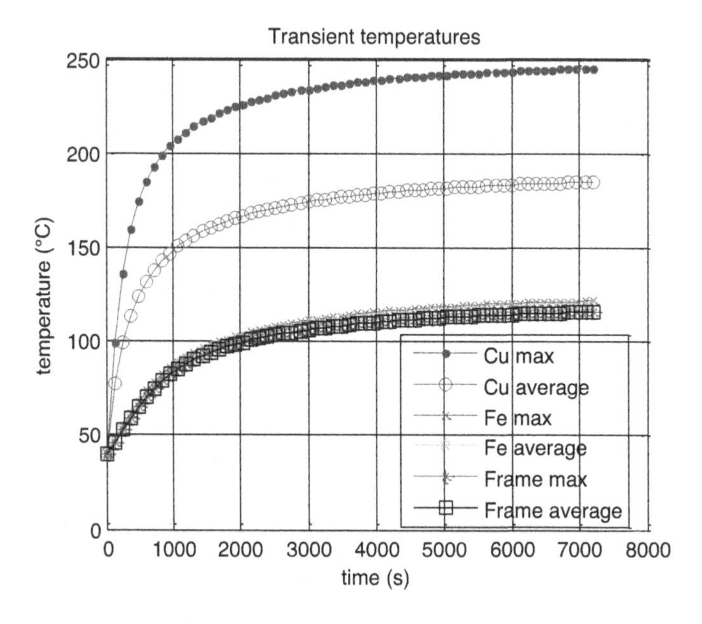

FIGURE 7.9 Transient temperature without impregnation.

The application are available online with the main file 'thermal.m' and the input file example 'mth5kWp10.m'.

7.5 SUMMARY

Chapter 7 introduces the thermal FEA of electric machines using the free FEMM4.2 software via a case study with a MATLAB program available online.

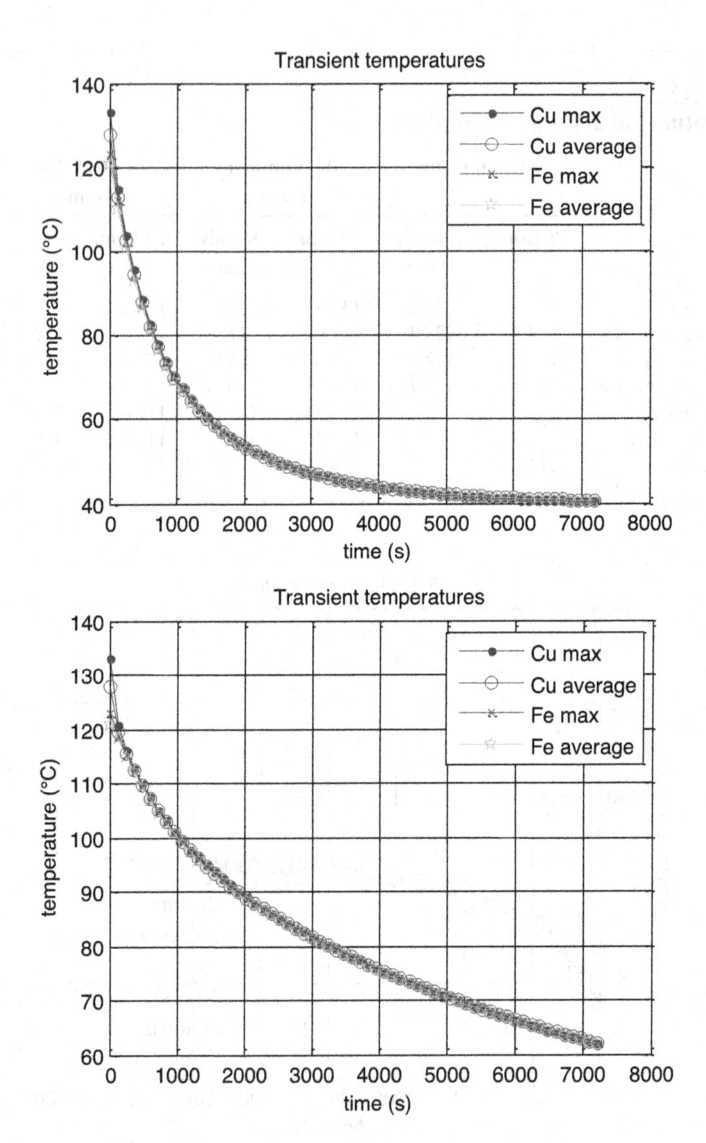

FIGURE 7.10 Cooling after shutdown of no frame machine (M1): C

It includes the presentation of:

- Single-body thermal model
- Two-body thermal model
- Equivalent thermal network
- FEA thermal electric machine analysis in terms of problem description and the extracting and interpretation of features of interest (boundary conditions, input

data file, steady state, and transients FEA thermal simulation of an electric machine with sample results from the dedicated computer program (thermal.m) available online.

Numerous important insights valuable to thermal design of electrical machines are presented in this chapter in a practical manner for easy assimilation by the new reader.

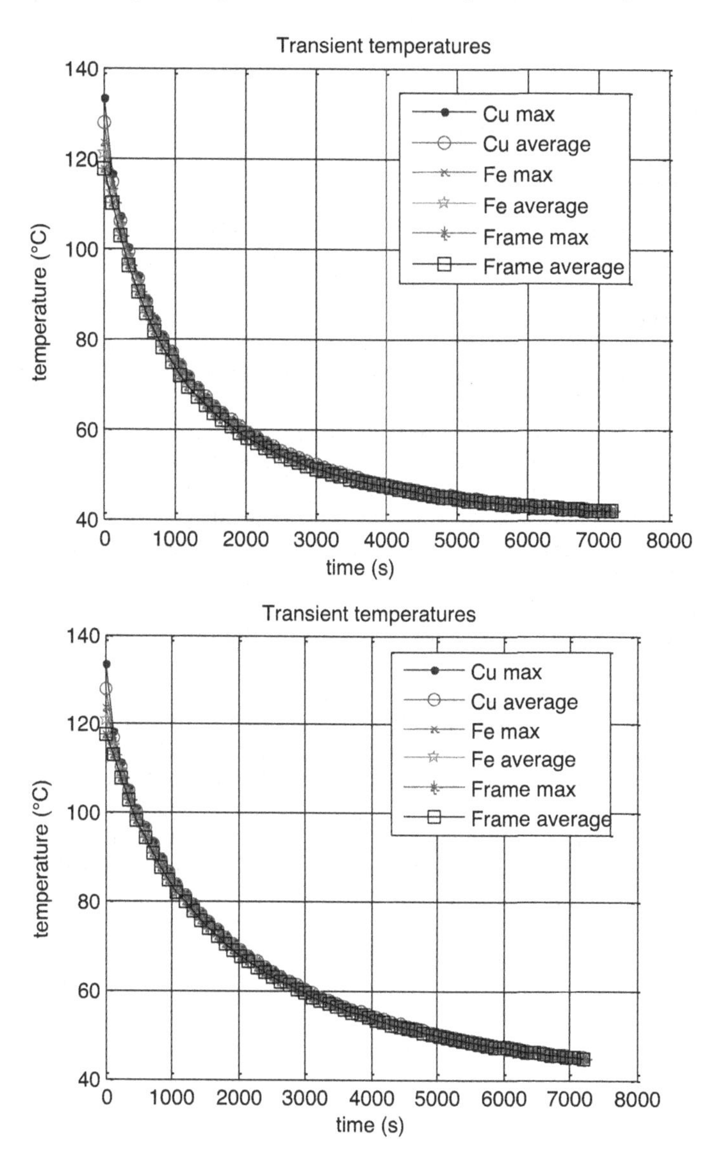

FIGURE 7.11 Cooling after shutdown of Aluminum frame machine (M2): Cooling after shutdown of no frame machine (M1).

TABLE 7.3
Temperatures at 2 Hours After Shutdown, –ambient Temperature 40°C

	M1—No Frame		M2 Aluminum Frame		Units
	Running Ventilation Kv = 5	Ventilation Shutdown Kv = 1	Running Ventilation Kv = 1.5	Ventilation Shutdown Kv = 1	
Tcu1—copper peak	40.6	62	42	45	°C
Tcu—copper average	40.6	62	42	45	°C
Tfe1—iron core peak	40.6	62.1	42	45.1	°C
Tfe—iron core average	40.6	61.9	41.9	44.9	°C
Tframe1—frame peak	–		41.8	44.8	°C
Tframe—frame average	–		41.8	44.8	°C
Simulation time	107.91	107.23	130.61	130.87	s

REFERENCES

1. https://www.marathongenerators.com/generators/docs/manuals/thermal-life.pdf (accessed 30 11.2020).
2. Luigi Alberti, Nicola Bianchi, Peter Baldassari, Ren Wang, Thermal Assisted Finite Element Analysis of Electrical Machines, *18th International Conference on Electrical Machines*, 2008, paper ID 900.
3. Shaoshen Xue, Z. Q. Zhu, Yu Wang, Jianghua Feng, Shuying Guo, Yifeng Li, Zhichu Chen, Jun Peng, Thermal-Loss Coupling Analysis of an Electrical Machine Using the Improved Temperature-Dependent Iron Loss Model, *IEEE Transactions on Magnetics*, 54, 11, November 2018, paper 8105005.
4. M. Rosu, P. Zhou, D. Lu, D. Ionel, M. Popescu, F Blaabjerg, V. Rallabandi, D. Staton, *Multiphysics Simulation by Design for Electrical Machines, Power Electronics, and Drives*, IEEE Press-Wiley, 2018.
5. https://www.femm.info/wiki/HomePage [last accessed 21.12.2020].
6. David Meeker, Finite Element Method Magnetics: Octave FEMM User's Manual, March 17, 2018, https://www.femm.info/wiki/HeatFlowTutorial [last accessed 21.12.2020].
7. Dawei Liang; Z.Q. Zhu; J.H. Feng; S.Y. Guo; Y.F. Li; J.Q. Wu; A.F. Zhao, Influence of Critical Parameters in Lumped Parameter Thermal Models for Electrical Machines, 2019 *22nd International Conference on Electrical Machines and Systems (ICEMS)*.

8 Optimal Electromagnetic Design of Electric Machines

The Basics

8.1 ELECTRIC MACHINE DESIGN PROBLEM

The design of electric machines consists of two distinct stages: the dimensioning calculus stage and the verification calculus stage.

Dimensioning includes the choosing of materials, the topology, and the calculus of all geometrical parameters starting from the machine design specifications. Because the relationships between the machine performance indexes and the geometrical dimensions are rather involved (and nonlinear), it is impossible to produce a set of inversion functions to distinguish the latter from the former:

$$p_i = f_i(X) \quad i = 1, n \tag{8.1}$$

where

p_i is the performance index from the design specifications,
X is the geometric variable and material properties vector, and
f_i is an involved (nonlinear) function.

In general, there is more than one variable vector X that satisfies p_i. Also, for some design specifications there may be no design solutions (e.g., unity power factor for an induction motor with a certain type of electric machine). By classical nonoptimal design, a possible solution X is found that satisfies most if not all design specifications. The optimal design selects, after mastering a good part of all feasible variable values, the solution that satisfies a given optimum criterion (e.g., minimum initial cost, weight, or maximum efficiency).

As the inverses of functions f_i are not admissible to determine the vector of geometric variables, the electric and magnetic stresses are chosen (current density, A/m^2; linear current loading, A turns/m; flux density, T; shear rotor stress N/m^2; etc.) based

DOI: 10.1201/9781003216018-8

on past experience. Also some detailed slot geometry parameters, such as slot opening at the airgap or slot wedge thickness, are given. Then, using the laws of electricity (Maxwell equations) and geometrical relationships, all geometrical parameters are calculated. As the electric and magnetic stresses have been chosen arbitrarily, there is no guaranty that the geometrical variable vector, X, will satisfy the design specifications, p_i. Consequently, the verification of performance is mandatory as part of the design process.

The verification calculus includes

- Electromagnetic verifications
- Thermal verifications
- Mechanical verifications

In general, to fulfill the performance p_i, the dimensioning calculus has to be redone a few times. So the geometrical dimensioning is found iteratively even within a classical design. The number of iterations depends on the strategy chosen to change the electric/magnetic stresses after each verification calculus routine. To reduce the number of iterations, a correlation between electric/magnetic stress changes and geometrical parameter variations has to be established. However, the number of iterations remains high, except for a highly experienced designer who "feels the way the design goes."

The stages of a classical design approach are shown in Figure 8.1. The dimensioning and verification calculi are computerized but the selection of electric/magnetic stresses and of initial geometric parameters is still done by the designer. If the design specifications are too restrictive, finding a solution can either be very tedious or even impossible. On the contrary, for low-level specification machines, configurations of lower cost or of good performance may be obtained at the same cost.

The stages of an optimal design are shown in Figure 8.2. The optimal design specifications contain, in general, minimal performance requirements and an objective function. The initial electromagnetic stresses and geometrical parameters are either specified by the designer or randomly generated through a dedicated routine within a specified domain. The dimensioning and parameter calculi are done as in classical designs, but, additionally, the objective function is also calculated. Then the electromagnetic, thermal, and mechanical verification calculi are performed, and for every performance index that is not suitable, a penalty is added to the objective function. If the penalty coefficients are high enough, then their requirements will be satisfied in the final design solution. Within this chapter, the thermal model is largely simplified while mechanically only the minimum shaft diameter is verified.

The detection of the minimum objective function value and the selection of the new values of electromagnetic stresses and initial geometrical variables before every iteration cycle are heavily dependent on the objective function optimum search method, as shown in the following text.

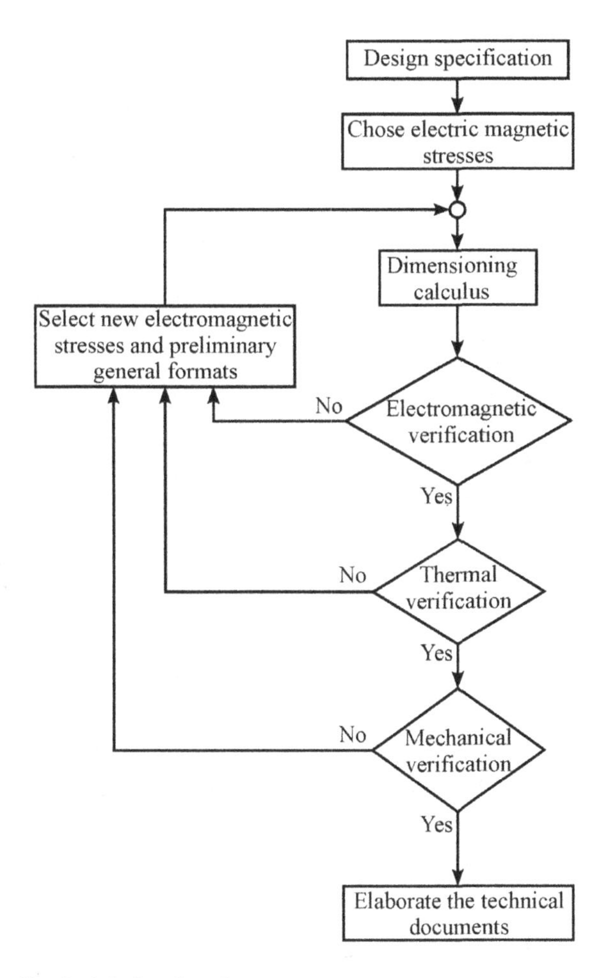

FIGURE 8.1 Classical design flowchart.

8.2 OPTIMIZATION METHODS

There are many definitions for optimization from the viewpoints of mathematics, science, and economics. A concise definition is given in [1]: "Optimization is the act to obtain the best results under given circumstance."

For electrical machines, the optimization could be applied in the machine (drive) design process in order to obtain the best machine (drive) performance according to a given criteria with minimum production cost, or it can be used in the drive/generator control. The first problem, designing and manufacturing the machine, is a large-scale optimization process. The optimization process has sense only associated with an optimization criterion. The optimization criterion for electric machine designs could be to reduce the production cost and to improve the electric machine

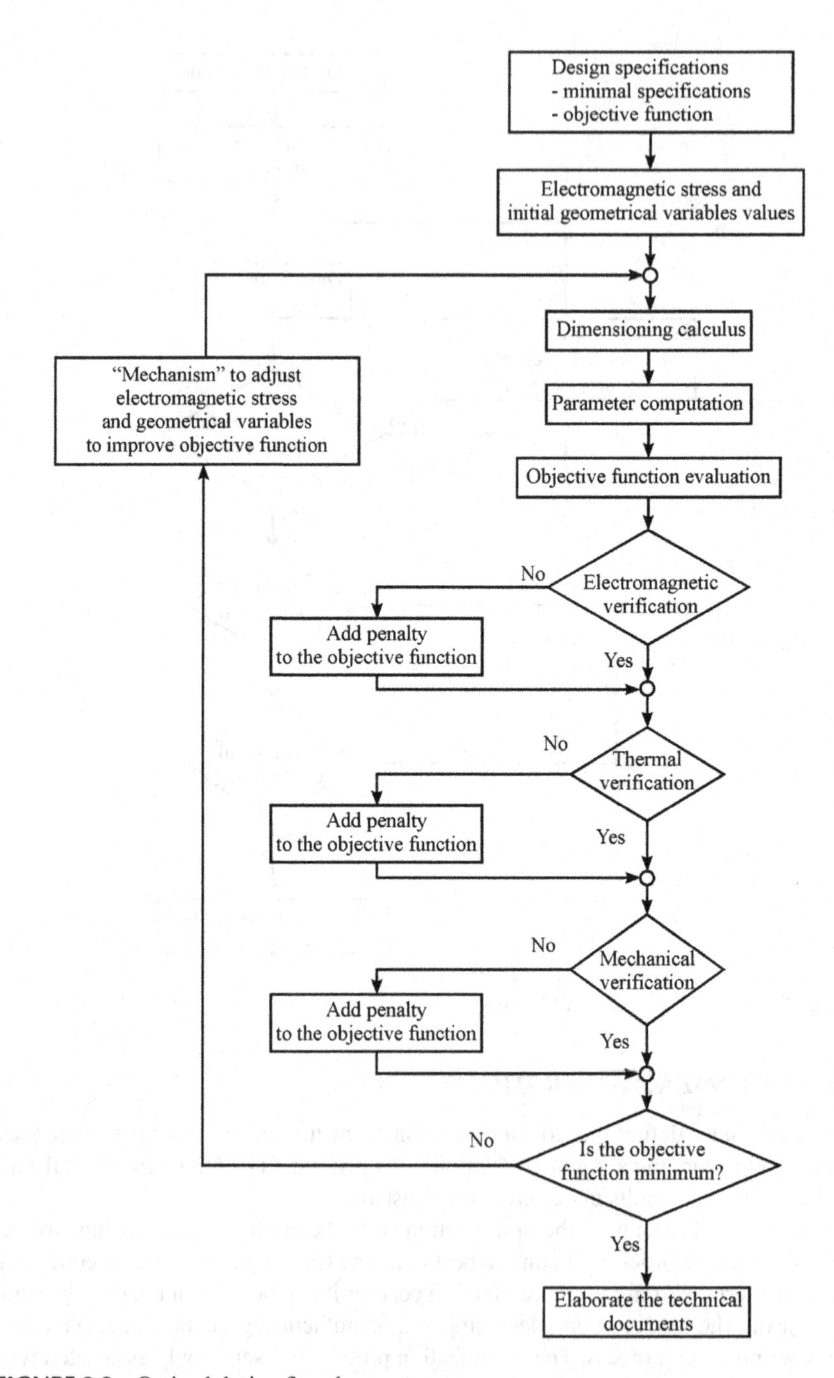

FIGURE 8.2 Optimal design flowchart.

performance, or to do both simultaneously. The production cost has a complex structure containing

- Materials used to manufacture the electric machine, which could be divided into active materials (e.g., stator and rotor coils, magnetic soft material cores, and permanent magnets) and passive materials (e.g., bearings, shafts, frames, ventilation (cooling) systems, winding terminals, and terminal boxes)
- Materials and energy used during electrical machine manufacturing
- Production facilities costs
- Designing, including optimization costs
- Promoting costs, marketing costs, etc.
- Loss capitalized and maintenance costs

The production cost optimization is a complex problem that entails managerial and engineering decisions. This book deals only with the active material cost optimization, which is a small but important brick in the global cost reduction. The influence of production facilities on active material costs is considered through technological limitations and technological factors that are constant inputs in active material optimization algorithms. For example, the slot-filling factor, which plays an important role in machine performance such as torque per volume (torque per unity mass), depends on the windings technology. The slot-filling factor is usually smaller than 0.35 for automatic winding machines for a series production but it could rise up to 0.6 for a highly skilled handworker. Increasing the winding slot-filling factor reduces the cost of the active material but it needs better production facilities. The optimization process produces new geometrical dimensions of magnetic core laminations. Introducing a new lamination geometry in production would be costly because it would require a new punching machine. It will be necessary to study how much the new machines would cost and how much time it would take to make returns on the investment. The optimal laminations dimension with the same criterion (let us consider, for simplicity, only the active material costs) depends on the prices of the active materials (i.e., copper, iron, aluminum, and permanent magnet). The geometric dimensions of magnetic laminations are changing with time due to the high dynamics of the changing prices of materials. The new optimized machine will be introduced in production only when the active material cost reduction is large enough to compensate for the new facilities cost, for a sustained period of investment recovery. A market price prediction is also useful for this purpose.

Often the minimum cost of a machine does not imply an optimum machine for the user. Users have other criteria, such as good efficiency and small initial price, by which to evaluate the optimum design of an electric machine. These two requirements may seem contradictory but there is an optimum somewhere between maximum efficiency and smaller machine cost. In many applications, the reliability of electric machine is more important than its cost.

The optimal design in mathematical formulation consists in the minimization of objective function, f_{ob}, which depends on optimization variable vector, X, and a set of constraints, ($g_i(X)$—inequality constraints and $h_i(X)$—equality constraints). Usually, the optimization variable domain is lower and upper bounded:

$$\begin{aligned} &\text{find } \bar{X} \text{ to minimize } f_{ob}\left(\bar{X}\right) \\ &g_i\left(\bar{X}\right) \le 0,\ 1 \le i \le m \\ &h_i\left(\bar{X}\right) = 0,\ \ 1 \le i \le k \\ &x_{i\ \min} \le x_i \le x_{i\ \max},\ 1 \le i \le n \end{aligned} \tag{8.2}$$

In electric machine design, the objective function could be the cost of active materials, the weight of active materials, the machine energy efficiency, or a combination of these. The usual constraints are machines over-temperature smaller than an acceptable temperature value, torque or power larger than the required value, starting current smaller than a maximum acceptable value, torque pulsation smaller than an acceptable value, and technological limitations. The objective function, f_{ob}, and constraint functions, g_i and h_i, are complex functions and do not usually have analytical expressions for their derivatives. Thus, analytical methods such as augmented Lagrange multipliers method cannot be applied. The equality constraints could be eliminated by choosing adequate optimization variables. Sometimes the objective function becomes more complex with the new optimization variables. There are several numerical methods [1] to solve the optimization problem described in Equation 8.2. The main drawback of numerical methods is the algorithm convergence to a local optimum, which could be far away from the global optimum. The optimization algorithm could fail in a local minimum even if the objective function is a unimodal function (i.e, a function that has a single minimum), due to the constraints. The probability of finding a global minimum increases by starting the optimization algorithm from several initial points. Another approach uses evolutionary algorithms, such as the genetic algorithm, which already works with a population of design vectors. The simulated annealing algorithm also has a better probability to reach the global optimum because it permits, sometimes, movement against the objective function gradient according to the virtual temperature.

A large class of optimization algorithms deals with search methods. Many search algorithms have been developed to solve the unconstraint optimization problem. The same methods could also be applied for the constraint problem, but it is necessary to check if the optimization vector belongs to the feasible domain (i.e., whether all constraints are satisfied). The penalty function is another category of methods that transforms the constraint problem into an equivalent unconstraint problem. The equivalent objective function becomes

$$F\left(\bar{X}\right) = f\left(\bar{X}\right) + r_k \sum_{i=1}^{m} G_i\left(g_i\left(\bar{X}\right)\right) \tag{8.3}$$

There are two categories of penalty functions, G_i: interior penalty functions and exterior penalty functions [1]. If the interior penalty functions are used, the optimal solutions locus changes but it remains in the feasible domain. When the r_k constant decreases, the vector that minimizes the equivalent function, $f(X)$, becomes closer to the vector that minimizes the initial objective function, $f(X)$. Extrapolating the series of vector $X(r_k)$ to $r_k = 0$ it is possible to obtain the vector that minimizes the initial objective function. The usual internal penalty functions are

$$G_i = -\frac{1}{g_i(\bar{X})}$$
$$G_i = \log\left(-g_i(\bar{X})\right) \tag{8.4}$$

If the exterior locus penalty function is used, then the vector locus which minimizes the equivalent objective function coincides with the vector which minimizes the objective function $f(X)$, when it is placed inside the feasible domain but it is placed outside of feasible domain when objective function, $f(X)$ has its minimum outside of feasible domain. The outside vector locus is moved closer to the feasible domain when the penalty factor, r_k, increases. When the penalty factor tends to infinity the solution locus tends to the feasible domain border. The usual external functions are

$$G_i = \max\left(0, g_i(\bar{X})\right)$$
$$G_i = \left\{\max\left(0, g_i(\bar{X})\right)\right\}^2 \tag{8.5}$$

In this book, the modified Hooke–Jeeves algorithm is used to illustrate the optimal design for induction machines and for permanent magnet synchronous machines. The external penalty function based on physical interpretation is used to transform the constraint system into an equivalent unconstraint system. It is well known that increasing the winding temperature over an admissible temperature reduces the life of the machine, which, in terms of optimization, means increasing cost in direct ratio with reducing life.

Considering the over-temperature constraint as in Equation 8.6, where θ is the machine temperature depending on the optimization variable and θ_{ad} is the maximum admissible temperature, the penalty function is computed considering the initial cost of the machine, $f_i(X)$, and the gain factor, k_t, which considers, for example, the number of times the machine life is reduced when the over-temperature increases with a given percent over the admissible temperature.

$$g_\theta(\bar{X}) = \frac{\theta(\bar{X}) - \theta_{ad}}{\theta_{ad}} \tag{8.6}$$

$$G_\theta = k_t \cdot \max\left(0, g_\theta(\bar{X})\right) \cdot f_i(\bar{X}) \tag{8.7}$$

The same judgment could also be considered for the demagnetization constraint of permanent magnet machines. There, a safety factor is considered to avoid the permanent magnet demagnetization for uncertain over-currents and uncertainties in material features. When the safety factor reduces, the probability of demagnetization increases and the machine life is reduced. Again, a penalty function could be built, considering the initial machine cost. Using the phenomenological penalty function, it will not be necessary to build the optimal solutions series for larger and larger penalty factor r_k as has been described for classical external penalty functions, in order to reach the feasible border. In fact, the feasible domain border is not a fixed border. The design optimization is usually done considering the rated load, though, usually, the machine runs at different loads. For some applications, the probability of load density is known and it could be used for optimal design. Also, the material features and their price evolution are known as their average and variance; thus, in fact, they are stochastic variables. The stochastic optimization methods are able to optimize the machine design using stochastic variables.

Today many electrical machines are designed for variable speed and one frequently constraint condition is to produce larger power or torque than minimum given power or minimum given torque versus speed as it is shown in Figure 8.3.

This is a parametric constraint. Its mathematical formulation is given in Equation 8.8 where T_{max} (X, n) is the maximum torque available for the optimization variable vector, X, at speed n, while T_d (n) is the minimum required torque at speed n:

$$-T_{max}\left(\bar{X},n\right)+T_d\left(n\right)\leq 0, \quad n_{min}\leq n\leq n_{max} \tag{8.8}$$

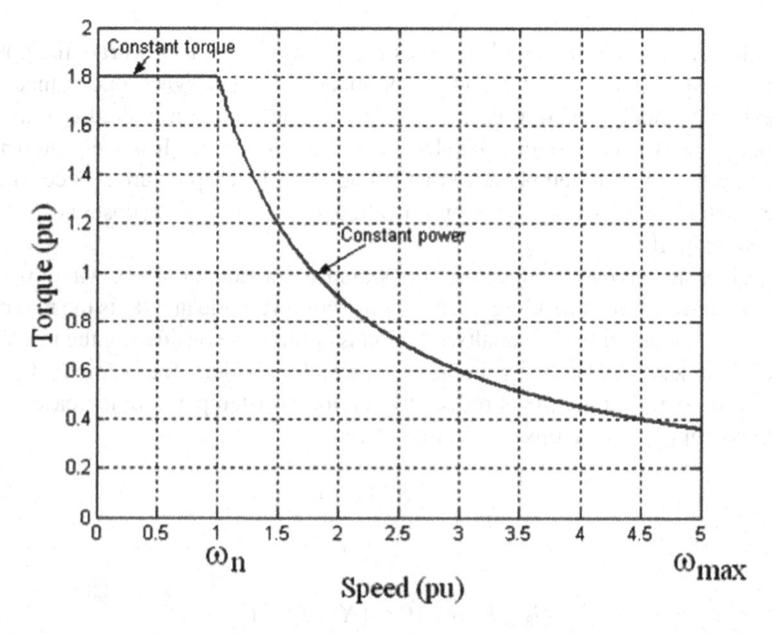

FIGURE 8.3 Parametric constraint—minimum torque versus speed for 5/1 constant speed and 1.8 torque overloading capability.

A simple method to solve the parametric constraint is by replacing it with several constraints. Generally, this method does not work efficiently and a large number of cases are necessary to replace the parametric constraint. Fortunately, for this particular case (Figure 8.3), only two or three particular constraints are necessary to replace the parametric constraint. This will start by introducing torque constraints at rated speed and at maximum speed. After optimal design, the parametric constraint will be checked and if the constraint requirements are satisfied for rated and maximum speeds but they are not satisfied for a region between these points, then it is necessary to add extra points to this region and repeat the design optimization. If there is a too restrictive constraint, such as maximum machine external diameter or machine length, maximum allowed current, and maximum permanent magnet emf voltage, then the designed machine does not satisfy the requirement torque even at base (rated) speed. It is then necessary to relax some of the constrains.

8.3 OPTIMUM CURRENT CONTROL

The torque of electrical machines is produced by a pair of magnetization and torque currents. One of the optimum control problems is to minimize the motor losses or the current. A case study is presented in the following text that considers the steady-state rotor field control of an induction motor. The optimization vector has two variables, magnetization current and torque current that are considered as phase current components. There are also two constraints: an equality constraint given through the torque equation and an inequality constraint, which states that the required stator voltage should be smaller than or equal to the available voltage. The stator voltage depends on the speed of the motor, so the second constraint is a parametric constraint. For small enough speeds, the voltage constraint does not influence the optimum current solution and in this case the optimum problem becomes

$$i_1 = \sqrt{i_{\mathrm{M}}^2 + i_{\mathrm{T}}^2} \tag{8.9}$$

$$T = 3p_1 \frac{L_{\mathrm{m}}}{L_{\mathrm{r}}} L_{\mathrm{m}} i_{\mathrm{M}} i_{\mathrm{T}} = 3p_1 L_{\mathrm{mr}} i_{\mathrm{M}} i_{\mathrm{T}} \tag{8.10}$$

We have here an optimization problem with two variables and an equality constraint. It is possible to eliminate one variable using the equality constraints. Usually, the magnetization inductance depends on the magnetization current but it could be considered independent of the torque current. From this reason, it is simple to compute the torque current from Equation 8.10 and replace its values in Equation 8.9.

$$i_1 = \sqrt{i_{\mathrm{M}}^2 + \left(\frac{T}{3p_1 L_{\mathrm{mr}}}\right)^2 \frac{1}{i_{\mathrm{M}}^2}} \tag{8.11}$$

A single variable (without constraint) optimum problem is thus obtained. Differential calculi will be used to find the solution but, before that, we can change the objective function from Equation 8.11 by a simple function that has the same extreme points:

$$f_{obj} = i_M^2 + \left(\frac{T}{3p_1 L_{mr}} \right)^2 \frac{1}{i_M^2} \tag{8.12}$$

The minimum current occurs for the solution of equation:

$$\frac{\partial f_{obj}}{\partial i_M} = 0 \tag{8.13}$$

If the magnetization inductance is constant, then Equation 8.13 becomes Equation 8.14 and it admits the analytical solution 8.15:

$$2i_M - 2\left(\frac{T}{3p_1 L_{mr}} \right)^2 \frac{1}{i_M^3} = 0 \tag{8.14}$$

$$i_M = \sqrt{\frac{T}{3p_1 L_{mr}}} \tag{8.15}$$

The torque current is computed from the constraint equation where the magnetization current is replaced with its optimal value. Finally, an important result should be noted: the minimum current for a constant magnetization inductance is obtained when the magnetization current is equal to the torque current.

$$i_T = \frac{T}{3p_1 L_{mr} i_M} = \sqrt{\frac{T}{3p_1 L_{mr}}} = i_M \tag{8.16}$$

The magnetization inductance is not constant in the real motor, so it is important to study the motor behavior when it is controlled with the magnetization and torque currents according to the optimization principle at constant inductance. Numerical results will be presented for an induction motor with the following parameters: rated power Pn = 1.1 kW, phase rated voltage Vn = 220V, phase rated current In = 2.77A, rated frequency fn = 50 Hz, rated speed nn = 1410 rpm, stator phase resistance $R1$ = 5.31 Ω, equivalent rotor resistance $R2$ = 5.64 Ω, short-circuit inductance L_{sc} =0.044 H at rated current and with the magnetization inductance versus magnetization linkage flux (peak value) from Figure 8.4.

The phase current versus torque is presented in Figure 8.5, where the motor is controlled at constant rated magnetization current and according to the optimum criterion of magnetization current equal to the torque current (i_M = i_T, L_{mr} = constant).

Also, $i_M = i_T$ for the real motor with magnetization inductance dependent on magnetization current produces largely different (larger) currents. We should note that for small torques there is a reduction in current even if the model is not exact but, for large torques, the phase current considering the optimization criterion is larger than for i_M = constant strategy. The optimal control current (curve "nlo") is somewhere

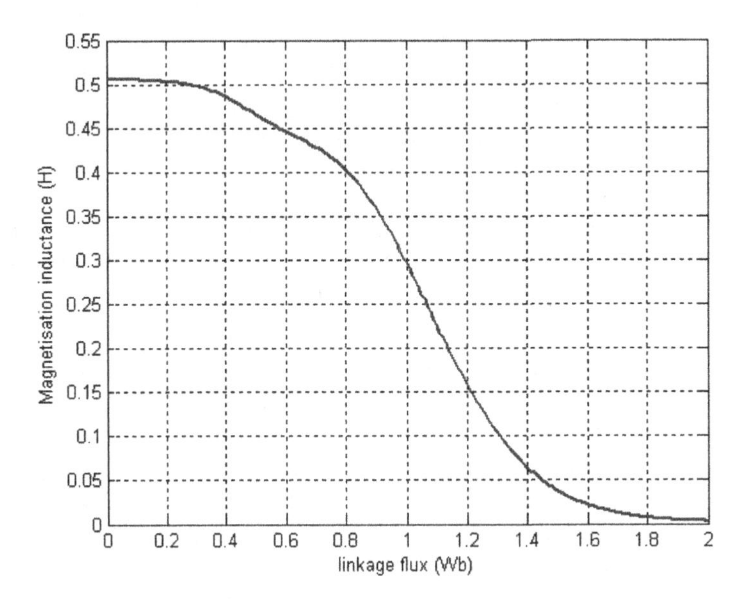

FIGURE 8.4 Magnetization inductance vs. linkage flux.

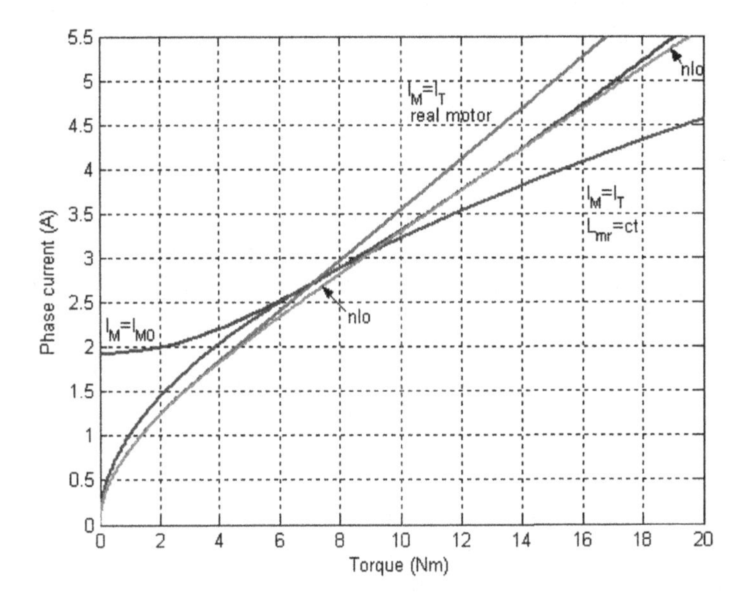

FIGURE 8.5 Phase current (rms value) vs. torque ($Tn = 7.45\text{Nm}$, $I_{M0} = 1.93\,A$, $L_{rmn} = 0.32$ H).

between, with good current prediction at low loads as expected. An important conclusion should be noted: when the process model is not accurate enough, the optimal solution could sometimes be worse than the nonoptimal solution. It is important to verify if each variable is reflected correctly in the objective function with the available model. The optimization principle shows better performance than the performance at constant magnetization current for torques that are smaller than rated torques but we should not jump to such conclusions because the studied motors are small motors with magnetization currents around 70% of the rated current. The positive effect of large magnetization currents is the motor capacity to produce 2.5 times larger torques than rated torques with only 2 times larger currents. At the rated load, the magnetization current is almost equal to the torque current.

The derivative of the objective function solution 8.12 is also the optimization key, if the dependence of magnetization inductance on magnetization current is considered, but in this case we get the following nonlinear equation:

$$i_{\mathrm{M}}^{4} = \left(1+\frac{i_{\mathrm{M}}}{L_{\mathrm{mr}}}\cdot\frac{\partial L_{\mathrm{mr}}}{\partial i_{\mathrm{M}}}\right)\left(\frac{T}{3p_1 L_{\mathrm{mr}}}\right)^2 \tag{8.17}$$

The L_{mr} inductance depends nonlinearly on the magnetization current and in many cases only a table with inductance values at several currents is available from measurements or from a numerical field model (FEM). The available values are not quite exact because they are affected by measurement errors or by numerical errors. Some methods to filter these errors use an analytical approximation of the inductance versus current based on least square error minimization. For example, the magnetization inductance curve presented in Figure 8.4 is based on several measured no-load current versus flux linkage and then an analytical approximation is found

$$L_{\mathrm{m}} = c_0 + \sum_{i=1}^{5} c_i e^{-2i\psi_{\mathrm{m}}^2} \tag{8.18}$$

where the series coefficients are $c_0 = 0.002093$ H, $c_1 = 3.3311$ H, $c_2 = -10.6093$ H, $c_3 = 17.1183$ H, $c_4 = -13.2716$ H, $c_5 = 3.93556$ H. It is possible to use many or fewer terms in the series but the coefficients should be recalculated again, because base functions of the exponential series are not orthogonal. Equation 8.17 becomes transcendental and only a numerical solution is possible. The iterative algorithm could be avoided if the torquc versus magnetization current is computed

$$T = 3p_1 L_{\mathrm{mr}} i_{\mathrm{M}}^2 \frac{1}{\sqrt{1+\frac{i_{\mathrm{M}}}{L_{\mathrm{mr}}}\cdot\frac{\partial L_{\mathrm{mr}}}{\partial i_{\mathrm{M}}}}} \tag{8.19}$$

The torque current is computed from the torque equation. The ratio between optimal torque current and magnetization current, k_{iopt}, depends only on the magnetization current and on the machine construction. This nonlinear function could be stored in a table and then it could be used in the optimal control:

$$i_{\mathrm{T}} = \frac{1}{\sqrt{1+\frac{i_{\mathrm{M}}}{L_{\mathrm{mr}}}\cdot\frac{\partial L_{\mathrm{mr}}}{\partial i_{\mathrm{M}}}}} i_{\mathrm{M}} = k_{iopt} i_{\mathrm{M}} \tag{8.20}$$

An example is given in Figure 8.6 where, based on Figure 8.4, L_m (i_m) and k_{iopt} are given. Returning to Figure 8.5 we can notice that the stator current for nonlinear optimization (nlo) is close to constant inductance IM control, which suggests that our machine is already saturated.

The energy performance using optimal current control is improving at low torque, (Figure 8.5) but the dynamic performance is poor when the machine is in flux weakening. The time constant to install the magnetic field in the motor is around 10 times larger than the time constant to install the torque. The minimum current optimization can be applied only when there is no sudden variation in the load torque and where fast torque response is not necessary. The stator current minimization reduces the losses but it is not equivalent to the minimum total losses optimization, which has the following objective function:

$$\begin{aligned} p_{loss} &= R_1\left(I_T^2 + I_M^2\right) + R_2 I_T^2 + \frac{\left(\omega_1 L_m\right)^2}{R_{Fe}} I_M^2 \\ &= \left(R_1 + \frac{\left(\omega_1 L_m\right)^2}{R_{Fe}}\right) I_M^2 + \left(R_1 + R_2\right) I_T^2 \end{aligned} \quad (8.21)$$

Decreasing the magnetization current at low load torque in order to reduce the stator current reduces the stator copper losses but the torque component increases and the rotor cage losses also increase. The iron losses decrease with the magnetization current but this also depends on speed. If the speed is small, the iron losses are negligible. A usual minimum losses optimum criterion gives a little larger magnetization current than the minimum current criterion. This text should serve only as a basis for evolved optimal design methods as those discussed in the following.

8.4 MODIFIED HOOKE–JEEVES OPTIMIZATION ALGORITHM

The Hooke–Jeeves optimization algorithm [2] is a pattern search algorithm that uses two kinds of moves: exploratory moves and pattern moves. In the following text, a modified version of Hooke–Jeeves algorithm is presented in order to allow constrained system optimization using external penalty functions.

The optimization algorithm contains the following steps:

1. Choose the optimized variables and the constant geometrical dimensions. The optimized variables are grouped in a vector. In our case (a PM brushless motor),

$$\bar{X} = \left(poles\; D_{ext}, w_{st}, w_c, h_c, h_{sy}, l_{pm}\right)^T$$

where

poles is the number of poles per primitive machine,
D_{ext} is the stator core outer diameter,
w_{st} is the stator teeth width,

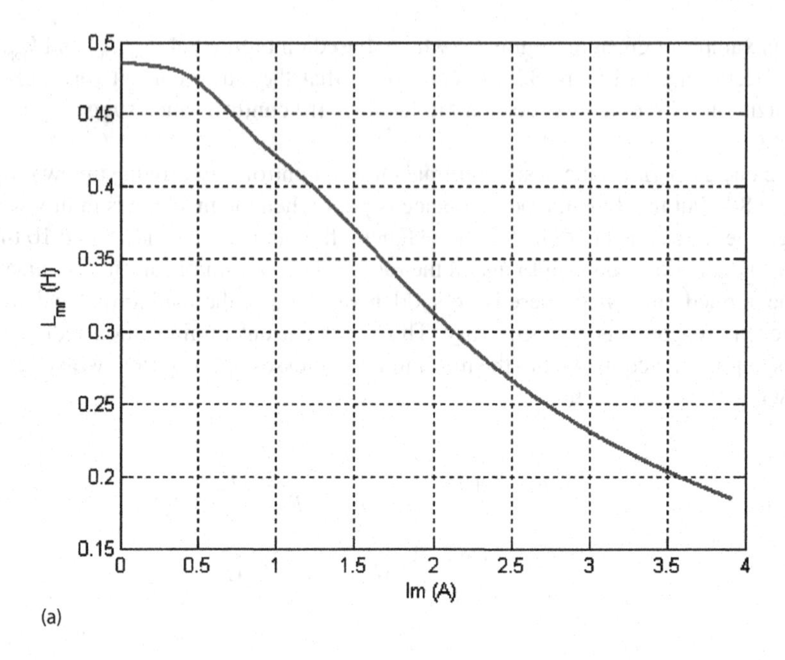

(a)

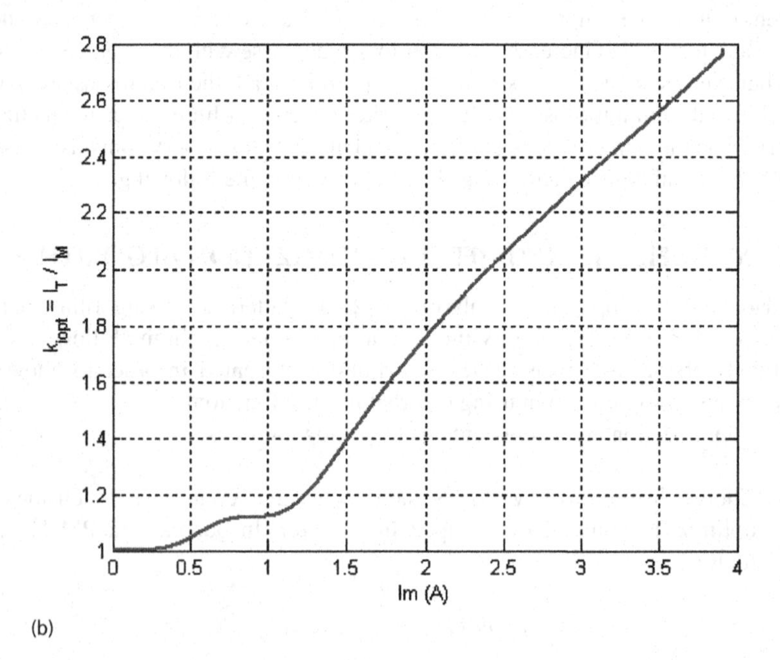

(b)

FIGURE 8.6 The magnetization inductance and the optimal ratio, k_{iopt}, between torque and magnetization current.

w_c is the coil width,
h_c is the coil height in the stator slot,
h_{sy} is the axial stator yoke length, and
l_{pm} is the permanent magnet radial length.

2. Find the technological limitations and geometrical constraints, and choose the optimized variable range $\overline{X}_{min}$ and $\overline{X}_{max}$. In a mathematical formulation,

$$\overline{X}_{min} \le \overline{X} \le \overline{X}_{max} \tag{8.22}$$

$$\overline{g}_1\left(\overline{X}\right) = \overline{0} \tag{8.23}$$

$$\overline{g}_2\left(\overline{X}\right) \le \overline{0} \tag{8.24}$$

where $\overline{g}_1 : \Re^n \to \Re^p$ and $\overline{g}_2 : \Re^n \to \Re^q$ are vector functions defined on variable space, considering "p" equality constraints and "q" inequality constraints.

3. Choose the objective scalar function, $f_1\left(\overline{X}\right), f_1 : R^n \to R$. The constraints could be included in the objective function considering the penalty function $f_p\left(\overline{X}\right)$ and $f_p : R^{p+q} \to R_+$:

$$f_p = \sum_{i=1}^{p+q} f_{pi}\left(g_i\left(\overline{X}\right)\right)$$
$$\text{with } f_{pi} = \begin{cases} 0 \text{ if } 1 \le i \le p \text{ and } g_i\left(\overline{X}\right) = 0 \\ 0 \text{ if } p+1 \le i \text{ and } g_i\left(\overline{X}\right) \le 0 \\ \text{monotonic positive } \forall \text{ otherwise} \end{cases} \tag{8.25}$$

Finally, the objective function is

$$f\left(\overline{X}\right) = f_1\left(\overline{X}\right) + f_p\left(\overline{X}\right) \tag{8.26}$$

4. Choose the initial values of optimized variable vector, $\overline{X}_0$, the initial step vector, $\overline{dX_0}$, the minimum step vector, $\overline{dX_{min}}$, and the step actualization ratio, r, with $0 < r < 1$.
5. Compute all necessary geometrical dimensions, the motor's performance from analytical models, and then evaluate the objective function f_0.
6. Make a research movement along each optimized variable in positive and negative directions (local grid search) using the initial step as has been shown for a two-dimensional vector in Figure 13.7. Compute the objective function and then compute the objective function gradient:

$$\overline{h} = \left(h_1, h_2, \ldots, h_n\right) = \left(\frac{\partial f}{\partial x_1}, \frac{\partial f}{\partial x_2}, \ldots, \frac{\partial f}{\partial x_n}\right) \tag{8.27}$$

The partial derivative is computed numerically using function evaluation in three points along the k-direction.

$$\frac{\partial f}{\partial x_k} = \begin{cases} \dfrac{f_k - f_0}{\mathrm{d}x_k} & \text{if} \quad f_{-k} \geq f_0 > f_k \\ \dfrac{f_0 - f_{-k}}{\mathrm{d}x_k} & \text{if} \quad f_{-k} < f_0 \leq f_k \\ \dfrac{f_k - f_{-k}}{2\mathrm{d}x_k} & \text{if} \quad f_{-k} \langle f_0 \rangle f_k \\ 0 & \text{if} \quad f_{-k} > f_0 < f_k \end{cases} \tag{8.28}$$

where f_k is the objective function evaluation in point $\overline{X}_k$ obtained from $\overline{X}_0$ by moving along the k-direction with $\mathrm{d}x_k$, and, respectively, $-\mathrm{d}x_k$ for f_{-k}. The point $\overline{X}_0$ is placed on the slope in the first and second case from Equation 8.28 ($\overline{X}_0$ is the worse point in case 3, but, is the better point in case 4). In the last two cases, the partial derivative is not mathematically correct but it provides information to leave the worse point or to stay near the better point. The step along gradient is computed as

$$\overline{\Delta} = \left(\Delta_1, \Delta_2, \ldots, \Delta_n\right) = \left(\frac{h_1 \cdot \mathrm{d}x_1}{\left\|\overline{h}\right\|}, \frac{h_2 \cdot \mathrm{d}x_2}{\left\|\overline{h}\right\|}, \ldots, \frac{h_n \cdot \mathrm{d}x_n}{\left\|\overline{h}\right\|}\right) \tag{8.29}$$

7. Make an optimized variable vector movement with step $\overline{\Delta}$ until the objective function decreases. Point P_i is reached in the two-dimensional example from Figure 8.7.
8. Repeat the research movement (step "6") in order to find a new gradient direction and then repeat the gradient movement (step "7") until the research movement is not able to find better points around the current point. This condition is equivalent to $\left\|\overline{h}\right\| = 0$, point P_j in Figure 8.7.
9. Reduce the variation step by ratio r and repeat the previous steps until the minimum of the variable variation is reached and the gradient norm $\left\|\overline{h}\right\|$ vanishes. When the research movement cannot find better points even using the

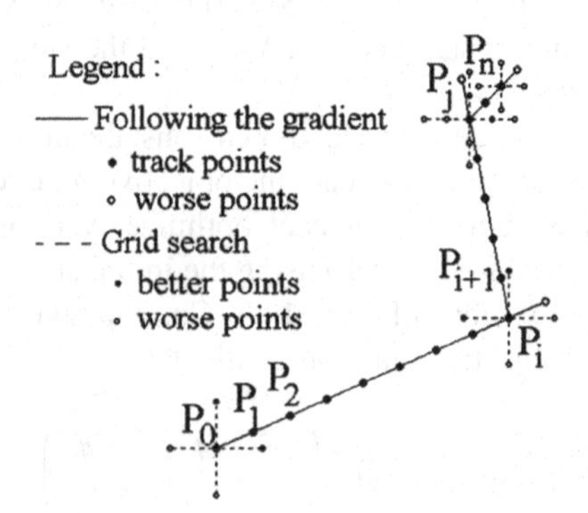

FIGURE 8.7 Search and gradient movement considering a 2D optimization problem.

smallest variable variation, then a minimum objective function is reached with the given resolution. This is probably not the global minimum. The algorithm should be run several times with different initial values of optimized variables in order to increase the probability to find the global minimum. The optimization algorithm may start with a 1.6 mm variation step for all geometrical variables and this is then reduced by ratio 2 until it reaches the minimum value equal by 0.1 mm. The minimum step value is usually chosen around the technological permissible variation. There is no reason to reduce the minimum step more than the permissible variation. The variation step should be a vector for a large machine because the ratio between the largest dimension and the smaller dimension of a variable could be hundreds or thousands.

The objective function could be considered as the initial cost of the machine plus the cost of the loss penalty plus penalty of the constraint function:

$$f = c_{\mathrm{i}}\left(\bar{X}\right) + c_{\mathrm{e}}\left(\bar{X}\right) + c_{\mathrm{p}}\left(\bar{X}\right) \tag{8.30}$$

where

c_{i} is the initial cost dependence,
c_{e} is the energy losses cost, and
c_{p}, is the penalty cost.

All of them must be in the same units; thus, for example, consider doubling the initial cost for every 10°C larger over-temperature than the acceptable maximum temperature or increase the cost in direct ratio to over-temperature when this becomes larger than the acceptable maximum.

Usually, the initial cost is the active material costs of the machine, and, for some applications, the required power electronics cost could be added:

$$C_{\mathrm{i}} = c_{\mathrm{Cu}} m_{\mathrm{Co}} + c_{\mathrm{lam}} m_{\mathrm{sFe}} + c_{\mathrm{Fe}} m_{\mathrm{rFe}} + c_{\mathrm{PM}} m_{\mathrm{PM}} + c_a m_t + c_{\mathrm{pe}} \frac{P_n}{pf} \tag{8.31}$$

where

c_{Cu} is the copper price, say, \$10/kg,
c_{lam} is the lamination price, say \$5/kg,
c_{Fe} is the rotor iron price, say, \$5/kg,
c_{PM} is the permanent magnet price, say, \$50/kg,
c_a is the extra price per total mass,
m_t, m_{c0} are the windings masses,
m_{sFe} is the stator lamination mass,
m_{rFe} is the rotor iron mass, and
m_{PM} is the permanent magnet mass.

The power electronics cost depends on the peak apparent power (in direct ratio to active power, P_n, and inverse ratio to power factor, *pf*) and power electronics price per VA - c_{pe}, say, \$0.025/VA for medium power converters. The machine design converges to an optimal power factor when the power electronics cost is added to the objective function because the power electronics cost depends to the apparent power and this is minimized when the power factor tends to unity. The grid connection equipment cost and reactive energy penalty cost could be used instead of the power electronics cost in order to optimize the power factor for grid-connected machines.

The energy efficiency optimization is introduced by considering the energy loss during the expected run time of the machine.

The energy losses cost expression is

$$c_e = P_n\left(\frac{1}{\eta_n(\bar{X})}-1\right)\cdot t_l\cdot p_e \tag{8.32}$$

where

P_n is the machine-rated power,
η_n is the rated efficiency,
t_1 is the operation time, and
p_e, energy price, is, say, \$0.1/kWh.

More sophisticated energy losses cost could be implemented, taking into account the loading factor, k_j, variation in time, and efficiency at that loading factor, η_j:

$$c_e = P_n\cdot t_l\cdot p_e\cdot\sum_j\left(\frac{1}{\eta_j(\bar{X})}-1\right)p_{kj}k_j$$
$$\sum_j p_{kj}=1 \tag{8.33}$$

The sum of probabilities to run at different loads, p_{kj}, should be equal by unity.

From many numerical examples, it could be observed that the loss energy cost is much larger than the initial cost. The optimization process is dominated by efficiency improvement rather than by initial cost reduction in these cases. The manufacturers and also many costumers are interested in the initial cost reduction with a minimum acceptable efficiency, given by governmental regulations or through free markets. It is then feasible to modify the objective function from Equation 8.32 according to this requirement:

$$c_e=\begin{cases}P_n\left(\dfrac{1}{\eta(\bar{X})}-\dfrac{1}{\eta_{\min}}\right)\cdot t_l\cdot p_e & \text{if } \eta(\bar{X})<\eta_{\min}\\ 0 & \text{if } \eta(\bar{X})\geq\eta_{\min}\end{cases} \tag{8.34}$$

Constraint penalty cost is composed from penalty cost of the over-temperature and penalty cost of the permanent magnet demagnetization due to peak or direct start current:

$$c_p = c_{pt}+c_{pdm} \tag{8.35}$$

The penalty cost for an over-temperature that is larger than the acceptable maximum is

$$c_{pt} = c_{pts} + c_{ptr} \tag{8.36}$$

$$c_{pts} = \max\left(0, T_w - T_{wad}\right) \cdot k_{ts} \cdot C_i \tag{8.37}$$

$$c_{pts} = \max\left(0, T_r - T_{rad}\right) \cdot k_{tr} \cdot C_i \tag{8.38}$$

where

c_{pts} is the penalty cost for stator (winding) over-temperature,
c_{ptr} is the penalty cost for rotor (PM) over-temperature,
T_w is the winding temperature,
T_{wad} is the maximum admissible winding temperature,
K_t is the a proportional constant,
c_i is the initial cost from Equation 13.31,
T_r is the rotor temperature, and
T_{rad} is the maximum admissible rotor (PM) temperature.

The penalty cost for permanent magnet demagnetization was computed considering that the "PM" is demagnetized when the total flux through the permanent magnet becomes negative due to the stator demagnetization current component:

$$c_{pdm} = \max\left(0, -\frac{\min\left(\Phi_{PMdm}\right)}{\Phi_{PM0}}\right) \cdot k_{dm} \cdot C_i \tag{8.39}$$

where

Φ_{PMdm} is the total flux trough PM at maximum demagnetization current equal to the rated, peak current multiplied by a demagnetization safety factor,
Φ_{PM0} is the total flux trough PM at no load, and
k_{dm} is the proportional constant.

8.5 ELECTRIC MACHINE DESIGN USING GENETIC ALGORITHMS

After a pattern search optimization algorithm such as that of Hooke–Jeeves we introduce here an evolutionary one.

The genetic algorithms consider a population of candidate solutions evolution under specific selection rules to a state that minimizes the cost function [3–5]. The genetic algorithms require a method to code the optimization variables, the objective function, and a set of selection rules for the evolution of the population's members. Developing a genetic algorithm for optimal design of electrical machines involves the following steps:

- Choosing the optimization variables, their minimum and maximum values, and their resolution. Some of the optimization variables should be integer numbers.

In this case, the resolution is equal by unity. It is not necessary to provide an initial value for the optimization vector.

- Choose a genetic codification of the optimization vector. There are two methods for the codification of the optimization variables as a genetic code: the binary code and the continuous (real) code.
- Choose the objective function and develop an analytical model in order to evaluate the objective function starting from the optimization variable vector. It is possible also to use FEM in order to compute the machine behavior, but, in this case, it is necessary to have a full parametric model to fully mechanize FEM (problem generation, solve the problem, and interpret the results). When such models are developed, it is necessary to evaluate thousands of candidate solutions. This method will be feasible by using parallel computations. However, the model should be accurate enough over the entire domain.
- Choose a selection algorithm in order to produce members for the next generation. Also, it is necessary to choose some criterion to stop the algorithm.

The number of bits required for the genetic code for every variable, n_{bit}, for binary codification is computed considering the number of distinct values for each variable, n_{vx}, and this depends on the variable range and its resolution:

$$n_{\text{vx}}(i) = \text{ceil}\left(\frac{X_{\max}(i) - X_{\min}(i)}{r_x(i)}\right) \tag{8.40}$$

$$n_{\text{bit}}(i) = 1 + \text{floor}\left(\log_2\left(n_{vx}(i)\right)\right) \tag{8.41}$$

where "ceil" is the round function to the nearest integer through plus infinity and "floor" is the round function through minus infinity.

For an induction machine with 16 optimization variables, it is necessary between two bits (number of slots per pole per phase is an integer number between 2 and 4 for small machines with four pairs of poles) and seven bits to code the variable that needs good resolution. In this particular example, the total number of bits required to store the whole vector information is 86, which means 2^{86} cases (about 7.73×10^{25}). Now, here is clear evidence that there is no chance for exhaustive search of an optimum solution.

The optimization vector computes for the genetic binary code with relation (8.42), and with relation (13.43) for continuous codification where g_{xb} is a integer between zero and n_{vx}, and g_{xc} is a real number between zero and unity:

$$\bar{X} = \bar{X}_{\min} + \bar{g}_{xb} \cdot \bar{r}_x \tag{8.42}$$

$$\bar{X} = \bar{X}_{\min} + \bar{g}_{xc} \cdot \left(\bar{X}_{\max} - \bar{X}_{\min}\right) \tag{8.43}$$

Many of the codes do not satisfy a minimum request condition, as is the case for geometric constraints, and they are not viable codes. The critical geometric dimensions are computed immediately and a member discards it if it does not observe the minimal requirement. Another random code will be generated until the initial population is complete. Usually, every variable is represented separately on the maximum number of bits. It is possible to store the entire information on a single "long integer" and execute the genetic operation simultaneously by the "all variables code" but the gain will not be important because the objective function evaluation is a very difficult operation.

The optimization design goal is to minimize the objective function while the genetic algorithm maximizes the fitness function. The genetic code is like an ascending order objective function. Thus, the better code is the first code. The Monte-Carlo roulette selection method selects the members to provide genetic codes for the next generation with a probability in direct ratio to the fitness function. A proper fitness function could be computed by dividing unity to the objective function. Then a per unit fitness function, f_f, is computed

$$f_g\left(\bar{X}_i\right)=\frac{1}{f\left(\bar{X}_i\right)} \tag{8.44}$$

$$f_r\left(\bar{X}_i\right)=\frac{f_g\left(\bar{X}_i\right)}{\sum_i^{ps} f_g\left(\bar{X}_i\right)} \tag{8.45}$$

$$r_k=\sum_{i=1}^{k} f_r\left(\bar{X}_i\right) \tag{8.46}$$

A random number, p, between zero and unity is generated. The index, k, of the smaller rank, r_k, larger than the random number, p, is chosen to be the first parent. For the second parents, another random number is generated and it is chosen using the same method. In this way, more adapted members have a chance to become parents and transmit the genetic information to the next generation. The first populations are very nonhomogeneous populations, and, using only this method to transmit genetic code through to the next generation, produces a rapid convergence where the generality is lost. In order to avoid a rapid convergence to a local solution, the fitness function of the members that are already parents is multiplied by an exclusion factor. In this way, the chance to become parents again is reduced. Two offsprings are produced by crossovers before they go to the next generation. Randomly, a part of the offspring suffers genetic mutation before it goes to the next generation. The members of the old generation are ranked again and then the process to produce new members is continued until a new, complete, generation is produced. The new generation will take the place of the old generation and the algorithm is repeated until the given number of generations is reached. The algorithm could be stopped before the maximum number of generations is reached if the fitness function difference between the best and the worst member becomes too small. This means that all population

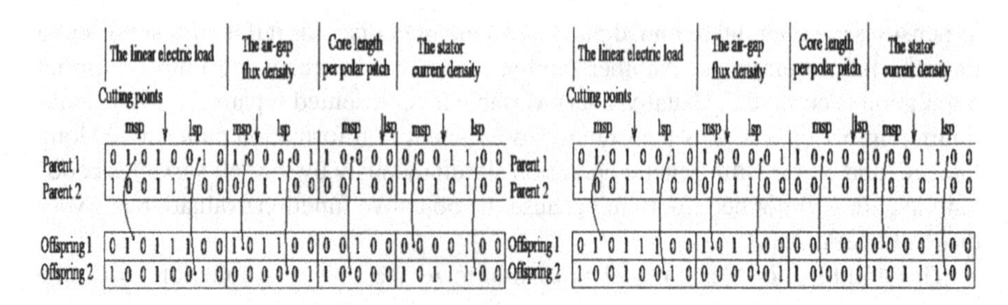

FIGURE 8.8 Offspring generation using crossover.

members are "close relatives" and probably many of them are perfect twins. It is possible to avoid twins in the population by discarding every new offspring that is identical with any already existing member from the next generation. The described principle could be applied for binary codification as well as for continuous codification.

The crossover is a technique to obtain two offspring from two parents by recombining the genetic code, as shown in Figure 8.8, for the binary codification of a few components of optimization vectors of induction machines such as specific linear current density (elsp—specific linear electric load), airgap flux density (Bagsp), the ratio between machine length and pole pitch (lcpertau), and the stator current density (Js). In this example, 8 bits are used to codify the specific electric load, 6 bits to codify the specific flux density in the airgap, 5 bits to specify the ratio between core length and pole pitch, and 7 bits to specify the stator current density. The whole information is stored on 86 bits. There are many possibilities for cutting and recombining the genetic code either as a single cutting or as multiple cuttings that could be in the same position or in variable positions. In this example, multiple cuttings, were chosen one for each piece of information coded and variable position cutting. The parts of the code available after cutting each variable region are assembled randomly in order to avoid sending the most significant code part (msp) from "parent 1" to "offspring 1" and least significant part (lsp) from "parent 2" to "offspring 1."

Mutation is another technique to produce offspring from a single parent. The parent code is copied identically to the offspring with an exception of one or two variables where the parent code is replaced with a random code as is observed in Figure 8.9 where a mutation occurs for "core length per pole pitch" variable.

A part of offspring code produced by crossover for continuous coding is computed by Equation 8.46 and the other part by Equation 8.47 where i and j are the optimization variable indexes. Applying Equation 8.47 or 8.48 is an arbitrary choice, to avoid a specific pattern for the crossover process.

$$
\begin{aligned}
g_{o1}(i) &= g_{p1}(i) + \alpha\left(g_{p2}(i) - g_{p1}(i)\right) \\
g_{o2}(i) &= g_{p2}(i) + \alpha\left(g_{p1}(i) - g_{p2}(i)\right)
\end{aligned}
\tag{8.47}
$$

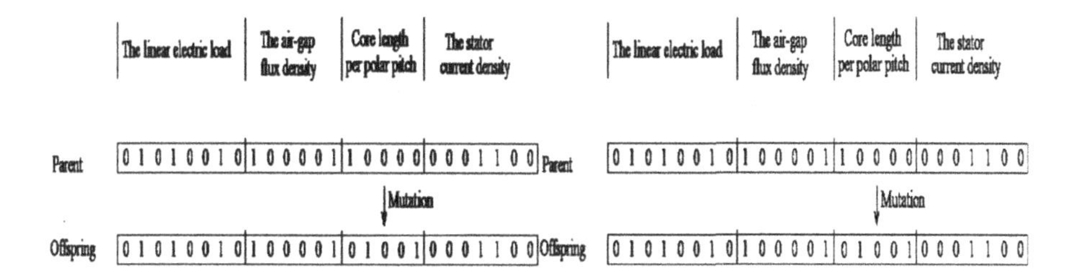

FIGURE 8.9 Offspring generation using mutation.

$$\begin{aligned} g_{o1}(j) &= g_{p2}(j) + \alpha\left(g_{p1}(j) - g_{p2}(j)\right) \\ g_{o2}(j) &= g_{p1}(j) + \alpha\left(g_{p2}(j) - g_{p1}(j)\right) \end{aligned} \tag{8.48}$$

The offspring optimization variables are always between the parent variables while for binary coding this happens only if the Gray code is used.

Examples of optimal design using genetic algorithms will be presented in the next chapters for permanent magnet synchronous machines and for induction machines. In the examples provided there the following parameters are used to control the genetic algorithm: population size; number of generations; elitism, that is, the number of the best members that are sent directly in the next generation; mutation rate, that is, the probability to have a mutation for each variable genetic code; and the excluding factor, that is, the rate of fitness function reduction for the parents after mating.

For more on optimal design of electric machines see reference [6].

REFERENCES

1. S.S. Rao, *Engineering Optimization: Theory and Practice*, 3rd edn., John Wiley & Sons, New York, 1996.
2. R. Hooke and T.A. Jeeves, Direct search solution of numerical and statistical problems. *J. ACM*, 8(2), 1961, 212–229.
3. R.L. Haupt, and S. Haupt, *Practical Genetic Algorithms*, 2nd edn., John Wiley & Sons, Hoboken, NJ, 2004.
4. D.E. Goldberg, *Genetic Algorithms*, Addison-Wesley, Reading, MA, 1989.
5. J.H. Holland, *Adaption in Natural and Artificial Systems*, The University of Michigan Press, Ann Arbor, MI, 1975.
6. G. Lei, J. Zhu, Y. Guo, "Multidisciplinary Design Optimization Methods for Electric Machines and Drive Systems", book, Springer Verlag, Berlin, Germany, 2016.

9 Optimal Electromagnetic Design of Surface PM Synchronous Machines (PMSM)

9.1 DESIGN THEME

In general, the following specifications make up the design theme of a variable speed PMSM:

- Base continuous power, P_b
- Base speed, n_b
- Maximum voltage, V_n
- Overload factor, k_1
- Maximum speed, n_{max}
- Power at maximum speed, P_{max}
- Number of phases, m

Also a few constraints are added:

- Efficiency at Pb and nb
- Material insulation class (allowed temperature)
- Protection degree against alien bodies
- Total initial cost of active materials

Flux weakening ($n_{max}/n_b > 1.5$) is not very feasible with surface PMSMs (SPMSMs)—(unless their synchronous inductance, L_s, is large in pu, such as for tooth-wound stator windings)—and, thus, $n_{max} = n_b$.

9.2 ELECTRIC AND MAGNETIC LOADINGS

The first stage in choosing the electric and magnetic loadings consists of general or optimization design of electric machines. For SPMSMs, these loadings are described as follows [1–6]:

- Specific electric loading, J_1 (A/m), represents the total effective value of ampere turns in stator slots per stator periphery length. J_1 relates to thermal loading and torque density. Large values of J_1 lead to large torque density, and consequently

DOI: 10.1201/9781003216018-9

to smaller machine size, which may end up in machine overheating and lower efficiency. When high torque density is targeted in the design, the specific tangential force, f_{tsp}, concept may be preferred. It is measured in N/cm^2 and varies from 0.1 N/cm^2 in micromotors to 10 N/cm^2 in larger torque density designs. J_l and f_{tsp} are complementary concepts.

- PM airgap flux density, B_{ag} (T), varies from 0.2 T in micromotors to 1 T in large torque density designs. Together B_{ag}, J_l, and ftp determine the volume of the machine for given base torque (Teb).
- Stator tooth flux density, B_{st} (T), determines the degree of magnetic saturation in the machine (Figure 9.1); it varies from 1.2 to 1.8 T, in general, for silicon-laminated stator cores.

As the machine's magnetic airgap (which includes the surface PM thickness) is large, the influence of the magnetic saturation of the teeth is rather small, and, thus, a large value of B_{st} may be chosen; in contrast, if B_{st} is large (and the fundamental frequency, f_{1b}, too is large) the core losses would also be large (Figure 9.2).

Smaller values of B_{st} lead to wider teeth, and thus thinner slots, that is, larger design current densities, and, consequently, larger copper losses. To prevent this, deeper slots are adopted, but then the machine's external diameter (and volume) is increased. Moreover, copper losses may not decrease notably by deepening the slot, as the average diameter (and length) of end connection of coils increases. So the design should start with applying moderate to large values of B_{st}, and then reducing or increasing them according to the performance of the machine based on these values, in comparison to its performance based on the design theme.

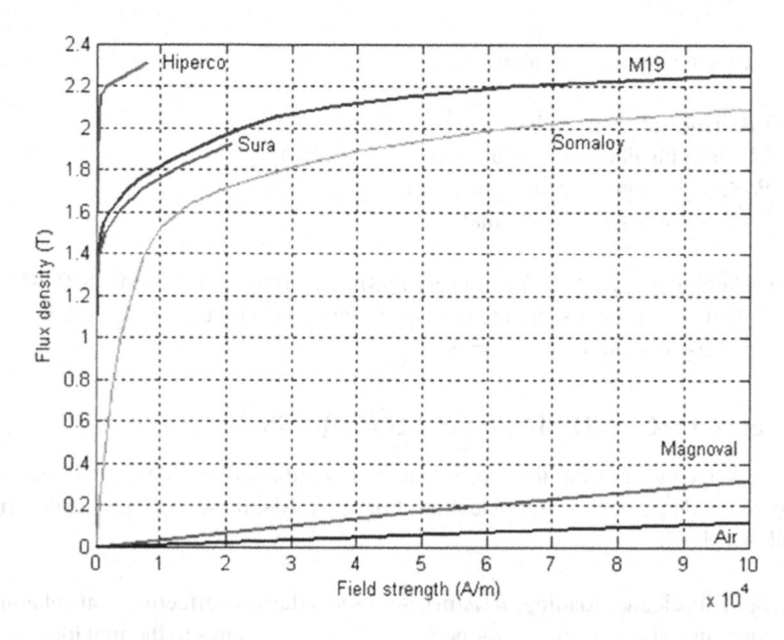

FIGURE 9.1 Magnetization curves for a few typical soft magnetic materials.

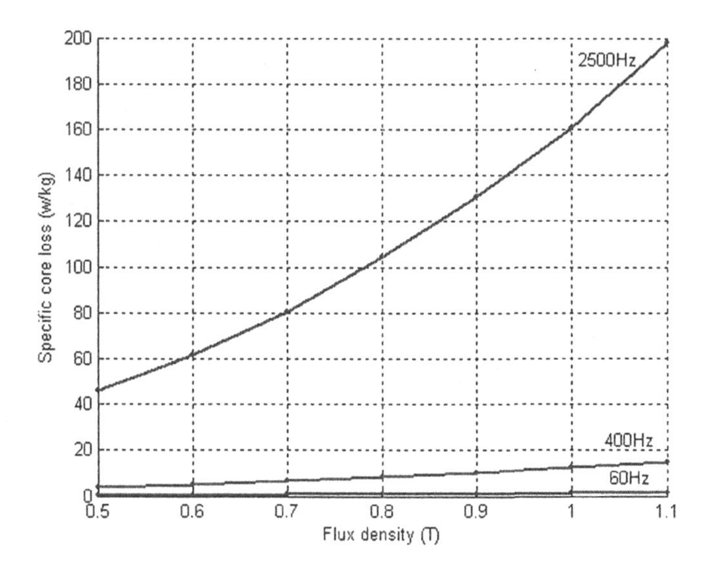

FIGURE 9.2 SURA007(0.18 mm) iron losses vs. flux density.

- Stator yoke flux density, B_{sy} (T), is chosen as a compromise between the level of magnetic saturation and the limitations due to core losses. Small values of B_{sy} may lead to a larger machine size and weight, especially if the number of poles is small ($2p = 2,4$).
- Current density, J_s (A/mm^2), determines the copper losses and the copper volume. Thin and deep slots that have small values of J_s (2–3.5 A/mm^2) may lead to a high leakage inductance and machine volume (and weight). On the other hand, high J_s values (>8 A/mm^2), in general, not only imply forced cooling but also lead to lower efficiency while reducing the machine volume.
- Rotor yoke flux density, B_{ry} (T), is important in the machine with a large number of poles (and a large diameter) when the PMs are not any more placed directly on the shaft.

9.3 CHOOSING A FEW DIMENSIONING FACTORS

- Machine shape factor, γ_c: the ratio between the axial (stack) length of the machine, l_{stack}, and the pole pitch, τ ($\gamma_c = l_{stack}/\tau$, usually between 0.3 and 3). The main geometry of an SPMSM is shown in Figure 9.3.
- PM pole span factor, γ_{PM}: the ratio between the PM width and the pole pitch. This factor influences the PM total weight, the emf space harmonics content, and, to some extent, the cogging torque.
- Number of slots per pole per phase, q_1.
- Number of winding layers, n_L ($n_L = 1$ for single-layer windings and $n_L = 2$ for two-layer windings).
- Stator winding current path count, a_1, is always a divisor of the number of poles for two-layer windings and of the number of pole pairs for single-layer

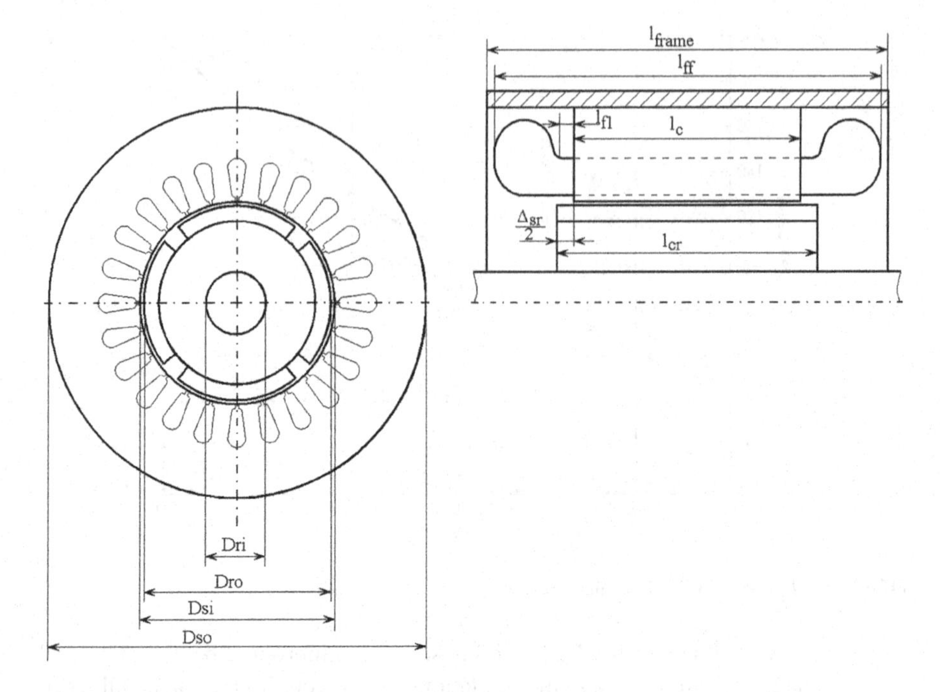

FIGURE 9.3 The main geometrical dimensions of a PMSM.

windings. To avoid the circulation current between current paths, due to inherent machine symmetry imperfection, $a_1 = 1$ whenever possible, with the exception of low-voltage, large-current (automotive) applications.

- Elementary conductors in parallel: For large currents (or for large fundamental frequency) the coil turns are made of multiple elementary conductors in parallel, with some degree of transposition, to reduce skin effects.
- Coil span, y_1, is the distance between the forward and return sides of the coil; it may be measured not only in mm but also in number of slot pitches: between $2q_1$ and $3q_1$ for three-phase distributed windings, but below 0.5 for tooth-wound windings.
- Slot opening, s_o (see Figure 9.4): The minimum value of s_o is limited by the possibility to introduce the coils, turn by turn, in the slot and by the increase of slot leakage inductance and PM flux fringing. Its maximum value is limited by the PM flux reduction, cogging torque increase, and torque ripple.
- Tooth top height, h_{s4}: The minimum value of h_{s4} is limited by technological (and magnetic saturation) constraints and its maximum value is limited by the increase the slot leakage inductance. (In high-speed PMSMs, with $f_{1b} > 1$ kHz, h_{s4} may be increased, to increase machine inductance L_s, and thus decrease current ripple).

9.4 A FEW TECHNOLOGICAL CONSTRAINTS

- Laminated-core-filling factor, $k_{stk} = 0.8–0.97$, is the ratio of the lamination's height to the total laminated core length (lamination's insulation coating means $k_{stk} < 1$).

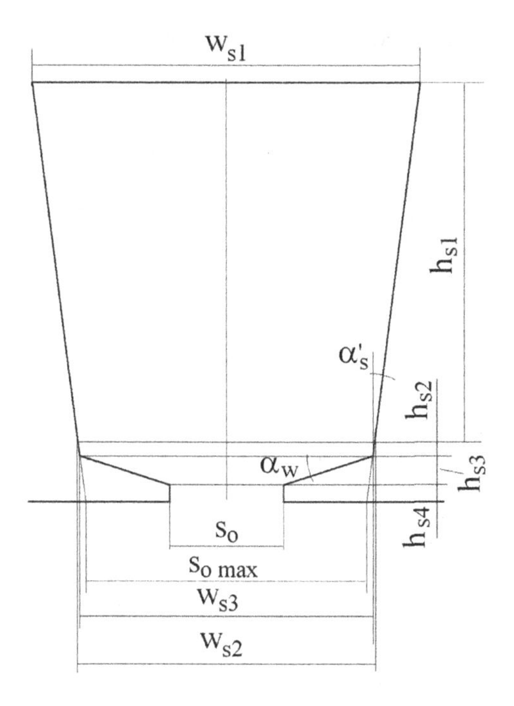

FIGURE 9.4 Stator slot geometry.

- Slot-copper-filling factor, k_{sf} = 0.33–0.7, has lower values correspond to semi-closed slots with coils introduced, turn by turn, in the slot, while larger values correspond to open slots and premade coils made of conductors having rectangular cross sections.
- Minimum airgap, g_{min}, from mechanical considerations in super high-speed machines is increased to reduce PM eddy current losses due to stator mmf space harmonics; a resin coating of PMs is also added for mechanical rigidity.
- Overload demagnetization safety factor, k_{sPM}: The maximum overload stator mmf that avoids PM demagnetization in the worst load scenario is limited by this factor.
- Slot geometry type: If more than one slot shape is used, a design expression for the chosen shape should be provided, and the computer code calls for the specific geometrical and specific magnetic permeance expressions on a menu basis.
- Angle of slot wedge, α_w.
- Slot insulation thickness.
- Straight-turn end connection, l_{f1}, that exits the slot.
- Difference between PM, l_{PM}, and stator stack length, l_{stack} ($l_{PM} - l_{stack} = (1-2)(g + h_{PM})$), where h_{PM} is the PM thickness (along the magnetization direction), in order to avoid axial forces on the bearings.
- Minimum shaft diameter, based on the maximum envisaged torque.

9.5 CHOOSING MAGNETIC MATERIALS

We refer here to permanent magnets and magnetic cores. For PMs, we choose the remnant flux density, B_r, the coercive field, H_c, and the maximum allowed rotor (PM) temperature, T_r. The choice of B_r depends on the airgap flux density: $B_{ag} < B_r$ for SPMSMs. The coercive field (force), H_c, depends on B_r and on the recoil pu permeability: $\mu_{rec} \approx (1.05–1.3)$ for NeFeB, SmCo5, and ferrite PMs ($B_r = (0.3–0.45)$ T); $\mu_{rec} \approx 1.05$ and $B_r = 0.6–0.8$ T for bonded PMs; and $B_r = 1.1–1.37$ T for sintered PMs at 20°C.

$$H_c \approx \frac{B_r}{\mu_{rec(pu)}\mu_0}(\text{A/m}) \tag{9.1}$$

The PM characteristics are obtained from the manufacturer's catalogs and they include

- B_r at 20°C
- H_c at 20°C
- B_r and H_c temperature coefficients
- Energy density
- Interior coercive field (force)
- PM magnetization field, $H_c \times (1.8–3)$ and $B_r \times (1.8–2.5)$
- Maximum operation temperature
- Electric resistivity of PM materials, to calculate the PM eddy current losses due to stator mmf space and time harmonics: much higher for bonded NeFeB than for sintered NeFeB
- Mass density (kg/m^3)

For up to 150 Hz maximum fundamental frequency, general silicon-laminated cores made of 0.5 mm (0.65 mm for increased productivity) nonoriented grain laminations are used. For fundamental frequencies above 150 Hz, thinner laminations (0.2 mm or less, but with stacking factor $k_{sk} > 0.8$) are to be used to reduce eddy current losses. The magnetization curve characterizes the silicon lamination (Figure 9.1).

Using soft magnetic materials with higher saturation flux densities ($B_{sat} = 2.35$ T at $H_{sat} = 10^4$ A/m for 50 Hz) leads to more compact PMSMs at a slightly higher initial cost. (Hyperco50 is more expensive than M19 silicon steel.) Other important characteristics of soft magnetic materials are the core losses that depend on the flux density amplitude and frequency. The core losses are segregated into eddy current and hysteresis losses. Some manufacturers offer core losses for a few frequencies (Figure 9.2). M19 and SURA are two widely used trade names for silicon laminations. Hyperco50 is a trade name for a high saturation flux density laminated material with up to 50% cobalt and moderate losses at frequencies below 500 Hz. Somaloy is the trade name of a soft material composite for $f_B > 500$ Hz (in general) and (or) where three-dimensional a.c. magnetic field lines have to be accommodated. Thin laminations such as SURA007 [6] (0.18 mm) have shown acceptable losses at 2500 Hz (Figure 9.2). Magnoval is a sintered material with low relative magnetic permeability ($\mu_{rec} = (3–5)\mu_0$) used for slot wedges to decrease PM flux density pulsations, and cogging and total torque pulsations, at the cost of larger leakage and synchronous inductances. Magnoval is also good to increase the machine inductance in high-frequency ($f_{1b} > 1\ kHz$) machines to reduce current ripple.

A good approximation of core power losses (in W/kg) for the practical frequency and the flux density is, still, the Steinmetz formula:

$$P_{tot} = k_h B_m^2 f + \pi^2 \frac{\sigma d^2}{6} \left(B_m f\right)^2 \tag{9.2}$$

k_h—material coefficient for hysteresis losses,
σ—lamination electric conductivity,
d—lamination thickness,
f—field frequency, and
B_m—a.c. flux density amplitude.

9.6 DIMENSIONING METHODOLOGY

First, we calculate the machine size constant, C_0, and the stator bore diameter, D_{si}, based on some already chosen parameters:

$$C_0 = \frac{\pi^2}{\sqrt{2}} B_{ag} J_1 k_w \tag{9.3}$$

where

B_{ag} is the PM airgap flux density (T),
J_1 is the linear specific electric loading (kA_{turns}/m) chosen from table 9.1,
k_w is the winding factor, which includes the zone factor, k_{ws}, and the chording factor, k_{chs}, and
y_s is the coil span in slot pitch count.

$$k_w = k_{ws} \cdot k_{chs}; \quad k_{ws} = \frac{\sin\left(\frac{\pi}{6}\right)}{q_1 \sin\left(\frac{\pi}{q_1 6}\right)}; \quad k_{chs} = \sin\left(\frac{y_1}{q_1 m} \cdot \frac{\pi}{2}\right) \tag{9.4}$$

Equation 9.4 is strictly valid for integer q_1. D_{si} in mm is

$$D_{si} = 1000 \cdot \sqrt[3]{\frac{60 p_s}{\pi n_n} \cdot \frac{P_n}{\lambda_c C_0}} \tag{9.5}$$

where

p_s is the number of rotor poles ($p_s = 2p_1$), which is either given or calculated for the base, speed and chosen frequency,

TABLE 9.1
Optimization Variable Bounds

Optimization Variable	Minimum Values	Maximum Values	Units	Comments
J_1	15	30	kA/m	Specific electric load specifications
Bagsp	0.45	0.75	T	Airgap flux density
Bst	1	2	T	Stator tooth flux density
Bsy	0.9	1.9	T	Stator yoke flux density
Bry	0.9	2.1	T	Rotor yoke flux density
J_s	3	8	A/mm²	Stator current density
lcpertau	0.5	3		Core stack length per pole pitch
q_1	2	4		Stator slots per pole per phase
so	1	5	mm	Stator slot opening width
sh4	0.5	2	mm	Stator tooth pole tip height
cSpan	0.66	1		Coil span per pole pitch
Alpm	0.5	1		Permanent magnet angle per pole angle

P_n is the base/rated electromagnetic power,
n_n—speed in rpm, and
λ_c—core length per pole pitch.

It is safe to use Equation 9.5 mainly if the reaction airgap flux density of the stator mmf is small (less than 25%) with respect to the PM airgap flux density, B_{ag}. An alternative equation uses the electromagnetic torque, T_{eb}, and the tangential specific force, f_{tsp} (in N/cm²):

$$D_{si}(\text{mm}) = 100 \cdot \sqrt[3]{\frac{2p_s}{10\pi^2} \cdot \frac{T_{eb}}{\lambda_c f_{tsp}}} \tag{9.6}$$

Figure 9.3 shows the main geometrical parameters of a PMSM.

Once D_{si} is established, the pole pitch, τ, and the core length stack, l_c, are calculated:

$$\tau_p = \pi \frac{D_{si}}{p_s} \tag{9.7}$$

$$l_c = \lambda_c \tau_p \tag{9.8}$$

Based on given $B_{ag} = (0.2\text{–}1\ \text{T})$ and $\lambda_{PM} = (0.6\text{–}0.9)$, the PM flux per pole, Φ_{PM}, is

$$\Phi_{PM} = \frac{2}{\pi} B_{ag} \tau l_c \sin\left(\lambda_{PM} \frac{\pi}{2}\right) \tag{9.9}$$

Now the stator and the rotor yoke thicknesses, h_{sy} and h_{ry}, may be calculated:

$$h_{sy} = \frac{B_{ag}}{B_{sy}} \cdot \frac{\tau_p}{\pi} \quad (9.10)$$

$$h_{ry} = \frac{B_{ag}}{B_{ry}} \cdot \frac{\tau_p}{\pi} \quad (9.11)$$

The number of stator slots, N_{ss}, the stator slot pitch, τ_{ss}, and the tooth width, w_{st}, are

$$N_{ss} = q_q \cdot m \cdot p_s \quad (9.12)$$

$$\tau_{ss} = \pi \frac{D_{si}}{N_{ss}} \quad (9.13)$$

$$w_{st} = \tau_{ss} \frac{B_{ag}}{B_{st}} \quad (9.14)$$

The geometrical stator parameters in Figure 9.4 may all be calculated, α_w is approximately 20°. But before that, the slot geometrical angle α_s, and the angle of tooth top (slice) α_{st} are determined:

$$\alpha_s = \frac{2\pi}{N_{ss}}; \quad \alpha_s' = \frac{\pi}{N_{ss}} \quad (9.15)$$

$$\alpha_{s0} = 2arc\sin\left(\frac{s_0}{D_{si}}\right) \quad (9.16)$$

$$\alpha_{st} = \alpha_s - \alpha_{s0} \quad (9.17)$$

We verify now if the tooth top (slice) is larger than the tooth width, w_{st} (Equation 9.12); if not, the slot opening, s_0, is reduced accordingly. At the limit, the slot may remain open when the central angle for a tooth is

$$\alpha_{stmin} = 2arc\sin\left(\frac{w_{st}}{D_{si}}\right) \quad (9.18)$$

For open slots,

$$s_{0\max} = D_{si}\sin\left(\alpha_s' - \alpha_{st\min}/2\right) \quad (9.19)$$

The main slot geometrical parameters (Figure 9.4) are

$$h_{s3} = \frac{\dfrac{s_{0\max} - s_0}{2} + h_{s4} \cdot tg\left(\alpha_s'\right)}{\dfrac{1}{tg\left(\alpha_w\right)} - tg\left(\alpha_s'\right)} \quad (9.20)$$

$$w_{s3} = s_{0\max} + 2\left(h_{s4} + h_{s3}\right)tg\left(\alpha_s'\right) \tag{9.21}$$

$$w_{s2} = w_{s3} + 2h_{s2}tg\left(\alpha_s'\right) \tag{9.22}$$

where h_{s2} is the slot classic insulation thickness.

The copper area in a slot, S_{Cu}, and the required slot area, S_{slot}, are

$$S_{\text{Cu}} = \frac{I_{\text{ts}}}{J_{\text{s}}} \tag{9.23}$$

$$S_{\text{slot}} = \frac{S_{\text{Cu}}}{K_{\text{sf}}} \tag{9.24}$$

where K_{sf} is the slot fill factor and I_{ts} is the total RMS ampere turn/slot, which is to be calculated from either Equation 9.23 or Equation 9.24:

$$I_{ts} = \frac{\pi D_{si}}{N_{ss}} J_l \tag{9.25}$$

$$I_{\text{ts}} = \frac{T_{\text{em}}}{\sqrt{2} p_1 \Phi_{\text{p}} N_{\text{ss}}} \tag{9.26}$$

Knowing the active slot area, S_{slot}, its main dimensions are

$$h_{s1} = \frac{-w_{s2} + \sqrt{w_{s2}^2 + 4S_{\text{slot}}\tan\left(\alpha_s'\right)}}{2\tan\left(\alpha_s'\right)} \tag{9.27}$$

$$w_{s1} = w_{s2} + 2h_{s1}\tan\left(\alpha_s'\right) \tag{9.28}$$

$$h_s = h_{s4} + h_{s3} + h_{s2} + h_{s1} \tag{9.29}$$

Then the outer stator diameter, D_{so}, is approximated to an integer in millimeters:

$$D_{s0} = \text{round}\left(D_{\text{si}} + 2h_s + 2h_{\text{sy}}\right) \tag{9.30}$$

Now the stator yoke thickness, h_{sy}, is

$$h_{\text{sy}} = \frac{D_{s0} - D_{\text{si}}}{2} - h_s \tag{9.31}$$

The end-connection coil length, l_{f}, the total half-turn length, l_{mc}, and the axial length of the stator between end connections, l_{ff}, are

$$l_{\mathrm{f}} = \frac{\pi}{2}\tau_{\mathrm{p}}\frac{y_1}{q_1 m}\left(1+\frac{h_{\mathrm{s}}}{D_{\mathrm{si}}}\right)+2l_{f1} \tag{9.31}$$

$$l_{\mathrm{mc}} = l_{\mathrm{c}} + l_{\mathrm{f}} \tag{9.33}$$

$$l_{\mathrm{ff}} = l_{\mathrm{c}} + 2l_{f1} + \tau_{\mathrm{p}}\frac{y_1}{q_1 m}\left(1+\frac{h_{\mathrm{s}}}{D_{\mathrm{si}}}\right) \tag{9.34}$$

The motor frame length can be calculated using l_{ff}.

9.6.1 Rotor Sizing

The PM thickness, h_{PM1}, is supposed to produce a certain PM flux density in the airgap, B_{ag0}, and also to avoid demagnetization under a given overload:

$$h_{\mathrm{PM1}} = \frac{B_{ag0}\delta_{\min}k_{\mathrm{s}}}{\mu_0 H_{\mathrm{c}}\left(1-\frac{B_{\mathrm{ag0}}}{B_{\mathrm{r}}}k_{\mathrm{s}}\right)} \tag{9.35}$$

$$h_{PM2} = 1000\frac{I_{ts}N_{ss}}{p_1\sqrt{2}}k_1 k_{\tau PM}\frac{1}{H_c} \tag{9.36}$$

where k_{s} is the saturation factor, k_{i}, the load factor, is considered 1.8 and $k_{\tau\mathrm{PM}}$ 1.5 representing the demagnetization safety factor.

In both cases, h_{PM} is in millimeters and the largest of h_{PM1} and h_{PM2} is chosen and rounded to tenths of millimeters. If h_{PM2} is larger than h_{PM1}, the airgap, g, has to be recalculated:

$$g = \mu_0 H_{\mathrm{c}} h_{\mathrm{PM}}\frac{1-\frac{B_{ag0}}{B_{\mathrm{r}}}k_{\mathrm{s}}}{B_{ag0}k_{\mathrm{s}}} \tag{9.37}$$

The gap g is rounded to a multiple of 0.05 mm:

$$g = 0.05\cdot\mathrm{round}\left(20g\right) \tag{9.38}$$

The external and interior diameters of the rotor, D_{ro} and D_{ri}, and the core length, l_{cr}, are

$$D_{r0} = D_{si} - 2g \tag{9.39}$$

$$D_{\mathrm{ri}} = D_{r0} - 2\left(h_{\mathrm{PM}} + h_{\mathrm{ry}}\right) \tag{9.40}$$

$$l_{\mathrm{cr}} = l_{\mathrm{c}} + \Delta_{\mathrm{sr}} \tag{9.41}$$

The interior rotor diameter, D_{ri}, is rounded to an integer in millimeters, and, thus, the rotor yoke thickness is slightly adjusted. If D_{ri} is smaller than the shaft diameter, either the dimensioning methodology is redone for a larger D_{si} (by reducing B_{ag}, J_1, and f_{tsp}) or shape factor, λ_c, or the PMs are placed on the shaft (for a massive rotor yoke).

9.6.2 PM Flux Computation

Failing to realize the required PM flux implies larger current (and copper) losses for the base speed (power) torque. For the preliminary design, we added a saturation factor, k_{sat}, and presupposed that it is feasible. This is needed to recalculate the PM flux with better precision, as the PMs have a linear demagnetization curve with $\mu_{rec} = (1.05\text{–}1.1)\mu_0$. Consequently, the Carter coefficient has to be calculated for the equivalent airgap, g_e:

$$g_e = g + \frac{B_r}{\mu_0 H_c} h_{PM} \tag{9.42}$$

$$k_c = \frac{\tau_s}{\tau_s - \gamma_s s_0} \tag{9.43}$$

where

$$\gamma_s = \frac{\left(\frac{s_0}{g_e}\right)^2}{5 + \frac{s_0}{g_e}} \tag{9.44}$$

Due to the nonlinearity of magnetic core materials, the PM airgap flux density has to be calculated iteratively after we use $k_s > 1$ for saturation from the preliminary design.

$$B_{ag0} = \mu_0 H_c \frac{h_{PM}}{g_e k_C k_s} \tag{9.45}$$

The PM flux per pole is calculated from Equation 9.9 and then we can recalculate B_{st}, B_{sy}, and B_{ry} from Equations 9.46 through 9.48:

$$B_{st} = B_{ag} \frac{\tau_s}{w_{st}} \tag{9.46}$$

$$B_{sy} = \frac{\Phi_p}{2 l_c h_{sy}} 10^6 \tag{9.47}$$

$$B_{ry} = \frac{\Phi_p}{2 l_c h_{ry}} 10^6 \tag{9.48}$$

Making use of the core magnetization curve through interpolation, the magnetic field in the stator teeth, H_{st}, in the stator yoke, H_{sy}, and in the rotor yoke, H_{ry}, are calculated. Based on this, the magnetization voltages, V_{mst}, V_{msy}, and V_{mry}, in the stator teeth, the yoke, and the rotor yoke are calculated:

$$V_{mst} = 0.001 H_{st} h_s \tag{9.49}$$

$$V_{msy} = 0.001 H_{sy} \frac{\pi \cdot D_x}{p_s C_x} \tag{9.50}$$

$$V_{mry} = 0.001 H_{ry} \frac{\pi \cdot D_y}{p_s C_y} \tag{9.51}$$

$$x = \pi D_x; \quad y = \pi D_y;$$

where

D_x and D_y are, respectively, the average diameters around which the average flux path closes in the two yokes

C_x and C_y are, respectively, two path length reduction coefficients that account approximately for the fact that nonuniform local saturation takes place in the stator and rotor yokes

$$D_x = D_{s0} - \frac{4}{3} h_{sy} \tag{9.52}$$

$$D_y = D_{ro} - 2h_{PM} - \frac{2}{3} h_{ry} \tag{9.53}$$

C_x and C_y depend on the flux density distribution and on the number of poles of the rotor. These may be calculated through interpolation from the C curves in Figure 9.5, obtained empirically or, better, by the FEM.

The magnetic voltage along the airgap, V_{mg}, is

$$V_{mg} = 0.001 g_e k_C \frac{B_{ag}}{\mu_0} \tag{9.54}$$

So the total mmf, V_m, required to produce B_{ag0} in the airgap is

$$V_m = V_{mst} + V_{msy} + V_{mry} + V_{mg} \tag{9.55}$$

The new value of the saturation factor, k_s, is calculated using

$$k_s = \frac{V_m}{V_{mg}} \tag{9.56}$$

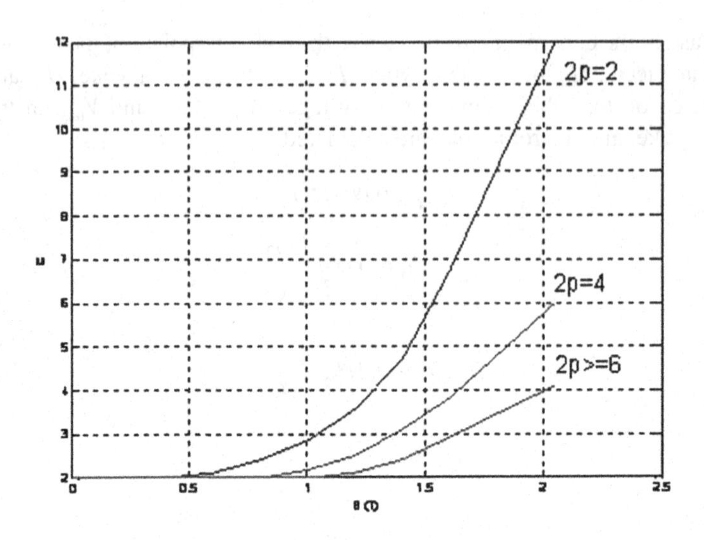

FIGURE 9.5 Flux path length reduction coefficient, *C*.

Based on this new value of k_s, the computation cycle of B_{ag0} is redone until an error of less than 1% between two successive values of k_s is obtained. High precision is required in the optimal design, but then a mix of an analytical and an FEM calculation has to be converged. To speed up the convergence of k_s (or B_{ag}) a relaxation coefficient may be used.

$$B_{\text{ag},k} = \left(1-r\right)B_{\text{ag},k-1} + r \cdot B_{\text{ag,new}} \tag{9.57}$$

where

$B_{\text{ag},k-1}$ is the old value of the airgap flux density
$B_{\text{ag,new}}$ is the value calculated with k_{sat} computed from Equation 9.56
$B_{\text{ag},k}$ the value of B_{ag} used in the next iteration cycle

A value of $r = 0.2–0.3$ should produce good convergence for saturated machines but slow convergence for less-saturated machines. The convergence problem may occur for very low saturation degrees, but then the trouble may be avoided by assigning a reasonable (safe) value to K_s from the start ($k_s = 1.05–1.1$). After B_{ag} is calculated with the desired precision, the PM flux per pole, Φ_{p}, is recalculated and, based on this, the PM flux linkage for a single-turn/coil, Ψ_{PM1}, is computed:

$$\Psi_{PM1} = q_1 \frac{n_L p_1}{a_1} \Phi_p k_w \tag{9.58}$$

Now the equivalent current through the one-turn coil winding is

$$I_{q1} = \frac{2T_n}{3p_1\Psi_{PM1}} \quad (9.59)$$

The stator phase resistance for the one-turn coil winding, R_{s1}, is

$$R_{s1} = 1000\rho \frac{n_{cs} l_{mc}}{a_1 S_{Cu}} \quad (9.60)$$

$$l_{mc} = 2(l_c + l_f)$$

where

ρ is the copper electric resistivity at the working temperature, T_{w1} (Equation 9.61) and

n_{cs} is the number of coils in series.

$$\rho = \rho_{20}\left(1 + (T_{w1} - 20)\alpha_{Cu}\right) \quad (9.61)$$

where

ρ_{20} is the copper electric resistivity at 20°C and
α_{Cu} is the resistivity coefficient with temperature.

The number of conductors in series per current path for a single-turn coil is

$$n_{cs} = \frac{2q_1 p_1}{n_L a_1} \quad (9.62)$$

where

a_1 is the number of current paths,
n_L is the number of winding layers ($n_L = 1,2$), and
$p1 = ps/2$ is pole pairs.

The cyclic magnetization inductance of the single-turn coil winding is

$$L_{m1} = 2m\mu_0 \cdot \frac{\left(\frac{n_{cs}}{2} \cdot f_{w1}\right)^2 \cdot l_c \cdot \tau}{\pi^2 \cdot p_1 \cdot k_C \cdot (1 + k_s) \cdot g_e} \quad (9.63)$$

The saturation factor, k_s, has been calculated on no load. For heavily saturated machines, k_s has to be calculated including the influence of the stator mmf. Equation 9.63 is valid for $q_1 > 1$. The leakage inductance is calculated using geometrical slot- and end-connection permeances, λ_{ss} and λ_{s0}.

$$k_2 = \begin{cases} \dfrac{3\left(2 - \dfrac{y_1}{mq_1}\right) + 1}{4} & \text{if} \quad \dfrac{y_1}{mq_1} > 1 \\ \dfrac{1 + 3\dfrac{y_1}{mq_1}}{4} & \text{if} \quad \dfrac{2}{3} < \dfrac{y_1}{mq_1} < 1 \\ \dfrac{6\dfrac{y_1}{mq_1} - 1}{4} & \text{if} \quad \dfrac{y_1}{mq_1} < \dfrac{2}{3} \end{cases} \tag{9.64}$$

$$k_1 = \frac{1 + 3k_2}{4} \tag{9.65}$$

$$\lambda_{ss} = k_1 \frac{h_{s1}}{b_1} + k_2 \left(\frac{h_{s2}}{b_2} + \frac{h_{s3}}{b_3} + \frac{h_{s4}}{s_0} \right) \tag{9.66}$$

where

b_1 is the active slot average width and

b_2 and b_3 are the widths corresponding to the trapezoidal-shape zone free of conductors,

$$b_1 = \frac{\left(w_{s1} + w_{s2}\right)^2}{0.25\left(3w_{s1} + w_{s2}\right) + 0.5w_{s1}^2 \dfrac{w_{s2} - 3w_{s1}}{\left(w_{s1} - w_{s2}\right)^2} + \dfrac{w_{s1}^4}{\left(w_{s1} - w_{s2}\right)^3} \log\left(\dfrac{w_{s1}}{w_{s2}}\right)} \tag{9.67}$$

$$b_1 \approx \frac{3}{2}\left(w_{s1} + w_{s2}\right)$$

$$b_2 = \frac{w_{s2} - w_{s3}}{\log\left(\dfrac{w_{s2}}{w_{s3}}\right)} \tag{9.68}$$

$$b_2 \approx \frac{w_{s2} + w_{s3}}{2}$$

$$b_3 = \frac{w_{s2} - s_0}{\log\left(\dfrac{w_{s3}}{s_0}\right)} \tag{9.69}$$

The analytical computation of the geometrical end-connection permeance, λ_{s0}, is difficult, but there are quite a few empirical approximate expressions that are widely used:

$$\lambda_{s0} = 0.34 q_1 \frac{l_f - 0.64\left(D_{si} + h_{sOA}\right)\dfrac{y_1}{2p_1 m q_1}}{l_c} \tag{9.70}$$

This expression is mainly valid for distributed windings ($q_1 > 1$), since for the tooth-wound coil λ_{s0} is larger as the conductors are closer to the core but the end connections are shorter.

Finally, the leakage inductance, $L_{\sigma 1}$, per phase for a single-turn coil is

$$L_{\sigma 1} = 2\mu_0 \left(\frac{n_{cs}}{2}\right)^2 \frac{\lambda_{ss} + \lambda_{s0}}{p_1 q_1} \cdot \frac{l_c}{1000} \tag{9.71}$$

The total cyclic inductance, L_{s1} (when all three currents are present) is

$$L_{s1} = L_{m1} + L_{\sigma 1} \tag{9.72}$$

Considering the total current in axis q and the parameters of the single-turn coil, we may calculate the number of turns per coil, s_{b1}, for operating at given voltage and speed (frequency):

$$s_{b1} = \frac{V_n}{\sqrt{\left(\omega_1 \Psi_{PM1} + R_{s1} I_{q1}\right)^2 + \left(\omega_1 L_{s1} I_{q1}\right)^2}} \tag{9.73}$$

$$s_{b1} \approx \frac{0.9 V_{nf} \sqrt{2}}{w_1 \Psi_{PM1}}$$

The number of turns per coil, s_{b1}, has to be an integer, so it has to be rounded to s_b. Consequently, the machine length has to be recalculated:

$$l_c = \frac{s_{b1}}{s_b} l_c' \tag{9.74}$$

where l_c' is the old machine length (Equation 9.8). The l_c value also is rounded to an integer in millimeters. The machine parameters, dependent on the machine length, such as R_s, Ψ_{PM1}, and L_{s1}, are recomputed. However, at least for $n_{s1} > 20$ turns, the use of Equation 9.74 only, is feasible.

Now we calculate the conductor area:

$$q_{Cu} = \frac{S_{Cu}}{s_b} \tag{9.75}$$

To reduce the skin effect, the use of n_{ce} elementary conductors of diameter d_{ce} in parallel may be required:

$$d_{ce} = 2\sqrt{\frac{q_{Cu}}{\pi n_{ce}}} \tag{9.76}$$

After norming d_{ce} (according to standards), the total final conductor area is recalculated. All machine parameters, N_1 (number of turns), Ψ_{PM}, R_s, L_m, and $L_{\sigma l}$ are recalculated for the final values of s_b and l_c calculated above:

$$N_1 = s_b q_1 \frac{n_L p_1}{a_1} \tag{9.77}$$

$$\Psi_{PM} = N_1 \Phi_p k_w \tag{9.78}$$

$$I_n = \frac{2\sqrt{2}}{m} \cdot \frac{T_{nem}}{p_1 \Psi_{PM}} \tag{9.79}$$

$$R_s = 1000\rho \frac{2N_1 l_{mc}}{a_1 q_{Cu}} \tag{9.80}$$

$$L_{\sigma 1} = 2m\mu_0 \frac{(N_1 k_w)^2 l_c \tau_p}{\pi^2 p_1 a_1 k_c k_s g_e}; \quad p_1 = p_s/2 \tag{9.81}$$

$$L_{\sigma 1} = 2\mu_0 \frac{N_1^2}{p_1 a_1 q_1} (\lambda_{ss} + \lambda_{s0}) \cdot \frac{l_c}{1000} \tag{9.82}$$

9.6.3 Weights of Active Materials

The weight of windings:

$$m_{cu} = 2ma_1 N_1 l_{mc} q_{Cu} \gamma_{Cu} 10^{-9} \tag{9.83}$$

The PM weight:

$$m_{PM} = \pi \lambda_{PM} (D_{r0} - h_{PM}) h_{PM} l_{cr} \gamma_{Fe} 10^{-9} \tag{9.84}$$

The rotor yoke weight:

$$m_{rFe} = \pi (D_{ri} + h_{ry}) h_{ry} l_{cr} \gamma_{Fe} 10^{-9} \tag{9.85}$$

The stator yoke weight:

$$m_{sy} = \pi (D_{s0} - h_{sy}) h_{sy} l_c k_{stk} \gamma_{Fe} \tag{9.86}$$

The stator teeth weight:

$$m_{\text{sth}} = N_{\text{ss}} w_{\text{st}} h_{\text{s}} l_{\text{c}} k_{\text{stk}} \gamma_{\text{Fe}} 10^{-9} \tag{9.87}$$

The stator core weight:

$$m_{\text{sFe}} = m_{\text{sy}} + m_{\text{sth}} \tag{9.88}$$

The active material weight:

$$m_{tot} = m_{rFe} + m_{PM} + m_{sFe} + m_{Cu} \tag{9.89}$$

As the laminations are stamped from squares, their weight, m_{uFe}, is in fact

$$m_{\text{uFe}} = D_{s0}^2 \cdot l_c \cdot k_{\text{stk}} \cdot \gamma_{\text{Fe}} \cdot 10^{-9} \tag{9.90}$$

These weights have to be used when the critical cost of the materials of the machine is calculated.

The rotor inertia, J_s, is

$$J_{\text{s}} = m_{\text{PM}} \cdot \frac{D_{r0}^2 + D_{r1}^2}{8} \cdot 10^{-6} + m_{\text{ry}} \cdot \frac{D_{r1}^2 + D_{\text{ri}}^2}{8} \cdot 10^{-6} \tag{9.91}$$

The shaft inertia is added to the value from Equation 9.91.

9.6.4 Losses

Winding losses:

$$P_{\text{Cu}} = m R_{\text{s}} I_{\text{n}}^2 \tag{9.92}$$

Core losses:

The core losses are calculated separately for the stator teeth and the stator yoke. Based on available core-loss data, we may calculate hysteresis and eddy current losses per unit volume either separately or together [4]:

$$p_{\text{ht}} = \frac{1}{\gamma_{\text{Fe}}} k_{\text{h}} f_{\text{n}}^{\text{efh}} B_{\text{st}}^{\text{eBh}} \tag{9.93}$$

$$p_{\text{hy}} = \frac{1}{\gamma_{\text{Fe}}} k_{\text{h}} f_{\text{n}}^{\text{efh}} B_{\text{sy}}^{\text{eBh}} \tag{9.94}$$

$$p_{\text{et}} = \frac{\pi^2}{6 \rho_{\text{Fe}} \gamma_{\text{Fe}}} f_{\text{n}}^{\text{efe}} B_{\text{st}}^{\text{eBe}} g_{\text{t}}^{\text{egt}} \tag{9.95}$$

$$p_{ey} = \frac{\pi^2}{6\rho_{Fe}\gamma_{Fe}} f_n^{efe} B_{sy}^{eBe} g_t^{egt} \quad (9.96)$$

$$p_{Fe_t} = p_{ht} + p_{et} \quad (9.97)$$

$$p_{Fe_y} = p_{hy} + p_{ey} \quad (9.98)$$

$$p_{Fe_t} = p_{Fe_t} m_{sth} \quad (9.99)$$

$$p_{Fe_y} = p_{Fe_y} m_{sy} \quad (9.100)$$

$$p_{Fe} = p_{Fe_t} + p_{Fe_y} \quad (9.101)$$

where
- efh is the frequency exponent for hysteresis losses,
- eBh is the flux density exponent for hysteresis losses,
- efe is the frequency exponent for eddy current losses,
- eBe is the flux density exponent for eddy current losses,
- egt is the thickness exponent, and
- gt is lamination thickness.

The segregation of yoke, p_{Fey}, and teeth, p_{Fet}, core losses allows to follow their evolution, and, thus, facilitates corrective measures to reduce them in the critical zone.

9.6.5 Thermal Verification

To roughly verify the winding over temperature of the machine, first the total frame area, A_{fr}, for heat transfer is calculated:

$$A_{fr} = 10^{-6}\left(\pi k_f D_{s0} l_f + \frac{\pi}{2} D_{s0}^2\right) \quad (9.102)$$

where k_f is the cooling surface increase ratio due to fins.

Then adding an equivalent heat transfer coefficient, α_t (in W/m °C), say from 14 (for unventilated frames) to 100 (for water-cooled frame jackets), the winding over temperature, T_{w1}, is

$$T_{w1} = \frac{P_{Cu} + P_{Fe}}{\alpha_t A_{fr}} \quad (9.103)$$

If the over temperature is high, while α_t has a realistic value in accordance with the adopted cooling system, then the machine design is redone from the start, with a smaller J_1 (or f_{tsp}) and/or longer λ_c (longer core length).

9.6.6 Machine Characteristics

In general, the machine characteristics for a PMSM include

- Base speed, n_b and maximum speed, n_{max}
- Continuous torque at base speed and base (maximum) voltage
- Peak torque at base speed
- Torque at maximum speed
- EMF at maximum speed
- Efficiency and power factor (or stator current) versus load and speed for given maximum fundamental inverter voltage, V_1

More on realistic modeling of SPMSMs is available in Refs. [2–4].

9.7 OPTIMAL DESIGN WITH GENETIC ALGORITHMS

For the optimization design of SPMSMs with genetic algorithms (GA), we introduce the following optimization variables:

- Linear electric loading, J_1 (A/m)
- Airgap flux density, B_{ag} (T)
- Stator teeth flux density, B_{st} (T)
- Stator yoke flux density, B_{sy} (T)
- Rotor yoke flux density, B_{ry} (T)
- Stator current density, J_s (A/mm^2)
- Machine shape factor, $\lambda_c = l_c/\tau$
- Slots per pole per phase, q_1
- Slot opening, s_o (mm)
- Tooth top height, h_{s4} (mm)
- Coil span, y_1 (mm)
- PM pole span coefficient, λ_{PM}

These 12 variables are grouped in a vector X_0. The variable vector X is not directly initialized by the designer, but he/she will attribute minimum and maximum values, X_{min} and X_{max}, together with the corresponding resolution ΔX. Also, all minimum values of geometrical variables that are imposed for technological or for mechanical reasons are grouped in G_{dmin}:

- Outer rotor diameter
- Interior rotor diameter

- Minimum active slot height
- Minimum total slot height
- Minimum slot width
- Minimum active slot area

The chosen GA operates the binary coding of optimization variables. As some of the optimization variables, such as the number of stator slots per pole per phase, q_1, and the coil span in slot pitches, are integers while others are rational numbers with extended dynamics, they are represented with different numbers of bits. Now, with the variable domain, $X_{max} - X_{min}$, and the representative resolution, the required number of bits, N_{bit}, is

$$N_{bit} = 1 + \text{int}\left(\log_2\left(\frac{X_{max} - X_{min}}{\Delta X}\right)\right) \tag{9.104}$$

To set the GA (detailed in Chapter 8) we give the population size, n_p, the number of individuals (in a generation), the number of generations, n_g, the elitism factor, k_{elit}, the mutation rate, r_m, and the occupation factor, k_{ex}, of those individuals who take part in producing genetic material for the next generation.

9.7.1 Objective (Fitting) Function

We have chosen a complex objective (fitting) function here: the initial cost of the motor plus the loss cost for the machine's active life plus additional framing and transportation costs due to its weight. The initial cost, C_i, includes only the costs of active materials, as the total manufacturing and selling costs depend highly on the fabrication technology and cost management of the particular manufacturer:

$$C_i = m_{cu} p_w + m_{Fe_u} p_{lam} + m_{PM} p_{PM} \tag{9.105}$$

where p_w, p_{lam}, and p_{PM} are the copper, the lamination, and the PM unitary prices, say in USD(or EU)/kg.

The cost of losses, C_E, is

$$C_E = P_N\left(1 - \frac{1}{\eta_N}\right) n_{hy} n_y p_E \tag{9.106}$$

where

n_{hy} is the annual operational hours,
n_y is the operational years, and
p_E is the energy cost (USD(EU)/kWh.

As a motor very rarely operates all the time at the rated power and speed, we may introduce an ideal operation cycle characterized by probability α_i to operate at different power loads, P_i, and efficiencies, η_i:

$$C_E = n_{hy} n_y p_E \sum_{i=1}^{n} \alpha_i P_i \left(1 - \frac{1}{\eta_i}\right) \tag{9.107}$$

where $\sum \alpha_i = 1, i = \overline{1,n}$.

Though C_E is in general larger than C_i, the value of C_E is diminished in the design as many buyers cannot afford a higher initial cost; in such cases, in fact, a kind of optimal efficiency in terms of the initial/total cost is adopted. Now the cost of losses, C_E, is

$$C_E = n_{hy} n_y p_E \sum_{i=1}^{n} \alpha_i P_i \tag{9.108}$$

where $P_i = \begin{cases} P_i \left(\dfrac{1}{\eta_{oi}} - \dfrac{1}{\eta_i}\right) & \text{if} \quad \eta_i < \eta_{oi} \\ 0 & \text{if} \quad \eta_i \geq \eta_{oi} \end{cases}$

In aircraft applications, low weight is a critical performance index. A similar situation occurs in wind generators when low-cost but poor-performance materials lead to a larger machine weight, which implies a heavy nacelle and tower, and their extra costs may offset the reduction in the generator initial costs. In order to force the algorithm to reduce the motor (generator) weight, we may add some additional costs, C_m, proportional to the motor weight, m_t:

$$C_m = m_t p_m \tag{9.109}$$

Also, surpassing the maximum allowable winding (as well as the core and the PM) temperature, leads to immature aging of the equipment. To avoid this, with the optimization algorithm we introduce a penalty cost for the over temperature, C_{temp}, in the objective function. This penalty cost, C_{temp}, may vary linearly or exponentially with the over temperature:

$$C_{temp} = \begin{cases} k_T \left(T - T_{max}\right) C_i & \text{if} \quad T > T_{max} \\ 0 & \text{if} \quad T_{max} \end{cases} \tag{9.110}$$

We may add penalty costs for any other technical constraints. For example, we may add a penalty for a lower power factor, which is related to the converter kVA rating costs, as also for the outer diameter or for the machine axial length, etc. The larger the specific costs of these constraints, the higher the probability that the optimization design will observe them.

Finally, the total cost, C_t, is

$$C_t = C_i + C_E + C_m + C_{temp} \tag{9.111}$$

9.7.2 PMSM Optimization Design Using Genetic Algorithms: A Case Study

An exercise design of a three-phase, four-pole synchronous motor with a surface PM rotor for a rated power of 2.2 kW and base frequency of 50 Hz is presented. The maximum and minimum values of optimization variables are presented in Table 9.1. The technological constraints used for this exercise design can be found in the input file (pmm1.m) of the PMSM optimal design MATLAB code, which is available online on the CD (which is attached to the book) in the "PMSMdesign" folder. The total cost, C_t, represents the objective function.

The evolution of the genetic algorithm is illustrated in the following figures. Every generation has a member that is best fitted for the required optimum criterion, which minimizes the cost function. The generation itself is characterized by the average cost function. For each generation, there is also a least-adapted member. Following the evolution of these three values, we may appreciate the GA convergence. At the beginning, there is a large difference between the cost function associated to the best member and the average value of the cost function. The value of the cost function of the least-adapted member is larger than 10 times the average value. A large gap between the minimum and the maximum cost function for members over a generation shows a population with a good evolution perspective. The statistics values (minimum, average, and maximum) of the cost function tend to decrease during the evolution. The minimum value of the cost function is decreasing monotonically along the generations while the average and the maximum values sometimes are increasing, especially for the first generations (Figure 9.6), where a large variation of the maximum cost can be observed.

In time, the members of the population begin to hold similar codes and the variation of the maximum and the minimum cost function from one generation to the next diminishes. Gradually, the minimum value of the objective function remains unchanged for many generations and the cost variation is small even if it exists.

When the members were chosen randomly, in pairs, to generate an offspring, a minimum difference in the cost function was imposed, in order to avoid a premature evolution. A premature evolution has produced numerous twins in the population, and the maximum and average costs collapsed to the minimum cost. The main method, crossover, used to generate an offspring is not able to bring about any improvement in this state. Then a new genetic code is produced only by mutation, but if the new members have cost functions higher than those of the majority of the population members, which are very similar, the new genetic code will be eliminated in a few generations.

The main geometric dimensions and the slot geometry evolution of the best member along generations are presented in Figure 9.7. It can be observed that geometric dimensions take discrete values and the same dimension can be arrived again after several generations.

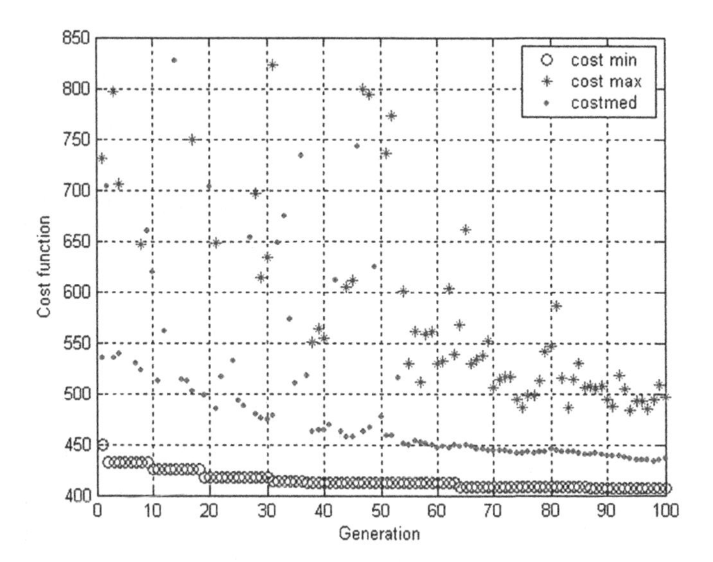

FIGURE 9.6 Minimum, maximum, and average cost function evolutions.

The power losses and the energy efficiency evolution of the best member are shown in Figure 9.8. The mechanical power loss is set constant, as 0.5% of the rated power, in the input power. A small efficiency improvement (around 1%) is observed in the first ten generations. The best-candidate solution from the first generation has already a good efficiency because it represents the best solution of a population of 150 members.

The machine components' weights and their cost evolutions are presented in Figure 9.9.

The laminations are punched from rectangular sheets and the minimum rectangular (square) surface that encloses the stator lamination is considered when computing the weight of used iron. When the rotor core is made of laminations, then it is punched from the same sheet as the stator lamination at no extra cost for the material. A solid rotor core is used in many cases for small PMSMs, and, in this case, the cost of the rotor core material is added.

The initial cost and energy cost evolutions for the best-candidate solution are shown in Figure 9.10. We can observe that their values are about the same, which is a well-known optimization "thumb rule." The sum of the initial cost and the energy cost is equal to the total cost, which is the objective function plotted in Figure 9.10 in a better scale. Usually, the over temperature penalty cost is equal to zero for the best-candidate solution.

The optimization variable evolutions for the best-candidate solution are presented in the following figures. The maximum and minimum values of the optimization variables as well as their resolutions are given in the input file. The number of discrete levels of each variable is usually not equal to powers of two. The required number of bits (Equation 9.102), necessary to cover the discrete levels of the optimization variable, could code a larger number of levels that is required from the initial file specification. Using the described crossover method, it is possible to generate an

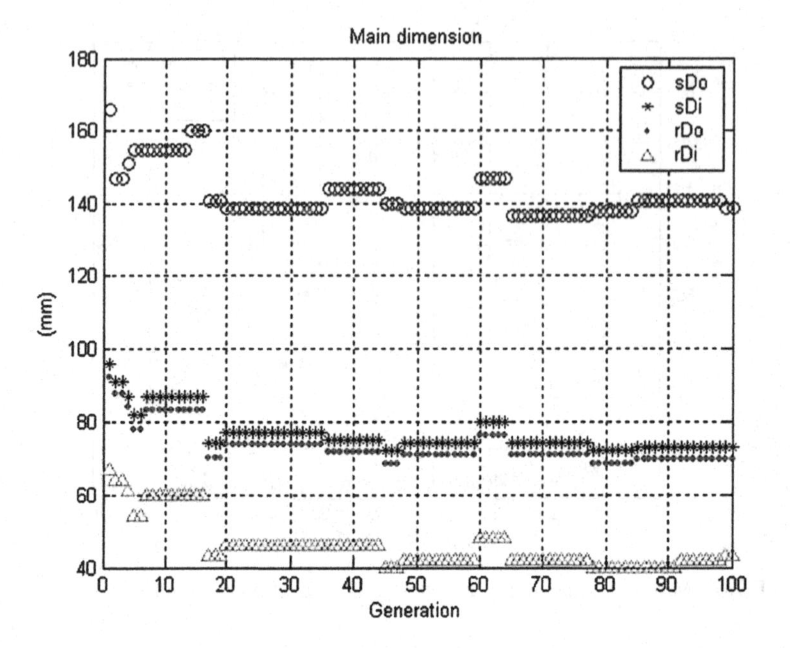

(a)

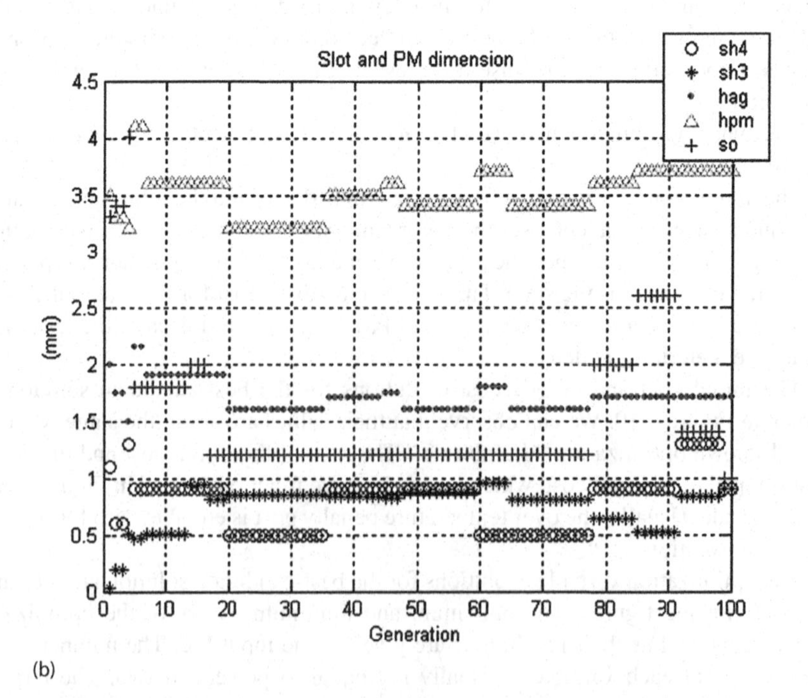

(b)

FIGURE 9.7 (a) Main geometric and (b) slot dimensions evolution of the best member.

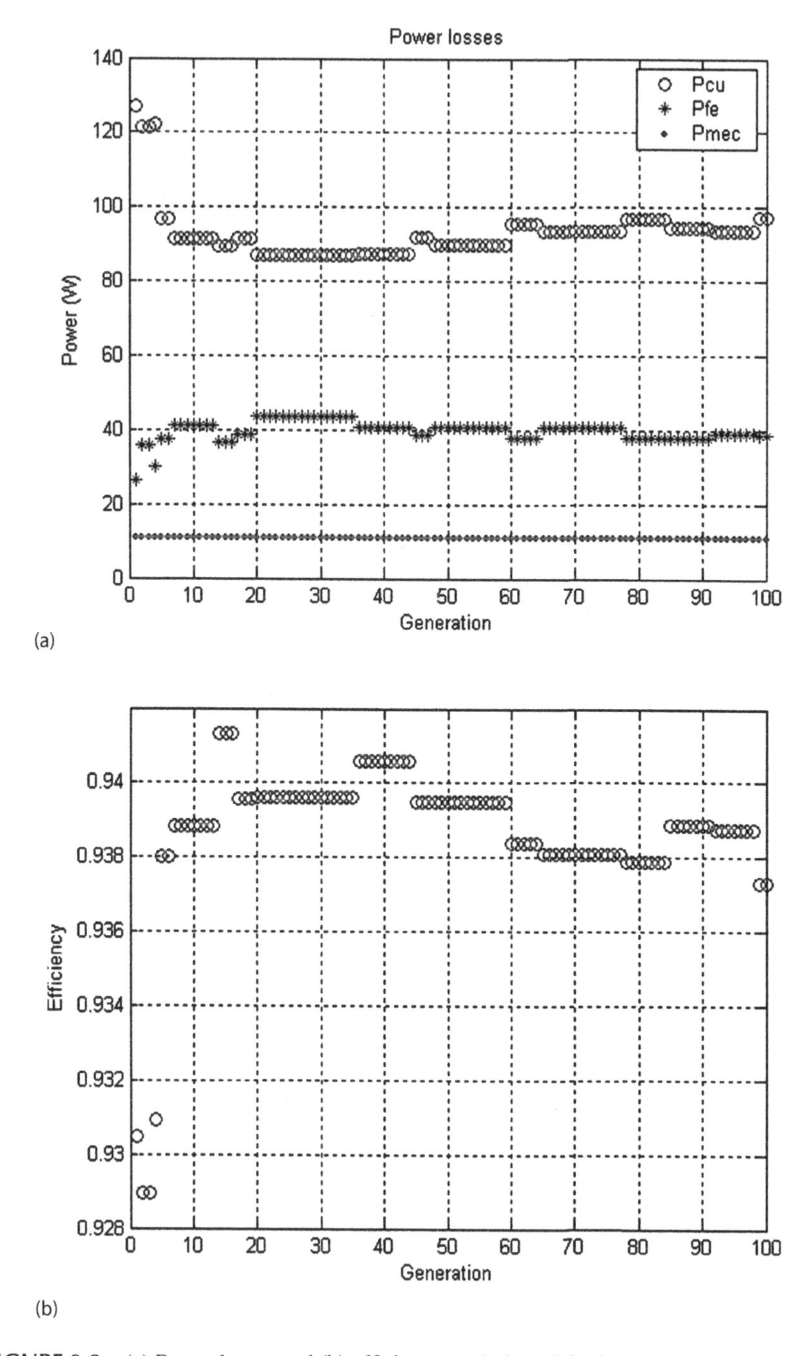

FIGURE 9.8 (a) Power losses and (b) efficiency evolution of the best member.

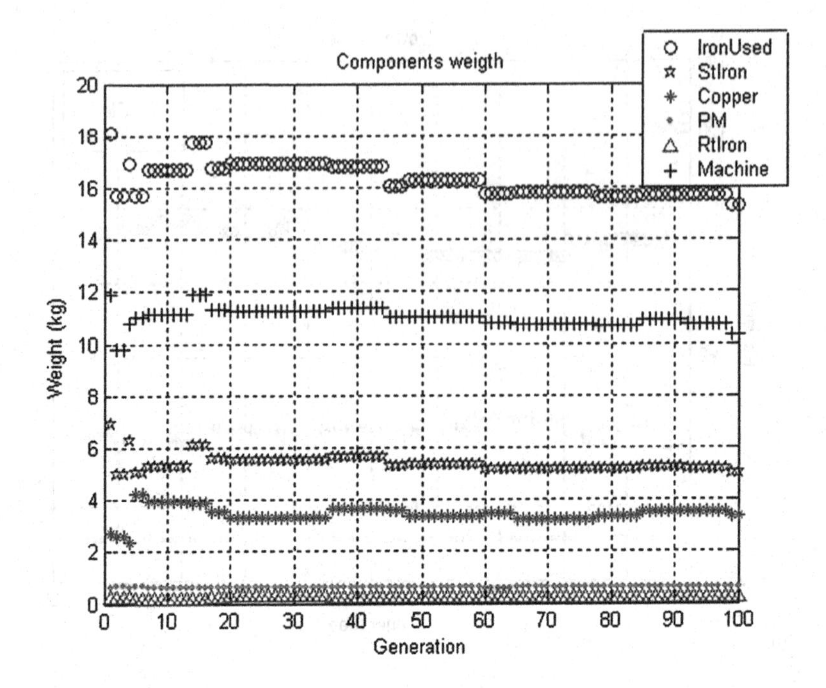

(a)

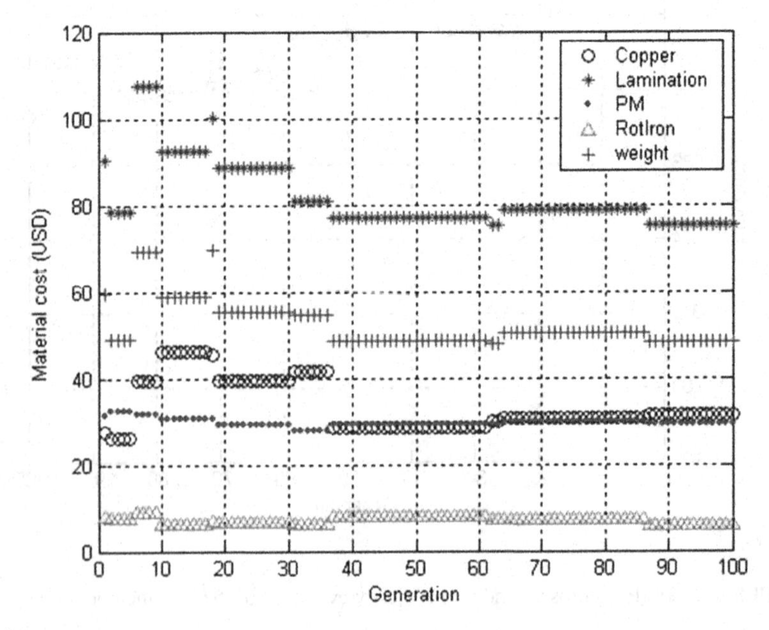

(b)

FIGURE 9.9 (a) The components' weights and (b) their costs for the best-candidate solution.

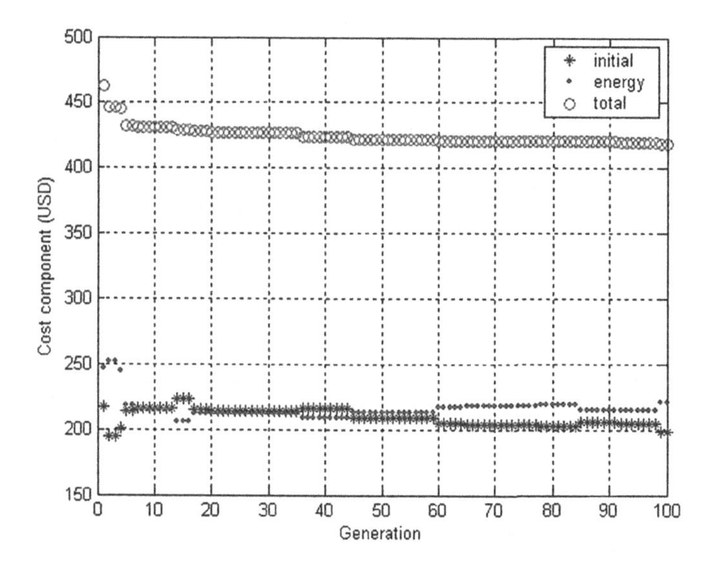

FIGURE 9.10 Evolution of the cost (objective) function of the best-candidate solution.

offspring that has one or more optimization variables larger than the upper bounds. One solution is to discard all offsprings that have their optimization variables outside the bounds and try again to generate viable offsprings. Another solution is to let the optimization variable be larger than the upper limits and let the optimization algorithm eliminate infeasible candidates. This method is applied when an optimization variable larger than the upper bounds has a physical meaning such as for electric and magnetic loads. One more method to solve this problem is to decrease the resolution in such a way that all binary codes are used to represent variables between given bounds. When a variable value out of its bounds is a nonsense, then a mathematical transformation is applied to force the optimization variable back between its bounds. For example, the ratio between the PM width and the pole pitch larger than one is nonsense. The following mathematical transformation is applied to avoid this nonsense:

$$\lambda_{PM} = \begin{cases} \lambda_{PM} & \text{if} \quad \lambda_{PM} \le 1 \\ 2 - \lambda_{PM} & \text{if} \quad \lambda_{PM} > 1 \end{cases} \tag{9.112}$$

Some features may have two distinct codes. For example, the code that is producing the number 1.1, is forced to produce a relative PM width equal to 0.9, and there exists also the code that produces directly 0.9.

The optimal value of linear electric load is larger than 40 kA/m (Figure 9.11), while the maximum value in the input file was set at 30 kA/m. The optimal value of the airgap flux also tends to its upper bound (Figure 9.12). Larger limits of these optimization variables will be set if there are no other restrictions.

The upper limit of the current density is large enough in our design, and, thus, there is no optimization value of the best-candidate solution larger than the upper

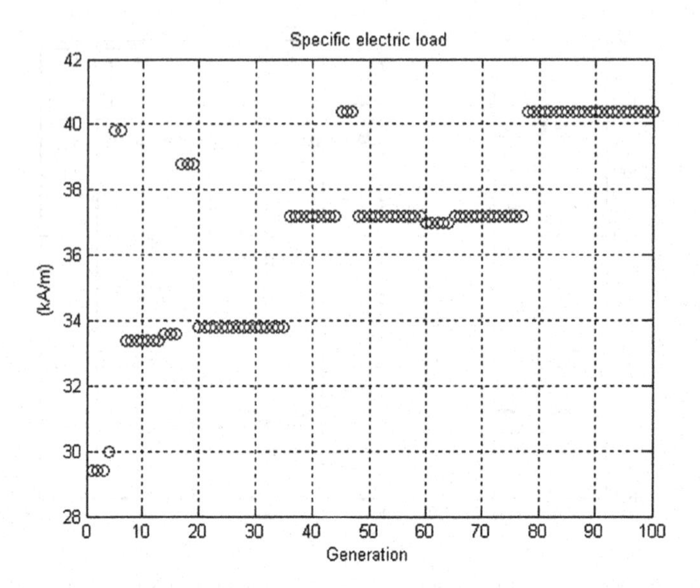

FIGURE 9.11 Specific electric load evolution of the best-candidate solution.

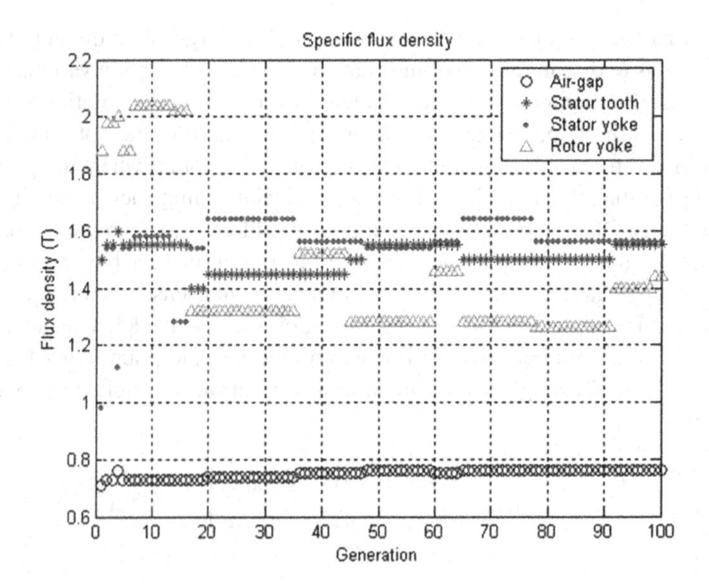

FIGURE 9.12 Flux density evolutions of the best-candidate solution.

limit (Figure 9.13). The codification method allows for larger current densities because the required domain needs 50 samples to be covered, and 6 bits are enough to code 64 samples.

The ratio of the core length to the pole pitch, the ratio of the coil span to the pole pitch, and the ratio of the PM width to the pole pitch are presented in Figure 9.14 for the best-candidate solution from each generation. The ratio of the coil span to the

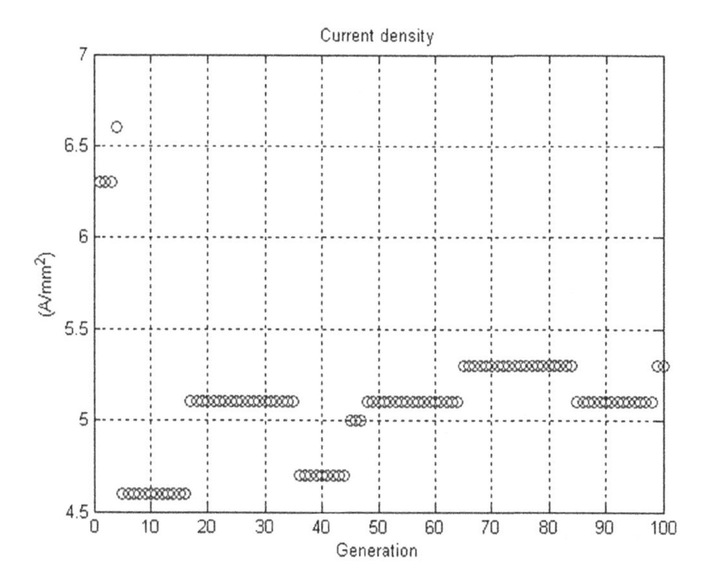

FIGURE 9.13 Current density evolution of the best-candidate solution.

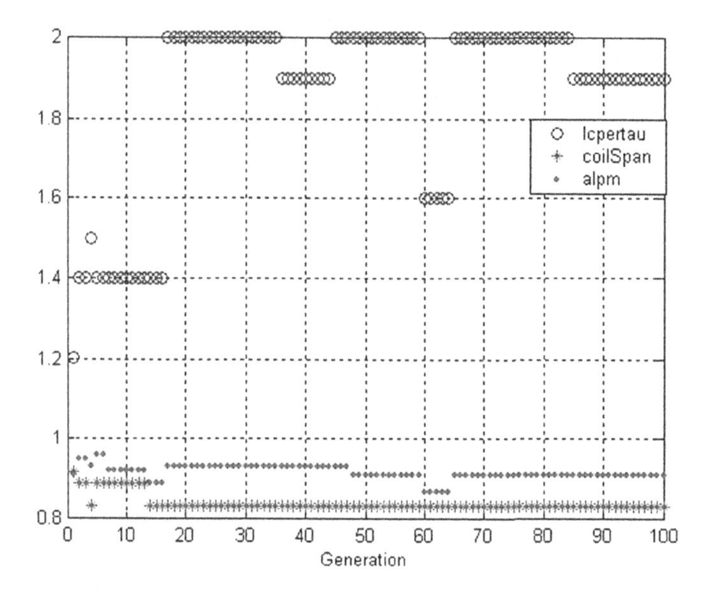

FIGURE 9.14 Evolutions of the ratios of dimensions of the best-candidate solution: core length per pole pitch (lcpertau), coil span per pole pitch (coilSpan), and PM width per pole pitch (Alpm) (all in pu).

pole pitch given by the GA is rounded to the nearest rational number, which has a denominator equal to the number of slots per pole.

The number of stator slots per pole per phase is always an integer and is shown in Figure 9.15 for the best-candidate solution from each generation.

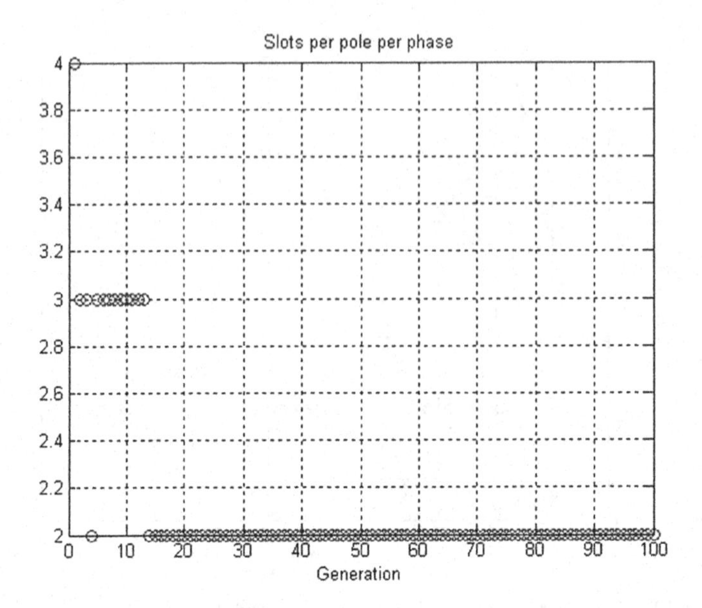

FIGURE 9.15 Evolution of the slots per pole per phase of the best-candidate solution.

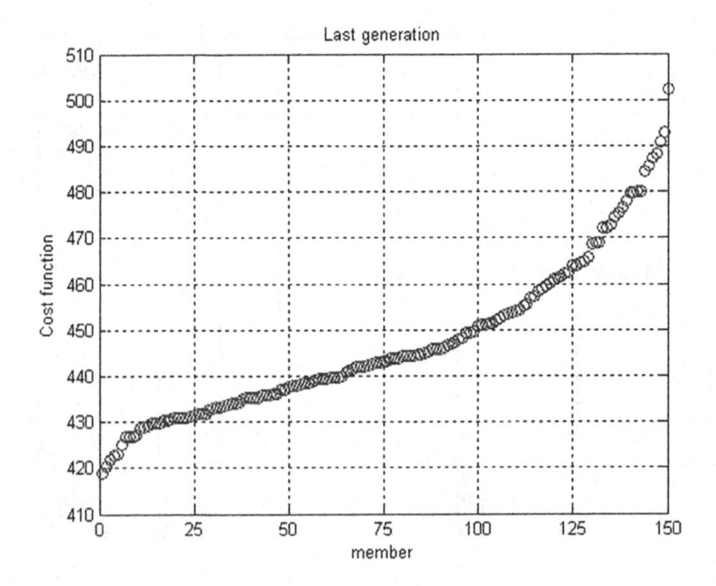

FIGURE 9.16 The cost function over the last generation.

All members of the last generation are obtained through a long selection process and it would be interesting to analyze the distribution of the optimization variables. At first sight the members look close to the objective function (Figure 9.16).

We may observe that not all possible levels are taken (Figures 9.17 through 9.20), and that there is no unambiguous rule on what values could minimize the optimization function.

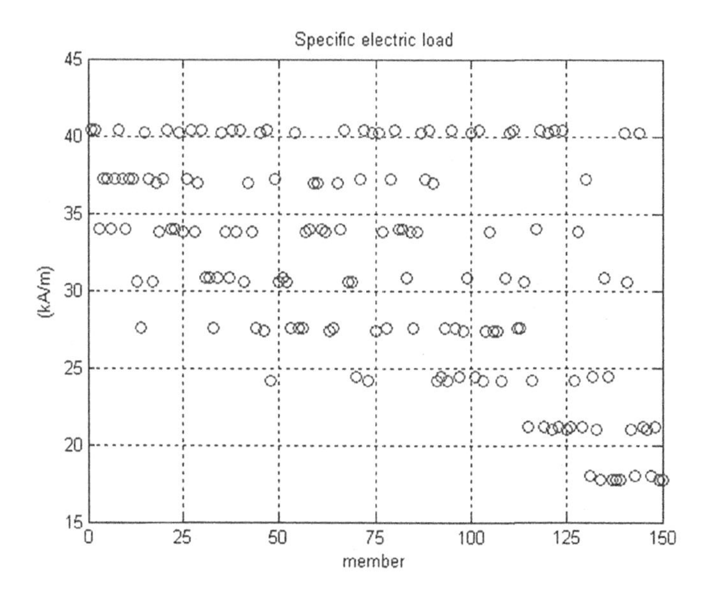

FIGURE 9.17 The specific electric load over the last generation.

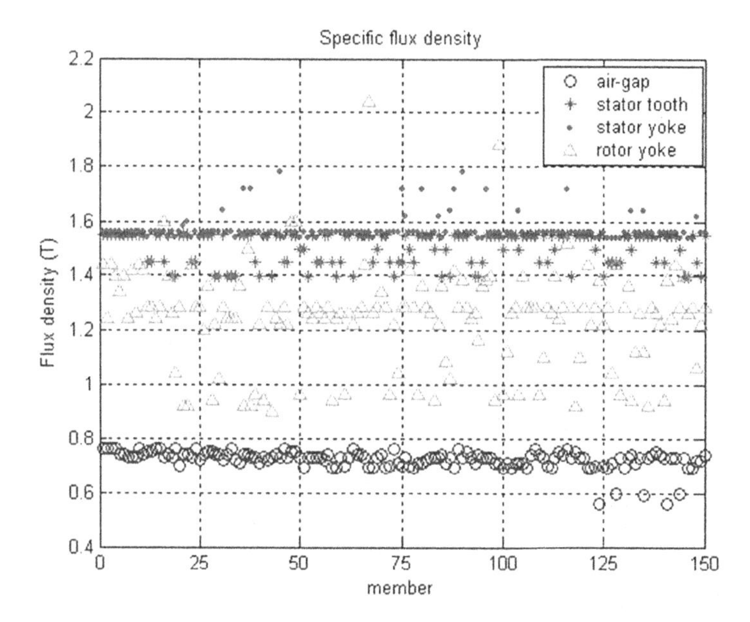

FIGURE 9.18 Flux densities over the last generation.

We are not able to assert that always the value of linear electric load that characterizes the best solution (40.4 kA/m) will produce good results (Figure 9.17).

We can state that probably a linear electric load smaller than 25 kA/m is too small to get an acceptable motor cost. Good solutions are possible with an airgap flux density around 0.75 T (Figure 9.18), respectively, with the stator yoke and stator tooth

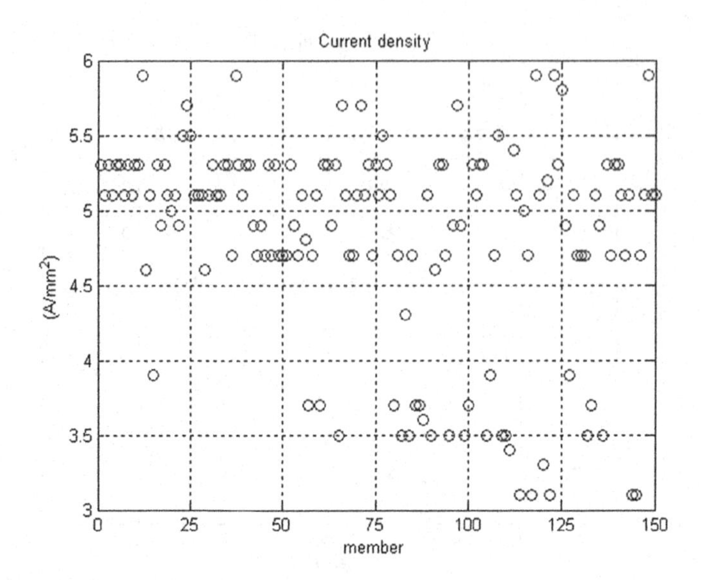

FIGURE 9.19 Current density over the last generation.

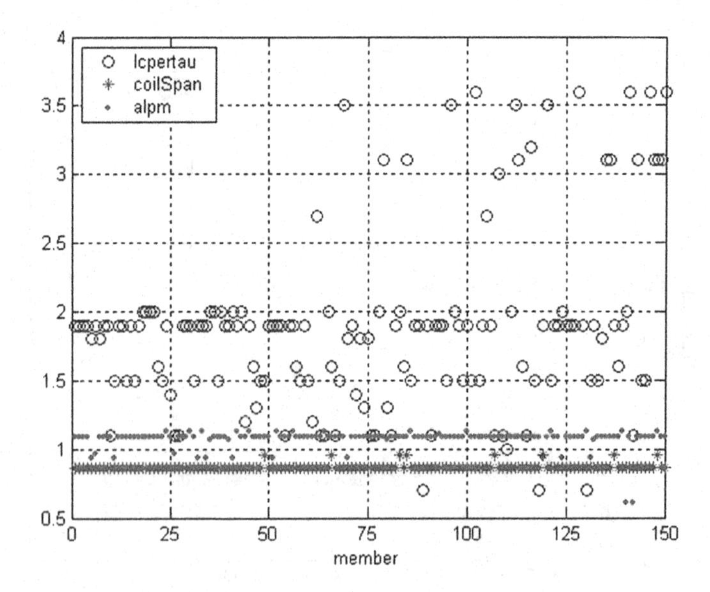

FIGURE 9.20 Ratio of dimensions over the last generation.

flux densities around 1.55 T. The rotor yoke flux density has a small influence on the objective function (Figure 9.18).

It is not possible, probably, to minimize the objective function when current densities become smaller than 4 A/mm^2. About the ratio-type optimization variables, we can state that a coil span per pole pitch equal to 0.83 and the PM width to pole pitch

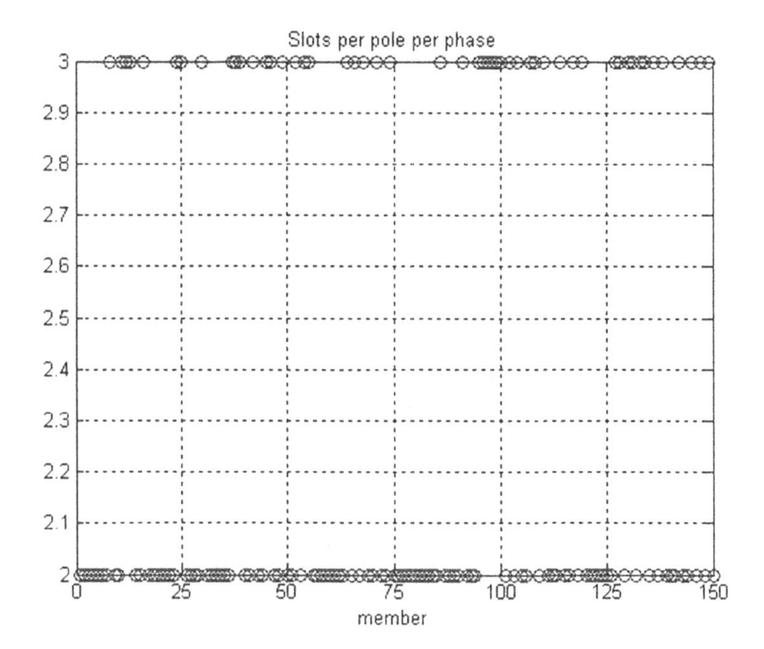

FIGURE 9.21 Slots per pole per phase over the last generation.

of around 0.91 could produce good solutions (Figure 9.20). A ratio of the core length to the pole pitch larger than 2.5 probably always produces bad results (Figure 9.20).

There are only two values selected for the number of slots per pole per phase, and we can remark that in the last generation two slots per pole per phase had higher occurrence frequencies than three slots per pole per phase, especially for high-performance members (Figure 9.21). The best-candidate solution for many populations has two slots per pole per phase, so, probably, this value will minimize the objective function.

Further improvement of the GA is possible by reducing the search area and even by setting some variables constant and then eliminating these from the optimization variable vector.

9.8 OPTIMAL DESIGN OF PMSMS USING HOOKE–JEEVES METHOD

The same optimization variable vector and objective function, as those presented for the GA method, are used also to implement the Hooke–Jeeves optimization algorithm (presented in detail in Chapter 8). The geometric dimensions as well as the optimization variables and cost function modifications look now as continuous functions, compared to the GA functions, where these functions look rather discontinuous. The evolutions of the main geometric and slot dimensions are presented in Figure 9.22.

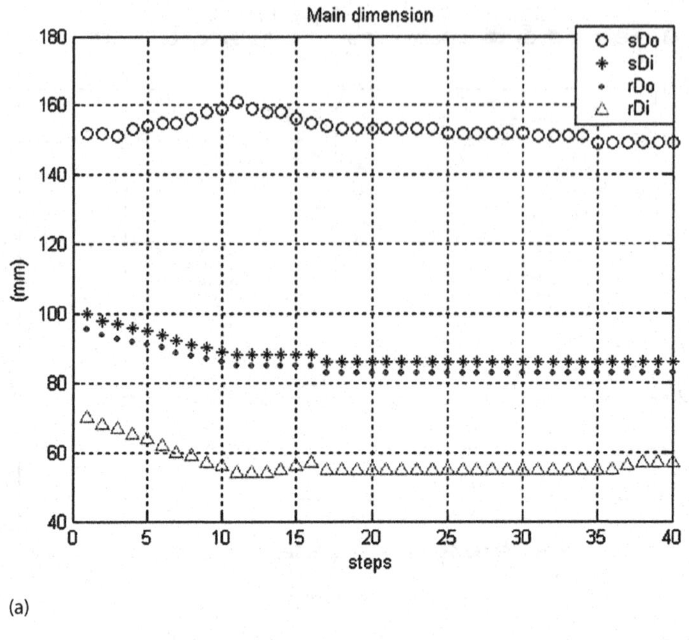

(a)

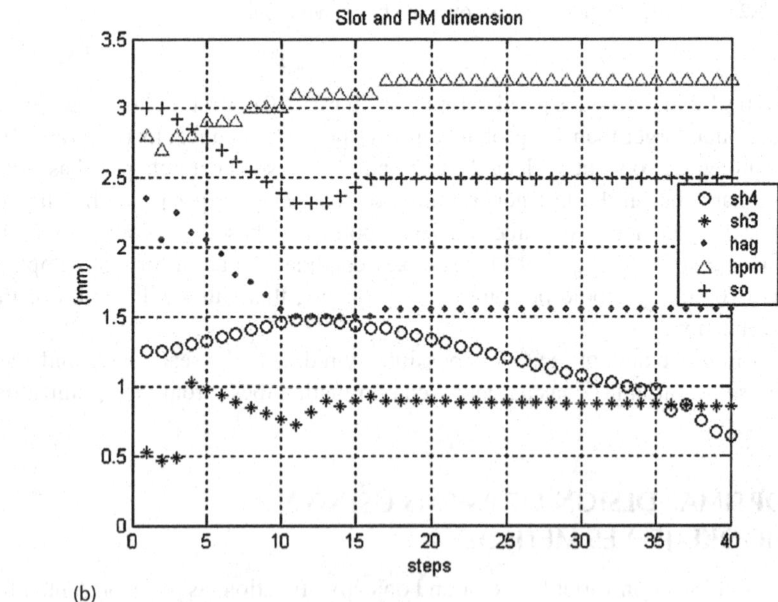

(b)

FIGURE 9.22 Geometric and slot dimensions evolution. (a) Main dimension and (b) slot and PM dimension.

The power losses and efficiency evolutions are shown in Figure 9.23, while the components' weights and costs are visible in Figure 9.24.

Cost function, specific (linear) electric load flux density, and current density evolutions are shown in Figures 9.25 through 9.28.

The ratio of the PM width to the pole pitch and the ratio of the coil span to the pole pitch are going to the same values as in the GA (Figure 9.29), while the core length per pole pitch ratio is going to a smaller value. The algorithm starts with three slots per pole per phase and after three steps it changes to two slots per pole per phase and remains at that value (Figure 9.30).

Comparative results of the optimal design using GAs and the Hooke–Jeeves method, respectively, are presented in Table 9.2.

The total cost is about the same for the two methods but the simulation time is about 75 times larger for GAs than for the Hooke–Jeeves algorithm. The GA chooses some optimization variables that are values outside the bounds, while the Hooke–Jeeves optimization variables remain strictly within the specified bounds. The motor efficiency designed by the Hooke–Jeeves algorithm is larger than the motor efficiency designed by the GA, but the initial cost is smaller for the GA and, probably, this is a consequence of a larger linear electric load allowed for the GA. The first variation step for each variable in the research process of the Hooke–Jeeves algorithm was 10% of the entire domain. Such a large initial step may help the algorithm examine a large area, and, thus, increase the chances to pass over local minima. Using Hooke–Jeeves methods repeatedly, from different (say 25) initial variable vectors (randomly chosen) would better the chance for global optimum, at one-third of GA computation effort.

9.9 FEM BASED OPTIMAL DESIGN OF A PM SPOKE MOTOR: A CASE STUDY

Optimal design requires a good mathematical model of electrical machine. Classical electrical machine as induction machine, d.c. excited synchronous machine, and surface permanent machine are well known and acceptable analytical models exist in the literature and industry practice. For new types of electrical machines a trusted analytical model does not exist and their development and validation is time-consuming and requires higher qualified experts. The increasing of computers performance allows to use an embedded FEM model to compute the machine performances in order to evaluate the objective function. The optimal design requires thousands of objective function evaluation and then the embedded FEM model has to be fast and it gives only key parameters as linkage flux, inductances at rated current and torque. The optimization variable should be adapted accordingly to the FEM opportunities but also considering the computation resource requirements of FEM. The FEM allows to analyze the influence of some geometric dimensions that in the analytical model are not faithfully reflected but, on the other hand, the number of

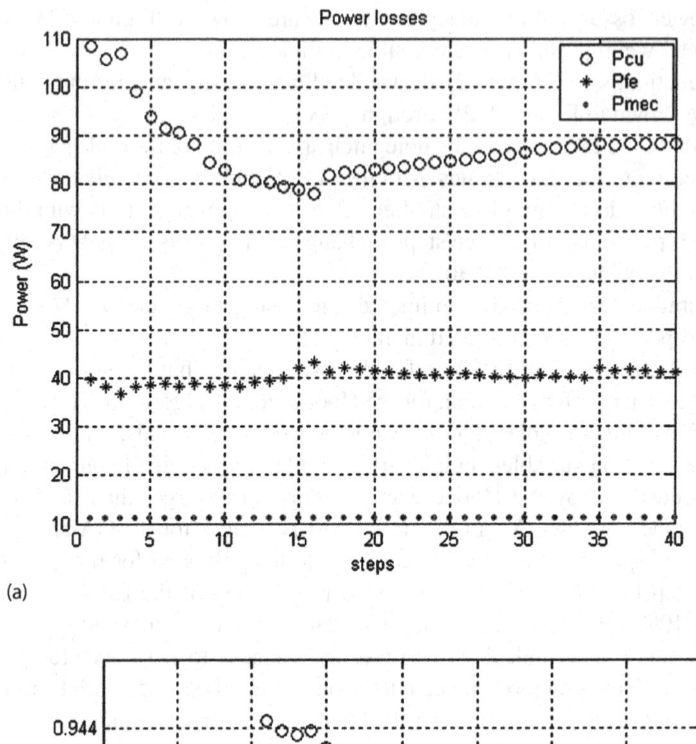

(a)

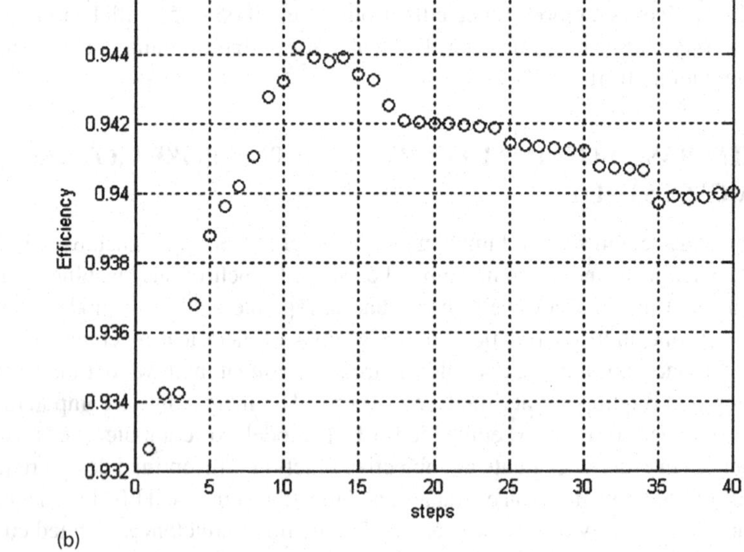

(b)

FIGURE 9.23 (a) Power losses and (b) efficiency evolutions.

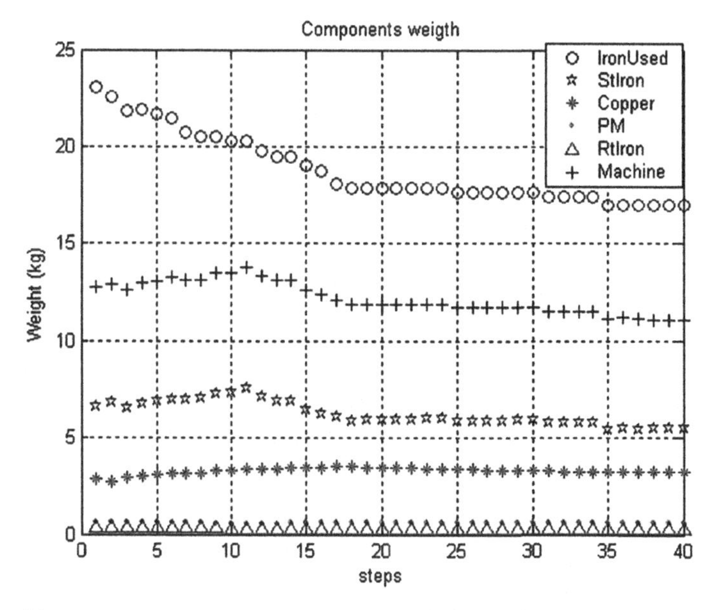

(a)

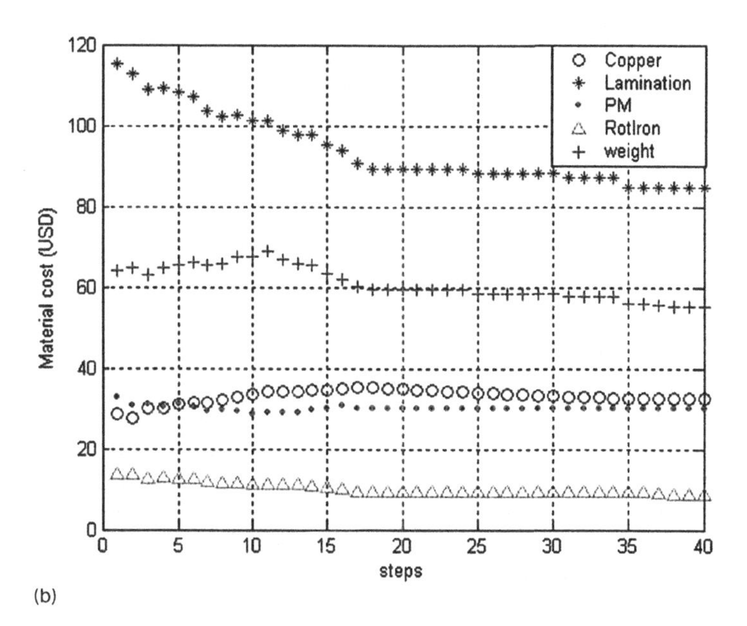

(b)

FIGURE 9.24 (a) Components' weights and (b) cost evolutions.

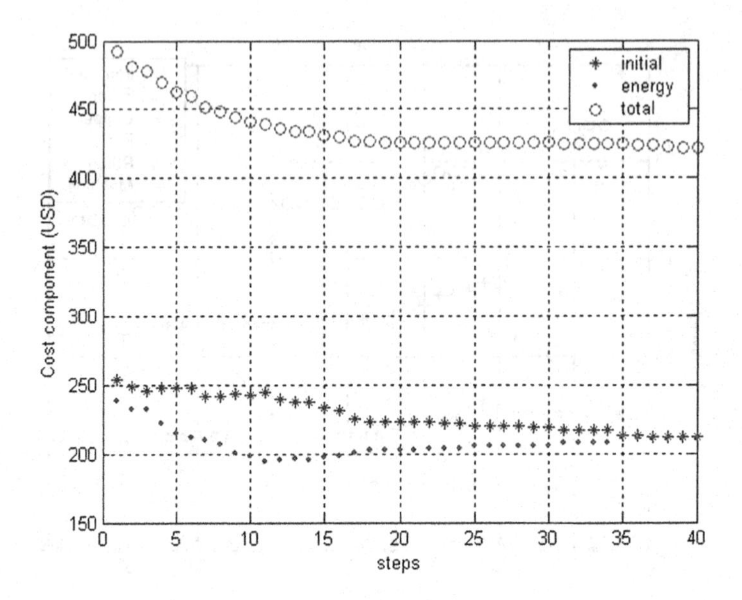

FIGURE 9.25 Cost (objective) function evolution.

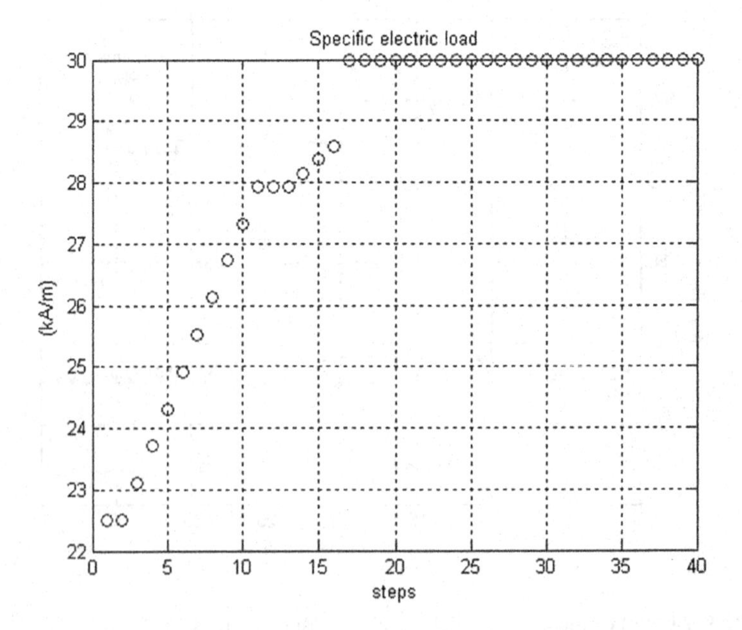

FIGURE 9.26 Specific load evolution.

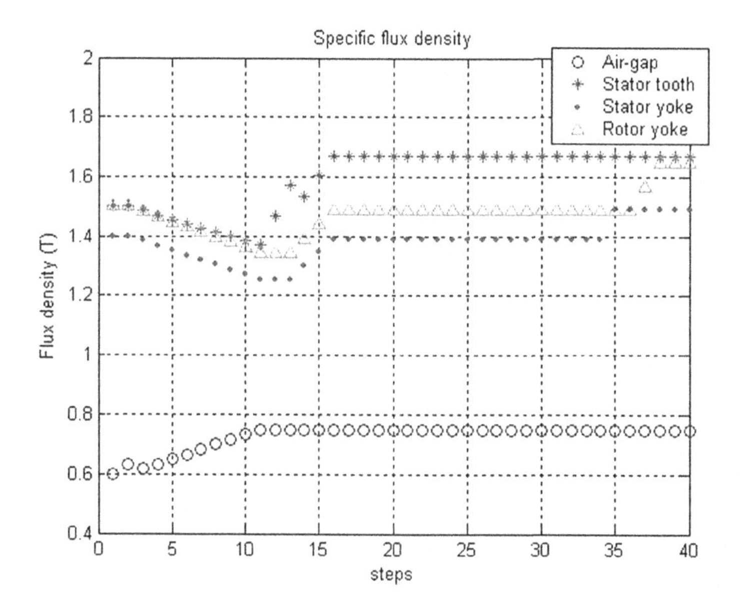

FIGURE 9.27 Flux density evolution.

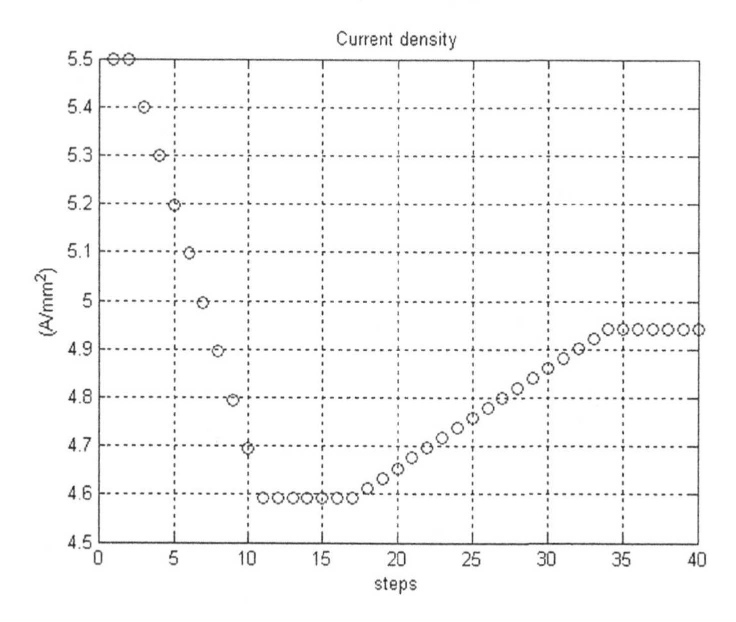

FIGURE 9.28 Current density evolution.

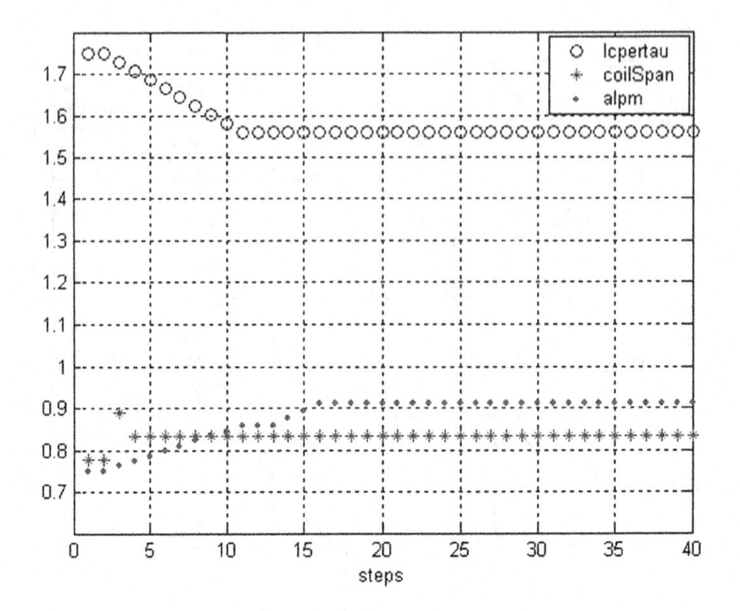

FIGURE 9.29 Dimensions ratios evolution.

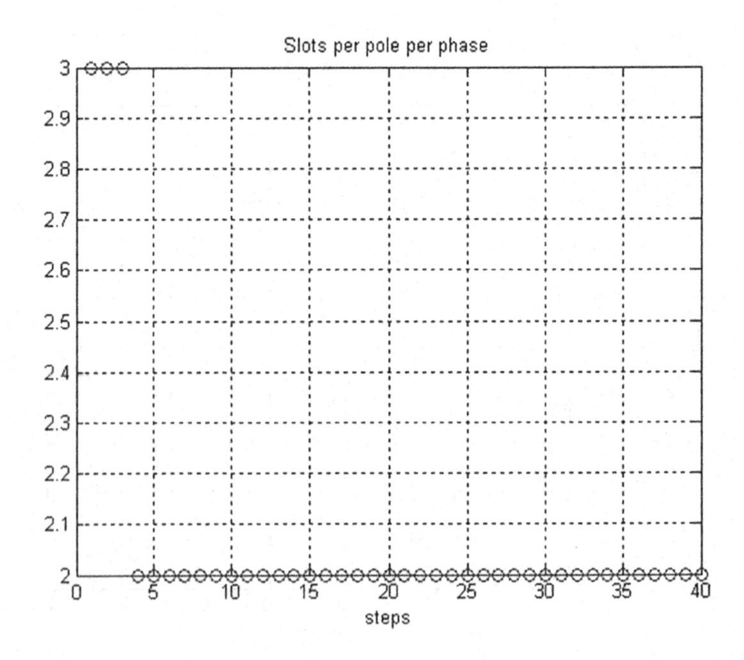

FIGURE 9.30 Slots per pole per phase evolution.

TABLE 9.2
Comparisons of Optimization Results

Parameter	GA	HG	Units	Comments
sDo	139	149	mm	Stator external diameter
sDi	73	86	mm	Stator inner diameter
sh4	0.9	0.6	mm	Stator tooth pole tip height
so	1	2.5	mm	Stator slot opening width
sh3	0.9	0.9	mm	Height of wedge for stator slots
shOA	24.2	20.7	mm	Total height of stator slot
sh1	21.7	18.5	mm	Main (coil) height of the stator slots
shy	8.8	10.8	mm	Stator yoke height
swt	4.7	5.1	mm	Stator teeth width
N_1	336	312		Number of turns per phase
rDo	69.6	82.9	mm	Rotor outer diameter
rDi	43	57	mm	Rotor inner diameter
hpm	3.7	3.2	mm	Permanent magnet height (thickness)
hag	1.7	1.55	mm	Airgap
J_1	40.4	30	kA/m	Linear electric load
Bagsp	0.76	0.75	T	Airgap flux density
sBtp	1.55	1.67	T	Stator tooth flux density
sByp	1.56	1.49	T	Stator yoke flux density
rByp	1.44	1.65	T	Rotor yoke flux density
J_s	5.3	4.94	A/mm^2	Stator current density
Lcpertau	1.9	1.6		Ratio of core stack length to pole pitch
q_1	2	2		Stator slots per pole per phase
cSpan	0.83	0.83		Ratio of coil span to pole pitch
alpm	0.91	0.91		Ratio of PM width to pole pitch
WeightStCu	3.365	3.255	kg	Coil weight
WeightIronUsed	15.268	16.961	kg	Used iron weight
WeightStIron	5.003	5.509	kg	Stator lamination weight
WeightPM	0.595	0.603	kg	PM weight
WeightRtIron	1.372	1.712	kg	Rotor iron weight
WeightMot	10.336	11.079	kg	Motor total weight
Iqn	3.9	3.81	A	Rated current
Vfn	220	220	V	Rated voltage
Pcu	97.39	88.15	W	Copper losses
Pfe	38.76	41.2	W	Iron losses
Pmec	11	11	W	Mechanical losses
Etan	93.73	94		Efficiency
Cu_c	33.649	32.547	USD	Copper cost
lam_c	76.339	84.806	USD	Used laminations cost
PM_c	29.772	30.173	USD	Permanent magnet cost
$rotIron_c$	6.8634	8.5575	USD	Rotor solid iron cost
pmw_c	51.679	55.395	USD	Penalty cost of active material weight
i_{cost}	198.30	211.48	USD	Initial material cost
$energy_c$	220.73	210.52	USD	Lost-energy cost during 10 years, 1500 h/year
t_{cost}	419.03	422.00	USD	Total cost (objective function)
sim_{time}	340.86	4.516	s	Simulation time on Pentium IV-2.6 GHz (with one initial variable vector) or 92 s = 4.516 × 20 (with 20 random initial variable vectors)

optimization variables should be reduced in order to reduce the computation effort. Embedded FEM optimal design was already used successfully to design single-phase PM motors [13], [14] where a cogging torque pattern could compensate the PM current interaction torque in a way that guarantees a resultant torque larger that a minimum required value for every rotor position. This is essential for single-phase motor starting from every rotor initial position. Evaluation of such constraint function is hard because the torque should be evaluated in many rotor position by FEM.

A spoke PM machine, Figure 9.31, case study is chosen to present the embedded FEM optimal design. In this case the number of FEM evaluations per designed motor is reduced drastically.

The first step to design an electrical motor is to choose the main specifications. Let us consider a motor with a rated power of Pn = 5 kW at base speed 2400 rpm with a range of constant power between base speed and maximum speed 4800 rpm and, to be more general, let us consider the most probable load P_1=5 kW at n_1=2000 rpm, and the minimum electrical efficiency at this test point η(P1,n1)=94%. In a spoke PM machine, a good PM flux concentration is obtained for a large number of poles, but there is constructive limitation of maximum number of poles by maximum frequency and, for small power machines, the maximum allowed outer diameter. Considering the maximum frequency of 400 Hz at maximum speed the poles numbers will be 10poles (5 pole pairs). The motor should be supplied from a standard power converter and then the number of phases is 3 and rated phase voltage V_n=230 V (RMS).

Having the main motor specifications, we can start the motor design. The design objective is to minimize the active material cost with an efficiency better than 94% in the test point, without PM demagnetization and avoid overheating.

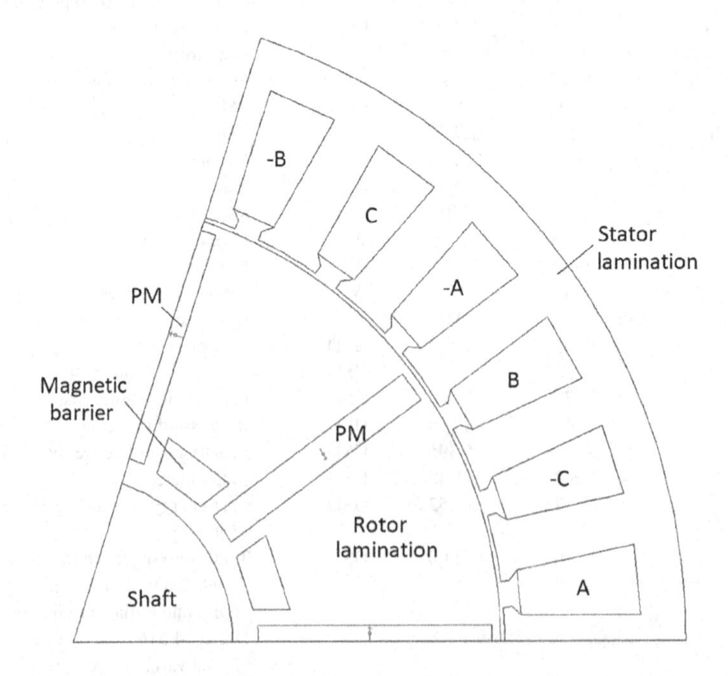

FIGURE 9.31 Spoke PM machine -cross sections (only 2 poles from 10 are presented).

It could be noticed that the number of poles is large and then one coil per pole per phase is chosen in order to keep a reasonable diameter at 5 kW rated power.

A FEM design is in fact the performance evaluation on an initial geometry and then iteratively adjusting the geometric dimensions to obtain the desired performance. For a given topology, all geometric dimensions could be computed from a few primary dimensions. The primary dimensions are given in the input file and all the other dimensions could be computed, let us name them secondary dimensions. The first option is to choose the primary dimensions, but some combination is better. A part of the primary dimension values will be changed during the optimization process (the optimization variables) and other parts are technological limitations and usually they are constant values. The optimization variables should influence the objective function and ideally, they are a convex set. In practice, it is difficult to prove that optimization variable vector is a convex set and thus, the optimization algorithm will detect unfeasible points and will continue to remain silent on the other branch. The initial optimization variable vector should belong to the feasible set. Sometimes is difficult to find a feasible initial vector if the search domain includes many or large unfeasible subdomains. Following the previous rules the optimization variables are chosen, X = (Dr0, lstack, hpm, lpm_pu, hsy_pu,wt_pu, hs_pu), where:

- Dro is rotor outer diameter,
- lstack is axial laminated stator core length,
- hpm is PM height
- lpm_pu is length of PM height in pu from maximum available space,
- hsy_pu is stator yoke height in pu from pole pitch,
- wt_pu is tooth width in pu (from slot pitch),
- hs_pu is slot height in pu (from pole pitch).

It is clear that optimization variables are the main constructive dimensions and many of them are given in per unit to avoid initialization troubles. The initial value of Dro and lstack could be a guess from a similar motor, desired values or they could be computed with (9.5) or (9.6) for rotor outer diameter and 9.8 for stator length. While the lpm_pu maximum value could be chosen 1 p.u. (its maximum theoretical values) the tooth width in per unit is usually between 0.3- 0.7. From a geometrical point of view there are upper limits for hsy_pu and hs_pu, but practically there is no reason to have upper limits larger than 0.5, respectively, 0.5 for hasy_pu, respectively, hs_pu.

The dimension constraints used for spoke PM motors (most of them given by technological limitations are as follows:

- hag—the air-gap height,
- Dri_min—the shaft minim diameter under rotor core,
- hry -the rotor height, the rotor yoke is required only for mechanical reasons,
- hrb—rotor bridge over PM, required only for mechanical reasons,
- hrb2- rotor base bridge,
- hrb2max—maximum allowed base bridge,
- hrbarrier—magnetic barrier height

- dhpm—distance between PM and lamination hole on PM smallest side,
- dlpm—minimum distance between PM and lamination hole on PM length (largest side),
- hs4—stator tooth pole tip height,
- hs3—stator tooth wedge height zone,
- sMs—stator slot mouth opening,
- ksfill—slot filling factor,
- ssf = 0.97—lamination stacking factor.

The primary geometric dimensions are shown in Figure 9.32.

For small power machines or when require PM width is larger, the magnetic barrier is eliminated and the rotor topology is that presented in Figure 9.33. When the distance between adjacent PMs becomes larger than hrb2max, the base bridge will be split and the magnetic barrier is introduced automatically by the design algorithm.

There is a mechanical alternative for spoke PM rotor, Figure 9.34, by using rotor detachable poles reinforced on a nonmagnetic ring with swallow tails. In order to reduce the q axis inductance additional flux barriers are used. This construction is not part of the present case study.

The second dimensions and parameters could be divided in two classes: a class of dimensions that are nondependent on the optimization variable like pole pairs, rated frequency, test frequency, rated torque, and test torque, respectively, dimensions that depend on the optimization variables. The nondependent variables are computed one

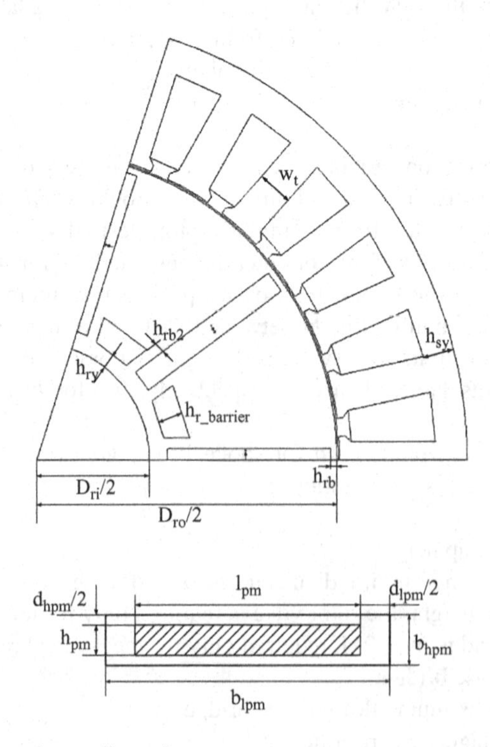

FIGURE 9.32 The primary dimensions.

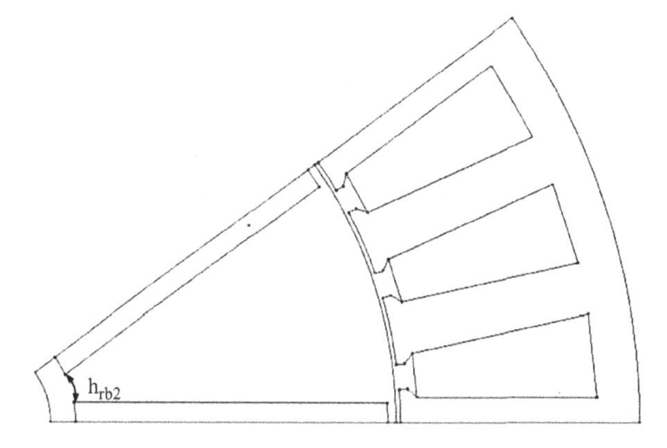

FIGURE 9.33 Rotor without magnetic barrier.

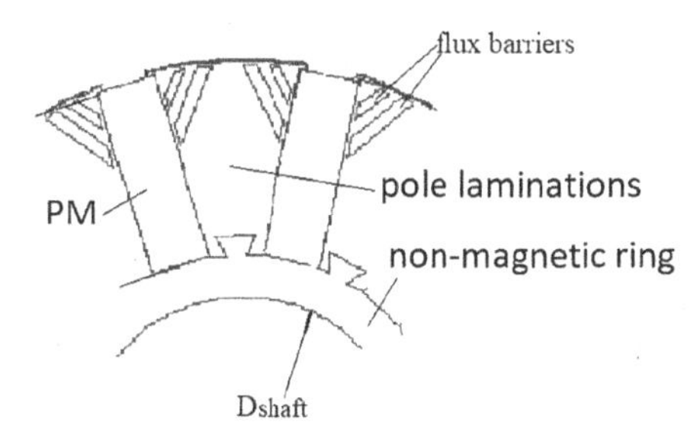

FIGURE 9.34 Mechanical alternative for PM spoke rotor.

time at initialization while the dependent variable will be computed repeatedly for each update of optimization variable. This way the computation speed is improved a little but more important is to organize the code to be readable and easy to check and test. If a not dependent variable is evaluated repeatedly, the computation time is increased a little, but if a dependent variable is not evaluated for each update of optimization variable, then it leads to a computation error difficult to troubleshoot.

After all geometric dimensions are available, the FEM model will be built by script, which includes preparing the used material, the electric circuit, boundary conditions, drawing the motor and then assign the materials, boundary conditions, and circuit to the geometry. If it is for the first time using FEMM please follow the Magneto-static tutorial [15] and then find the script command in [16].

Preparing the materials means to add the material to the model from the Library Materials if it exists, Figure 9.35 "Air", "M19"—for laminations, "1006 Steel" for shaft, or to define a new material as in Figure 9.36 for permanent magnets and conductors. The Library Materials has a large variety of conductors with normalized diameters that are recommended to use when the FEM is used to compute the parameters for a designed machine. The embedded FEM design assumes windings with a single turn per coil while the number of turns per coil will be computed in post-processing when the permanent flux, winding resistance and inductances are available. Since the initial conductor diameter is not the final diameter it is a nonsense to

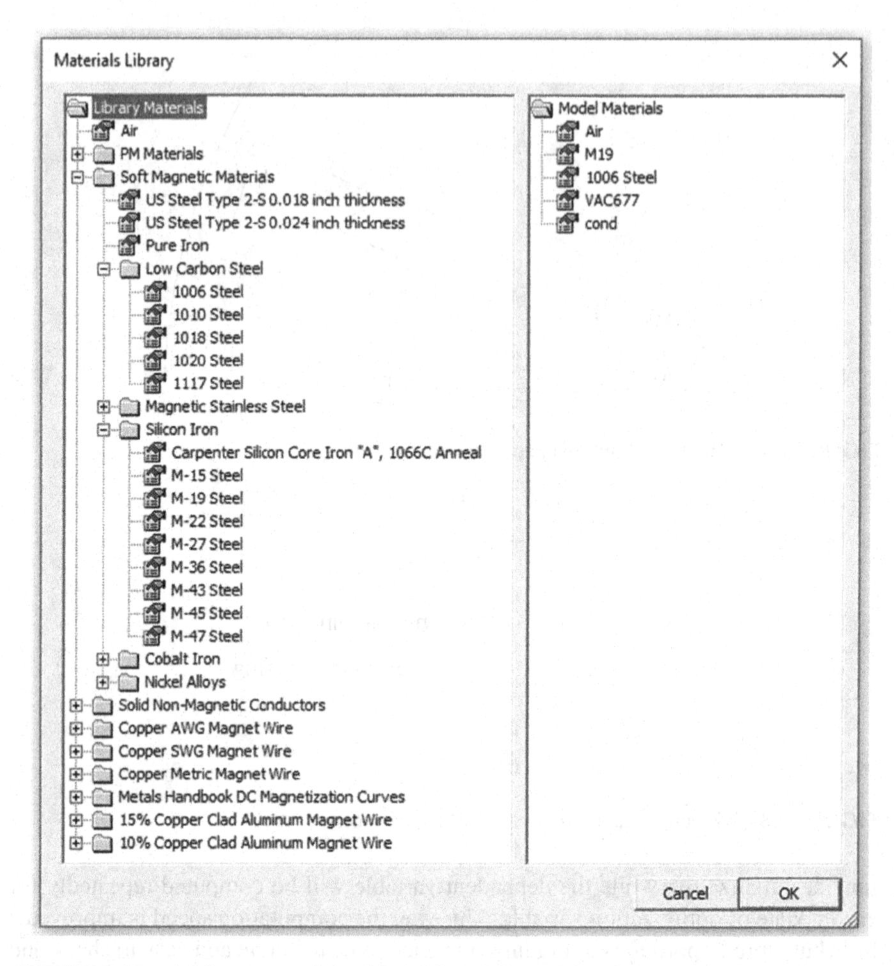

FIGURE 9.35 Adding a material to Model from Library Materials.

normalize and often there is no normalized diameter close to the initial conductor diameter.

The distance beetwen PM and lamination is very small and if this is introduced in the model then the number of the mesh nodes will be increased and consequently the required computation time is increased. If the airgap between PM and laminations is not considered the error of PM flux could be up to 10%. A solution to reduce the error is to replace the real PM with an echivalent PM that fill all laminations hole but will produce the same magnetic field. The equivalent coercitive magnetic strength, H_{ce}, and relative permeability μ_{re} are computed analytically.

$$k_{hPM} = \frac{h_{PM}}{b_{hpm}}; k_{wPM} = \frac{l_{PM}}{b_{lPM}} \tag{9.113}$$

$$H_{ce} = \frac{k_{hPM}}{\mu_r \left(2 - k_{wPM} + k_{hPM}\left(1 - k_{wPM}\right)\right)} \cdot H_c \tag{9.114}$$

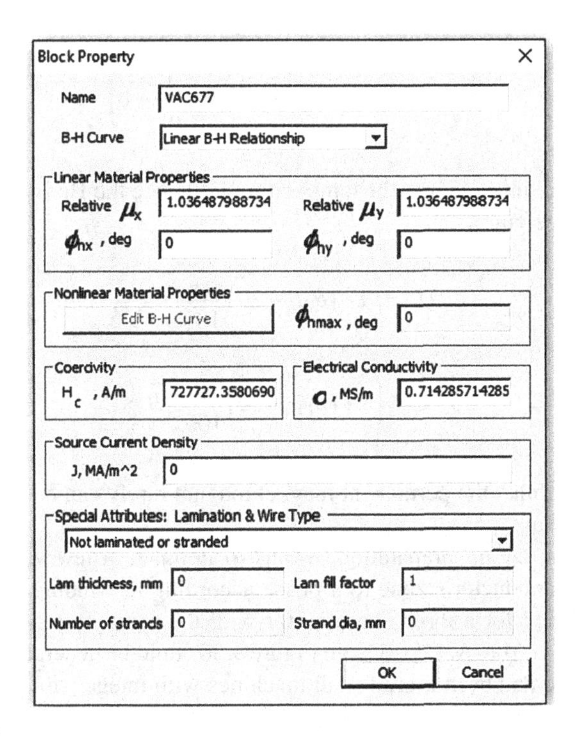

(a)

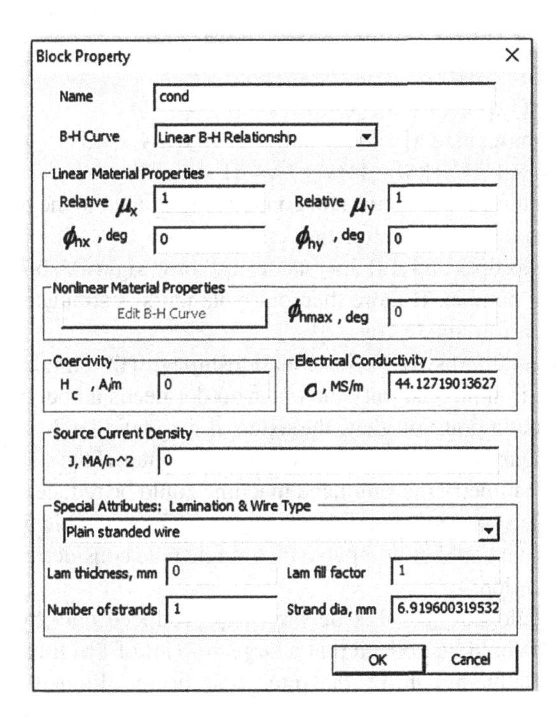

(b)

FIGURE 9.36 Defining new materials.

$$\mu_{re} = 1 - k_{wPM} + \frac{k_{wPM}}{1 + k_{hPM}\left(\frac{1}{\mu_r} - 1\right)} \tag{9.115}$$

In order to take into account the temperature influence the Hc is computed for a PM expected temperature.

$$H_c = \left(1 + \left(\theta_{PM} - 20\right)\frac{k_{\theta Hc}}{100}\right)H_{c20} \tag{9.116}$$

$$B_r = \left(1 + \left(\theta_{PM} - 20\right)\frac{k_{\theta Br}}{100}\right)B_{r20} \tag{9.117}$$

For a real machine, the permanent magnet magnet rarely can be taken directly from Library Material.

The electric circuit preparation, means to initialize a few vectors that allow to associate the conductor's zone to a phase according to winding scheme. Three circuits are defined for a three-phase motor with the usual name A, B, C. The phase sequence A, – C, B, – A, C, – B from Figure 9.30 could be generalized to simulate the entire machine and even more for all machines with integer slot per pole per phase, q1. The circuit name could be stored in a string. The following Matlab code will produce the slot allocation vectors, "w_name"—store an index to extract the phase name from "phase_name" string while the "w_sign" stores the current conventional sense.

```
phase_name='ABC';
w_q1=ones(1,q1);
w_name=phase_name([w_q1 3*w_q1 2*w_q1 w_q1 3*w_q1 2*w_q1]);
w_sign=[w_q1 -w_q1 w_q1 -w_q1 w_q1 -w_q1]
```

After drawing the geometry, the slot could be associated to the phase simultaneously with the material by a single function call.

mi_setblockprop('cond',1,0.5,w_name(ii),0,0,w_sign(ii)*sb0)where "ii" is the slot associated number. If more than one pole pairs is simulated ii will be replaced with remainder of ii divided by 3 q1.

Boundary conditions are in close relationship with the simulated number of poles. If all machine is simulated only the outer border needs a boundary condition. From energy conversion point of view, the external magnetic field could be neglected, so the Dirichlet boundary condition is applied on the outer of the stator core. Considering the magnetic symmetry the simulated machine could be reduced at a single pole pair. In this case, the symmetric condition is applied. The magnetic field produced by PM and d axis current could be computed on a single pole considering the anti-symmetric boundary condition.

Magnetic field line and flux density map produced only by PMs are shown in Figure 9.37. It could be noticed that a large amount of PM flux is lost through rotor base bridge. Reducing, -hrb2, the rotor base bridge, limited by mechanical constraints; it is possible to produce more magnetic fields by adding a radially magnetized PM in the magnetic barrier.

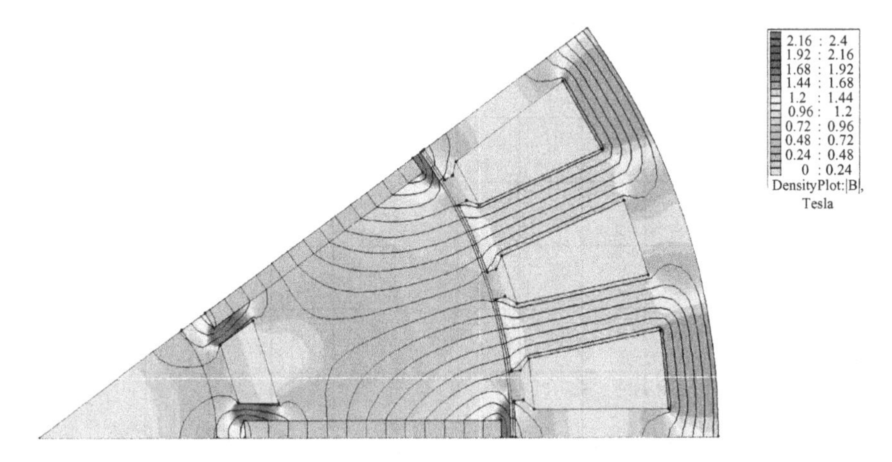

FIGURE 9.37 PM Field lines and flux density map.

The only existing flux is PM flux along d axis and, as a circuit value for the current rotor position, it could be computed as

$$\psi_{PM0} = \frac{\psi_{PMB} - \psi_{PMA}}{\sqrt{3}} \tag{9.118}$$

The stator rated Ampere-turns (rms) are computed:

$$I_{n0} = \frac{T_n\sqrt{2}}{m\, p\psi_{PM0}} \tag{9.119}$$

In the same way, the Ampere turns is computed for test torque.

The objective function is evaluated without computing the numbers of turns.

The optimization variables evolution during optimization process are shown in Figure 9.38. The slot relative height is increasing by about 40% in the first 2 steps. This reduces the copper losses Figure 9.39 and increases the efficiency Figure 9.40 with the price of increasing the total active mass, Figure 9.41. The iron losses in the test point are smaller and efficiency is a little larger than required. The over-temperature is computed at rated power and the over-temperature penalty function imposes the maximum rated losses. The objective function is still very large for the first two steps (2673 USD, respectively, 980.3 USD) and they are not presented in the objective function graphics, Figure 9.42. Increasing of stator slot height is reducing the copper losses but at the same time increases the outer diameter that means increasing cooling surface; so, from the third step, all penalty functions become zero and the objective function is equal with initial cost, Figure 9.42.

The last value of the objective function is 60.1 USD and it is equal with the active material cost.

The final motor parameters are computed in the post-optimization where two more FEM evaluations are performed. In order to compute d axis inductance, the phase current is set to get rated demagnetization current.

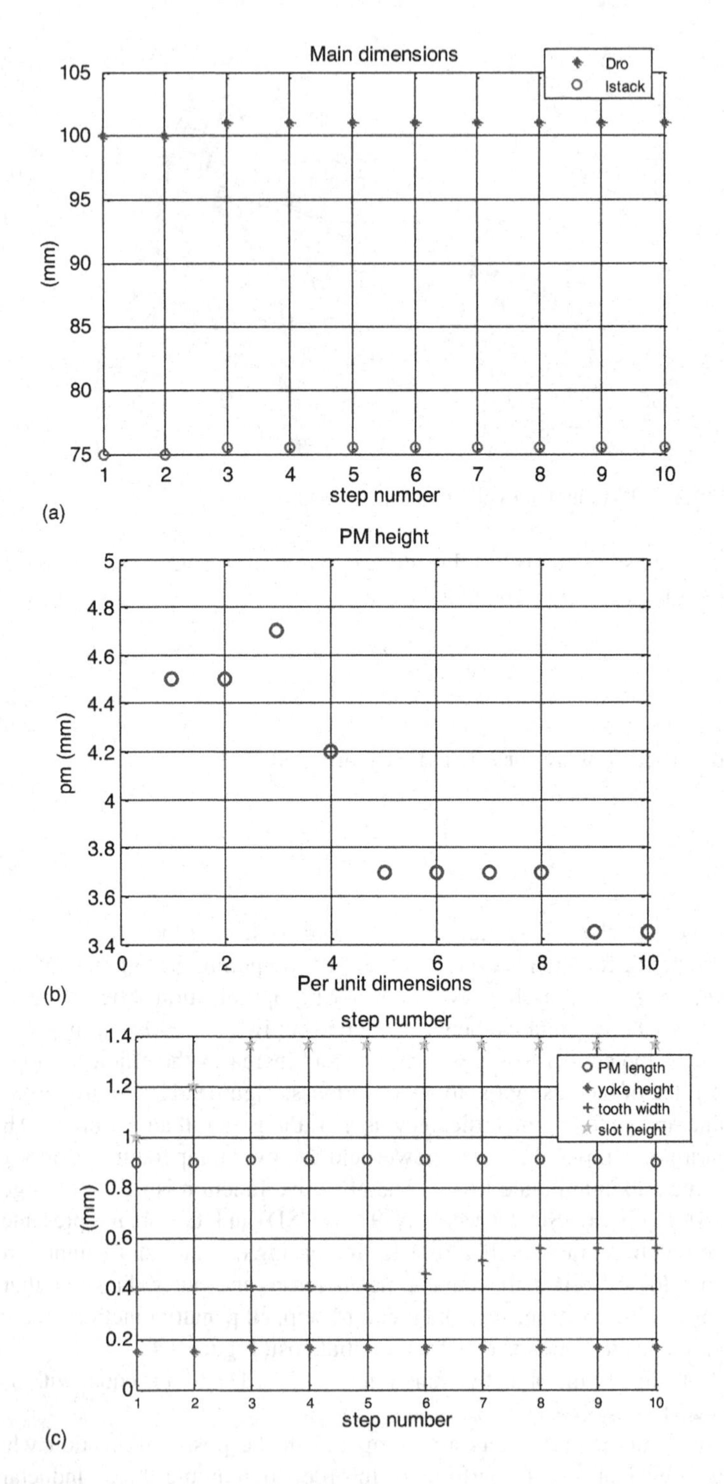

FIGURE 9.38 Optimization variables evolutions.

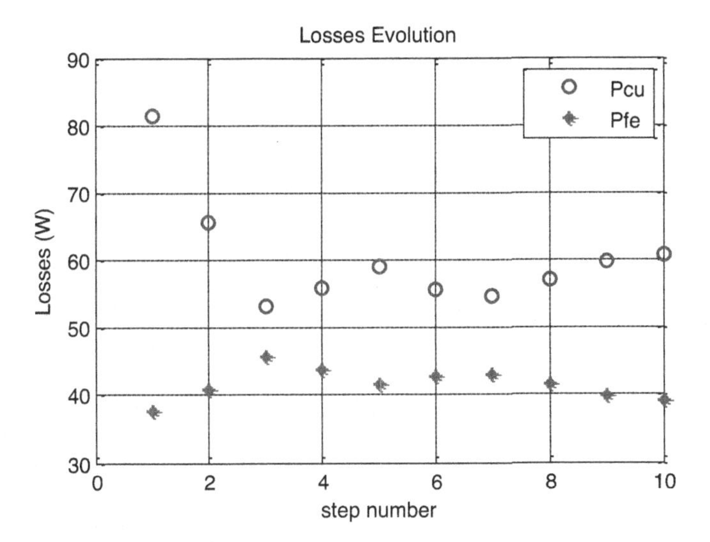

FIGURE 9.39 Copper and Iron losses evolutions.

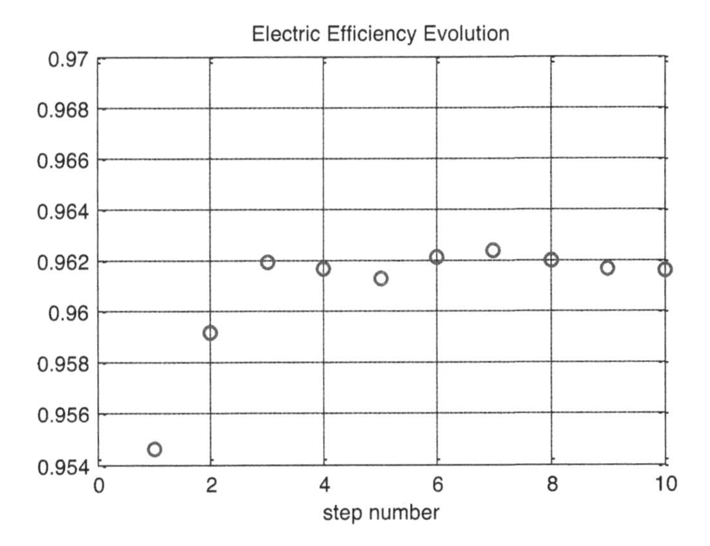

FIGURE 9.40 Efficiency evolutions.

$$i_A = -i_B = I_{n0}\sqrt{\frac{3}{2}}, \quad i_C = 0 \tag{9.120}$$

The field line and flux density are shown in Figure 9.43.

The compound flux is along d axis, ψ_d will be computed with (9.118).

The q axis inductance is computed by setting the phase current to produce magnetic field in q axis. For this position and phase distribution as in Figure 9.31 the phase Ampere-turns is:

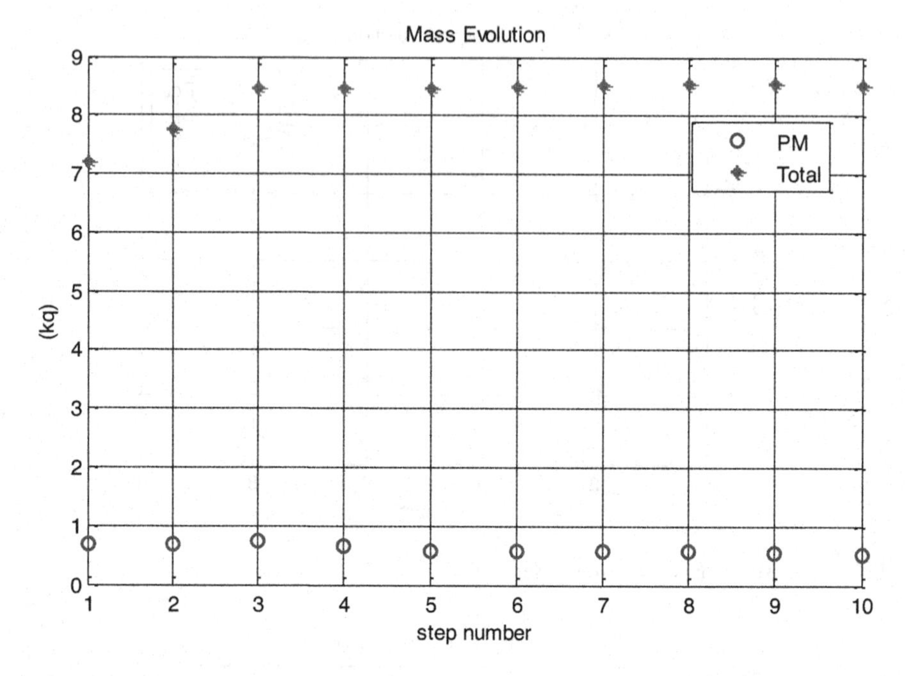

FIGURE 9.41 Total and OM mass evolutions.

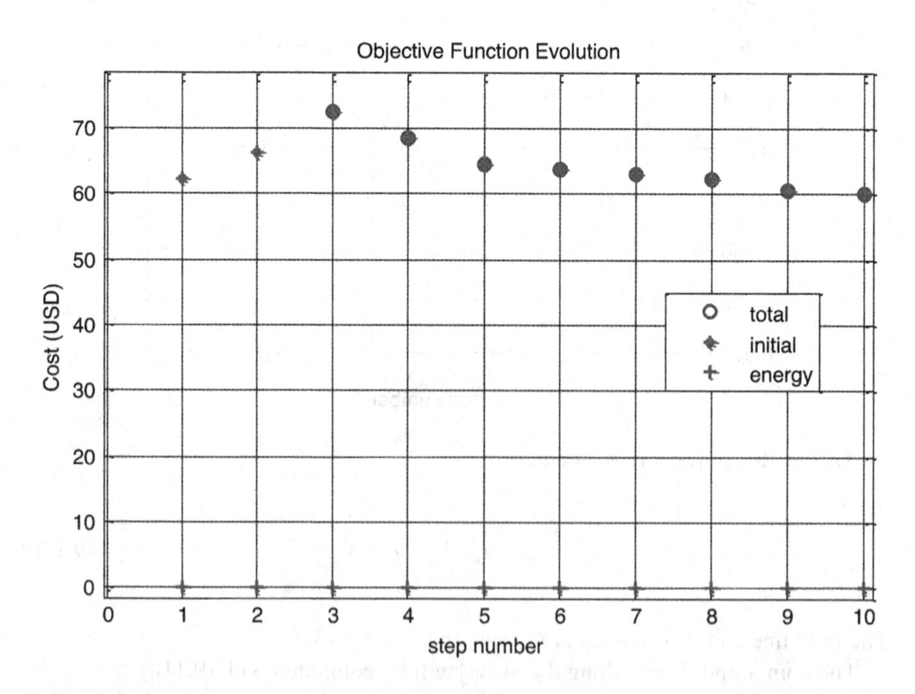

FIGURE 9.42 Objective function and its main components.

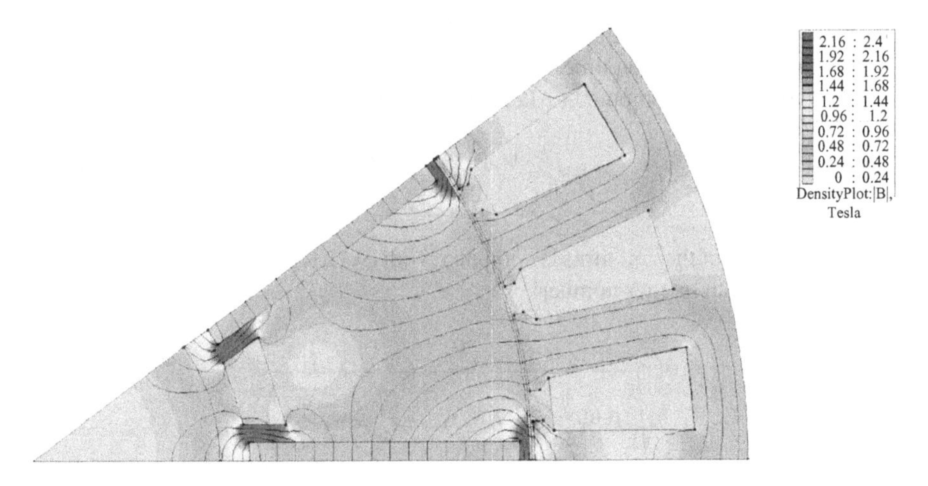

FIGURE 9.43 Field lines and flux density: PM and rated d axis demagnetization current.

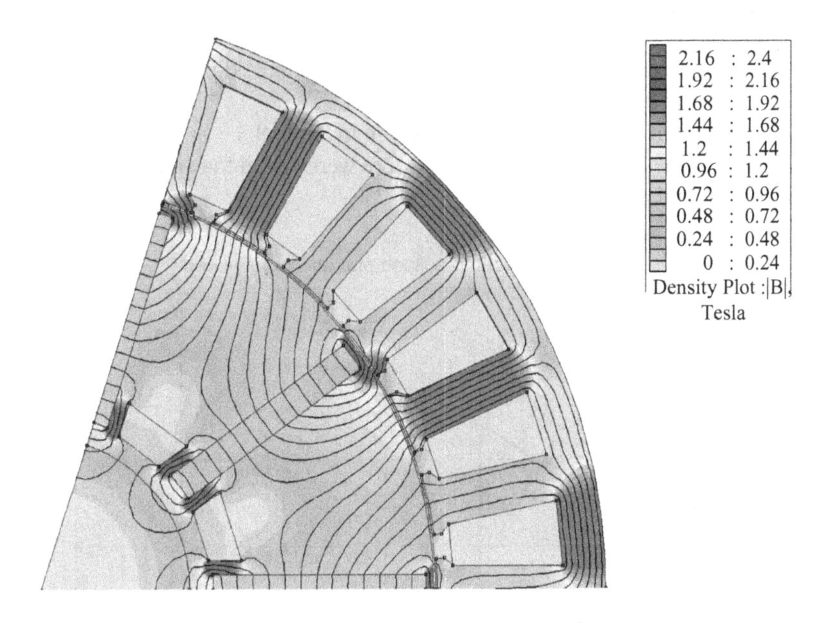

FIGURE 9.44 Field lines and flux density: PM and rated q axis current.

$$i_a = i_b = -\frac{I_{n0}}{\sqrt{2}} i_c = I_{n0}\sqrt{2} \tag{9.121}$$

The FEM is evaluated on a pole pair Figure 9.44.

The d axis flux is the PM flux, but in presence of the q axis current, ψ_{PMiq}. However, it is computed again with (9.118). It could be noticed that cross saturation is decreasing the PM flux from 7.1 mWb to 6 mWb, (for a winding with a single turn per coil) which is about 15%. When a large cross-coupling saturation is observed introducing the PM flux evaluation during the optimization process should consider it despite of increasing the computation time.

The q axes flux is computed (9.122) and then the q axis inductance is computed.

$$\psi_q = \frac{\psi_A + \psi_B}{2} - \psi_C \tag{9.122}$$

Now the number of the coil turns is computed with (9.123) and then it will be rounded (usual down) to an integer number:

$$s_b = \frac{V_{fn}}{\sqrt{\left(p_1 w_n \frac{\psi_{PMiq}}{\sqrt{2}} + R_s I_{n0}\right)^2 + \left(p_1 w_n \frac{\psi_q}{\sqrt{2}}\right)^2}} \tag{9.123}$$

The permanent magnet flux, stator resistance, and inductance will be recomputed considering the real number of turns per coil.

The Electromagnetic power versus current at constant rated voltage and at maximum speed, Figure 9.45, proves the ability to produce rated power at rated current, even larger power due to negative i_d current.

The main parameters of the motor are given in Table 9.3.

The slot volume (coil part) and of stator core were added to parameters, to be used as input in thermal analysis by FEM.

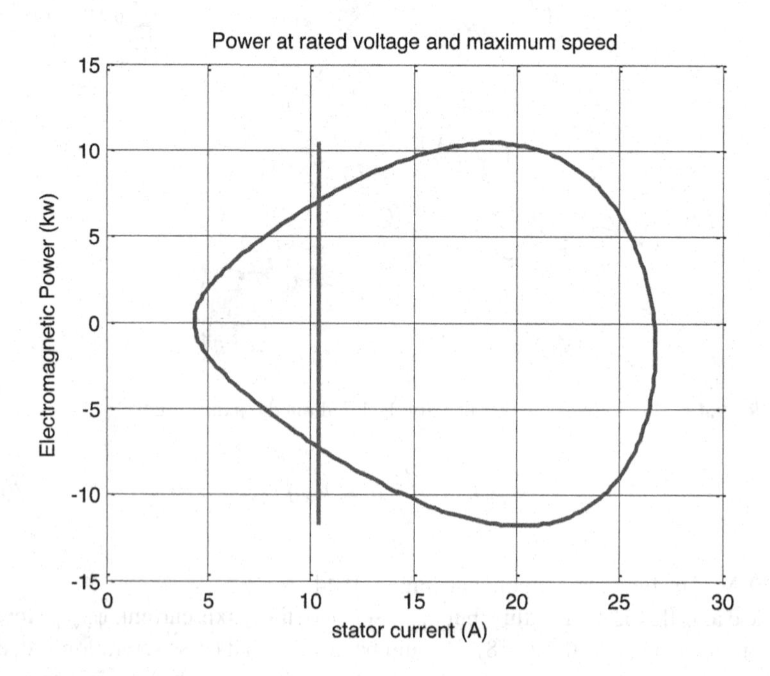

FIGURE 9.45 Electromagnetic power versus phase current at max. speed.

TABLE 9.3
Main parameters

No.	Parameter	Value	UM	No.	Parameter	Value	UM
1	Rated power	5	kW	18	Pcun	220	W
2	Rated speed	2,400	rpm	19	Pfen	50.62	W
3	poles	10		20	etan	94.87	%
4	Vf	230	V	21	N1	145	turns
5	In	10.42	A	22	dco	1.29	mm
6	Dso	145	mm	23	Mcu	1.353	kg
7	Dsi	102	mm	24	Mst iron	3.13	kg
8	lstack	75.5	mm	25	Mrot iron	3.51	kg
9	hpm	3.45	mm	26	Mpm	0.53	kg
10	lpm	26.45	mm	27	Mmot	8.51	kg
11	wtooth	5.9	mm	28	cost	60.14	USD
12	hag	0.5	mm	29	Vol slot	7.098	cm^3
13	Rated torque	19.89	Nm	30	Vol st core	400.86	cm^3
14	PM linkage flux	0,2062	Wb	31	Tn/Vol	15.96	Nm/liter
15	Rs	0.675	Ohm	32	Tn/weight	2.33	Nm/kg
16	Ld	8.2	mH	33	T_optim	403.85	sec
17	Lq	13.3	mH	34	T_total	412	sec

Observing the acceptable optimization time of about 404 seconds, the embedded FEM optimization could be considered feasible. A single FEM evaluation per tried machine is used during optimization loop and then a new FEM evaluation in the post-optimization. The PM flux is decreased in presence of the pure Iq current (due to cross saturation) and thus the electrical recomputed efficiency, decreases from 96.16% to 95.79% in the test point. Introducing more FEM evaluations in the optimization will avoid some systematic errors, with the price of increasing several times the optimization duration. Computation time could be reduced by using better the multicore processor in parallel computation.

A Matlab design code, ipmsg.m was developed and it is available online. An example input file is m5 kW.m for a 5 kW power is also available.

9.10 CONCLUSION

This chapter may be considered a solid introduction to the analytical optimization design of PMSMs. The results of the full analytical model should help to grasp better the principles and some subtleties.

Though magnetic saturation was considered in the analytical model, the cogging torque was not taken into account. Total torque pulsations, mainly due to magnetic saturation, or due to static or dynamic eccentricities, were not considered either.

Tying up analytical optimization design methodologies (as in this chapter) and integrating them into finite element software, for validation and exploration of the cogging torque, total torque ripple, etc., are, perhaps, the next steps. Interested readers are invited to follow them up on their own [7,8].

Finally, FEM—only—based optimal design codes for EMs have been introduced recently but the total computation time is higher [9–12]. However, a simplified such code was introduced here for a spoke-PM-rotor PMSM.

REFERENCES

1. D.C. Hanselman, *Brushless Permanent Magnet Motor Design*, 2nd edition, The Writers Collective, Cranston, RI, 2003.
2. J.R. Hendershot and T.J. Miller, *Design of Brushless Permanent-Magnet Motors*, Oxford University Press, Oxford, U.K., 1995.
3. W. Ouyang, D. Zarko, and T.A. Lipo, Permanent magnet machine design practice and optimization, *Industry Applications Conference, 41st IAS Annual Meeting. Conference Record of the 2006 IEEE*, Tampa, FL, Vol. 4, pp. 1905–1911, 2006.
4. C.B. Rasmusen, Modeling and simulation of surface mounted PM motors, PhD thesis, Institute of Energy Technology, Aalborg University, Aalborg, Denmark, 1996.
5. IEC/EN 60034.
6. Thin Non-Oriented Electrical Steel—SURA 007, Products Information, Surahammars Bruk, Sweden.
7. L. Tutelea, I. Boldea, Surface permanent magnet synchronous motor optimization design: Hooke Jeeves method versus genetic algorithms, *Industrial Electronics (ISIE), 2010 IEEE International Symposium on*, pp.1504–1509, 2010.
8. V. Gradinaru, L. Tutelea, I. Boldea, Hybrid analytical/FEM optimization design of SPMSM for refrigerator compressor loads, *International Aegean Conference on Electrical Machines and Power Electronics (ACEMP) and Electromotion Joint Conference*, pp. 675–680, 2011.
9. Y. Duan, D. M. Ionel, "A review of recent developments in electric machines design optimization methods with a PMSM benchmark study", *IEEE Trans*, Vol. IA-49, No. 3, 2013, pp.1268–1275.
10. B. Ma, G. Lei, J. Zhu, Y. Guo, C. Liu, "Application oriented robust design optimization method for butch production of permanent magnet motors", *IEEE Trans*, Vol. IE-65, No. 2, 2018, pp.1728–1739.
11. D. -K. Lim, S-Y Jung, K.-P. Yi, H.-K. Jung, "A novel sequential—stage optimization strategy of an IPM synchronous generator design", *IEEE Trans*, Vol. IE-65, No. 2, 2018, pp.1781–1790.
12. D. Zarko, S. Stipetic, M. Martinovic, M. Kovacic, T. Jercic, Z. Hanic, "Reduction of computational effort in Finite element-based PM traction motor optimization", *IEEE Trans*, Vol. IE-65, No. 2, 2018, pp.1799–1807.
13. F. J. H. Kalluf; A. D. P. Juliani; I. Boldea; L. Tutelea; A.-S. Isfanuti; A. R. Laureano, "Single-phase stator-ferrite PM double saliency motor performance and optimization", ACEMP—OPTIM 2015.
14. A.-S. Isfanuti, L. N. Tutelea, I. Boldea, T. Staudt, "Small-power 4 stator-pole stator-ferrite PMSM single-phase self-starting motor drive: FEM-based optimal design and controlled dynamics", OPTIM ACEMP 2017.
15. David Meeker, *FEMM 4.2 Magnetostatic Tutorial*, https://www.femm.info/wiki/MagneticsTutorial.
16. David Meeker, *Finite Element Method Magnetics: OctaveFEMM User's Manual*, March 17, 2018, dmeeker@ieee.org, http://www.femm.info.

10 Optimization Design of Induction Machines

10.1 REALISTIC ANALYTICAL MODEL FOR INDUCTION MACHINE DESIGN

In this chapter, we introduce for convenience a rather complete analytical realistic alternative model of the sub 100 KW induction machine and use it in an optimal design computer code based on the GA and the Hooke–Jeeves algorithms via a numerical example.

The general design of induction motor is similar to the general design of the PMSM, with minor differences. For induction machine design, there are specific design themes and a dedicated set of chosen variables and technological constraints [1–10].

10.1.1 Design Theme

In the design theme of the induction machine, the following parameters are always given:

- Base continuous power, P_b, or rated power, P_n
- Base speed, n_b
- Rated voltage, V_n
- Number of phases, m
- Overload factor, k_1

When the motor is designed to run without power electronics, the minimum starting torque and the maximum direct grid connection starting current are required in the specifications list.

For variable-speed applications, the maximum speed, n_{max}, and the power at maximum speed, P_{max}, are added to the specifications.

Other frequent constraints are

- Efficiency at P_b and n_b
- Power factor at rated power and speed, pf_{spa}
- Material insulation class (allowed temperature)
- Protection degree against alien bodies
- Total initial cost of active materials

DOI: 10.1201/9781003216018-10

10.1.2 Design Variables

The design variables are chosen and iteratively changed directly by the designer or by an optimization algorithm in order to achieve the design theme requirements. The main design variables are

Linear current density (electric linear load), J_l (kA/m)

- Airgap flux density, B_{agsp} (T)
- Machine shape factor—ratio of core length stack to pole pitch, λ_c
- Stator current density, J_s (A/mm²)
- Rotor current density, J_r (A/mm²)
- Stator teeth flux density, B_{st} (T)
- Stator yoke flux density, B_{sy} (T)
- Rotor teeth flux density, B_{rt} (T)
- Rotor yoke flux density, B_{ry} (T)
- Number of stator slots per pole per phase, q_1
- Stator slot opening width, s_{os} (mm)
- Rotor slot opening width, s_{or} (mm)
- Stator tooth top height, h_{s4} (mm)
- Rotor tooth top height, h_{r1} (mm)
- Stator coil span, $\text{cSpan} = \dfrac{\text{y1}}{\tau\text{m}}$

The following parameters are chosen according to the available technology and motor size:

- Current parallel path count, a_1
- Number of winding layers, n_L
- Airgap, g
- Difference between rotor core length and stator core length, dl_r
- Straight-turn end connection length, l_{f1}, that exits the slot
- Axial distance between end-coil connection and frame, l_{f2}
- Core-stacking factor, k_{stk}
- Slot-filling factor, k_{sf}
- Ratio between rotor end ring and rotor bar current densities, $k_{Jendr2Jr}$
- Rotor slot slant in number of rotor slot pitches, r_{ss}
- Angle of slot wedge, a_w
- Number of rotor slots
- Slot insulation thickness, slotInsulThick
- Thickness of slot closure (wedge), slotClosureThick
- Number of elementary conductors per turn, n_{ce}
- Assumed average temperature of the stator winding, T_{w1}
- Assumed average temperature of the rotor winding (cage), T_{w2}
- Maximum admissible winding temperature, T_{wmax} (it depends on the insulation class)
- Maximum temperature of the cooling fluid (or ambient), T_{amb}

- Thermal transmission coefficient, α_t
- Increasing factor of cooling surface from frame fins and cooling fluid velocity, k_{ff}
- Iron loss factor (the iron loss is larger due to field nonuniformity), k_{pfe}

The soft magnetic material for stator and rotor magnetic cores and winding material (copper or aluminum for the rotor) are also chosen at this time. The mechanical losses estimation is also performed.

For the induction machine, the minimum mechanical feasible airgap is used. For the computation to start, the following values could be chosen:

$$g = \begin{cases} 0.1 + 0.2\sqrt[3]{P_n} & \text{if} \quad p_1 = 1 \\ 0.1 + 0.1\sqrt[3]{P_n} & \text{if} \quad p_1 \geq 2 \end{cases} \tag{10.1}$$

The airgap, g, is obtained in millimeters and the rated power, P_n, is given in kilowatts in Equation 10.1.

10.1.3 Induction Machine Dimensioning

The induction machine design starts with several preparation calculations, such as grid apparent power S_n, using chosen variables and constraints. Then, the machine constant and stator inner diameter are computed:

$$S_n = \frac{P_n}{\eta_{\text{spec}} \cdot pf_{\text{spec}}} \tag{10.2}$$

The winding factor k_w is computed from the zone factor, k_{ws}, and the chording factor, k_{chs}, using the same relation 9.4 as for the permanent machine (see also Ref. [3]):

$$C_0 = \frac{\pi^2}{\sqrt{2}} B_{agsp} J_l k_w \tag{10.3}$$

$$D_{si} = 1000 \cdot \sqrt[3]{\frac{60 p_s}{\pi n_n} \cdot \frac{S_n}{\lambda_c C_0}}; \quad (\text{mm}) \tag{10.4}$$

The pole pitch, τ, and stator core length, l_{c1}, are computed in the same way as for PMSM (Equations 9.7 and 9.8) using the inner stator diameter, D_{si}.

The main geometric dimensions are shown in Figure 10.1. The stator slots geometry is shown in Figure 6.42 and the rotor slots geometry is shown in Figure 6.43.

The stator and rotor yoke widths are

$$h_{sy} = \frac{\tau_p B_{agsp}}{\pi B_{sy}} \tag{10.5}$$

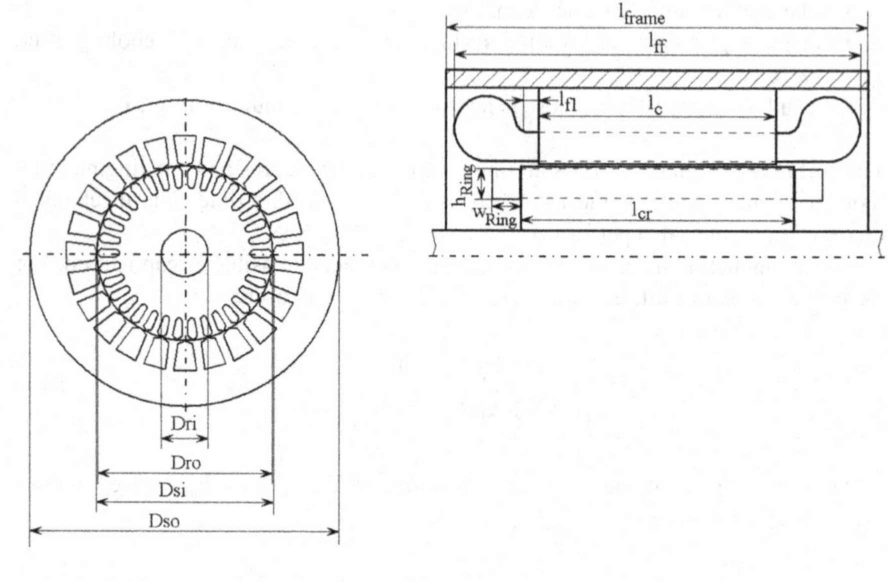

FIGURE 10.1 Induction machine geometry.

$$h_{ry} = \frac{\tau_p B_{agsp}}{\pi B_{ry}} \tag{10.6}$$

The number of stator slots and the stator slots pitch and tooth width are

$$N_{ss} = 2q_1 m p_1 \tag{10.7}$$

where p_1 is the number of pole pairs,

$$\tau_{ss} = \frac{\pi D_{si}}{N_{ss}} \tag{10.8}$$

$$w_{st} = \frac{B_{agsp}}{B_{st}} \tau_{ss} \tag{10.9}$$

Now, the polar flux, Ψ_p, and the first approximation of the number of turns per phase, N_1, and the number of turns per phase (or current path), s_{b1}, are computed:

$$\Psi_p = \frac{2}{\pi} B_{agsp} \tau_p l_c \tag{10.10}$$

$$N_1 = \frac{V_{nf}\left(0.93 - 0.97\right)}{\pi\sqrt{2} f_n k_w \Psi_p} \tag{10.11}$$

$$s_{b1} = \frac{N_1 a_1}{p_1 q_1 n_L} \tag{10.12}$$

where a_1 represents the number of current paths in parallel. The number of turns per coil should be an integer number so the computed value, s_{b1}, is rounded to the nearest integer, s_b. If the nearest integer is zero, then one turn per coil will be used. The number of turns per phase, N_1, and the machine core length, l_c, will be recomputed:

$$N_1 = \frac{s_b q_1 p_1 n_L}{a_1} \tag{10.13}$$

$$l_c = \frac{s_{b1}}{s_b} l_{c1} \tag{10.14}$$

l_{c1}—initial value of l_c

Now, the stator slot dimensions are computed in the same way as for PMSMs (Equations 9.15 through 9.22).

10.1.3.1 Rotor Design

The number of rotor slots is chosen according to the number of poles and the number of stator slots (slots per pole per phase) in such a way as to avoid large synchronous parasitic torques and radial forces. Table 10.1 gives the most used pairs of stator and rotor slot counts.

Now, the rotor outer diameter, D_{ro}, is computed followed by the rotor slots pitch, τ_{rs}, and the rotor tooth width, w_{rt}:

$$D_{ro} = D_{si} - 2g \tag{10.15}$$

$$\tau_{rs} = \frac{\pi D_{ro}}{N_{rs}} \tag{10.16}$$

$$w_{rt} = \frac{B_{agsp}}{B_{rt}} \tau_{rs} \tag{10.17}$$

The rotor slot opening is compared with the maximum feasible rotor slot opening, and if it is larger, it will be adjusted to a maximum feasible value.

The rotor slots geometry is computed next. At first the current, I_{rb}, through the rotor bars and then the requested area, A_{rb}, for the rotor bar are calculated. The coefficient 0.95 accounts for the fact that a part of the stator current is the magnetization current:

$$I_{rb} = 0.95 \frac{\pi D_{si} J_l}{N_{rs}} \tag{10.18}$$

TABLE 10.1
Stator and Rotor Number Turns

Pole Pairs	q_1	Number of Rotor Slots
1	4	18 20 22 28 30 33 34
	6	25 27 28 29 30 43
2	2	16 18 20 30 33 34 35 36
	3	24 28 30 32 34 45 48
	4	30 36 40 44 57 59
	5	36 42 48 50 70 72 74
	6	42 48 54 56 60 61 62 68 76 82 86 90
3	2	20 22 28 44 47 49
	3	34 36 38 40 44 46
	4	44 46 50 60 61 62 82 83
4	2	26 30 34 35 36 38 58
	3	42 46 48 50 52 56 60
6	2	69 75 80
	2.5	86 87 93 94

$$A_{\mathrm{rb}} = \frac{I_{\mathrm{rb}}}{J_{\mathrm{r}}} \tag{10.19}$$

The rotor tooth is assumed to have parallel walls.

The radius, R_{r1}, of the circular upper part of rotor slot is

$$R_{r1} = \left(\frac{a'}{\cos\left(\frac{\alpha_{rs}}{2}\right)} - \sqrt{\left(a' \tan\left(\frac{\alpha_{rs}}{2}\right)\right)^2 - \left(\frac{o_{rs}}{2}\right)^2} \right) \tan\left(\frac{\alpha_{rs}}{2}\right) \tag{10.20}$$

where

$$a' = D_{r0} - h_{r1} - \frac{\sqrt{D_{r0}^2 - o_{rs}^2}}{2} - \frac{w_{rt}}{2\sin\left(\frac{\alpha_{rs}}{2}\right)} \tag{10.21}$$

$$\alpha_{rs} = \frac{\pi}{N_{rs}}$$

$h_{r1} \in [0.5, 1.5]$ mm and $o_{rs} \in [0.3, 2]$.

If the required rotor bar area, A_{rb}, is larger than area of a circle with R_{r1} radius, then the rotor bare slots dimensions are computed as

$$w_{r1} = 2R_{r1} \cos\left(\frac{\alpha_{\mathrm{rs}}}{2}\right) \tag{10.22}$$

$$h_{r2} = \sqrt{R_{r1}^2 - \left(\frac{o_{rs}}{2}\right)^2} \tag{10.23}$$

The required and the maximum possible area of the trapezoidal part of the rotor slot are now computed:

$$A_{r2} = A_{rb} - o_{rs}h_{r1} - \frac{1}{2}\left(\pi + \alpha_{rs} - 2a\sin\left(\frac{o_{rs}}{2R_{r1}}\right) + \sin\left(\alpha_{rs}\right)\right)R_{r1}^2 + \frac{1}{2}o_{r1}h_{r2} \tag{10.24}$$

$$A_{r2\max} = \frac{w_{r1}^2}{4\tan\left(\frac{\alpha_{rs}}{2}\right)} \tag{10.25}$$

A maximum available area smaller than the required area means a smaller diameter, or a smaller rotor current density or rotor tooth flux density. When the computation is done automatically using a computer code, it is necessary to solve this problem in the same way, and, in our example, the required area is reduced to the available area. This implies an increase in the rotor current density. A warning message will be prompted on the display.

The height of the trapezoidal part of the rotor slots, h_{r3}, is

$$h_{r3} = \frac{w_1}{4\tan\left(\frac{\alpha_{rs}}{2}\right)}\left(1 - \sqrt{1 - \frac{\frac{A_{r2}}{A_{r2\max}} - a''}{1 - a''}}\right) \tag{10.26}$$

where

$$a'' = \frac{\pi - \alpha_{rs} - \sin\left(\alpha_{rs}\right)}{2\left(\cos\left(\frac{\alpha_{rs}}{2}\right)\right)^2}\tan\left(\frac{\alpha_{rs}}{2}\right) \tag{10.27}$$

The other rotor slot dimensions are

$$w_{r2} = w_{r1} - 2h_{r3}\tan\left(\frac{\alpha_{rs}}{2}\right) \tag{10.28}$$

$$R_{r2} = \frac{w_{r2}}{2\cos\left(\frac{\alpha_{rs}}{2}\right)} \tag{10.29}$$

$$h_{r4} = R_{r2}\left(1-\sin\left(\frac{\alpha_{rs}}{2}\right)\right) \tag{10.30}$$

$$h_{rOA} = h_{r1} + h_{r2} + h_{r3} + h_{r4} \tag{10.31}$$

If the required rotor bar area is smaller than the area of a circle with R_{r1} radius, then the rotor bare slots become a circle with radius

$$R_{r1} = R_{r2} = \sqrt{\frac{A_{rb} - o_{rs}h_{r1}}{\pi}} \tag{10.32}$$

Now, the rotor inner diameter is computed:

$$D_{ri} = D_{r0} - 2\left(h_{rOA} + h_{ry}\right) \tag{10.33}$$

with h_{ry} to be found in 9.11

The rotor inner diameter is rounded to an integer value in millimeters.

10.1.3.2 Stator Slot Dimensions

The stator slot area, A_{Cu}, to host the bare coils, the slot area, A_{ss}, the slot useful height, h_{s1}, the total height, h_{sOA}, and the slot width, w_{s1}, are

$$A_{Cu} = \frac{\pi D_{si} J_1}{N_{ss} J_s} \tag{10.34}$$

$$A_{ss} = \frac{A_{Cu}}{k_{sf}} \tag{10.35}$$

k_{sf}—slot filling factor ϵ [0.33, 0.5] (see 9.24)

$$h_{s1} = \frac{-w_{s2} + \sqrt{w_{s2}^2 + 4A_{ss}\tan\left(\frac{\alpha_{ss}}{2}\right)}}{2\tan\left(\frac{\alpha_{ss}}{2}\right)} \tag{10.36}$$

for w_{s2} and α_{ss} see chapter 9 (9.22). α_s' can be used instead of $\frac{\alpha_{ss}}{2}$.

$$h_{sOA} = h_{s1} + h_{s2} + h_{s3} + h_{s4} \tag{10.37}$$

h_{s2}—insulation thickness between 0.2 and 1 mm

$$w_{s1} = w_{s3} + 2\left(h_{s1} + h_{s2}\tan\left(\frac{\alpha_{ss}}{2}\right)\right) \tag{10.38}$$

The stator outer diameter becomes

$$D_{s0} = D_{si} + 2\left(h_{sOA} + h_{sy}\right) \tag{10.39}$$

In the previous geometric computation we met different mathematical operations including square root extraction and subtraction, and, sometimes for a bad correlation of input parameters, it could be possible to obtain a negative or quite complex value, which does not make sense. If the computation steps are made one by one, it is possible to stop the computation and choose other initial variables immediately when such a useless value is observed. The same path could also be followed when the computation is done by a computer code but the computer code is simpler when the useless geometric values are verified in groups. At this stage, it is possible to verify if R_{r1}, h_{r2}, and h_{r3} are real and D_{ri}, $sTeethAlpha$, h_{s3}, w_{s3}, R_{r1}, h_{r2}, h_{r1}, R_{r2}, w_{r2} are larger than some minimum values. If these verifications fail, new variables are produced by genetic optimization algorithms in the previous step is repeated until all parameters pass the test. For the Hooke–Jeeves algorithm, the highest value is attributed to the objective function (all parameter computation is skipped). It is important to have compatible initial values of optimization variables for the Hooke–Jeeves algorithm; otherwise, the computation code result will be an error message. All stator and rotor geometric dimensions are computed after all checked values pass the test.

10.1.3.3 Winding End-Connection Length

The end-connection length of the stator winding, l_f, is computed considering the end coil as a semicircle with an average diameter equal to the coil average opening. The right parts of the coil end connection, l_{f1}, are added.

$$l_f = \frac{\pi}{2} k_y \tau_p \left(1 + \frac{h_{sOA}}{D_{si}}\right) + 2l_{f1} \tag{10.40}$$

$$k_y = \frac{y_1}{q_1 m}$$

y_1—coil span
q_1—number of slots per pole per phase
m—number of phases
$l_{f1} \approx [3,8]$ mm

The wound stator length, l_{ff}, is approximated

$$l_{ff} = l_c + l_{f1} + k_y \tau_p \left(1 + \frac{h_{sOA}}{D_{si}}\right) \tag{10.41}$$

The rotor end-connection axial length computed considering its height, which is equal to the total rotor slot height is h_{rOA}, the ring current, I_{er}, and the current density in the rotor ring (in direct ratio with the rotor bar current density, J_r, with the ratio factor, k_{Jr}) are

$$I_{\text{er}} = \frac{I_{\text{rb}}}{2 \cdot \sin^2\left(\frac{p_1 \cdot \pi}{N_{\text{rs}}}\right)} \tag{10.42}$$

$$A_{\text{Ring}} = \frac{I_{\text{er}}}{k_{J_r} J_{\text{r}}} \tag{10.43}$$

$k_{Jr} \approx [0.7, 0.9]$ and represents the ration between endring current density and rotor bar current density.

$$h_{\text{Ring}} \approx h_{\text{rOA}} \tag{10.44}$$

$$w_{\text{Ring}} = \frac{A_{\text{Ring}}}{h_{\text{Ring}}} \tag{10.45}$$

The inertia of the rotor without shaft is

$$\begin{aligned} J_{\text{ir}} = \frac{1}{8}\Big(&\left(D_{\text{ri}}^2 + D_{\text{rc}}^2\right) m_{\text{ry}} + \left(D_{\text{rc}}^2 + D_{\text{ro}}^2\right)\left(m_{\text{rt}} + m_{\text{rb}}\right) + \Big(\left(D_{\text{ri}} - 2h_{r1}\right)^2 \\ &+ \left(D_{\text{ri}} - 2h_{r1} - 2h_{\text{Ring}}\right)^2\Big) m_{\text{ring}}\Big) \end{aligned} \tag{10.46}$$

$$m_{ry} = \rho_{Fe} l_{cr} \frac{\pi}{4}\left(D_{rc}^2 - D_{ri}^2\right); D_{rc} = D_{ri} + 2h_{ry};$$

10.1.4 Induction Machine Parameters

The stator windings resistance is

$$R_{s0} = \frac{s_b q_1 p n_L \left(l_c + l_{soh}\right)}{A_{sb} a_1^2} \rho_{Tw} \tag{10.47}$$

ρ_{Tw}—stator winding resistivity at Tw temperature (see 9.61)
A_{sb}—cross area of stator bar
n_L—number of winding layers

$$n_L = \begin{cases} 1, \text{ for one layer } \dfrac{y_1}{q_1 m} = 1 \\ 2, \text{ for two ore more layers } \dfrac{2}{3} \le \dfrac{y_1}{q_1 m} \le 1 \end{cases}$$

The rotor resistance is computed separately for the rotor bar and short-circuit ring and then it is reduced to the stator winding:

$$R_{rb} = \rho_{2Tw} \frac{l_{rc}}{A_{rb}}; \rho_{2Tw} - rotor\ resistivity \tag{10.48}$$

$$R_{\text{Ring}} = k_{\text{rRing}} \cdot \rho_{2Tw} \frac{\pi D_{\text{mRing}}}{A_{\text{Ring}} N_{\text{rs}}} \tag{10.49}$$

$$D_{mRing} \approx D_{ro} - h_{rOA}$$

where

D_{mRing} is the average diameter of the short-circuit ring
k_{rRing} is the reduction factor of the rotor ring to the rotor bar

The expression of the rotor ring resistance was computed considering equivalent losses.

$$k_{rRing} = \frac{1}{2\sin^2\left(\frac{p_1\pi}{N_{rs}}\right)} \tag{10.50}$$

The dimension of the cross area of the short-circuit ring is comparable with that the rotor slot pitch area. The short-circuit ring is rather a massive conductor and its real resistance is not equal to the computed resistance. In some books, this difference is included in the reduction factor, which is considered an empirical coefficient.

Finally, the rotor equivalent resistance (reduced to stator) is

$$R_{\text{r}} = \left(R_{\text{rb}} + R_{\text{rRing}}\right)\left(N_1 \frac{k_{\text{w}}}{k_{\text{wr}}}\right)^2 \frac{m}{N_{\text{rs}}} \tag{10.51}$$

Usually, the rotor bars are skewed in order to reduce the torque pulsations and noise. The rotor winding factor, k_{wr}, is computed from the skewing ratio, r_{sk} (rotor slot skewing divided by rotor slot pitch), given by the designer as an initial parameter:

$$k_{wr} = \frac{2\tau_p \sin\left(r_{sk} \frac{\pi\tau_{rs}}{2\tau_p}\right)}{r_{sk}\pi\tau_{rs}} \tag{10.52}$$

The magnetization inductance depends on the magnetization current. It is simple to compute the magnetization current versus the magnetization flux considering the magnetic curve of the magnetic core. At the beginning, one value (or several values for a curve) of the airgap flux density is considered. The flux density in the stator tooth, stator yoke, rotor tooth, and rotor yoke are computed using the magnetic tube circuit rule. The magnetic field is computed by interpolation using the magnetization curve and then the magneto-motive force necessary to produce that field is

$$V_m = V_{mag} + V_{mst} + V_{msy} + V_{mrt} + V_{mry} \tag{10.53}$$

$$V_{mag} = \frac{B_{ag}}{\mu_0} k_C g \tag{10.54}$$

where k_C is the Carter factor and it is given as product of Carter factors k_{C1} and k_{C2} computed separately for the stator and the rotor as in Equation 9.43. The magneto-motive force for the stator and the rotor teeth and yokes are also computed in a similar way as for PMSM.

The polar flux, Φ, is given by Equation 10.56 where k_f is the shape factor of the airgap flux density distribution. The shape factor is equal to $\pi/2$ for unsaturated core and it decreases as is shown in Figure 10.2. when the magnetic core is saturated.

The shape factor, k_f, is shown versus stator tooth saturation factor, k_{ts}:

$$k_{ts} = 1 + \frac{V_{mst} + V_{mrt}}{V_{mag}} \tag{10.55}$$

$$\Phi = \frac{\tau_p}{k_f} l_c B_{ag} \tag{10.56}$$

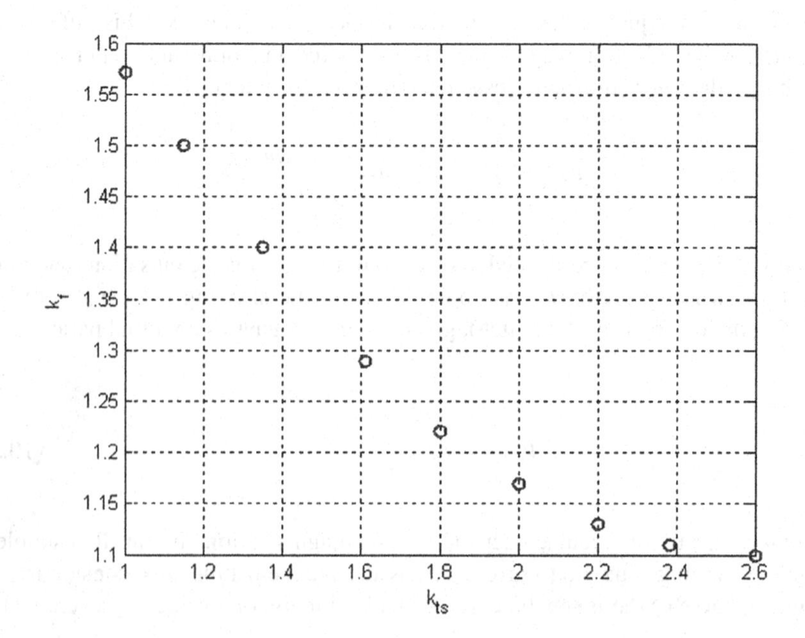

FIGURE 10.2 The shape form factor of the flux density distribution vs. stator tooth saturation factor k_{ts}.

The phase airgap flux linkage Ψ_m, magnetization current I_m, and inductance L_m are

$$\Psi_m = N_1 k_w \Phi \tag{10.57}$$

$$I_m = \frac{\pi \cdot p_1 \cdot V_m}{m \cdot N_1 \cdot k_w} \tag{10.58}$$

$$L_m = \frac{\Psi_m}{I_m} \tag{10.59}$$

It is also possible to compute nonsaturated values of the cyclic magnetization inductance, L_{m0}:

$$L_{m0} = 2\mu_0 m \frac{(N_1 k_w)^2}{\pi^2} \frac{\tau_p l_c}{p_1 k_C g} \tag{10.60}$$

The stator leakage inductance is computed considering the slot, end connection, and differential permeances. If the stator slots have the same shape form and the same end-connection distributions as for the PMSM stator, then the same expressions are used for stator slots permeance (Equations 9.63 through 9.69), respectively, Equation 9.70, for end connection.

The stator differential permeance is [1]:

$$\lambda_{sd} = 0.3\rho_{d1} k_{01} \sigma_{d1} (q_1 k_w)^2 \frac{\tau_{ss}}{k_C g} \tag{10.61}$$

where

$$k_{01} = 1 - 0.033 \frac{o_{ss}^2}{g \tau_{ss}} \tag{10.62}$$

$$\sigma_{d1} = \left(\frac{\sin\left(\frac{\pi}{6q_1}\right)}{\sin\left(y_c \frac{\pi}{2}\right)} \right)^2 \cdot \sum_{u=6k\pm1, k\geq1} \left(\frac{\sin\left(\nu y_c \frac{\pi}{2}\right)}{\sin\left(\nu \frac{\pi}{6q_1}\right)} \right)^2 \tag{10.63}$$

And ρ_{d1} is given in Figure 10.3a when the rotor bars slant is zero and in Figure 10.3b, when the rotor bar slant is equal to the rotor slot pitch.

Finally, the stator leakage inductance, $L_{s\sigma}$, is

$$L_{s\sigma} = 2\mu_0 \frac{N_1^2}{p_1 q_1} (\lambda_{ss} + \lambda_{sd} + \lambda_{s0}) l_c \tag{10.64}$$

for λ_{ss} see 9.66

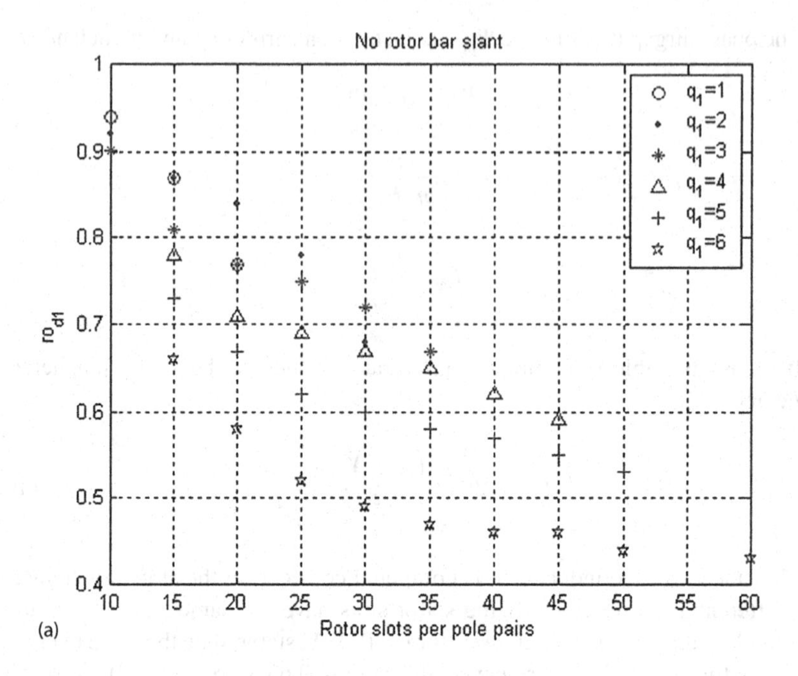

(a)

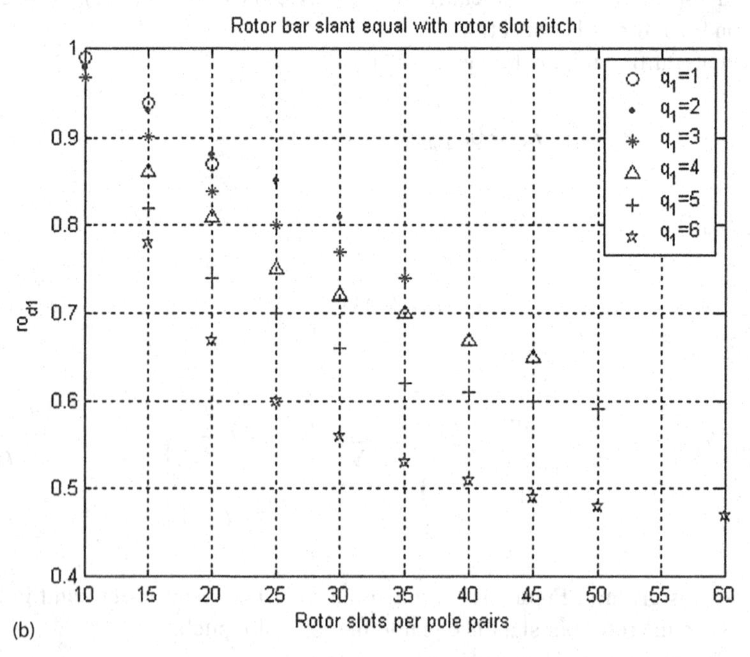

(b)

FIGURE 10.3 The ρ_{d1} factor. (a) No rotor bar slant and (b) rotor bar slant equal with rotor slot pitch.

The rotor slot permeance for the current rotor slot shape, Figure 10.1, is

$$\lambda_{rs} = \frac{\pi}{6} + \frac{2h_{r3}}{3(w_{r1} + w_{r2})} + \frac{h_{r1}}{o_{rs}} \tag{10.65}$$

The rotor end ring permeance is

$$\lambda_{r0} = k_{rRing}^2 \cdot \frac{D_{wRing}}{l_{rc} \cdot N_{rs}} \cdot og\left(4.7 \frac{D_{wRing}}{h_{Ring} + 2w_{Ring}}\right) \tag{10.66}$$

The rotor differential permeance is [1]

$$\lambda_{rd} = 0.3k_{02} \cdot \tau_{d2} \cdot \left(\frac{N_{rs}}{6p_1} \cdot f_{wr}\right)^2 \cdot \frac{\tau_{rs}}{k_c \cdot g} \tag{10.67}$$

where

$$k_{02} = 1 - 0.033 \cdot \frac{o_{rs}^2}{g \cdot \tau_{rs}} \tag{10.68}$$

The σ_{d2} coefficient is given in Figure 10.4 versus q_2, the average number of slots per pole and stator phase in the rotor:

$$q_2 = \frac{N_{rs}}{6p_1} \tag{10.69}$$

Finally, the rotor leakage inductance is

$$L_{r\sigma} = 4\mu_0 m \frac{(N_1 \cdot k_w)^2}{N_{rs}} \cdot (\lambda_{rs} + \lambda_{rd} + \lambda_{r0}) l_{rc} \tag{10.70}$$

The machine performance is computed using the standard steady-state circuit model.

10.2 INDUCTION MOTOR OPTIMAL DESIGN USING GENETIC ALGORITHMS

After the rather complete model of induction machine was introduced earlier, an exercise design of three-phase two-pole induction motor with short-circuit rotor at rated power 22 kW and 50 Hz base frequency is presented here. The total motor cost

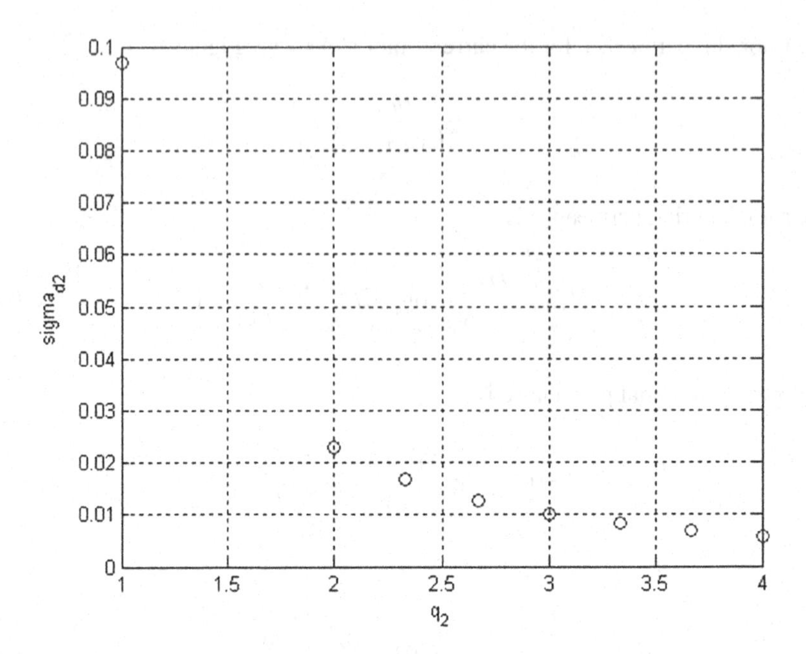

FIGURE 10.4 The σ_{d2} factor.

is the objective function. The maximum and the minimum of optimization variables are presented in Table 10.2.

The technological constraints used for this exercise design could be found in the input file (im1.m) of the induction machine optimal design MATLAB code, which is available online. Table 10.2 also contains the final values of optimization design exercise for three cases: GA—genetic algorithm with 100 members in the population and 200 generations evolution; GA1—genetic algorithm with 30 members of the population and 50 generations evolution; and the Hooke–Jeeves algorithm (HG). The optimization objective cost function is the motor life cost, which includes the initial active material cost and the energy loss cost.

The main optimization variables, specific electric load, J_1, and airgap flux density, B_{agsp}, have similar values for the three compared cases. For GA1, the specific electric load is smaller but 50 generations are too small a number to find the optimum motor. The optimization variable code used in genetic algorithms allows larger values than maximum bounds as can be observed for rotor and stator yoke flux densities. The specific flux densities in the stator and rotor teeth seem to be too large for the usual magnetic core like "M19." In fact, the real flux density is smaller because, at such large magnetic saturation, the shape form factor k_f of the airgap flux density distribution decreases from the ideal value 1.57 T to 1.1 T for high saturation. The same polar flux produced for ideal (sinusoidal) flux distribution with a peak of 2.5 T flux density in the teeth needs less than 1.8 T in the teeth. Every generation is characterized by the average, the minimum, and the maximum values of the cost function. Their evolution is shown in Figure 10.5.

The minimum value, which belongs to the best machine from the population, decreases monotonically. The average value of the cost function also tends to decrease.

TABLE 10.2
Optimization Variables

Optimization Variable	Minimum Values	Maximum Values	GA	GA1	HG	Units
J_1	12	40	28	24.2	27.31	kA/m
B_{agsp}	0.4	0.85	0.63	0.64	0.63	T
B_{st}	1.2	2.1	2.46	1.66	2.1	T
B_{sy}	1	2	1.74	1.62	1.7	T
B_{rt}	1.2	2.1	2.34	2.42	1.75	T
B_{ry}	1	2	1.84	1.88	1.56	T
J_s	2.5	8	2.8	3.2	3.52	A/mm^2
J_r	2	7.5	2.2	2	3.62	A/mm^2
sh_4	0.3	2	0.8	2.1	1.1	mm
s_{os}	0.5	2	1.4	0.8	1.2	mm
rh_1	0.3	2	0.5	0.9	1.2	mm
s_{or}	0.3	2	1.2	0.3	1.1	mm
Lcpertau	0.5	2	0.9	1	1	
q_1	2	6	5	5	4	
cSpan	0.66	1	0.73	0.8	0.83	
rSlots			18	18	18	

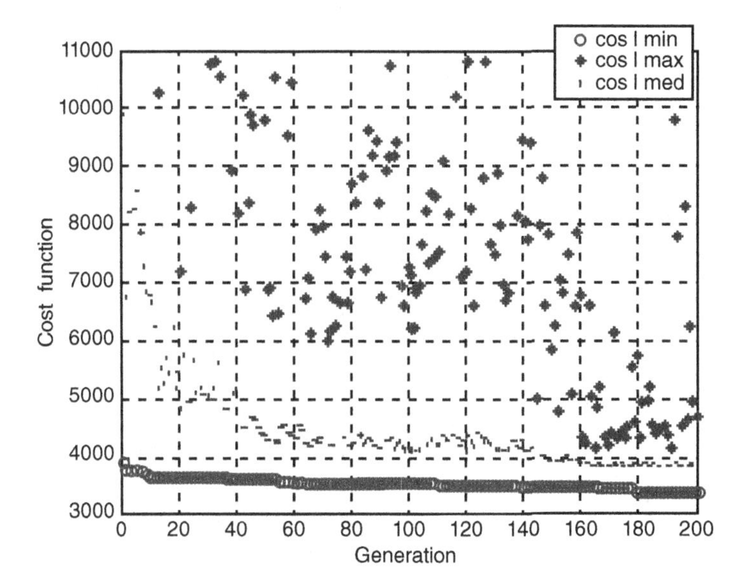

FIGURE 10.5 Minimum, maximum, and average cost function evolutions.

The evolution of representative optimization variables of the best member along the 200 generations is shown in Figures 10.6 through 10.11. The linear electric load and the current density in the stator and rotor have several discrete values but we cannot observe any such tendency when the performance of the best member is improved, Figure 10.6. Moreover, for the rotor current only three discrete values appear but the first one shows up only at the first generation.

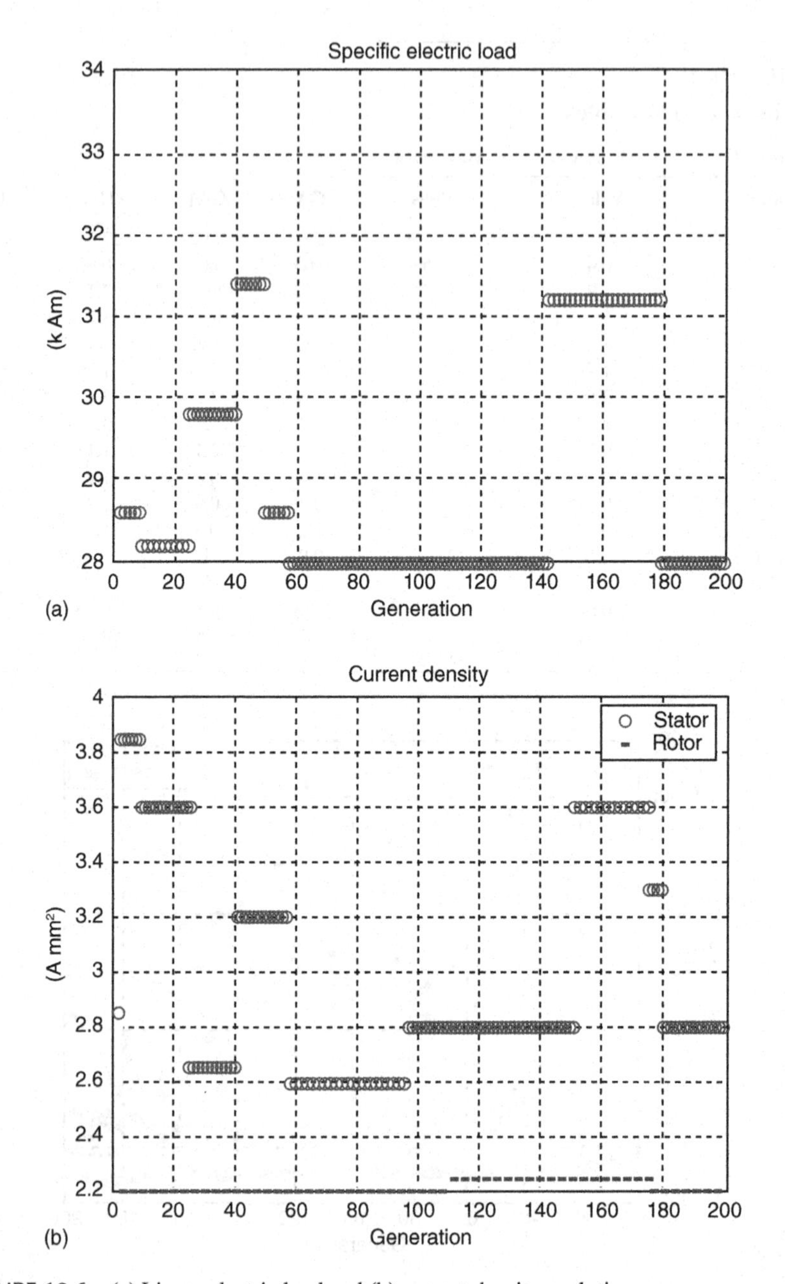

FIGURE 10.6 (a) Linear electric load and (b) current density evolutions.

The airgap flux density also shows discrete values but it increases slowly from the first generation to the last generation, Figure 10.7. There seems to be a good correlation between the decreasing objective function and the increasing airgap flux density. The flux densities in the stator and the rotor teeth also increase when the best-candidate solution improves.

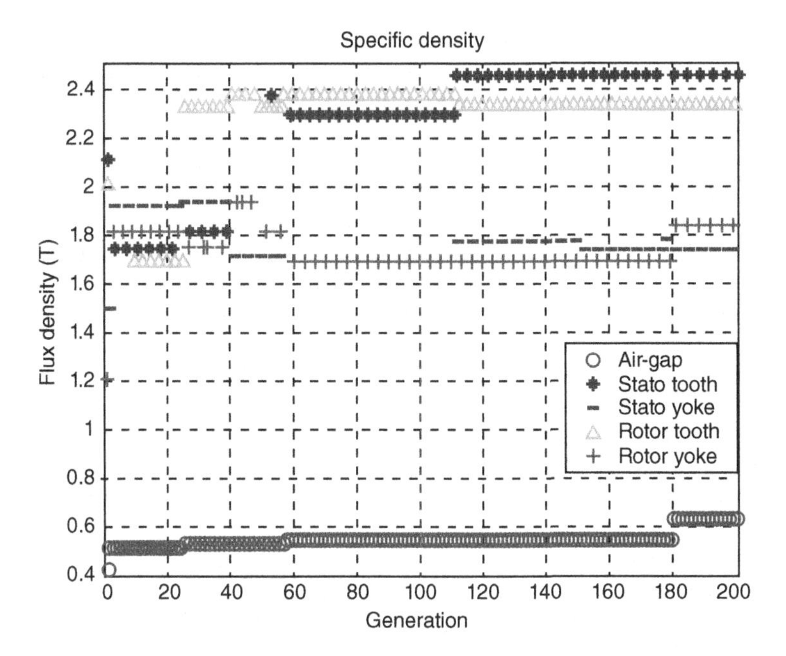

FIGURE 10.7 Airgap and magnetic core flux density evolutions.

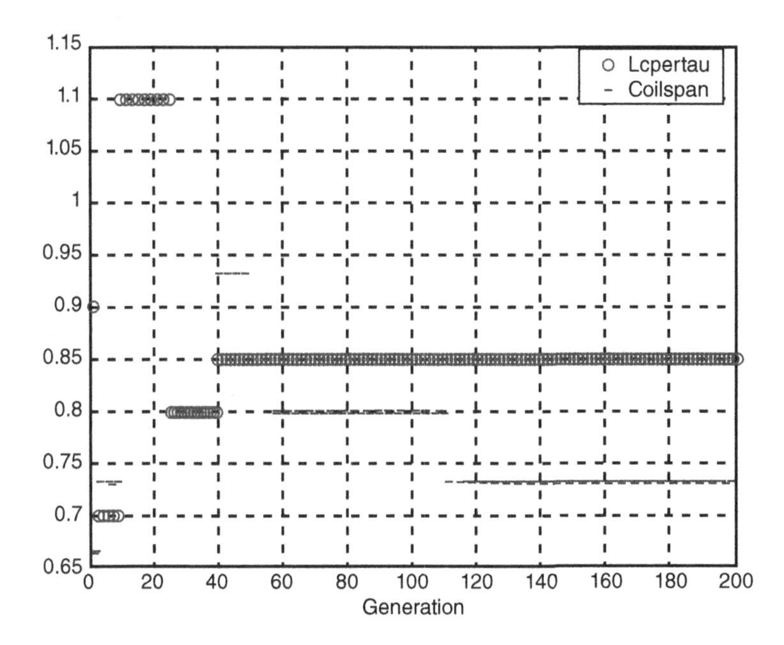

FIGURE 10.8 Dimensions of ratio evolutions of the best-candidate solutions: core length per pole pitch (lcpertau) and coil span per coil pitch (coilSpan).

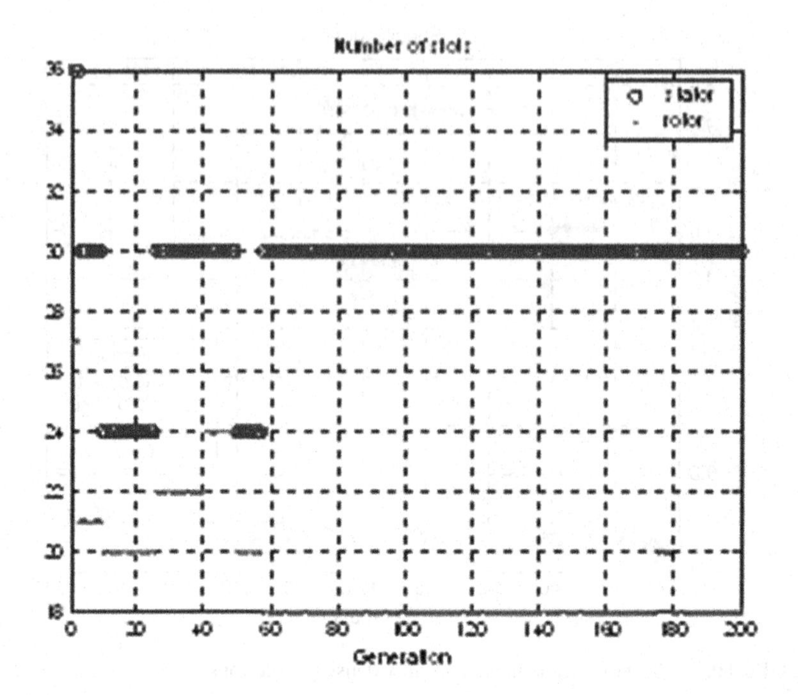

FIGURE 10.9 Number of stator and rotor slots evolutions.

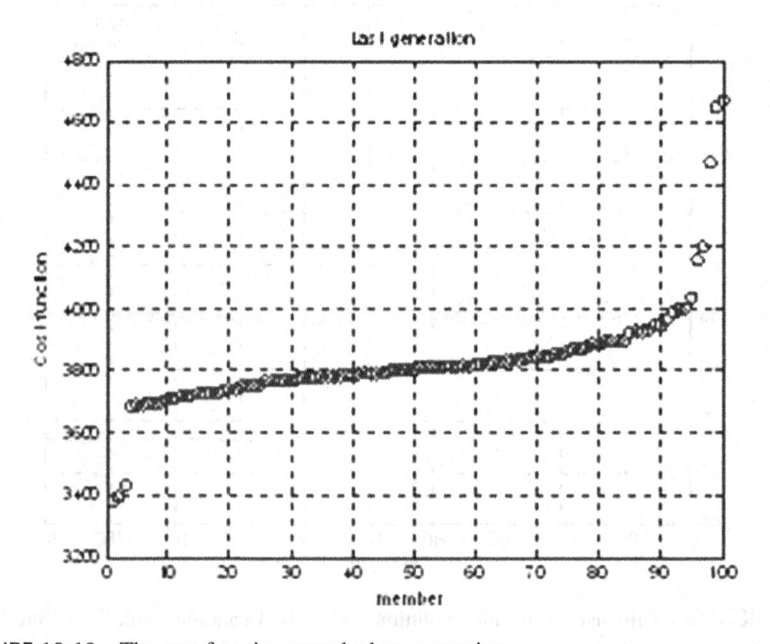

FIGURE 10.10 The cost function over the last generation.

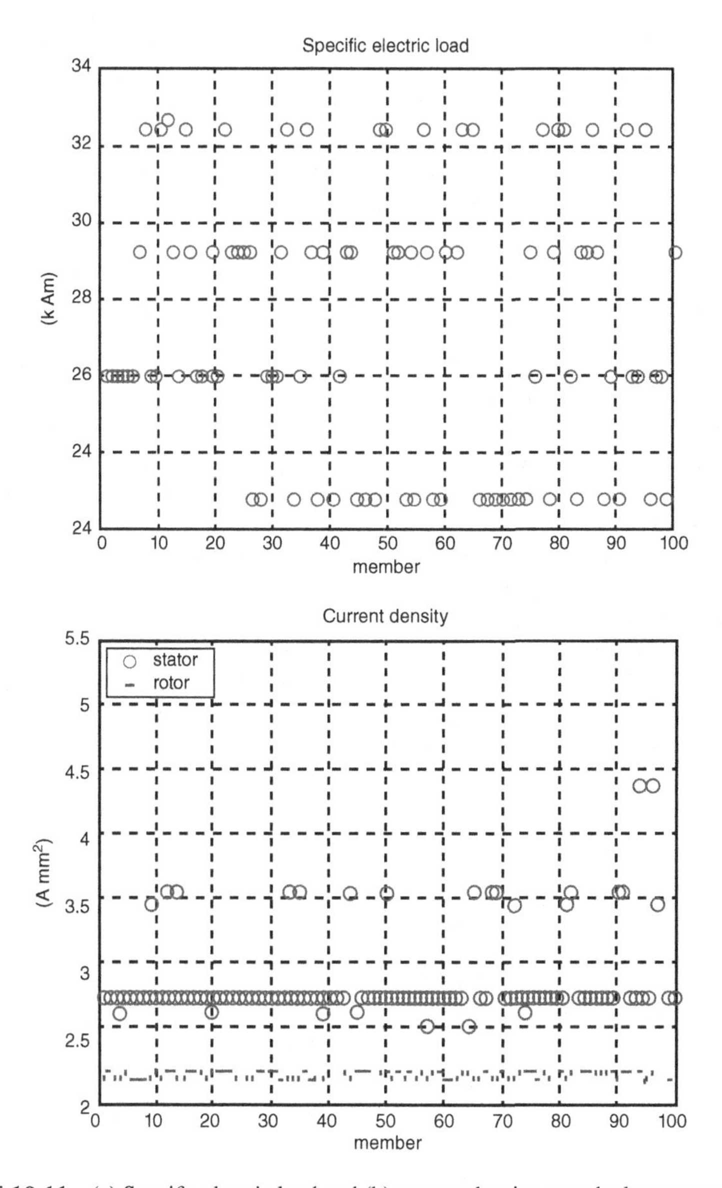

FIGURE 10.11 (a) Specific electric load and (b) current density over the last generation.

The ratio between core length and pole pitch, respectively, the coil span and polar pitch, shows a random variation, with several discrete values, Figure 10.8, and, then, for the last generation, they remain unchanged. The numbers of stator and rotor slots, Figure 10.9, also have only a few discrete values and they remain unchanged for the last generations.

The cost function for the members of the last generation and the optimization variable distribution could show the evolution potential of the genetic algorithm if it would be continuous. In our case, we can observe in Figure 10.10 that the cost function still has a large variation. The members were sorted in the order of increasing cost function. We can observe that there is no correlation between the specific electric load and the cost function, Figure 10.11. Only several discrete values of the specific electric load were retained in the last generation, but their distribution still covers a large region. The stator current density distribution contains only a few values and their distribution is not correlated with the cost function. They are placed at the lowest limit of the current density. The rotor current, though, is an exception; for the three members, the rotor current density is larger but the cost function also has larger values. The airgap flux density shows a correlation with the cost function for the last generation, Figure 10.12.

The largest values correspond to small values of the cost function. There are only a few discrete values of the airgap flux density and the cover range is small. If the global minimum of the objective function is placed at a larger value of the airgap flux density, then further evolution for this member could be very slow because only a good mutation could produce larger airgap flux. We can see also that a large teeth flux density could produce a better objective function.

The ratio between the core length and the pole pitch, and the ratio between the coil span and the pole pitch are reduced to a few values and their distribution seems to be uncorrelated with the cost function, Figure 10.13. For the chosen number of stator slots, only two from five possible values were retained as is shown in Figure 10.14. These are combined with only three values for the rotor slots number. There is no visible correlation between the number of the slots and the cost function. However, we can conclude that 24 or 30 slots in the stator are better than 18, which probably is

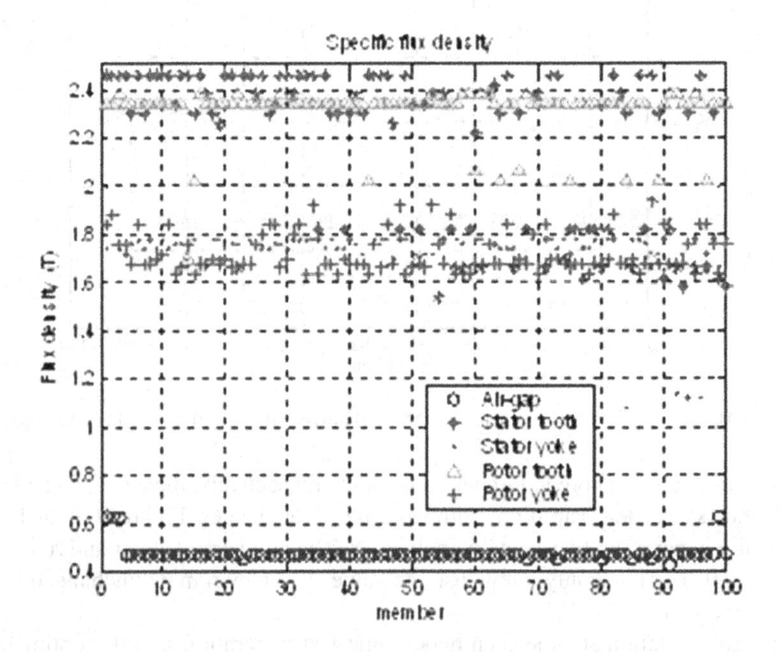

FIGURE 10.12 Flux densities over the last generation.

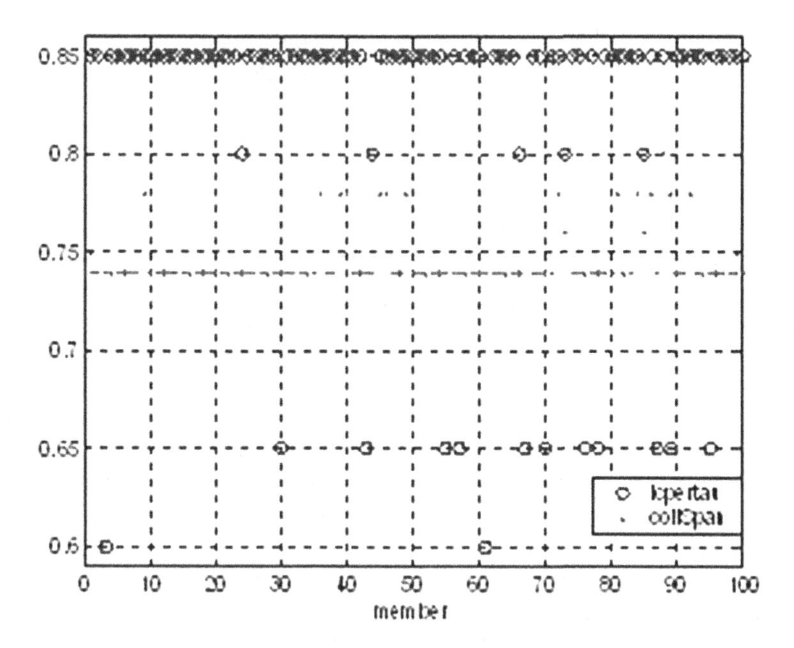

FIGURE 10.13 Dimension ratios over the last generation.

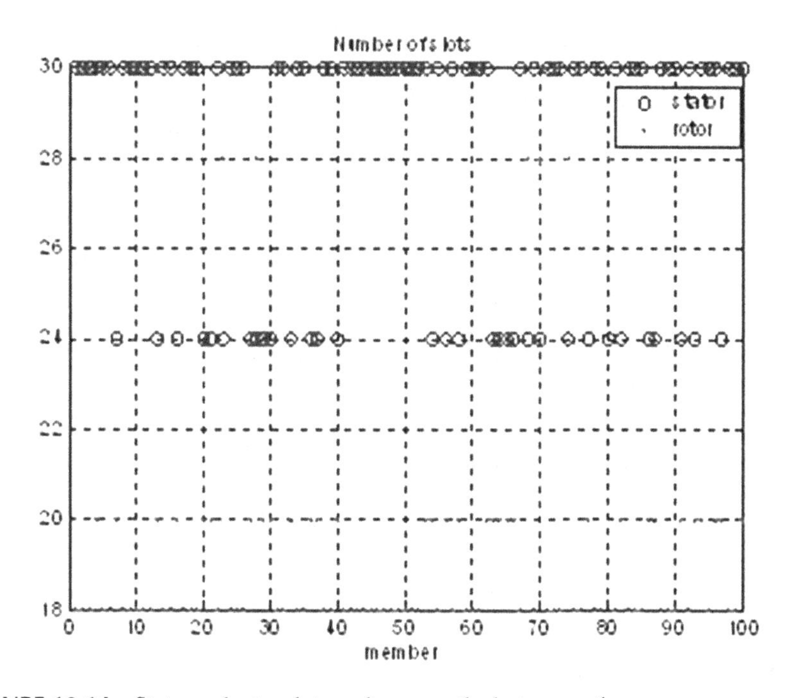

FIGURE 10.14 Stator and rotor slot numbers over the last generation.

too small a number, and 36 which probably is too large a number, because these have no selection in the last generation, which has better performance than in the first generation.

The rotor and stator diameter evolution is shown in Figure 10.15. There are several discrete values that cover a small variation domain. The stator and rotor slots dimension variation is shown in Figure 10.16. The variation domain of the slot height is moderate compared with its values.

The power losses and efficiency evolutions are shown in Figure 10.17. The copper losses show a tendency to decrease, while the motor efficiency shows a tendency to increase. The variation is discrete and not monotonous. The power factor evolution is also presented in Figure 10.17. This one too shows a nonmonotonic tendency to increase.

The motor and its stator and rotor weights evolution are shown in Figure 10.18. The weight of the windings and magnetic core components evolution are shown in Figure 10.19. We cannot observe any direct correlation between the weight evolution and the cost function evolution, which decreases monotonically, as shown in Figure 10.20.

The components of the cost function such as copper, laminations, cage, and undesired loss energy are also shown in Figure 10.20. The initial cost contains the cost of active materials such as copper, lamination, and cage as the weight cost. The total cost (objective function) contains the initial cost and the energy loss cost. It may also

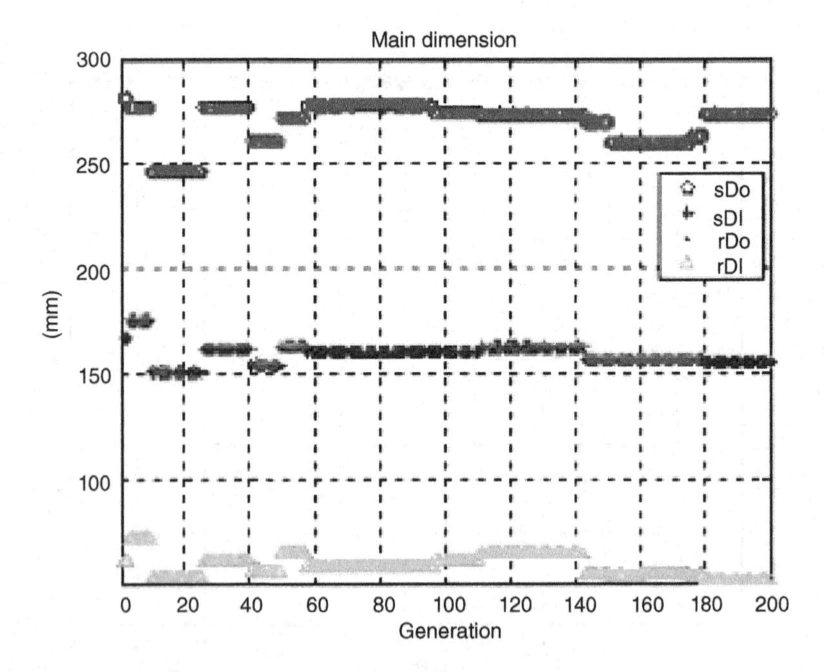

FIGURE 10.15 Rotor and stator diameters evolution.

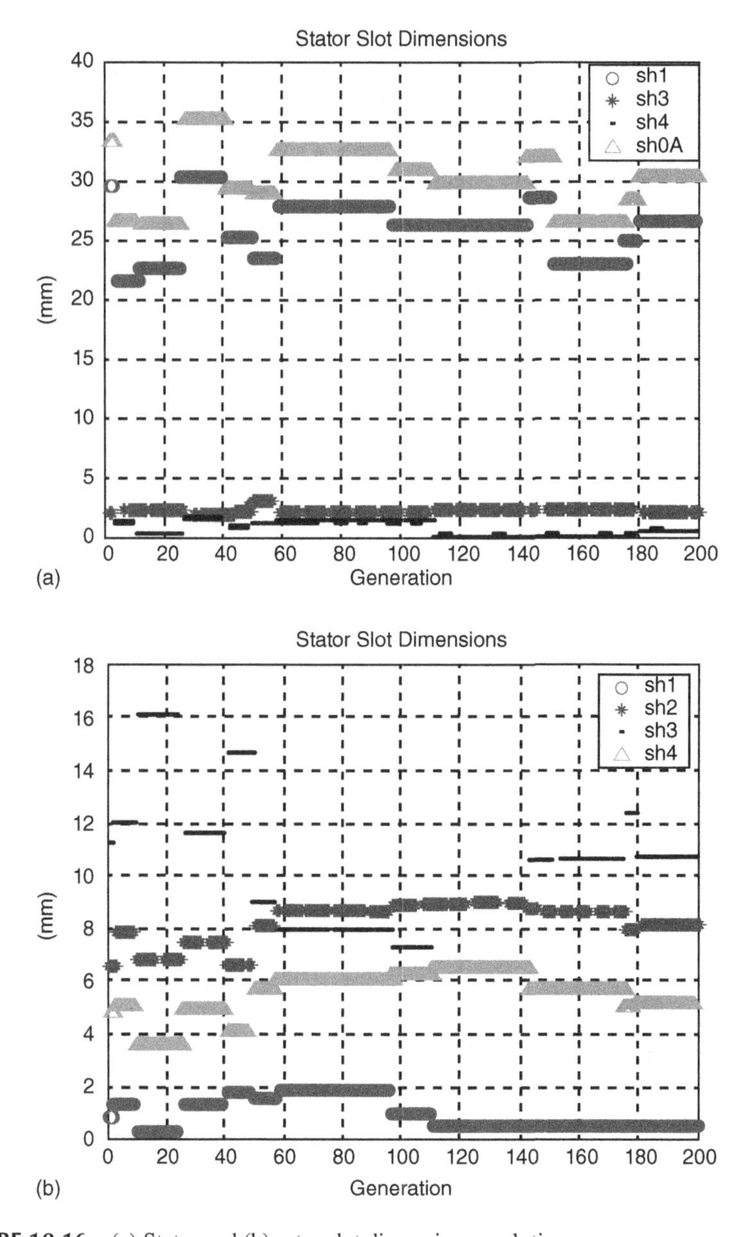

FIGURE 10.16 (a) Stator and (b) rotor slot dimensions evolutions.

contain a penalty cost for an over-temperature larger than the maximum admissible, and a starting current penalty cost if the latter is larger than the maximum allowed (in our case, six times larger than the rated current). Note that no penalty cost is present in Figure 10.20, which shows the best member from every generation.

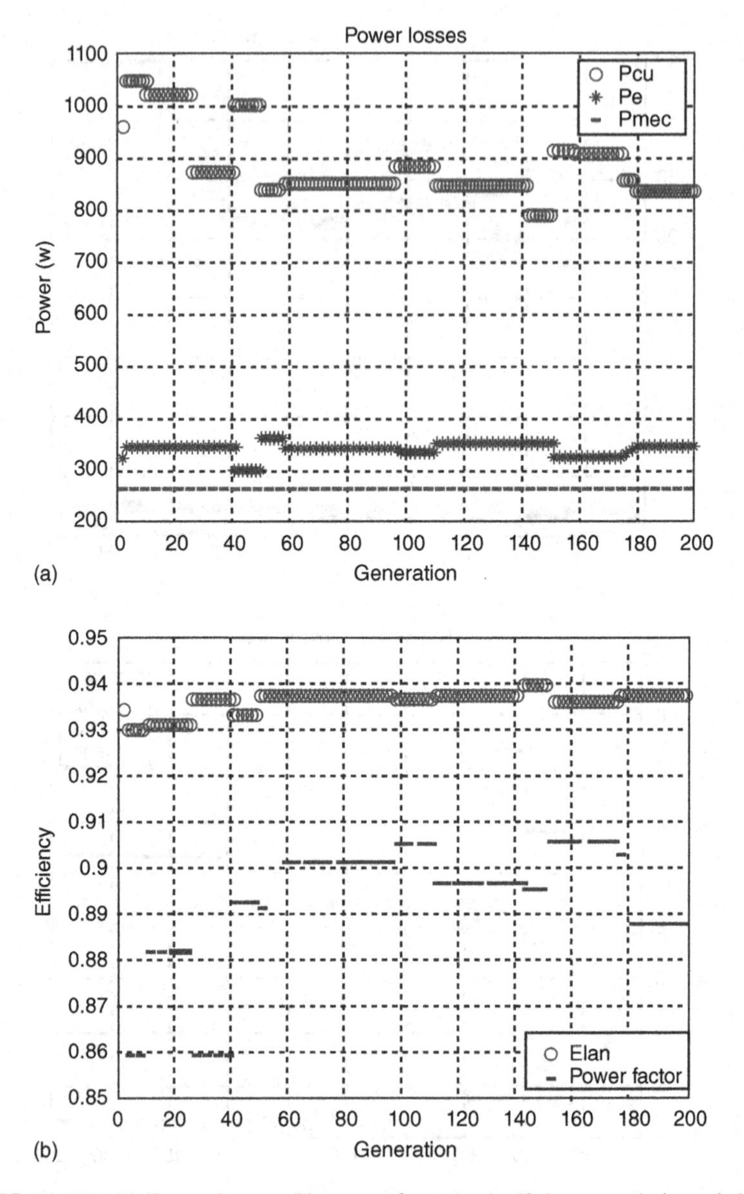

FIGURE 10.17 (a) Power losses, (b) power factor and efficiency evolution of the best member.

10.3 INDUCTION MOTOR OPTIMAL DESIGN USING HOOKE–JEEVES ALGORITHM

The same optimization variable vector and objective function, total cost C_E, as that presented for genetic algorithm method is also used to implement the Hooke–Jeeves optimization algorithm. The specific electric load has a small step variation during the optimization process, Figure 10.21.

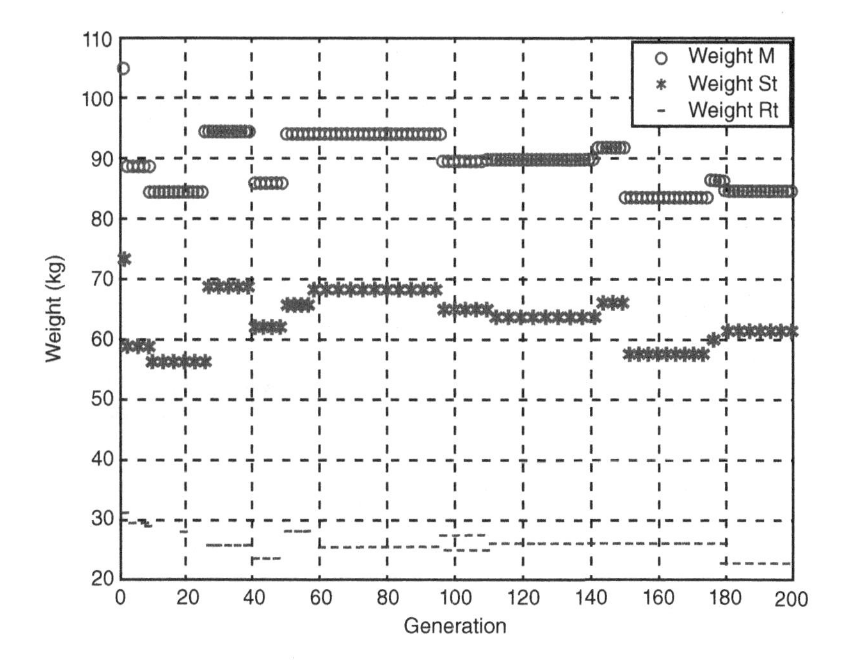

FIGURE 10.18 Motor weight evolution.

The final value is close to the final value obtained from the genetic algorithm method. It can also be seen from this figure that stator and rotor current densities decrease.

The airgap flux density remains constant during the optimization process, Figure 10.22, but its initial value is around the optimum value from the genetic algorithm optimization. The stator tooth flux increases toward its maximum value. The other core flux densities also have a tendency to increase, but they do not reach the upper bounds.

The ratio between the core length and the pole pitch decreases by about 25%, while the coil span ratio remains constant, Figure 10.23. The stator slots number remains constant at 24 slots, Figure 10.24, the same as that which appears in the last generation of the genetic algorithm optimization. The number of rotor slots decreases at the first optimization step from 34 to 18 slots, which was also found as optimum with the genetic algorithm.

The rotor and stator diameter evolutions are shown in Figure 10.25. The stator inner diameter, sDi, and the rotor outer diameter, rDo, increase a little, while the stator outer diameter, sDo, increases a little more. Therefore, the stator slot height also increases, Figure 10.26.

The copper losses decrease while the iron losses increase a little, so the efficiency, etan, increases, Figure 10.27. The power factor shown in Figure 10.27 also decreases to acceptable values.

The motor active weight increases by about 10kg due to the increase in stator weight, Figure 10.28.

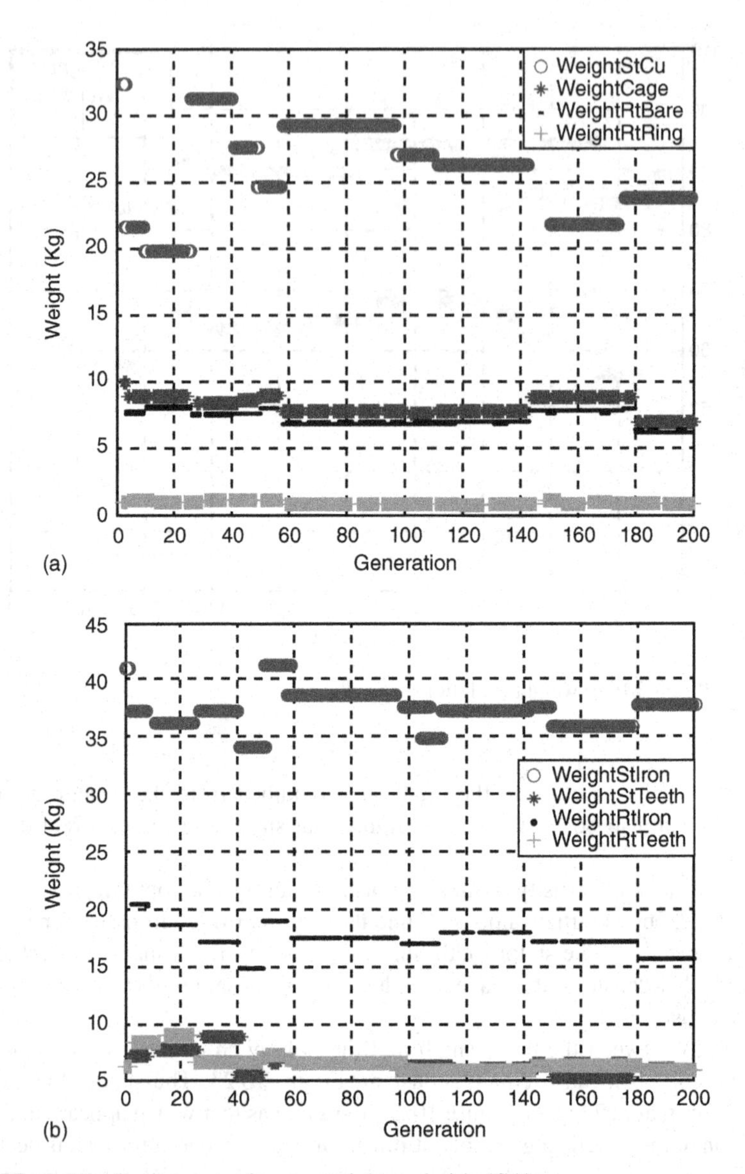

FIGURE 10.19 Components of motor weight evolution (GA).

The increase in stator weight is produced especially by additional stator windings (copper), Figure 10.29.

The objective function decreases consistently, Figure 10.30. The first decreasing step is produced by the elimination of penalty cost. The current density and the machine diameter are not changed between the first and the second solution, and the stator teeth top, hs 4, increases significantly; thus we can conclude that the start current penalty has been eliminated. Thereby the cost function continues to decrease due to efficiency improvement despite the initial increase.

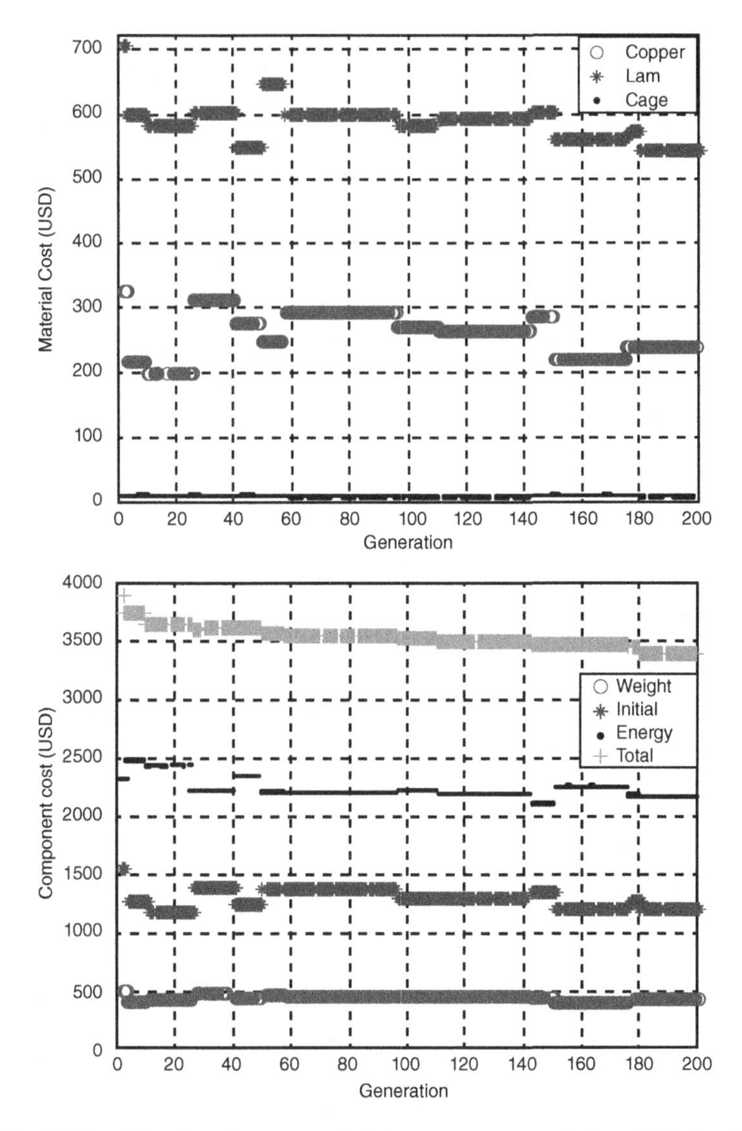

FIGURE 10.20 Objective function and their component evolution (GA) for the best member.

10.4 MACHINE PERFORMANCE

The machine performance was computed using the circuit model but the magnetization inductance was computed considering magnetic saturation. The performance characteristics of the motor designed with genetic algorithm, GA, are compared with the motor performance designed by the Hooke–Jeeves optimal algorithm, HG.

The linkage flux versus the magnetization current is presented in Figure 10.31 for both motors. It can be seen that at small currents (5 A), the linkage flux of GA motors

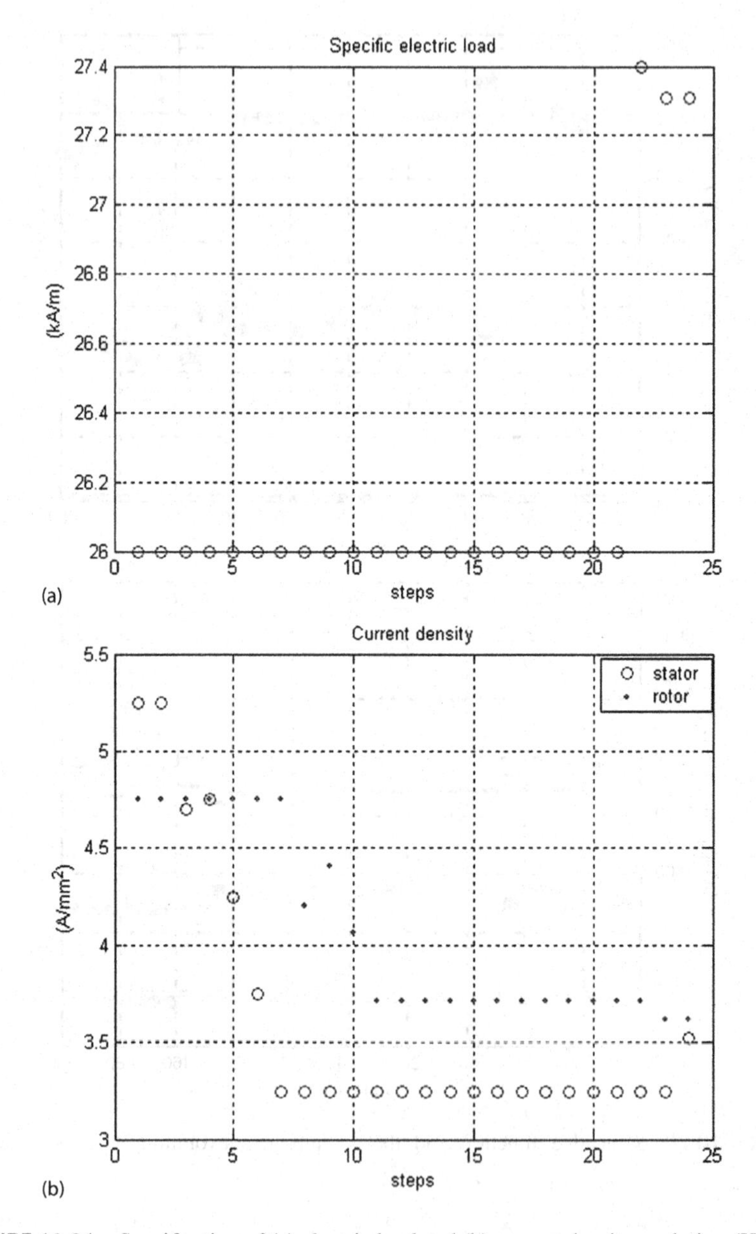

FIGURE 10.21 Specification of (a) electric load and (b) current density evolution (HJ).

is a little larger than HG machines while at medium and large currents, the flux of HG motor is larger than for GA motors. Therefore, HG motor is more saturated than the GA motor.

The same conclusion can also be drawn from magnetization inductance versus magnetization current, Figure 10.32, and from airgap flux density versus

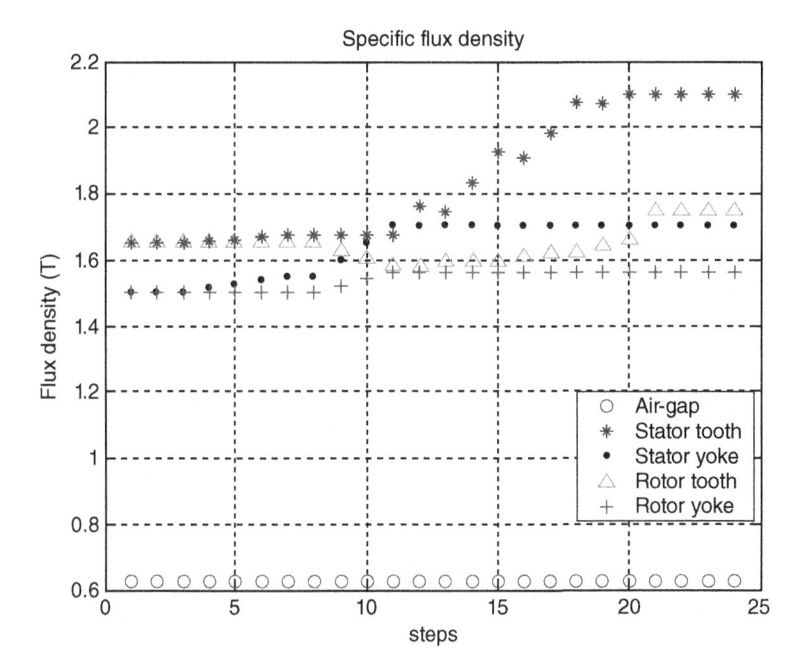

FIGURE 10.22 Flux density evolution (HJ).

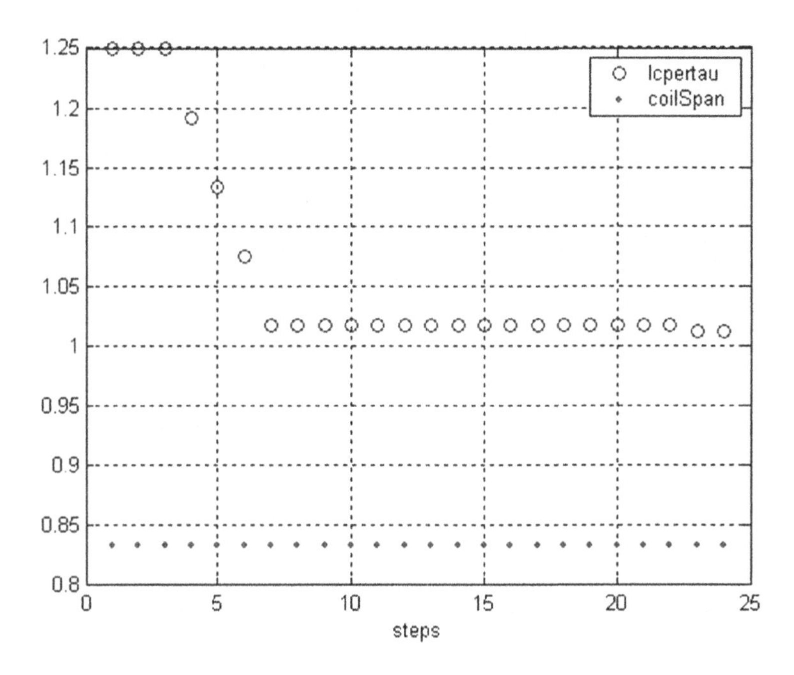

FIGURE 10.23 Dimension ratios evolution (HJ).

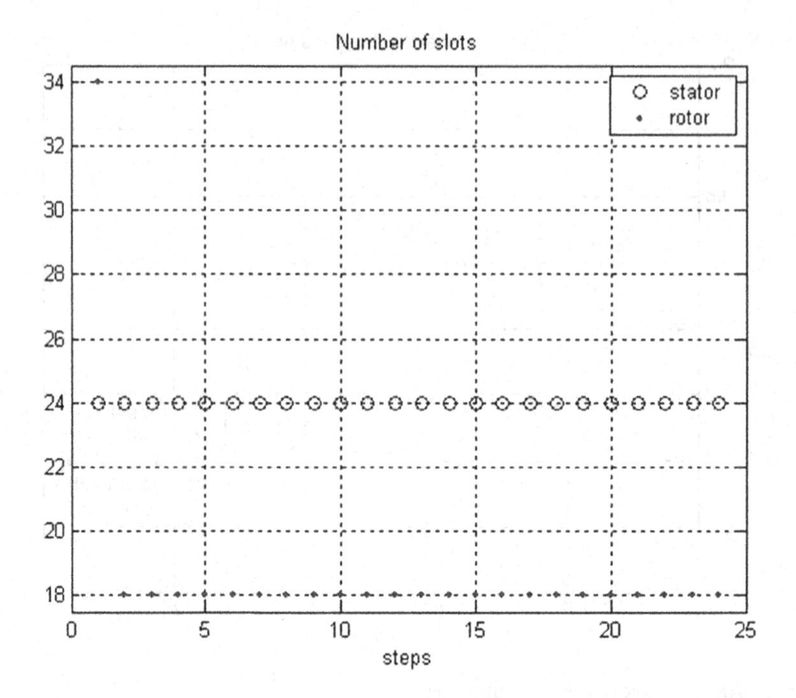

FIGURE 10.24 Stator and rotor slots evolution (HJ).

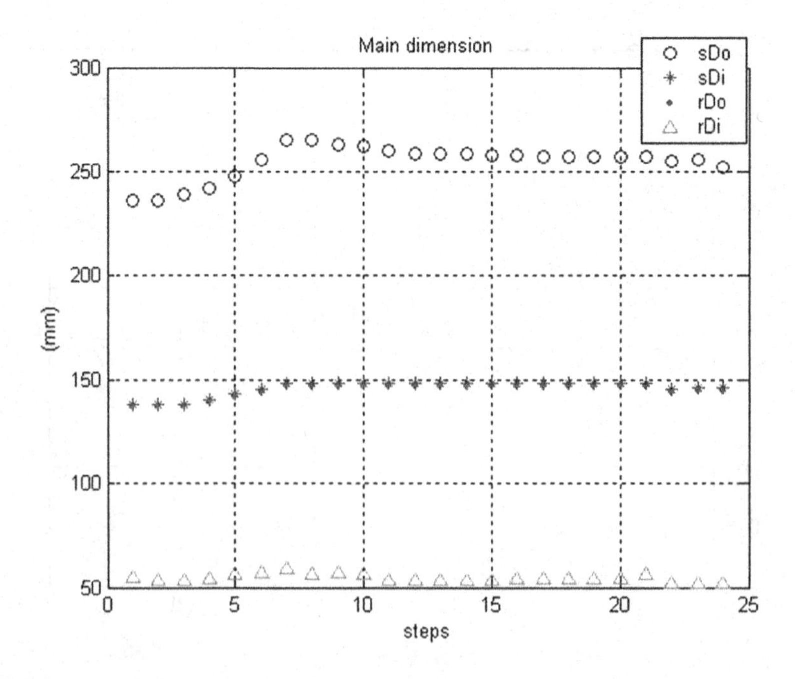

FIGURE 10.25 Rotor and stator diameters evolution (HJ).

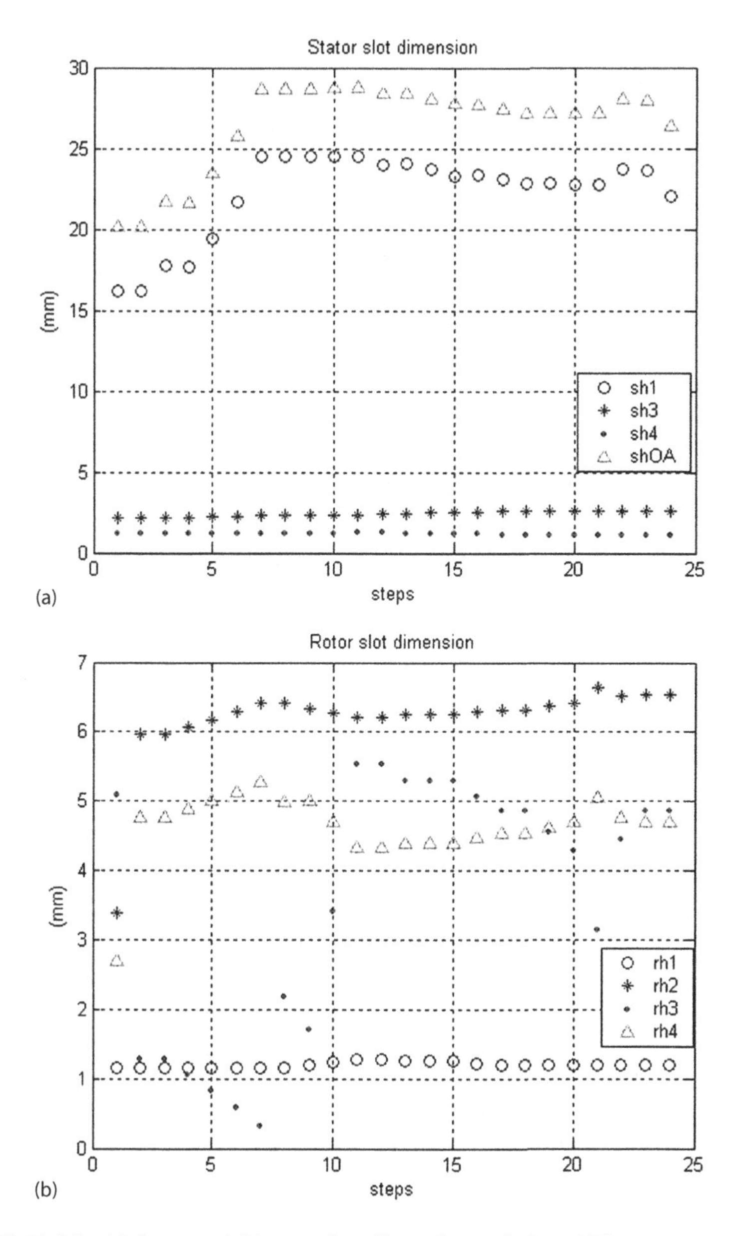

FIGURE 10.26 (a) Stator and (b) rotor slots dimension evolutions (HJ).

magnetization current, Figure 10.33. The saturation factors versus the magnetization current are shown in Figure 10.34.

The stator yoke has the largest contribution to the saturation factor despite the large specific flux density in the stator and the rotor teeth. This is the same as in two-pole machines that have a long magnetic line path through the stator yoke.

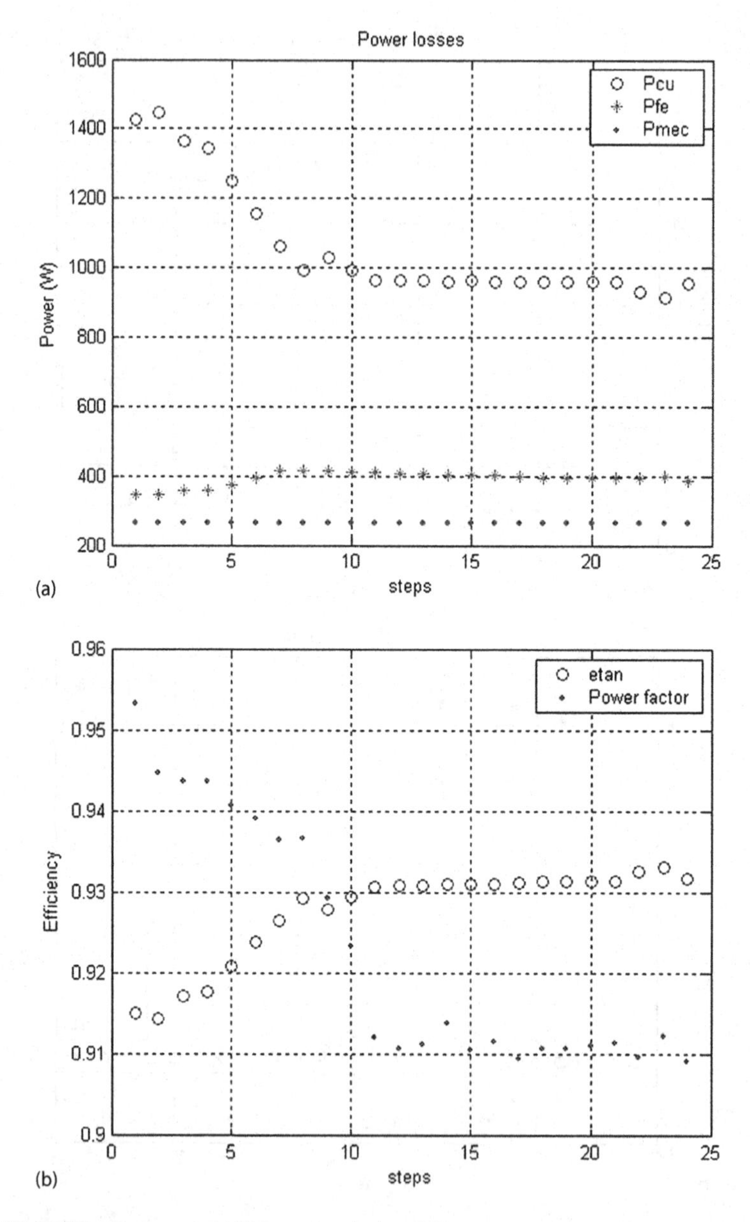

FIGURE 10.27 Losses and efficiency evolution (HJ).

The torque versus slip for both machines is shown in Figure 10.35. The GA motor has a smaller peak torque and a smaller critical slip than the HG machine. Also, the starting torque of the GA motor is smaller than that of the HG motor. The smaller rotor resistance and leakage inductance of the GA motor (Table 10.3) explains these features.

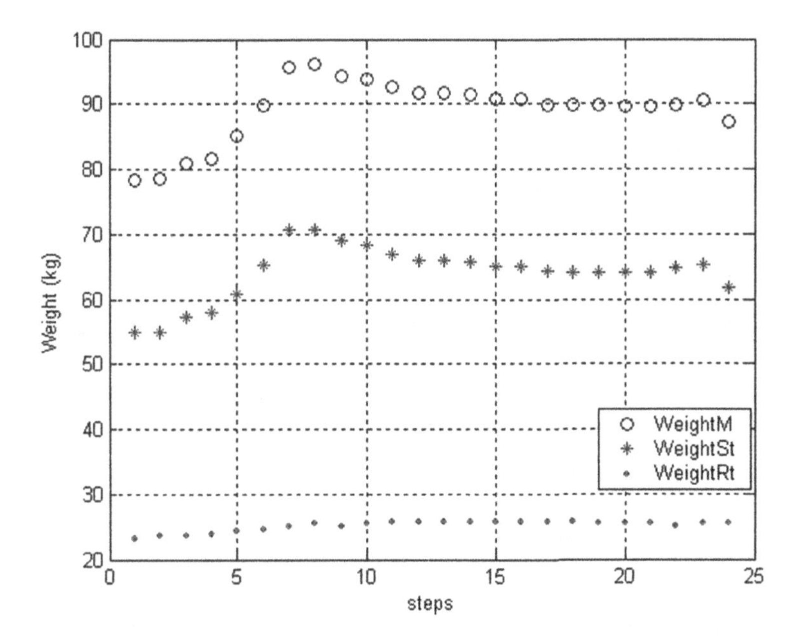

FIGURE 10.28 Motor weight evolution (HJ).

The starting current of the HG motor is a little smaller than for a GA motor, Figure 10.36.

The acceptable ratio of the starting current to the rated current was obtained by imposing a penalty cost for the motors with the starting current larger than allowed for. Without this penalty, the optimization algorithm produces motors with a starting current up to 12 times larger than rated currents. Such kinds of motors are not suitable to be directly connected to the grid. However, they have a large peak torque, which implies a large speed range at constant power, in variable speed drive.

The input and output powers versus the slip are shown in Figure 10.37, while the efficiency and power factors versus the slip are shown in Figure 10.38.

The current versus torque for both motors is shown in Figure 10.39, while the efficiency and power factors versus torque are shown in Figure 10.40.

The main dimensions and parameters of the designed motors are presented in Table 10.3.

The optimization steps of the Hooke–Jeeves algorithm are smaller and the required computation time is also much smaller as is evident in Table 10.3 It is only 5.73 s for the Hooke–Jeeves algorithm, while for the genetic algorithm with a population of 100 members and 200 generations, the simulation time reaches 286 s. The Hooke–Jeeves algorithm falls into a local minimum with the objective function equal by $3635 while the genetic algorithm finds an objective function of $3381, which is about 5% smaller. The initial cost seems to be similar for both HG and GA. Using a small number of population members (30) and a small number of generations, the

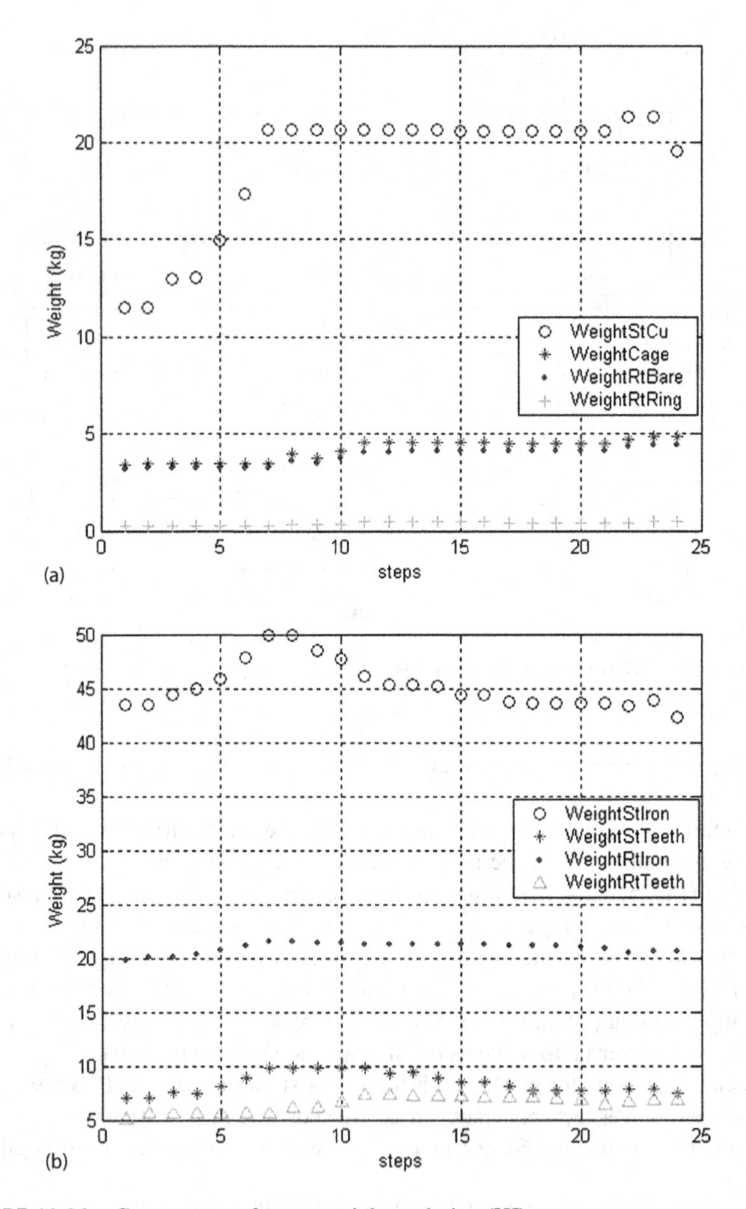

FIGURE 10.29 Components of motor weight evolution (HJ).

optimization time with GA is reduced but the objective function is comparable with the Hooke–Jeeves method. Starting HJ method with a few random initial variable vectors should increase probability for a global optimum, with still notably lower computation time than for GA.

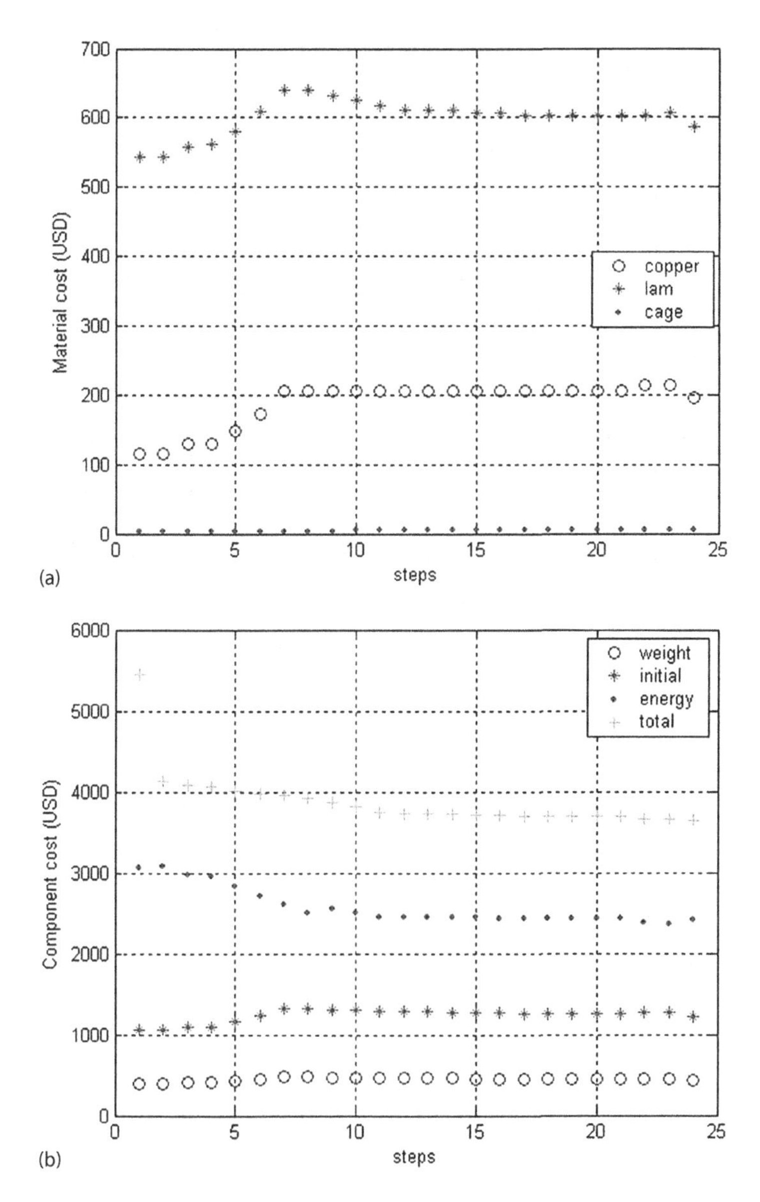

FIGURE 10.30 The objective (cost) function and their components evolutions.

10.5 CONCLUSION

To summarize, this chapter has proceeded in the following manner:

- A rather complete analytical model of IM, including magnetic saturation, was provided; for alternative IM models see Refs. [4, 6]

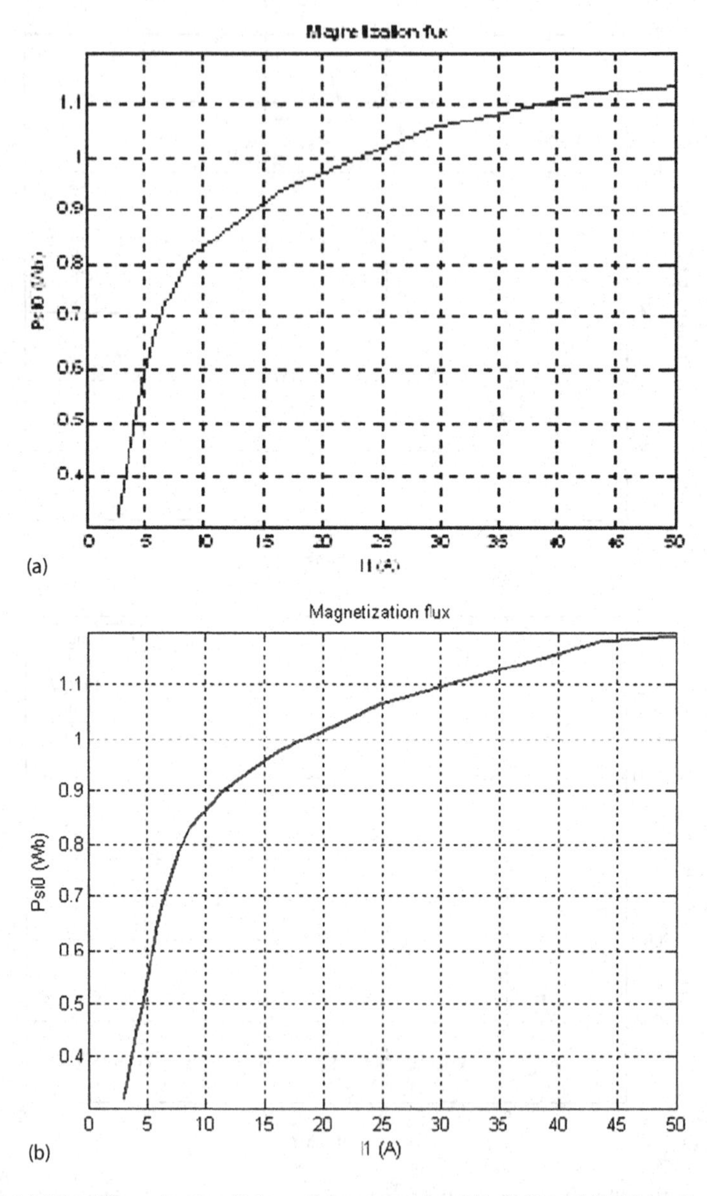

FIGURE 10.31 Magnetization linkage flux vs. current: GA (left) and HG (right).

- The analytical model was then incorporated into an optimal design computer code based on GA, and, respectively, the Hooke–Jeeves algorithm
- For the total motor cost as the cost function, the same machine was optimally designed by GA and HJ methods and the results were thoroughly illustrated and discussed

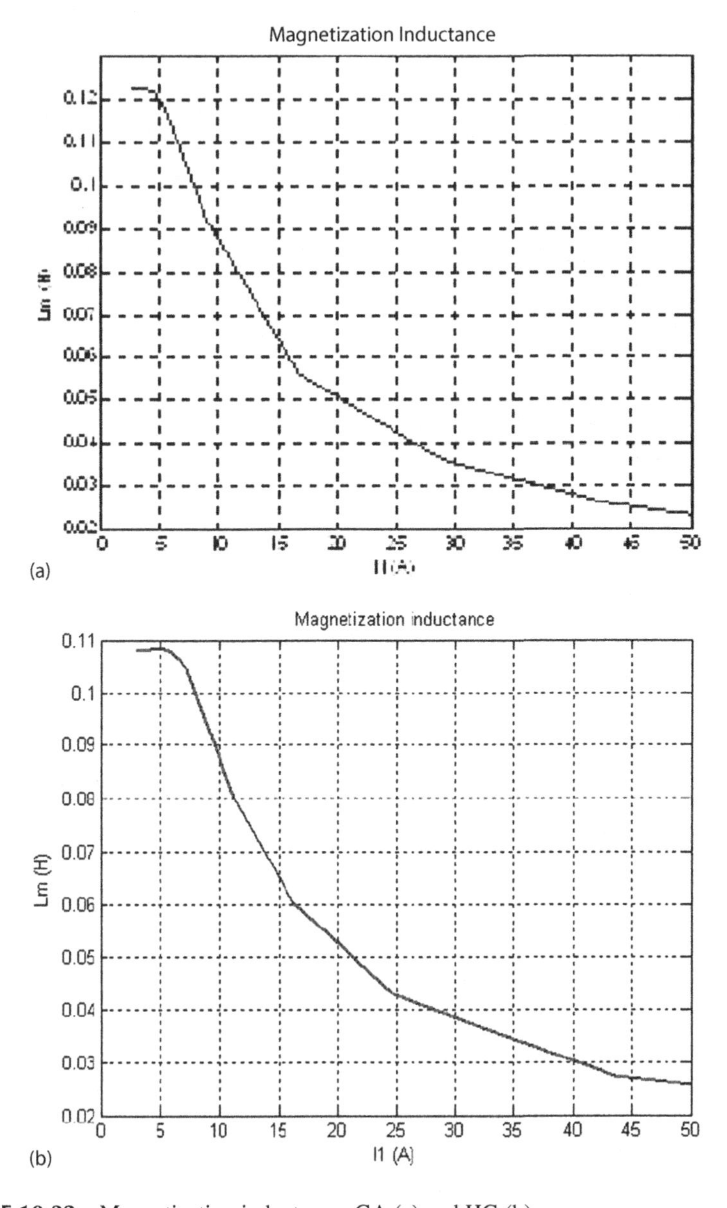

FIGURE 10.32 Magnetization inductance: GA (a) and HG (b).

- Both optimal design algorithms produced good results but the HG algorithm required 50 times less time for computation (for one initial variable vector)
- To avoid falling into local optimum, the HG algorithm has to be started from a few (15–20) initial sets of variables
- After optimal design, the use of FEM would be in place for trial verifications

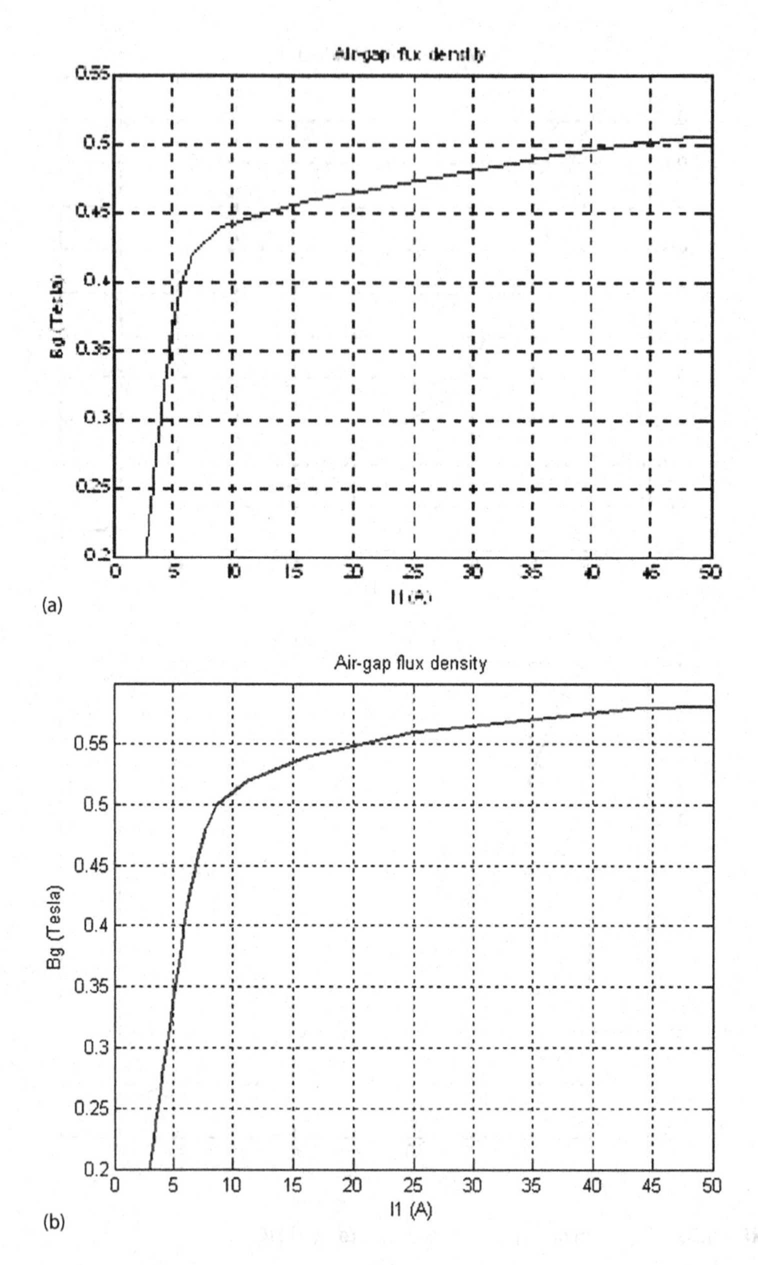

FIGURE 10.33 Airgap flux density vs. magnetization current: GA (a) and HG (b).

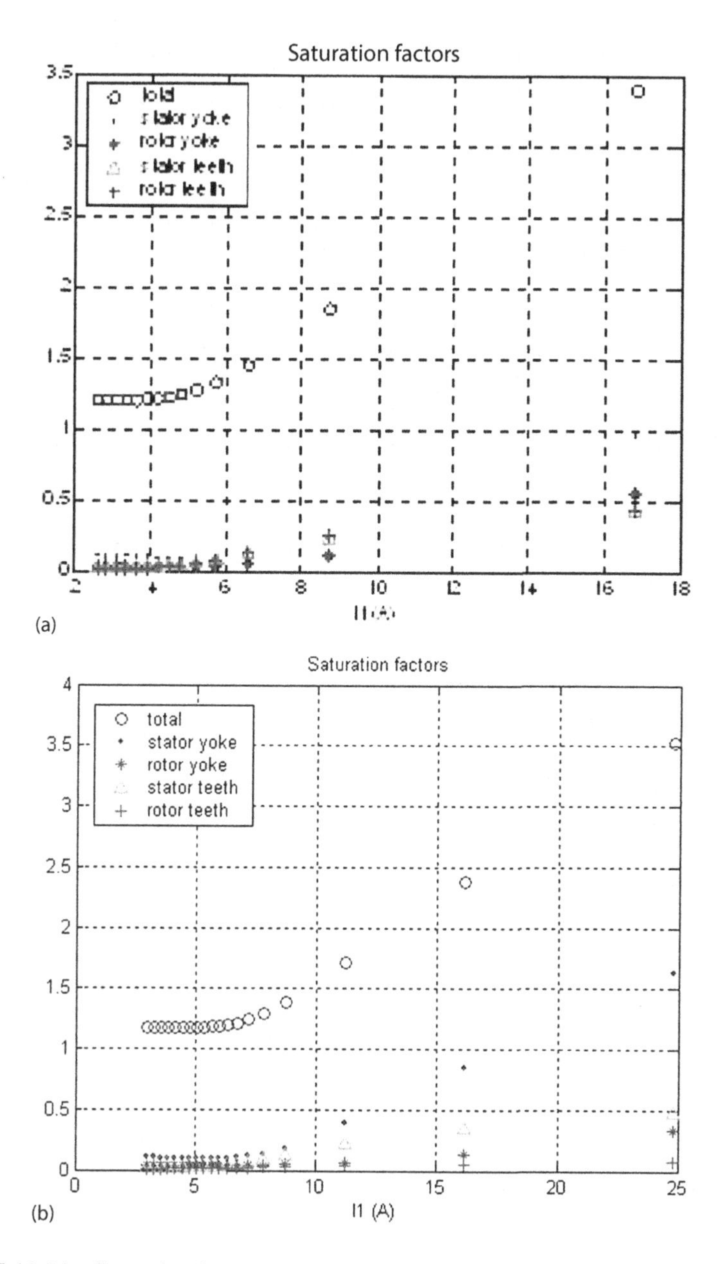

FIGURE 10.34 Saturation factors vs. magnetization current: GA (a) and HG (b).

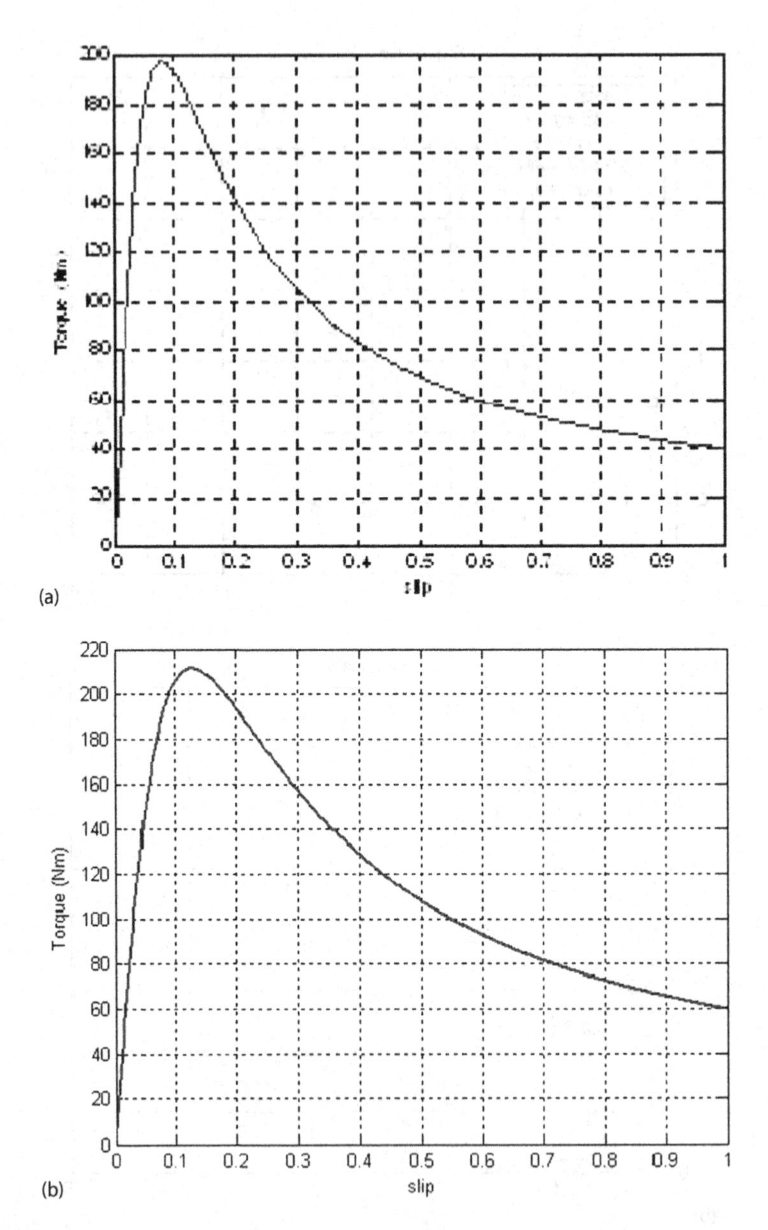

FIGURE 10.35 Torque vs. rotor slip: GA (a) and HG (b).

TABLE 10.3
Parameters of Optimized Induction Motor

Parameter	GA1	GA2	HG	Units	Comments
D_{so}	273	269	252	mm	Stator outer diameter
D_{si}	156	152.2	146	mm	Stator inner diameter
D_{ri}	52	52	52	mm	Rotor inner diameter
L_c	192.2	223.5	244.5	mm	Stator core length
g	0.65	0.65	0.65	mm	Airgap
hs_{OA}	30.5	28.2	26.4	mm	Stator slot height
h_{s1}	26.7	23.5	22.1	mm	Stator slot main height
h_{s3}	2.3	2	2.6	mm	Stator slot wedge height
h_{sy}	28.1	30.2	26.6	mm	Stator yoke height
w_{st}	4.2	6.1	5.7	mm	Stator teeth width
$N1$	60	50	48		Number of turns per phase
h_{rOA}	24.5	23.7	17.3	mm	Rotor slot height
w_{r1}	16.1	15.7	12.9	mm	Rotor slot width
w_{r2}	12.3	12.3	11.2	mm	Rotor slot width
h_{ry}	26.7	25.9	29.2	mm	Rotor yoke height
b_{rt}	7.5	7.2	9.3	mm	Rotor teeth width
WeightStCu	23.868	19.234	19.536	kg	Stator winding weight
WeightIronUsed	108.391	122.337	117.475	kg	Processed iron weight
WeightStIron	37.58	47.671	42.308	kg	Stator iron weight
WeightRtIron	9600.13	10.722	13.76	kg	Rotor iron weight
WeightCage	7.161	7.58	4.861	kg	Rotor cage weight
WeightM	84.31	92.07	87.261	kg	Motor weight (active material)
I_{1n}	39.49	40.24	38.58	A	Rated current (rms)
V_{fn}	220	220	220	V	Phase rated voltage (rms)
P_{cu}	837.56	845.16	956.75	W	Copper losses
P_{fe}	346.54	382.59	386.81	W	Iron losses
P_{mec}	264	264	264	W	Mechanical losses (1.2% of Pn)
E_{tan}	93.82	93.65	93.19	%	Rated efficiency
Cosphin	0.89	0.87	0.91		Rated power factor
sR	0.113	0.116	0.112	Ω	Stator d.c. resistance
rR	0.08	0.071	0.12	Ω	Rotor d.c. resistance
RFe	372	332	340	Ω	Equivalent iron losses resistance
$Lm0_{sat}$	53.17	54.959	62.07	mH	Magnetization inductance at rated voltage
Lsl	1.632	1.774	1.449	mH	Stator leakage inductance
Lrl	1.544	2.287	1.537	mH	Rotor leakage inductance
rJ	0.075	0.079	0.074	Kgm^2	Rotor inertia
Cu_c	238.677	192.338	195.363	USD	Stator winding cost
lam_c	541.957	611.685	587.374	USD	Iron cost
$cage_c$	7.161	7.58	4.861	USD	Processed iron cost
pmw_c	421.552	460.352	436.304	USD	Cage cost
i_{cost}	1209.348	1271.955	1223.903	USD	Weight penalty cost
$energy_c$	2172.148	2237.62	2411.335	USD	Initial cost
t_{cost}	3381.496	3509.575	3635.238	USD	Total cost (objective function)
sim_{time}	286.047	23.343	5.735	s	Simulation time (Intel dual core)

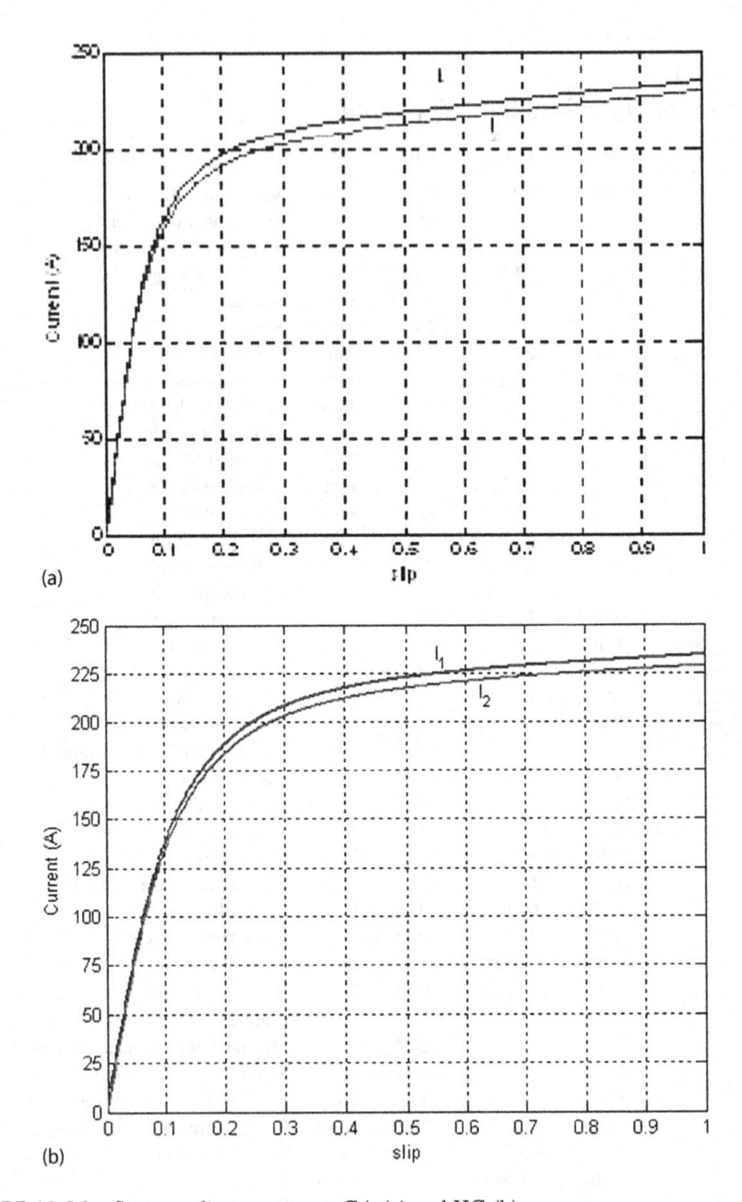

FIGURE 10.36 Stator and rotor current: GA (a) and HG (b).

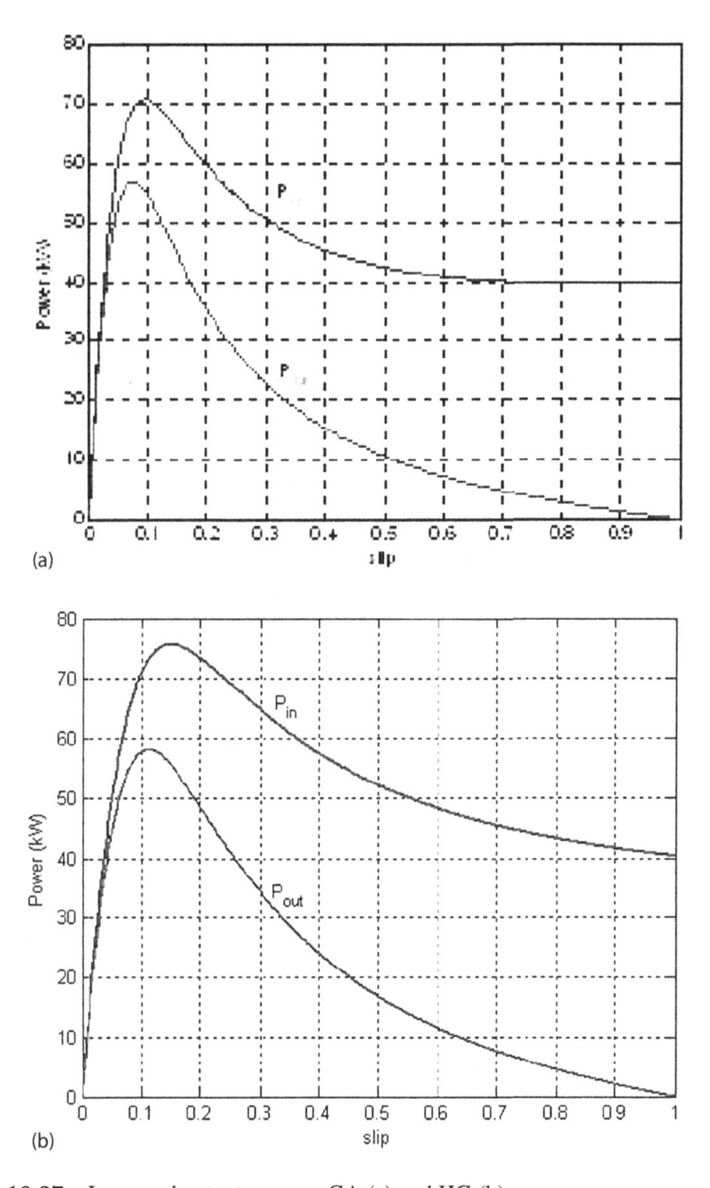

FIGURE 10.37 Input and output powers: GA (a) and HG (b).

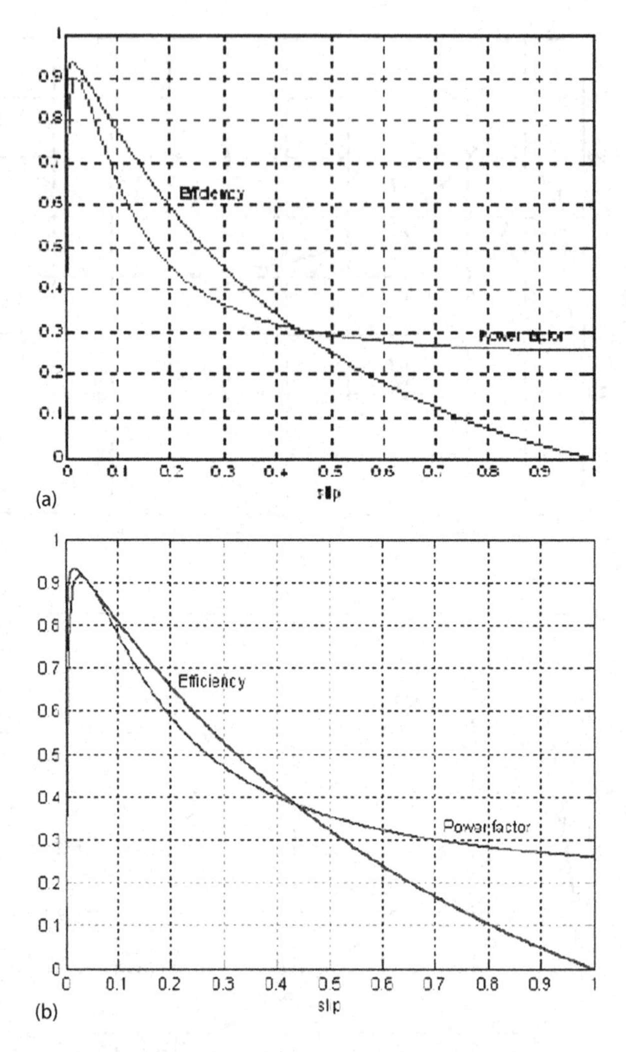

FIGURE 10.38 Efficiency and power factor vs. slip: GA (a) and HG (b).

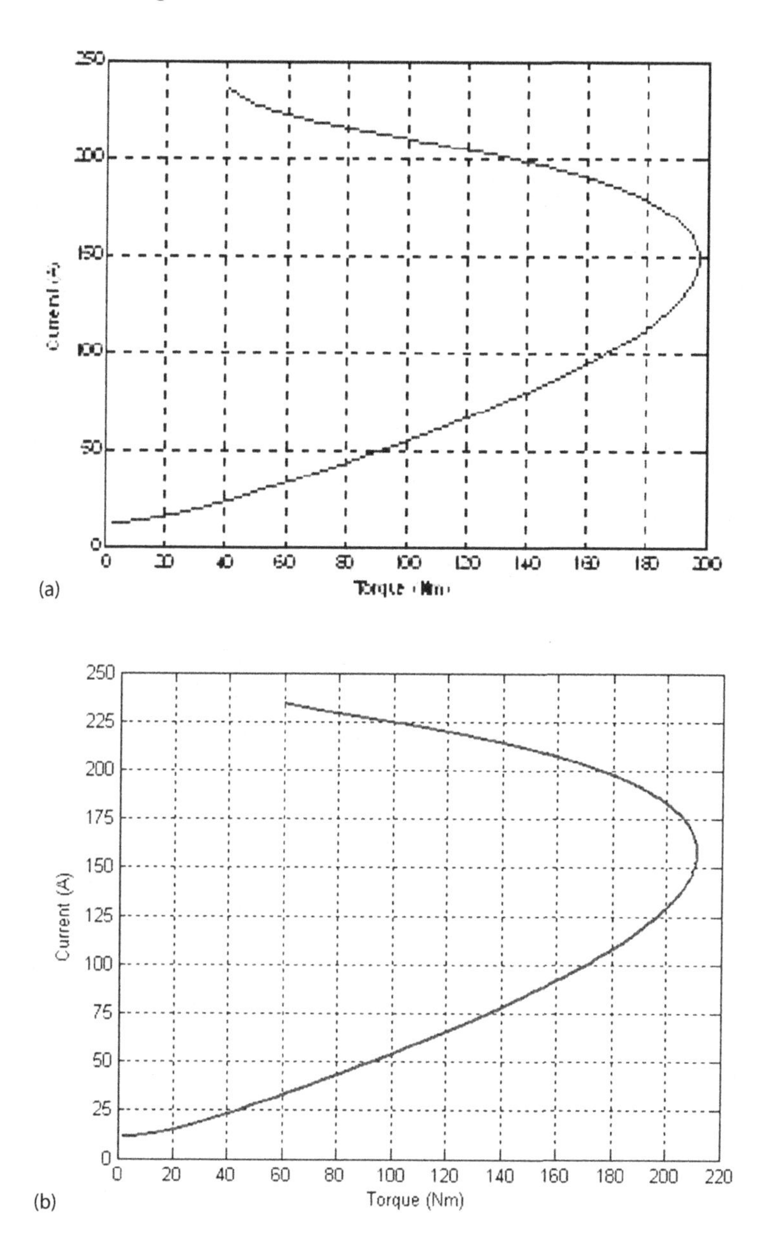

FIGURE 10.39 Current vs. torque: GA (a) and HG (b).

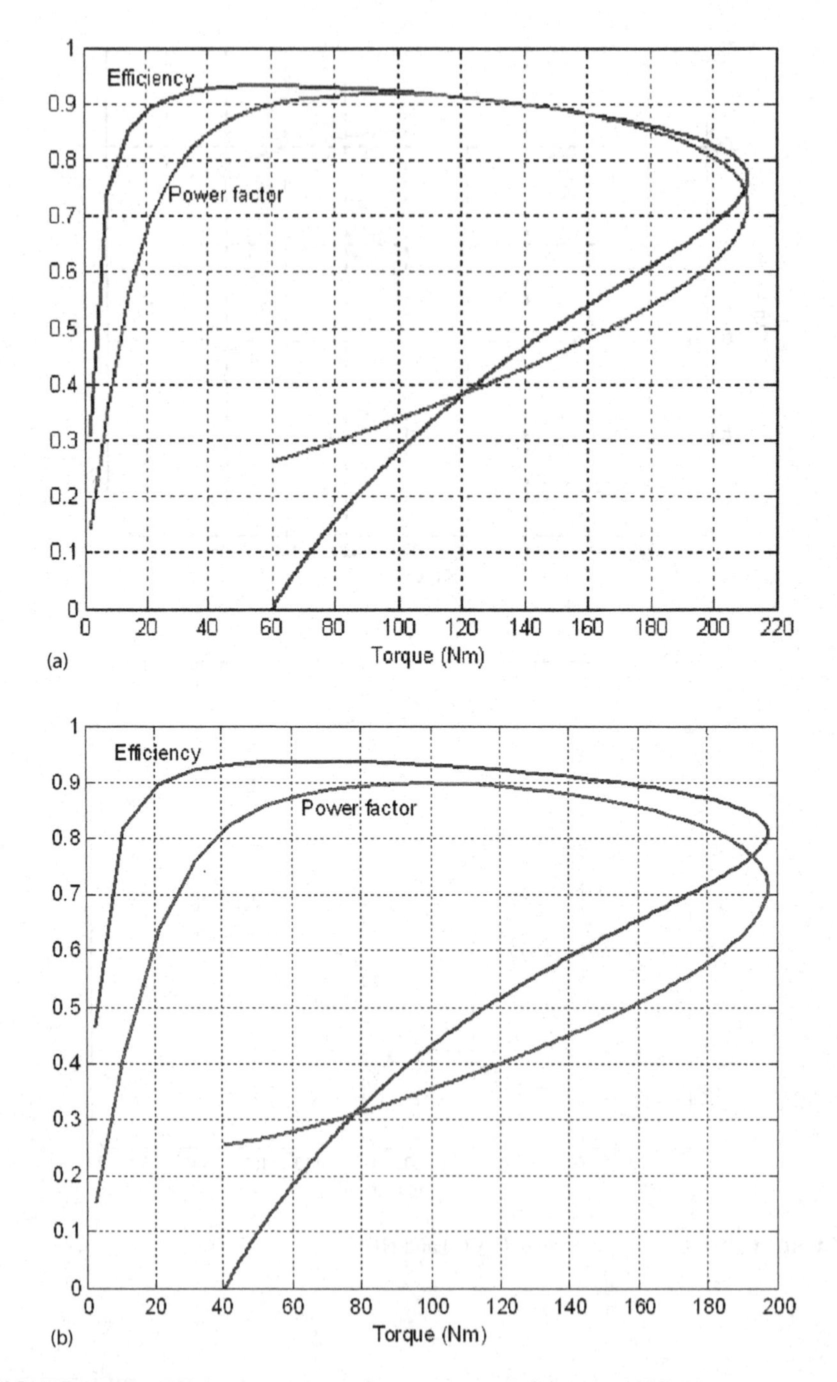

FIGURE 10.40 Efficiency and power factor vs. torque: GA (a) and HG (b).

- Integration of the optimal design codes into a complete design code, as house based on GA and HG, is yet to follow but some early attempts appeared recently [11].
- For more on IM optimization design see *IEEE Trans*. vol. EC & MAG, IE and References [12–14].

REFERENCES

1. I. Boldea and S.A. Nasar, *Induction Machine Design Handbook*, CRC Press, New York, second edition, 2010.
2. M.A. Awadallah, Parameter estimation of induction machines from nameplate data using particle swarm optimization and genetic algorithm techniques, *Elec. Power Compon. Syst.*, 36(8), 2008, 801–814.
3. E.S. Hamdi, *Design of Small Electric Machines*, John Wiley & Sons, Chichester, U.K., 1994.
4. V. Ostovic, *Dynamics of Saturated Electric Machines*, Springer-Verlag, New York, 1989.
5. B. Heller and V. Hamata, *Harmonic Field Effects in Induction Machines*, Elsevier, Amsterdam, the Netherlands, 1977.
6. G. Madescu, I. Boldea, and T.J.E. Miller, The optimal lamination approach to induction, machine design global optimization, *IEEE Trans. Ind. Applic.*, 34(3), 1998, 422–428.
7. Z.M. Zhao, S. Meng, C.C. Chan, and E.W.C. Lo, A novel induction machine design suitable for, inverter-driven variable speed systems, *IEEE Trans. Energy Convers.*, 15(4), 2000, 413–420.
8. T.C. O'Connell and P.T. Krein, A preliminary investigation of computer-aided Schwarz-Christoffel transformation for electric machine design and analysis, *2006 IEEE COMPEL Workshop*, Rensselaer Polytechnic Institute, Troy, NY, July 16–19, 2006, pp. 166–172.
9. M.B. Norton and P.J. Leonard, An object oriented approach to parameterized electrical machine design, *IEEE Trans. Magn.*, 36(4), 2000, 1687–1691.
10. J. Avery and R. King, Effects of specification requirements on induction machine design and operation, Paper No. PCIC-2003-11, pp. 111–120.
11. W. Li, P. Wang, D. Li, X. Zhang, J. Cao, J. Li, "Multiphysical field collaborative optimization of premium induction motor based on GA", *IEEE Trans*, IE-65(2), 2018, 1704–1710.
12. D. Mingardi, N. Bianchi, "Line start PM-assisted synchronous motor design optimization and tests", *IEEE Trans*, IE-64(12), 2017, 9739–9747.
13. L. N. Tutelea, T. Staudt, A. A. Popa, W. Hoffman, I. Boldea, "Line start 1 phase-source split phase capacitor cage-PM rotor-Relsyn motor: modeling, performance and optimal design with experiments", *IEEE Trans*, IE-65(2), 2018, 1772–1780.
14. Y. Wang, S. Niu, W. Fu, "Sensitivity analysis and optimal design of a dual mechanical port bidirectional flux-modulated machine", *IEEE Trans*, IE-65(1), 2018, 211–220.

Index

A

B

C

D

E

F

G

K

L

P

R

S

Printed in the United States
by Baker & Taylor Publisher Services